T0344684

Nanotechnology Commercialization

Nanotechnology Commercialization

Manufacturing Processes and Products

Edited by

Dr. Thomas O. Mensah, Editor in Chief
Georgia Aerospace Systems, Nano technology Division, Georgia Aerospace, Inc., Atlanta, GA

Dr. Ben Wang, Editor
School of Industrial and Systems Engineering, Georgia Tech Manufacturing Institute, Georgia Institute of Technology, Atlanta, GA

Dr. Geoffrey Bothun, Editor
Department of Chemical Engineering, University of Rhode Island, Kingston, RI

Dr. Jessica Winter, Editor
William G. Lowrie Department of Chemical and Biomolecular Engineering, The Ohio State University Columbus, OH

Dr. Virginia Davis, Editor
Department of Chemical Engineering, Auburn University, Auburn, AL

A Joint Publication of the American Institute of Chemical Engineers and John Wiley & Sons, Inc.
Published by John Wiley & Sons, Inc., Hoboken, New Jersey.

Registered Office(s)
John Wiley & Sons, Inc., 111 River Street, Hoboken, NJ 07030, USA

Editorial Office
111 River Street, Hoboken, NJ 07030, USA

For details of our global editorial offices, customer services, and more information about Wiley products visit us at www.wiley.com.

Wiley also publishes its books in a variety of electronic formats and by print-on-demand. Some content that appears in standard print versions of this book may not be available in other formats.

Library of Congress Cataloging-in-Publication Data Applied for
Hardback ISBN: 9781119371724

Cover Design: Wiley
Cover Image: © davidf/Gettyimages

Set in 10/12pt WarnockPro by SPi Global, Chennai, India

Printed in the United States of America

10 9 8 7 6 5 4 3 2 1

Contents

List of Contributors

Oluwamayowa Adigun
School of Chemical Engineering
Purdue University
West Lafayette, IN
USA

Geoffrey D. Bothun
Department of Chemical Engineering
University of Rhode Island
Kingston, RI
USA

Bryan W. Boudouris
School of Chemical Engineering
Purdue University
West Lafayette, IN
USA

Christina Brantley
U. S. Army Research, Development, and Engineering Command
Redstone Arsenal
Alabama
USA

Daniel C. Davis
Georgia Aerospace Systems
Nano technology Division
Georgia Aerospace, Inc.
Atlanta, GA
USA

Virginia A. Davis
Department of Chemical Engineering
Auburn University
Auburn, AL
USA

Souvik De
Artie McFerrin Department of Chemical Engineering
Texas A&M University
College Station, TX
USA

Cerasela Z. Dinu
Department of Chemical and Biomedical Engineering
West Virginia University
Morgantown, WV
USA

Eugene Edwards
U. S. Army Research, Development, and Engineering Command
Redstone Arsenal
Alabama
USA

Alexander S. Freer
School of Chemical Engineering
Purdue University
West Lafayette, IN
USA

Christopher Gilpin
Life Sciences Microscopy Facility
Purdue University
West Lafayette, IN
USA

Micah J. Green
Artie McFerrin Department of Chemical Engineering
Texas A&M University
College Station, TX
USA

Rakesh Gupta
Department of Chemical and Biomedical Engineering
West Virginia University
Morgantown, WV
USA

Michael T. Harris
School of Chemical Engineering
Purdue University
West Lafayette, IN
USA

Joseph H. Koo
Department of Mechanical Engineering
The University of Texas at Austin
Austin, TX, USA

Kil Ho. Lee
William G. Lowrie Department of Chemical and Biomolecular Engineering
The Ohio State University
Columbus, OH
USA

Yongye Liang
Department of Materials Science and Engineering
Southern University of Science and Technology
Shenzhen
China

Jodie Lutkenhaus
Artie McFerrin Department of Chemical Engineering
Texas A&M University
College Station, TX
USA

Micah C. McCrary-Dennis
High-Performance Materials Institute, FAMU-FSU College of Engineering
Tallahassee, FL
USA

Thomas O. Mensah
Georgia Aerospace Systems
Nano technology Division
Georgia Aerospace, Inc.
Atlanta, GA
USA

Laurie Mueller
Life Sciences Microscopy Facility
Purdue University
West Lafayette, IN
USA

Gauri Nabar
William G. Lowrie Department of Chemical and Biomolecular Engineering
The Ohio State University
Columbus, OH
USA

Okenwa O. Okoli
Intel Corporation, Ronler Acres
Hillsboro, OR
USA

David O. Olawale
R.B. Annis School of Engineering, Shaheen College of Arts and Sciences, University of Indianapolis
Indianapolis, IN
USA

Vinka Oyanedel-Craver
Department of Civil and Environmental Engineering
University of Rhode Island
Kingston, RI
USA

Mihail C. Roco
National Science Foundation
Arlington, VA
USA

and

National Nanotechnology Initiative, U.S. National Science and Technology Council
Washington, DC
USA

Paul B. Ruffin
Alabama A&M University
Normal, Alabama
USA

Matthew Souva
William G. Lowrie Department of Chemical and Biomolecular Engineering
The Ohio State University
Columbus, OH
USA

Arda Vanli
Department of Industrial and Manufacturing Engineering
Florida State University
Tallahassee, FL
USA

Alixandra Wagner
Department of Chemical and Biomedical Engineering
West Virginia University
Morgantown, WV
USA

Ben Wang
School of Industrial and Systems Engineering
Georgia Tech Manufacturing Institute, Georgia Institute of Technology
Atlanta, GA
USA

Kan Wang
School of Industrial and Systems Engineering
Georgia Tech Manufacturing Institute, Georgia Institute of Technology
Atlanta, GA
USA

Jessica O. Winter
William G. Lowrie Department of Chemical and Biomolecular Engineering
The Ohio State University
Columbus, OH
USA

and

Department of Biomedical Engineering
The Ohio State University
Columbus, OH
USA

Barbara Wyslouzil
William G. Lowrie Department of Chemical and Biomolecular Engineering
The Ohio State University
Columbus, OH
USA

and

Department of Chemistry and Biochemistry
The Ohio State University
Columbus, OH
USA

Chuck Zhang
School of Industrial and Systems Engineering
Georgia Tech Manufacturing Institute, Georgia Institute of Technology
Atlanta, GA
USA

Xiao Zhang
Department of Materials Science and Engineering
Southern University of Science and Technology
Shenzhen
China

Xing Zhang
Department of Materials Science and Engineering
Southern University of Science and Technology
Shenzhen
China

Preface

The Frontiers of Nanotechnology Book Series has the objective of advancing techniques for scale-up and transition of nanotechnology processes to industry. This book provides insight into the current status of advanced nanotechnology processes and their scale-up to semi-industrial and full-scale industrial levels, while addressing key scale-up challenges.

The impact such understanding has on full-scale nanotechnology manufacturing on business and marketing strategy, including expansion and execution, is necessary after years of major investments in the technology worldwide.

This book has the objective of also providing relevant technical and engineering framework and the latest innovative work in the area of nanotechnology manufacturing and scale-up. Chemical engineering and industrial engineer methods were adopted in addressing the manufacturing challenges.

In this first volume of the Frontiers of Nanotechnology Series, authors from leading US agencies, including Department of Defense, the National Science Foundation, National Laboratories, private companies, and leading US universities as well as international experts, have examined challenges in transitioning this technology from research to large-scale manufacturing environment. I want to express my gratitude to all of the authors for tackling such an important but complex engineering subject.

Researchers from US Army ARDEC Huntsville, national agencies, and leading universities and engineering departments, such as Georgia Institute of Technology, University of Texas, Austin, Purdue University, Auburn University, University of Rhode Island, University of West Virginia, Ohio State University, Florida State University, as well as the South China University and others have contributed significantly to this book.

In this first book of the Frontiers Series, authors have focused on the chemical engineering aspects of nanotechnology scale-up such as the chemistry and nanocatalyst applications in commercial processes, mixing and integration into solutions, analyzing interfacial aspects of nanotube dispersion, a critical step in nanomanufacturing, and an important challenge in scale-up. Statistical analysis for controlling continuous processing and predicting nanomaterials performance of sheets of nanostructures, fundamentals of nanomanufacturing using spray techniques are also covered.

Also presented in this book are high-temperature ablative materials for rocket motors and reentry vehicles for space applications, including finite element modeling of transport phenomena in ablative materials performance, advanced missile shell structures and nanocomposites incorporating nanosensors for advanced military applications, and vacuum-assisted resin transfer molding processing of nanocomposites including finite element analysis of processes. Also examined is the use of mechanical properties of fabricated composites as a method of evaluating process control and product performance. The authors have also explored applications of bioinspired approaches for fabricating nanocircuits, and finally toxicity, environmental, and safety issues regarding nanomaterials processing are presented in the last two chapters.

Environmental, safety, and toxicity of carbon nanotubes is important in the commercialization process since workers can be exposed to these nanoparticles, with serious health implications and adverse economic impact on the profitability of nanomanufacturing companies. EHS (Environmental Health and Safety) area must be addressed through engineering methods for this industry to thrive and be sustainable.

I want to thank all my coeditors, professors Ben Wang, Georgia Tech Manufacturing Institute; Jessica Winters, Ohio State University; Virginia Davis, Auburn University; and Geoffrey Bothun, University of Rhode Island, for assisting me as chapter contributors and reviewers of the technical manuscripts for this important book.

I want to express my special gratitude to Mike Roco at the National Science Foundation, a champion of the National Nanotechnology Initiative, NNI, in the United States for authoring the overview chapter for the book and giving me insights into critical gaps that exist in the commercialization of nanotechnology.

There is a paradigm shift in engineering design of processes as demonstrated by the US National Materials Genome project and key parts of this approach were employed in some of the work presented in this book, and it is our hope that this approach will continue to guide all future research into the scale-up of nanotechnology.

Thomas O. Mensah

Editor in Chief

Dr Thomas Mensah is a fellow of the National Academy of Inventors (NAI), fellow of the American Institute of Chemical Engineers (AIChE), and associate fellow of the American Institute of Aeronautics and Astronautics (AIAA). He holds seven US patents in fiber optics awarded in a 6-year time frame.

He worked at AT&T Bell Laboratories and Corning Glass Works and was one of the key innovators and inventors of large-scale processes that moved fiber optics from the research laboratories to manufacturing and commercial environments in the United States. He is currently the president of Georgia Aerospace Systems and served as a coprincipal investigator on the Carbon Nanotube Weapons Platform Development for the US Department of Defense. He also served as the director at large of the Nanoscale Engineering Forum at AIChE. This forum continues to organize feature conferences and technical sessions and symposium in nanotechnology at AIChE Annual Meetings around the country.

1

Overview: Affirmation of Nanotechnology between 2000 and 2030

Mihail C. Roco[1,2]

[1] *National Science Foundation, Arlington, VA, USA*
[2] *National Nanotechnology Initiative, U.S. National Science and Technology Council, Washington, DC, USA*

1.1 Introduction

In the nanoscale domain, nature transitions from the fixed physical behavior of a finite number of atoms to an almost infinite range of physical–chemical–biological behaviors of collections of atoms and molecules. The fundamental properties and functions of all natural and man-made materials are defined and can be modified efficiently at that scale. The unifying definition of nanotechnology, based on specific behavior at the nanoscale and the long-term nanotechnology research and education vision, was formulated in 1997–1999, and its implementation begun with National Nanotechnology Initiative (NNI) in 2000. We have estimated that it would take about three decades to advance from a scientific curiosity in 2000 to a science-based general purpose technology with broad societal benefits toward 2030 [1–3] (see www.wtec.org/nano2/).

A long-term strategic view is needed because nanotechnology is a foundational general purpose field. *Three development stages* of nanotechnology, corresponding to the level of complexity of typical outcomes, have been envisioned: passive and active nanostructures in the first stage of development (*Nano 1*), nanosystems and molecular nanosystems in the second stage (*Nano 2*), and converging technology platforms and distributed interconnected nanosystems in the last stage (*Nano 3*).

We use the *definition of nanotechnology* as set out in *Nanotechnology Research Directions* [2]. Nanotechnology is the ability to control and restructure matter at the atomic and molecular levels in the range of approximately 1–100 nm, and exploiting the distinct properties and phenomena at that scale as compared to those associated with single atoms or bulk behavior. The aim is to create materials, devices, and systems with fundamentally new properties

Nanotechnology Commercialization: Manufacturing Processes and Products, First Edition.
Edited by Thomas O. Mensah, Ben Wang, Geoffrey Bothun, Jessica Winter, and Virginia Davis.

and functions for novel applications by engineering their small structure. This is the ultimate frontier to economically change materials and systems properties, and the most efficient length scale for manufacturing and molecular medicine. The same principles and tools are applicable to different areas of relevance and may help establish a unifying platform for science, engineering, and technology at the nanoscale. The transition from the behavior of single atoms or molecules to collective behavior of atomic and molecular assemblies is encountered in nature, and nanotechnology exploits this natural threshold.

This chapter describes the timeline and affirmation of nanotechnology, its three stages, key challenges, and discusses nanotechnology return on investment.

1.2 Nanotechnology – A Foundational Megatrend in Science and Engineering

Nanotechnology is a foundational, general purpose technology for all sectors of the economy dealing with matter and biosystems, as information technology is a general purpose technology for communication and computation. Biotechnology and cognitive technologies are two other foundational technologies growing at the beginning of the twenty-first century (Figure 1.1). Table 1.1

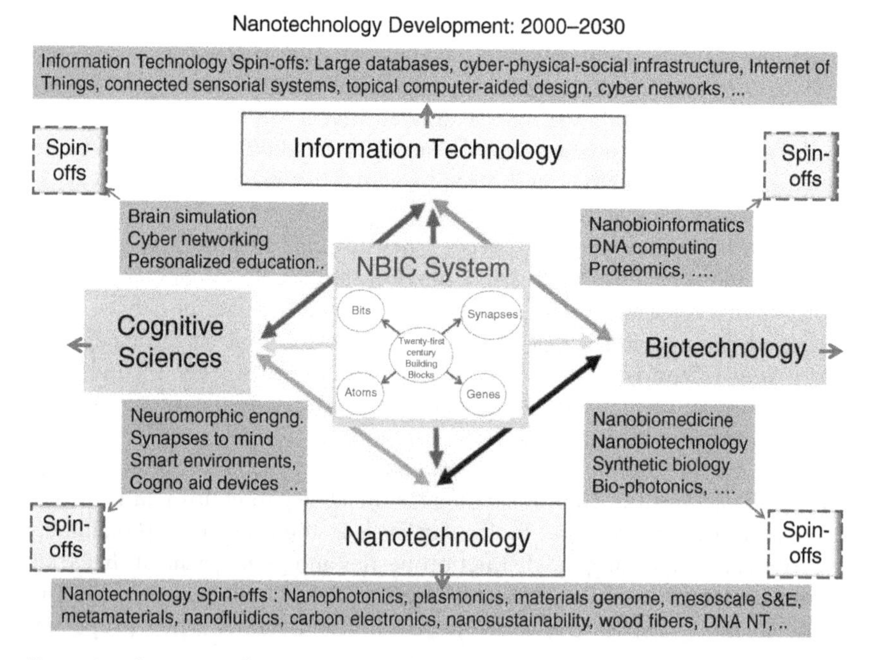

Figure 1.1 Converging foundational technologies, and their interdisciplinary and spin-offs subfields. Modified from Roco and Bainbridge [4].

Table 1.1 Proposed classification of science and technology platforms.

Category	I. *Foundational* S&T platform (system architecture)	II. *Topical* S&T platform (hierarchical system from I)	III. *Application field* platform (branched, inter- and recombination)	IV. *Product and service* platform (spin-off, inter- and recombination)
S&T Platforms	• **Nanotechnology**: (atom architecture) • **Information S&T** (bit architecture) • **Modern bio S&T** (gene architecture) • **Cognitive S&T** (synapsis architecture) • **Artificial Intelligence S&T** (system design)	***Essential:*** Photonics Semiconductors Genomics Biomedicine ***Contributing:*** Synthetic biology Neuromorphic eng Proteomics Nanofluidics Metamaterials	Cell phone system Transportation Medicine Energy conversion and storage Agriculture Space exploration	Car components Medical devices Nano coatings LEDs Nano lasers ....
Typical timescales	25–50 years	10–25 years	5–10 years	3–5 years
One-step investment amplification factor	$k_{f(\text{undamental})}$	$k_{t(\text{opical})}$	$k_{a(\text{pplication})}$	$k_{p(\text{product and service})}$
Cumulative investment amplification factor	$k_f\,k_t\,k_a\,k_p$	$k_t\,k_a\,k_p$	$k_a\,k_p$	k_p
Game changer for:	Knowledge	Technology approach	Application field	User consumption

shows several category levels of science and technology (S&T) platforms according to their level of generality and societal impact: foundational S&T, topical S&T, application domain, and product/service platform. While there are only five foundational S&T platforms most dynamic at this moment (Figure 1.1), the number of topical S&T platforms increases with the number of spin-offs, interplatform and further recombination growth. Each topical S&T platform has several application domains, which at their turn each have a series of products and related services. The importance of foundational platforms – and in particular its most exploratory component part at this moment, nanotechnology – is underlined by the cumulative investment amplification factor by developing the respective S&T platform that is a product of the foundational platform investment amplification factor, with the topical, application area and product amplification factors.

Nanotechnology continues exponential growth by vertical science-to-technology transition, horizontal expansion to areas as agriculture/textiles/cement, and spin-off areas (~20) as spintronics/metamaterials/..., progressively penetrating in key economic sectors. The number of World of Science publications on nano-extended 20 new terms between 1990 and 2014 that now represent over ¼ of the total publications (Figure 1.2). For this reason, it is increasingly difficult to identify the R&D programs around the word supporting nanotechnology because they are called after an activity that

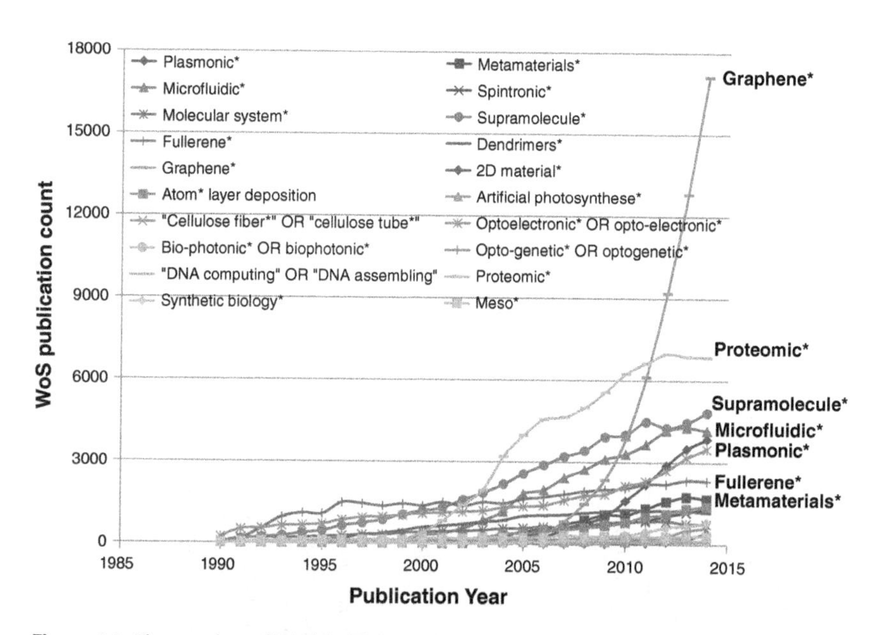

Figure 1.2 The number of World of Science (WoS) publications on nano-extended 20 new terms between 1990 and 2014.

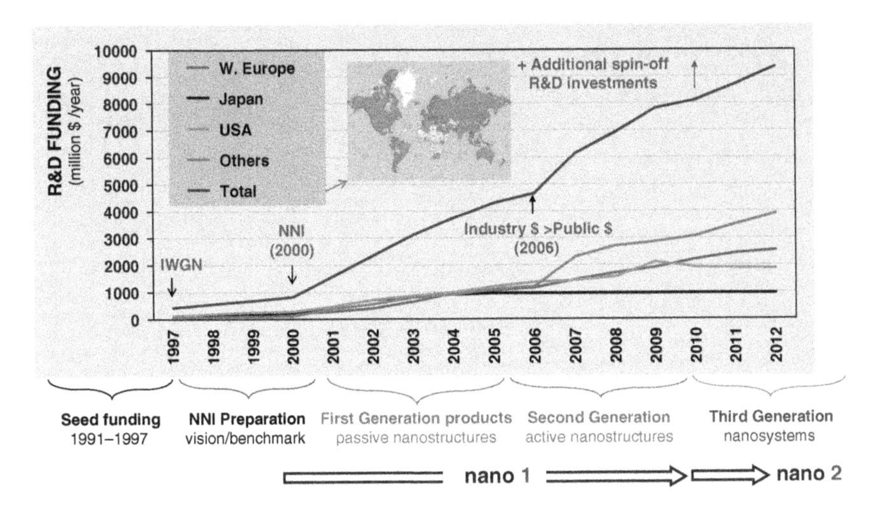

Figure 1.3 International government R&D funding the interval 2000–2012, after 2013 – increase use of new terms and platforms (using NNI definition, 81 countries, MCR direct contacts).

branched out of the foundational field. Figure 1.3 illustrates international government R&D funding the interval 2000–2012 [9].

Most of the larger science and technology initiatives have been justified in the United States and abroad mainly by application-related and societal factors. For example, the Manhattan Project during World War II (with centralized, goal-focused, and simultaneous approaches), the Apollo Space Project (with a centralized, focused goal), and Networking and Information Technology Research and Development (top-down initiated and managed, and established when mass applications justified the return of investment). The initiation of the NNI was motivated primarily by its long-term science and engineering goals and general purpose technology opportunity, and has been managed using a bottom-up approach combined with centralized coordination. A few comments underlying this characteristic are as follows:

> Charles Vest, President National Academy of Engineering (PCAST meeting, White House, 2005): *"NNI is a new way to run an initiative"*

> Steve Edwards, "Hall of Fame for Nanoscale Science and Engineering" (Jan. 1, 2006): *"...persuading the U.S. government, not to mention the rest of the world, to support nanotechnology. It was a masterful job of engineering the future"*

> Tim [5], President of the European Nanobusiness Association, and Cientifica Co. (2015): *"nanotechnology [is] the first truly global scientific revolution."*

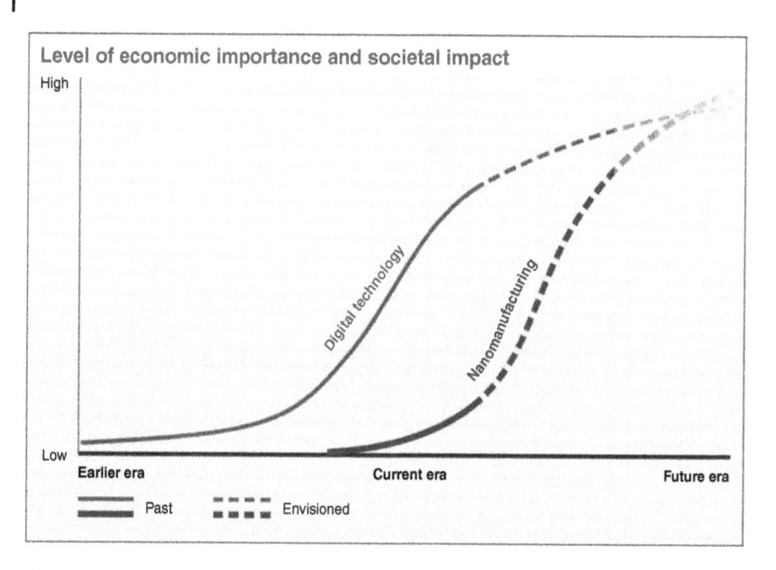

Figure 1.4 S-curves for two science and technology megatrends: past and envisioned conceptualization of "Nanomanufacturing" and "Digital Technology" [6].

Nanotechnology promises to become a general purpose technology with large-scale applications similar to digital technology. It could eventually match or outstrip the digital revolution in terms of economic importance and societal impact once the methods of investigation and manufacturing are developed and the underlying education and infrastructure are established. During about 2020–2030, nanotechnology could equal and even exceed the digital revolution in terms of technology breakthroughs, investments, and societal importance (Figure 1.4) [6].

The nanotechnology development S-curve shown in Figure 1.4 is supported by the data in Table 1.2 showing an increase of the world annual rate of revenues growth from 25% in 2000–2010 to 44% in 2010–2013.

Examples of S&E generic platforms with areas of high impact

Nanotechnology-enabled products by sectors with the most traded nanoproducts in 2014 according to Lux Research [7] and other industry sources:

- Materials and manufacturing: Fiber-reinforced plastics, nanoparticle catalysts, coatings, insulation, filtration, transportation (cars, trucks, trains, planes, and ships), and robotics (actuators and sensors). For example, Exxon-Mobil has multibillion dollar applications on nanostructured catalysts. TiO_2, MWCNTs, and quantum dots are some of the most frequently encountered nanocomponents. Nanoscale coatings, imprinting, and roll-to-roll are three most common manufacturing processes.

Table 1.2 Global and US revenues from products incorporating nanotechnology.

Revenues all in $ billions	2000	2010	Annual rate 2000–2010	2011	2012	2013	2014	Annual rate 2010–2013	Estimation 2020	Estimation 2030
World	30	335	***25%***	514	852	1190	1620	***48%***	*3000*	*30,000*
United States	13	110	***24%***	170	213	284	370	***38%***	*750*	*7,500*
United States/World	43%	33%	***−1%***	33%	25%	24%	23%	***−10%***	*25%*	*25%*

Source: Data from *Roco et al.* [1] for 2000–2010, est. 2020, and est. 2030; and from Lux Research [7] for 2011–2014.

- Electronics and IT: Semiconductors, mobile electronics and displays, packaging, thermal management, batteries, supercapacitors, paint, and integration with nanophotonics
- Healthcare and life sciences: Diagnostic and monitoring sensors (cancer), cosmetics, food products and packaging, personal care products, sunscreen, packaging, surgical tools, implantable medical devices, filtration, treatments (cancer radiation therapies) and medications formulations, contrast agents, quantum dots in lab supplies such as fluorescent antibodies, and drug delivery systems.
- Energy and environment: Fuel cells, hydrolysis, catalysts, solar cells, insulation, filtration, supercapacitors, grid storage, monitoring equipment (sensors), water treatment, and purification

Nanotechnology development between about 2000 and 2030 is an example of the *convergence–divergence process* for science and technology megatrends. The convergence process has four phases:

a) First is the "*creative phase*" (Figure 1.5). Confluence of knowledge of bottom-up and top-down disciplines, of various sectors from materials to medicine, and of various tools and methods of investigation and synthesis have led to an increasing control of matter at the nanoscale.
b) The convergence enables the creation/integration of successive generations of nanotechnology products and productive methods ("*Integration/fusion phase*").
c) At the beginning of the divergent stage, the spiral of innovation ("*Innovation phase*") leads to new products and applications that are estimated to reach $3 billion in 2020.
d) The spiral of innovation branches out thereafter into new activities, including spin-off disciplines and productive sectors, business models, expertise, and decision-making approaches ("*Spin-off phase*"). An essential element of progress is the emergence of completely new skills in nanotechnology, some of which have resulted from the interface with other fields and others resultant as a spin-off element of nanoscience, such as spintronics, plasmonics, metamaterials, carbon nanoelectronics, DNA nanotechnology, optogenetics, and molecular design to create hierarchical systems. New technology platforms will be created such as in nanosensors systems, components in robotics, nanoelectronics in cars, cyber manufacturing, unmanned vehicles, and light space crafts.

The exponential growth of nanotechnology through discoveries, technological transition, horizontal expansion, and its spin-off areas is expected to continue at high rates through 2030. The economic estimations currently made by

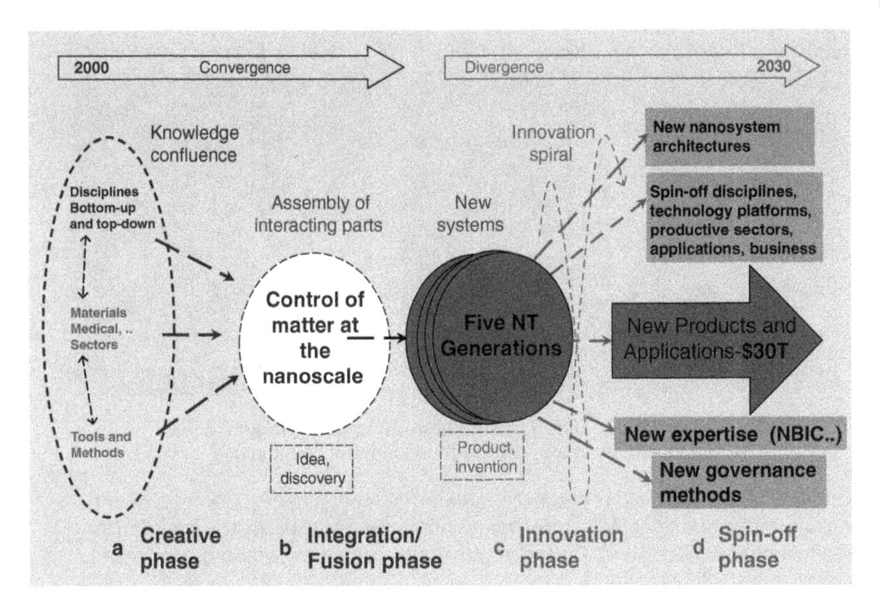

Figure 1.5 2000–2030 convergence–divergence cycle for global nanotechnology development. Modified from Roco and Bainbridge [4].

valuing the commercialization of particular products (e.g., single-wall carbon nanotube of single-sheet graphene) are not the most representative for effective products because the most valuable developments are the know-how and technological capabilities that are changing fast toward composite and modular nanosystems. This is particularly true if the material has mostly a "schooling" role ("model nanotechnology" vs "economic nanotechnology").

1.3 Three Stages for Establishing the New General Purpose Technology

We have estimated that about 30 years are needed for nanotechnology development from a scientific curiosity to mass use in economy that may be separated into three stages. Each stage is defined by its investigative methods and synthesis/assembling techniques, level of nanoscale integration and complexity of the respective products, typical application areas, education needs, and risk governance. These characteristics are documented in four reports (Figure 1.6). A schematic of the three successive stages and corresponding nanotechnology generation products are shown in Figure 1.7. Each stage differs by the types

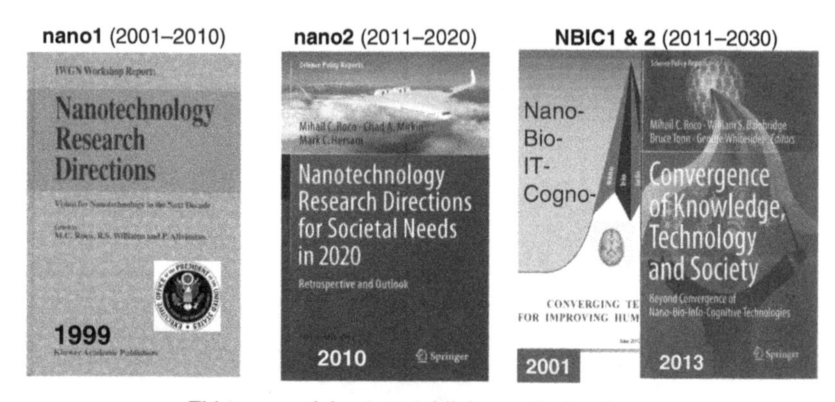

Figure 1.6 Thirty-year vision to establish nanotechnology: changing the research and education focus and priorities in three stages from scientific curiosity to immersion in socioeconomic projects [1–4]. The reports are available on www.wtec.org/nano2/ and www.wtec.org/NBIC2-report/.

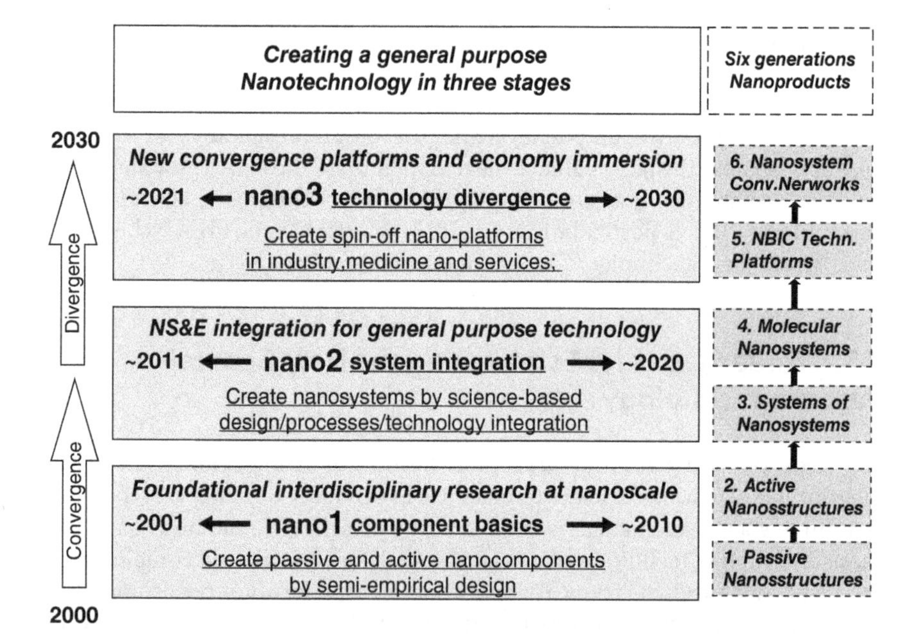

Figure 1.7 Creating a general purpose technology in three stages between 2000 and 2030. Each stage includes two generations of nanotechnology products. Modified from Roco *et al.* [1].

of new measurements and simulation methods, advanced processing/synthesis methods, integration and complexity levels, use and application, as well as risk governance [1].

The R&D focus in nanotechnology is evolving: from synthesizing components by semiempirical processing and creating a library of nanostructures (such as carbon nanotubes, quantum dots, and sheets of graphene) in 2000–2010, to science-based design and manufacturing of nanoscale devices and systems (such as in biomedicine and nanoelectronics) in 2010–2020, and establishing economic nanotechnology applications in various technology platforms and immersion in socioeconomic projects in 2020–2030.

The planning in **the first stage (*Nano 1*)** was focused on discovery of individual phenomena and semiempirical synthesis of nanoscale components. For example, quantum effect repulsion forces between surfaces at small interfaces have been identified, and the first quantum device was built. In another example, quantum dots, nanoparticles, nanotubes, and nanocoatings were created from a majority of elements in the periodic table, and their atom and molecular assembling mechanism explored for generalizations. Typical measurements of the atomic structure with femtosecond changes were indirect, using time and volume averaging. The nanocomponents were used to improve the performance of existing products, such as nanoparticle-reinforced polymers. The main penetration of nanotechnology was in advanced materials, nanostructured chemicals, electronics, and pharmaceutics. For example, Moore's Law for semiconductors has continued because of nanoscale components added in their fabrication. Education transitioned from fixed properties at the microscale to engineered properties based on the nanoscale understanding of nature and technology. Governance was mostly science-centric with a focus on nanotechnology environmental, health and safety aspects. National and international organizations have formulated the basics for nomenclature to be used in scientific publications and for standards classifications. An international, multidisciplinary nanoscale science and technology community has been established.

During **the second decade (*Nano 2*)** of nanotechnology development, the focus is on integration at the nanoscale and science-based creation of devices and systems for fundamentally new products – including self-powered nanodevices, self-assembling systems on multiple scales from the nanoscale, and nano-bio assemblies. Examples include nanofluidics systems, integrated sensorial systems, and nanoelectronics and display systems. Direct measurements and simulations with atomic and femtosecond resolutions have been undertaken for many-atom systems encountered in biological and engineering applications. There is an increased focus on new performance in new domains of application and on innovation methods. Nanotechnology penetration is faster in nanobiotechnology, energy resources, food and agriculture, forestry, and cognitive technologies, as well as in nanoscale simulation-based design methods. In education, we see more attention to cross-disciplinary "T" or "reverse T" learning including general nanotechnology education in the horizontal component. Societal aspects are increasingly on expanding sustainability, exploiting the potential for increased productivity and addressing socioeconomic issues with a focus on healthcare. Governance is increasingly user-centric and multiple player participatory. Global implications are seen on economy, sustainability, and balance of forces.

After 2010, there is an increased focus on nanoscale science and engineering integration with other knowledge and technology domains and their applications [3] that will continue through the end of **the third decade (*Nano 3*)** of nanotechnology development. After about 2000, convergence of nanotechnology with other key technologies will subsequently lead to bifurcation into emerging and integrated technology platforms. NBIC-based measurements will be needed in these new technology platforms. Integration of foundational and general technologies will branch out to new fields of research and production. Education will need to be more focused on unifying concepts and connecting phenomena, processes, and technologies. The role of bottom-up and horizontal interactions will increase in importance as compared to top-down measures in the S&T governance. The risk analysis will expand to hybrid bio-nano systems and human–technology coevolution. New competencies, socioeconomic platforms, and production capabilities will be taking a significant role in the economy. The international community will be more connected through the scientific and technological developments. It will create opportunities for new models of collaboration and competition.

New generations of nanotechnology products and productive processes are timed with the introduction of new prototypes of nanotechnology products and with the successive increases in the degree of control, integration, complexity, and risk. Table 1.3 defines the estimation for introduction of various generations along the three conceptual stages of nanotechnology development.

A dominant trend in the interval 2020–2030 is envisioned to be immersion of nanotechnology with other emerging and established technologies, in industry, medicine and services, and in education and training for societal progress to become the largest technology driver in most economical sectors together with information technology.

Priorities of the U.S. NNI at the beginning of *Nano 2* [1] were grouped and are funded under several "Nanotechnology Signature Initiatives (NSI)" since 2011: (i) Nanoelectronics for 2020 and Beyond; (ii) Sustainable Nanomanufacturing; (iii) Nanotechnology for Solar Energy; (iv) Nanotechnology Knowledge Infrastructure, and (v) Nanosensors. In 2016, the NSI on "Nanotechnology for Solar Energy" was completed and replaced by the NSI on "*Water Sustainability through Nanotechnology*" (www.nano.gov/node/1577). PCAST [8] has recommended for new set of grand challenges for Nano 2 looking 15 years ahead. The first has been dedicated to "*Nanotechnology-Inspired Grand Challenge*

Table 1.3 Generations of nanotechnology products and productive processes, and the corresponding interval for beginning commercial prototypes.

Stage	Generation	Main characteristics
Nano 1 *Component basics*	**G1:** **Passive nanostructures** (~2000–2005)	The nanostructures have stable behavior during their use. They typically are used to tailor macroscale properties and functions a) Dispersed nanostructures, such as aerosols, colloids, and quantum dots on surfaces b) Contact nanostructures, such as in nanocomposites, metals, polymers, ceramics, and coatings
	G2: **Active nanostructures** (~2005–2010)	The nanostructures change their composition and/or behavior during their use. They typically are integrated into microscale devices and systems and used for their biological, mechanical, electronic, magnetic, photonic, and other effects a) Bioactive with health effects, such as targeted drugs, biodevices, and artificial muscles b) Physico-chemical active, such as amplifiers, actuators, adaptive structures, and 3-D transistors
Nano 2 *System integration*	**G3:** **System of nanosystems** (~2010–2015)	Three-dimensional nanosystems frequently incorporated into other systems and using various syntheses and assembling techniques such as bio-assembling, robotics with emerging behavior, and evolutionary approaches. A key challenge is networking at the nanoscale and hierarchical architectures. Research focus will shift toward heterogeneous nanostructures and supramolecular system engineering. This includes directed multiscale self-assembling, artificial tissues and sensorial systems, quantum interactions within nanoscale systems, processing of information using photons or electron spin, and assemblies of nanoscale electromechanical systems (NEMS)

(Continued)

Table 1.3 (Continued)

Stage	Generation	Main characteristics
	G4: Molecular nanosystems (~2015–2020)	Heterogeneous molecular nanosystems, where each molecule in the nanosystem has a specific structure and plays a different role. Molecules will be used as devices and from their engineered structures and architectures will emerge fundamentally new functions. Designing new atomic and molecular assemblies is expected to increase in importance, including macromolecules "by design", nanoscale machines, and directed and multiscale self-assembling, exploiting quantum control, nanosystem biology for healthcare, and human–machine interface at the tissue and nervous system level. Research will include topics such as atomic manipulation for design of molecules and supramolecular systems, controlled interaction between light and matter with relevance to energy conversion among others, exploiting quantum control mechanical–chemical molecular processes, nanosystem biology for healthcare and agricultural systems, and human–machine interface at the tissue and nervous system level
Nano 3 Technology divergence	**G5: NBIC integrated technology platforms** (~2020–2025)	Converging technology platforms from the nanoscale based on new nanosystem architectures at confluence with other foundational emerging technologies. This includes converging foundational technologies (nano-bio-info-cogno) platforms integrated from the nanoscale
	G6: Nanosystem convergence networks (~2025–2030)	Distributed and interconnected nanosystem networks, across domains and interacting at various levels (foundational, topical, application, products/service), for health, production, infrastructure, and services. This includes networks of foundational technologies (nano-bio-info-cogno) platforms and their spin-offs including for emerging nano-biosystems

for Future Computing" (http://www.nano.gov/grandchallenges). A focus in 2011–2020 is research on the third generation of nanotechnology products including nanosystems, self-powered nanodevices, and nano-bio assemblies. There is an increased focus on nanoscale science and engineering integration with other knowledge and technology domains to create new nanosystem architectures and corresponding technology platforms by 2030 ("Converging knowledge, technology and Society: Beyond Nano-Bio-Info-Cognitive Technologies" [3]).

1.4 Several Challenges for Nanotechnology Development

Several key challenges for nanotechnology development in the next 15 years are identifying and realizing:

- *Path to new technological platforms.* Identifying and supporting technology platforms exploiting the main features at the nanoscale such as self-assembling, quantum materials and devices, nanofluidics fluid-based manufacturing, new logic and memory paradigms, nanophotonics, and plasmonics.
- *Path to economicity.* Increasing productivity and economical production are the main reasons for advancing the tools of nanoscale science and engineering. In the first stage, the challenge was to prove the ability to create and change nanostructures by control no matter the cost or sustainability. Now, we increasingly aim at minimizing the cost and resources. For example, a general approach is nanomodular materials and systems by design that would allow efficient assembling of nanomodules that individually maintain their nano-specific properties [9]. The purpose is to build economical and versatile products at relatively low temperatures and pressures, as well as to create things that are not possible otherwise such as nanosensors systems and synthetic organs.
- *Path to sustainability.* An initial goal of nanotechnology has been manufacturing using less materials, energy, and water while reducing the waste. An approach was leaving the molecules to do what "they like," such as self-assembling at low temperature and pressures using noncovalent molecular interactions. This may be associated with the term green nanomanufacturing. Another goal was to develop new methods for efficient energy conversion and storage, economic filtration, and computing, to name several important pathways to sustainable society.
- *Path to creativity, inventions, and innovations in nanotechnology applications.* Because nanoscale science and engineering is at the confluence of multiple disciplines and methods and there are multiple application areas including by convergence with other technologies, there is a good potential for

innovation. For example, the technology and business branching out in the nanotechnology convergence–divergence cycle created multiple opportunities for value add, such as nano-biomedical devices and synthetic biology, nanosensors for medical and industrial use, and nanotechnology for robotics and autonomous vehicles.

- *Path to wellness.* Nanotechnology enables replacing clinical medicine with individual treatment based on conditioning the cell at the subcellular level, and treating chronic diseases with molecular medicine. Physical and mental wellness is at the core of quality of life.
- *Path to almost zero-power internet of things.* The connectivity between efficient logic and memory devices, sensors, robotics, and distributed energy harvesting enabled by nanotechnology
- *Path to a comprehensive, flexible, and connected infrastructure* responding to the requirements of the three nanotechnology stages (Figure 1.7). Examples of the connected R&D infrastructure in the United States in 2015 are shown in Table 1.4.
- *Path to responsible governance and public acceptance.* There is an increased role of ethical, legal, and other social issue (ELSI) and of emergent nano-biosystems, nano-neurotechnology, and overall converging technologies effects.
- *Nanotechnology growth potential in the future.* It depends on setting the proper vision and goals, planning R&D to reach the vision, and assuring R&D funding under public scrutiny, and competition with other sectors and other economies.

1.5 About the Return on Investment

Foundational science and technology fields have a larger impact as compared to topical technologies they generate, but need longer time to be developed and implemented. *Nanotechnology is a foundational technology with implications on knowledge, productivity, health, sustainability, security, and overall wellness that is establishing its methods, transformative approach, and infrastructure between about 2000 and 2030.*

Still in its infancy in creating novel nanosystems, nanotechnology in 2015 has already shown its promise to society. The United States invested about $20 billion in nanotechnology R&D through the NNI in the fiscal years 2001–2015. The cumulative US nanotechnology commitment since 2000 places the NNI second only to the space program in terms of civilian science and technology investment [10]. Overall, it has been estimated that NNI spent about one fourth of the global government funding since 2001.

Table 1.4 Examples of main NNI R&D centers, user facilities, and networks sponsored by NSF.

Name	Institution(s)
NSF – 10 Networks	
National Nanofabrication Infrastructure Network (NNIN) – 15 nodes (user facilities) (www.nnin.org)	Cornell University – main node (NNIN recompetition in 2015)
National Nanotechnology Coordinated Infrastructure (NNCI) after September 2015 – 16 nodes	
National Nanofabrication Infrastructure Network (NNIN) – 15 nodes (user facilities) (www.nnin.org)	Cornell University – main node (under recompetition in 2013–2014)
Network for Computational Nanotechnology (NCN) (nanoHUB.org)	Purdue University – main node
National Nanomanufacturing Network (NNN) (www.internano.org)	University of Massachusetts, Amherst – main node
Centers for Nanotechnology in Society (CNS) (cns.asu.edu)	Arizona State University and University of California, San Diego
Nanoscale Informal Science Education (NISE) Network (www.nisenet.org)	Museum of Science, Boston – main node
Nanoscale Science and Engineering Centers (NSEC)	Distributed centers
Materials Science and Engineering Centers (MRSECs)	Distributed centers
Nanosystems Engineering Research Centers (NERC)	Distributed centers
Centers for the Environmental Implications of Nanotechnology (CEIN) (www.cein.ucla.edu)	University of California, Los Angeles, and Duke University
Center for National Nanotechnology Applications and Career Knowledge (NACK) (nano4me.org)	Pennsylvania State University
NSF – Three Science and Technology Centers	
Center for Energy Efficient Electronics Science (nanoelectronics) (www.e3s-center.org)	University of California, Berkeley
Emergent Behaviors of Integrated Cellular Systems (nanobiotechnology) (hhttp://ebics.net/about)	Massachusetts Institute of Technology
Center for Integrated Quantum Materials (ciqm.harvard.edu)	Harvard University

One successful case study at the beginning of the NNI about 2000 was for giant magnetic resonance (GMR) that brought a significant performance improvement in computer hard disks and had an annual economic impact of few billions dollars.

An estimation about the long-term impact of nanotechnology was presented in September 2000 during the review of NNI and published in 2001 [4]. The estimation was based on consultation with experts and technical executives in industry in 10 sectors. The level of penetration of nanotechnology was estimated in each sector per group of products and then multiplied by the total production levels. The findings lead to the overall conclusion that nanotechnology will bring revenues of about $1 trillion by 2015 increasing with an annual rate of about 25%. In 2004, Lux Research adopted a similar approach of evaluation using direct surveys in industrial units. The respective survey data on past intervals have a good international database and were used for comparison. Other topical studies on market size based on the collection of information from a single sector and partial databases have been less useful for comparison.

In 2014, the US annual public R&D investment (from federal, state, and local sources) has been estimated to about $1.8 billion while the private investment was about $4.0 billion [7]. In 2013, in the United States alone, according to the same industry surveys, there are more than about $370 billion in revenues from products incorporating nanotechnology as a key functional and competitive component. In several areas, nanotechnology has become a large part of the market. For example, around 60% of semiconductors and over 40% of manufactured catalysts have some form of nanotechnology involved. The technology has also shown a footprint in emerging research, with approximately 70% of energy-related proposals submitted to National Science Foundation having a basis in nanotechnology. The numbers are significant considering the variety of proposals and ideas. If we consider that for each $0.5 million annual production is needed a nanotechnology worker, then this would require about 740,000 nanotechnology jobs (Figure 1.8). The revenues from products incorporating nanotechnology are about 150 (~318/2.1) times larger than public R&D funding in 2013 that have catalyzed such output. The *investment amplification factors* from foundational nanotechnology to the topical technologies and then to application technology platforms leading to final product and service delivery (Table 1.1) support such a large return.

Nanotechnology is rapidly growing in a field still in formation, at the beginning of the S curve of development (Figure 1.4). Only relatively simple nanostructures are found in applications, such as nanolayers in semiconductor industry and coatings, dispersions in catalyst industry and paints, and nanomodules with surface molecular recognition for targeted medical therapeutics, to name a few. Despite the fact that nanotechnology is still in the formative phase of development, if one would consider an average tax of 20% and apply this to about $318 billion US market incorporating nanotechnology in 2013, the result would be

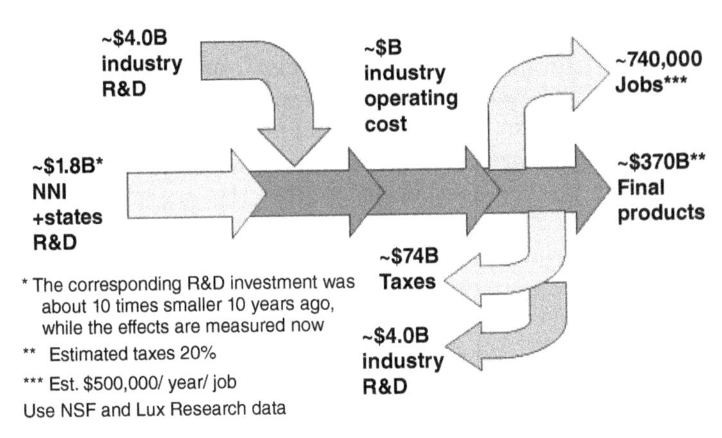

Figure 1.8 The flow of nanotechnology R&D investments and outcomes in the United States in 2014. Roco 2016; updated data and Figure 1.4 from Roco *et al.* [1].

$64 billion in tax revenue in 1 year (2013) that exceeds the total 13 year R&D investment of NNI by a factor of almost four.

The revenues from products incorporating nanotechnology in the world based on direct survey in industry [7] have reached $1014 billion in 2013, with 31.8% of this amount in Europe, 31.3% in the United States, 30.5% in Asia, and 6.4% in the rest of the world. The main sectors identified in the survey are materials and manufacturing with 59%, electronics and information technology with 29%, healthcare and life sciences with 10%, and energy and environment with 2%.

Key long-term qualitative and quantitative targets set up in 2000 for this interval have been realized and even some exceeded such as in memory devices, nanoelectronics, molecular detection of cancer, and market impact. Several illustrations are as follows:

- In his presentation made on January 21, 2000, at Caltech, President Clinton stated in his announcement of NNI [11]: "Just imagine materials with 10 times the strength of steel and only a fraction of the weight; shrinking all the information at the Library of Congress into a device the size of a sugar cube; detecting cancerous tumors that are only a few cells in size. Some of these research goals will take 20 or more years to achieve. But that is why – precisely why – ....there is such a critical role for the federal government." All these targets are currently more advanced than initially envisioned with significant progress on structural metals, foams and composites, creating addressable nanostructures for memory devices that go to 12 atoms and even one atom, and methods for targeted detection and treatment of cancer [1].

- The original report with a 10-year vision [2] recommended progress from fundamental concepts to applications in 10 sectors, and establishing a flexible infrastructure and education community. The progress reported after 10 years shows that a multidisciplinary community has been established using a suitable specialized infrastructure and the progress in various sectors generally has reached the core objectives [1]. The exploratory concepts outlined in the 1999 report are still valid. Nanotechnology as a foundational S&T field promised to branch out in various technology platforms. Large impact illustrations of the potential of nanotechnology with multibillion dollar revenues have been the applications GMR at the beginning of NNI 1998–000 (about $2 billion in direct production in 2000, R&D lead by IBM), targeted drug delivery (PHARMA), semiconductors (SRC), and nanostructured catalysts (Exxon-Mobil) [1]. The promise that nanotechnology will penetrate several key industries has been realized. Catalysis by engineered nanostructured materials impacts 30–40% of the US oil and chemical industries (Chapter 10 in the Nano 2020 report); semiconductors with features under 100 nm constitute over 30% of that market worldwide and 60% of the US market (Chapter on Long View in Nano 2020 report); molecular medicine is a growing field, and only in 2010 about 15% of advanced diagnostics and therapeutics are nanoscience based. These and many other examples show nanotechnology is well on its way to reaching the goal set in 2000 for it to become a "general-purpose technology" with considerable economic impact.
- We have estimated that nanotechnology funding and production will grow by about 25% in the first 10 years after 2000. Global nanotechnology development has shown significant and consistent increases by high annual rates from 2000 to 2013 of about 25% for primary workforce (about 600,000 in 2010) and market value of final products incorporating nanotechnology (about $300 billion in 2010). The nanotechnology labor and markets are estimated to continue to double every 3 years, reaching about $1 trillion market with 2 million jobs around 2015, and $3 trillion market encompassing 6 million jobs by about 2020. The rate of increase for venture capital investment is about 30% reaching about $1.3 billion in 2010. As shown earlier, in 2000, it was estimated that the products incorporating nanotechnology will bring world revenues of $1 trillion by 2015 [4]. The industry survey of Lux Research [7] has reported that this target has been realized in 2013.

The nanotechnology R&D investment has significant returns despite the relatively short term for fundamental discoveries of a foundational field in science and engineering to find the way to applications. Nanotechnology already has a major and lasting impact that promises to be more relevant for healthcare, environment, and manufacturing here on Earth than the Space program. The R&D

challenges for the future in establishing nanotechnology in the economy are the creation of science-based nanosystems, new nanosystem architectures for new technology platforms, emergence of nano-bio hybrid systems, and convergence with other foundational technologies for societal benefit. The nanotechnology revenues are estimated to reach about 5% of the Gross Domestic Product (GDP) in developed economies by 2020 and, respectively, over 10% of GDP by 2030.

Nanoscale science and engineering in the past 10 years is a springboard for future nanotechnology applications and other emerging technologies. We estimate that introduction of nanotechnology in various economic sectors such as electronics and pharmaceutics will lead to at least 1% increase annually in productivity during the 2020s in a similar manner as another general purpose technology – information technology – did in the 1990s.

R&D investments for various topical S&T platforms have been recognized to have significant returns in long term (with an average investment amplification factor for topical platforms of $k_t \sim$ 3–5 times; Table 1.1) as they support creation and use of new products, services, and tools with higher efficiency. Nanoscale science and engineering, like information and communication technology, is a foundational, general purpose technology that supports the topical S&T platforms. By assuming a similar fractal effect in improving basic methods and tools in foundation and topical platforms ($k_f \sim k_t$), then the cumulative return on investment may be roughly estimated as being ($k_f \times k_t$) $\sim$ 10–20 times.

Nanotechnology would reach economic, large-scale application including convergence with other foundational technologies and their spin-offs domains by 2030, at a similar level of development of infrastructure, tools, and design methods that information technology and the Internet achieved by about 2000.

1.6 Closing Remarks

Nanotechnology development has become an international scientific and technological endeavor with focused R&D programs in over 80 countries after the announcement of the NNI in the United States in 2000. The long-term vision and collaboration among the national and international programs are essential factors in this global development. Across the major developments in science and technology at present, nanotechnology positions itself as the most exploratory of them. With much of the research in many sectors, such as in information technology for example, focused on applications, in nanoscale science and engineering there are less established methodologies – giving it perhaps the greatest scope for discovery, manipulation, and diversification in coming years.

Nanotechnology is still in the formation phase of creating nanosystems by design for fundamentally new products. As we are at the beginning of the S-curve of nanotechnology development [6], a main challenge is to prepare the knowledge base, manufacturing, people, physical infrastructure, and anticipatory governance for nanotechnology of tomorrow. In this phase of development of nanotechnology, there is a need to invest in dedicated R&D programs on new nanotechnology methods and system architectures suitable for various economy sectors, adapt education programs and physical infrastructure to its unifying and integrative concepts, and institutionalize nanotechnology in the respective societal institutions. The global nanotechnology-based labor and markets are estimated to double about every 3 years, reaching over $3 trillion market encompassing 6 million jobs by 2020, and about one order of magnitude more by 2030 when nanotechnology will become a key competency and a significant component of GDP in developed countries.

Acknowledgments

This chapter is based on the author's experience in the nanotechnology field, as founding chair of the NSET subcommittee coordinating the NNI and as a result of interactions in international nanotechnology policy arenas. The opinions expressed here are those of the author and do not necessarily reflect the position of U.S. National Science and Technology Council (NSTC)'s Subcommittee on Nanoscale Science, Engineering and Technology (NSET) or National Science Foundation (NSF).

References

1 Roco, M.C., Mirkin, C., and Hersam, M.C. (2011) Nanotechnology research directions for societal needs in 2020. *Journal of Nanoparticle Research*, **13** (**3**), 897–919, (Full report with the same title was published by Springer, Dordrecht, 2011, 670p). Available on www.wtec.org/NBIC2-Report/.
2 Roco, M.C., Williams, R.S., and Alivisatos, P. (1999) *Nanotechnology Research Directions: Vision for the Next Decade*, Springer (formerly Kluwer Academic Publishers) IWGN Workshop Report 1999, Washington, DC: National Science and Technology Council. Also published in 2000 by Springer, Dordrecht. Available on http://www/wtec.org/loyola/nano/IWGN.Research.Directions/.
3 Roco, M.C., Bainbridge, W.S., Tonn, B., and Whitesides, G. (2013) *Converging Knowledge, Technology and Society: Beyond Nano-Bio-Info-Cognitive Technologies*, Springer, available on www.wtec.org/NBIC2-Report/.
4 Roco, M.C. and Bainbridge, S.M. (2003) *Converging Technologies for Improving Human Performance: Nanotechnology, Biotechnology, Information Technology and Cognitive Sciences*, Springer.
5 Harper, T. (2015) AzoNano Reviews.

6 GAO (Government Accountability Office) (2014) Forum on Nanomanufacturing. Report to Congress, GAO-14-181SP.
7 Lux Research (2016) Revenue from nanotechnology (Figure 11), in *Nanotechnology Update: U.S. Leads in Government Spending Amidst Increased Spending Across Asia* (eds I. Kendrick, A. Bos, and S. Chen), Lux Research, Inc. report to NNCO and NSF, New York, p. 17.
8 PCAST (2014) *Report to the President and Congress on the Fifth Assessment of NNI*, White House, Washington, DC, http://www.whitehouse.gov/sites/default/files/microsites/ostp/PCAST/pcast_fifth_nni_review_oct2014_final.pdf.
9 Roco, M.C. (2014) *Nanotechnology Tomorrow: to System Integration and Societal Immersion*, Nano Monterrey, November 10, Keynote presentation.
10 Lok, C. (2010) Nanotechnology: small wonders. *Nature*, **467**, 18–21.
11 Clinton (2000) Caltech, "Presidential archives". http://www.gpo.gov/fdsys/pkg/WCPD-2000-01-24/html/WCPD-2000-01-24-Pg122-3.htm

2

Nanocarbon Materials in Catalysis

Xing Zhang, Xiao Zhang, and Yongye Liang

Department of Materials Science and Engineering, Southern University of Science and Technology, Shenzhen, China

2.1 Introduction to Nanocarbon Materials

As a basic element, carbon plays an essential role in the Earth's civilization. The unique ability of carbon atoms to form robust covalent bonds within themselves or with other nonmetallic atoms in diverse hybridization states (sp, sp^2, and sp^3) enables them to afford a wide range of structures, from small molecules to long-chain polymers and three-dimensional macrostructures [1]. In the past few decades, many new carbon structures have been discovered and explored to fill the gap between small organic molecules and macroscopic natural carbon materials (such as graphite and diamond) [2]. Especially, the discovery of fullerenes in 1985 opened the door to a novel group of carbon allotropes in the nanoscale [3]. Subsequently, a variety of carbon nanostructures were discovered and studied to continuously push the development of carbon-based materials to a new horizon. These carbon nanostructures can be roughly divided into two general groups based on the predominant types of covalent bonds linking between carbon atoms. The first group is characterized by its graphitic structure, which is primarily made up of sp^2 carbon atoms that are densely packed in a hexagonal honeycomb crystal lattice. The representative members in this group are fullerenes, carbon nanohorns, carbon nanotubes (CNTs), graphene, graphene nanoribbon, carbon dots, and carbon nanofibers. Graphene is regarded as the most representative because it can serve as the "building block" to form other graphenic/graphitic nanoallotropes from a theoretical perspective. Another group of carbon nanostructures is mainly constructed with sp^3 carbon atoms and may be incorporated with slight amorphous and graphitic regions. Nanodiamond is a representative member in this group. Besides, there are some special carbon nanostructures existing in

Nanotechnology Commercialization: Manufacturing Processes and Products, First Edition.
Edited by Thomas O. Mensah, Ben Wang, Geoffrey Bothun, Jessica Winter, and Virginia Davis.

a mixed hybridization form such as sp–sp^2 and sp–sp^2–sp^3, with graphdiyne and amorphous carbon as the typical representatives, respectively [4, 5].

Nanocarbon materials have been extensively studied for decades for utilizations in various fields including electronics, optoelectronics, energy conversion and storage, catalysis, biomedical materials, and structural materials [6–10]. The reason for the intense research interest in nanocarbon materials is due to their appealing characteristics, such as high electroconductivity, large surface area, excellent mechanical properties, high thermal conductivity, unique optical properties, and tunable biocompatibility [11, 12]. In addition, nanocarbon materials can be easily synthesized, functionalized, processed, and hybridized with other materials through general physical or chemical methods, enabling them to be star materials in research and industry areas [7, 13, 14]. Some of the superior properties of nanocarbon are highly desirable in catalysis. In this chapter, we briefly give an overview of the recent advancements of nanocarbon materials in electrocatalysis and photocatalysis applications.

2.2 Synthesis and Functionalization of Nanocarbon Materials

Nanocarbon materials can be synthesized through two kinds of strategies: bottom-up and top-down routes [2, 9]. The bottom-up route is more general, which constructs carbon nanostructures from small organic molecules as the carbon source. This method is convenient for tuning the morphology, composition, and sizes. In contrast, the top-down strategy is appealing as it is suitable for large-scale production at low cost. In this section, we focus on the synthesis and functionalization of carbon nanostructures such as CNTs, graphene/graphene oxide, carbon nanodots, and mesoporous carbon, which have been extensively studied in the catalysis area.

2.2.1 Synthesis and Functionalization of Carbon Nanotubes

CNTs are a representative one-dimensional structure allotrope of nanocarbon materials, which can be considered as a hollow cylinder formed through scrolling single- or multilayered of graphene [1]. Currently, CNTs are generally synthesized by three methods including arc discharge deposition, laser ablation, and chemical vapor deposition (CVD) [2, 6, 15]. CNTs were first discovered accidentally when characterizing the products of fullerene synthesized by arc discharge deposition [16]. However, CNTs obtained from this method had many disadvantages such as low yield, low selectivity, and difficult quality control. Later, Smalley *et al.* developed a new method in which CNTs were synthesized by heating the graphitic carbon with laser ablation instead of arc discharge [17]. Both these methods suffered from low yield and high operation temperature, which impeded the extensive application of these methods. CVD is widely accepted as the most promising route to synthesize CNTs with high yield and excellent selectivity [18, 19]. The CVD method generally uses gaseous carbon sources, such as methanol, ethanol, and hydrocarbons, and can be performed at relatively lower temperature (500–1200 °C) with the assistance of transition metal catalysts, such as Fe, Ni, and Co [20–23]. In a CVD process, the diameter, wall number, and length of CNTs can be tuned by varying the structure and composition of the catalysts [20]. Recently, Wei *et al.* synthesized high-quality CNTs with lengths up to half meter by suspending the catalysts in the reaction chamber during the CVD growth process of CNTs [24]. Through about two decades' exploration, CNTs can now be massively synthesized at low cost, which lays a good foundation for the large-scale applications of CNTs in all kinds of fields [25–27].

Despite owning intriguing electrical, mechanical, and thermal properties, intrinsic CNTs are generally inert in chemical catalysis [28]. The intercalation of heteroatoms such as boron, nitrogen, oxygen, sulfur, selenium, and phosphorus into the graphitic lattice of CNTs could induce electronic modulation to alter their optoelectronic properties and/or chemical activities. Heteroatom doping can be achieved either by *in situ* doping during the synthesis process of CNTs or by posttreatment of CNTs with heteroatom-containing species [29–35]. In addition, CNTs functionalized with inorganic or organic species through covalent and noncovalent methods could induce synergistic effects to improve their chemical catalytic activities [12, 36–38].

2.2.2 Synthesis and Functionalization of Graphene and Graphene Oxide

Graphene is the most basic building block in the family of graphitic nanoallotropes. With the unique atom-thick two-dimensional structure, it exhibits good interactions with other materials besides superior electrical, mechanical, optoelectronic, and thermal properties [1, 14]. Compared with CNTs, the synthesis methods for graphene are more flexible. High-quality graphene could be obtained either by top-down or bottom-up routes. Graphene was first discovered by Novoselov and Geim through a micromechanical exfoliation method. The graphite crystal was repeatedly stripped until a monoatomic sheet was maintained [39]. Since then, several bottom-up approaches have been developed for synthesizing graphene through chemical reactions. Among these methods, CVD methods are regarded as the most prevalent ones, which could epitaxially grow graphene on insulated SiC substrate [40] or catalytically grow graphene on metal substrates such as Cu and Ni [41, 42]. Nowadays, methods for scalable synthesis of graphene with low cost CVD are in progress.

Graphene can also be derived from the reduction of graphene oxide (GO), but its electrical performance is generally inferior to the less-defective CVD graphene [43]. In catalysis applications, GO can be a better candidate for the reason that GO has many oxygen-containing groups (such as hydroxyls, epoxides, carbonyls, and carboxyl acid groups), which are beneficial for further functionalizations through covalent and noncovalent methods [12]. Further, GO could be synthesized through low-cost and scalable solution chemistry methods such as the modified Hummers' method [43]. Recently, the electrochemical exfoliation methods has been regarded as a promising alternative to the conventional Hummers method for large-scale production of high-quality and nonmetal contaminated GO [44]. Currently, most graphene-based materials used in catalysis stem from GO. The functionalizations of defect-less CVD graphene often adopt noncovalent interactions such as $\pi-\pi$ stacking and hydrophobic wrapping [45]. While for GO, in addition to the noncovalent interactions, covalently bonding with heterogeneous species through the defective sites such as oxygen-containing groups could be more effective. It has been demonstrated that the defective sites in GO or oxidized CNTs are the preferred nucleation sites for heterogeneous metal oxides or other compounds, resulting in high-performance hybridized materials between them (Figure 2.1) [12].

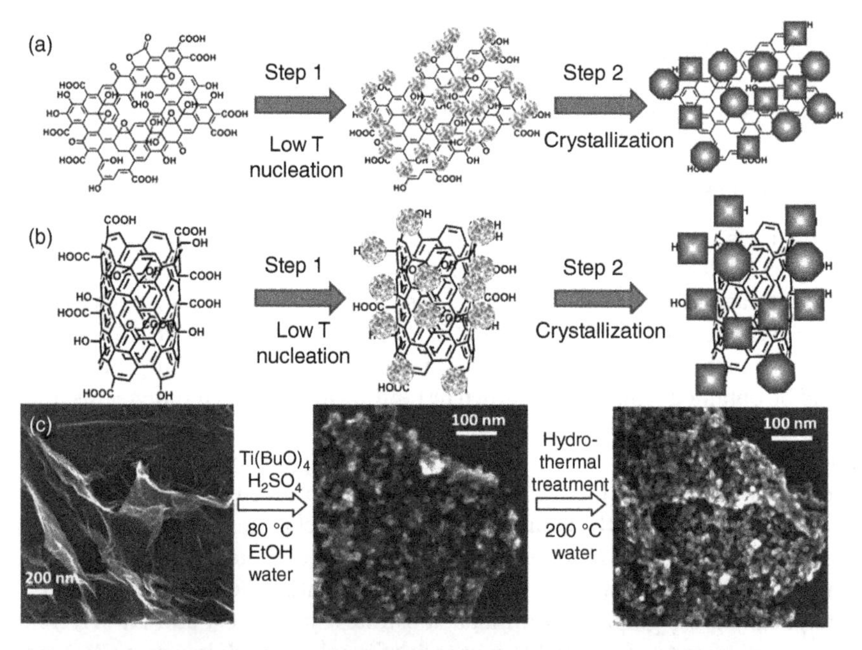

Figure 2.1 Schematic illustration for the construction of inorganic nanoparticle/nanocarbon hybrids. The cases of hybridizing with (a) GO and (b) CNT. (c) Transmission electron microscopy (TEM) images to show an example of TiO_2/GO hybrid. Liang *et al.* 2013 [12]. Reproduced with permission of American Chemical Society.

Similar to CNTs, doping with heteroatoms into the graphitic lattice or on the edge sites of graphene or GO is an alternative route to improve the chemical activities for catalysis applications. The doping strategies also mainly include *in situ* doping during the synthesis of graphene or GO and posttreatment of graphene or GO by treating with heteroatom-containing species as the doping source. A variety of nonmetal elements such as boron, nitrogen, sulfur, phosphorus, and fluorine have been used as the heteroatoms for doping graphene or GO [46–51].

2.2.3 Synthesis and Functionalization of Carbon Nanodots

Carbon nanodots constitute a fascinating class of newly discovered nanocarbons that comprises discrete, quasi-spherical nanoparticles with sizes below 10 nm [52]. Initially, carbon nanodots attracted considerable attention mainly due to their unique photoluminescence properties, which are strongly related to their sizes, excitation wavelengths, and surface terminations [53, 54]. Carbon nanodots were first discovered as a byproduct in preparing single-walled CNTs (SWCNTs) by arc discharge methods [55]. Later, numbers of methods have been developed to fabricate carbon nanodots, including arc discharge [55], laser ablation [56], electrochemical exfoliation [57], hydrothermal carbonization [58], ultrasonic-assisted carbonization [59], oxidization etching of CNT or graphene [60], microwave-assisted carbonization [61], and electrochemical carbonization [54]. Among these methods, the obtained carbon nanodots are generally terminated with oxygen-containing groups on the outer surface of graphitic core except arc discharge and laser ablation. It is widely suggested that the oxygen-containing groups endow carbon nanodots with excellent photoluminescence properties [52, 53].

In addition to the superior photoluminescence properties and morphology, carbon nanodots also exhibit similar properties with respect to GO. Therefore, the methods for functionalization of GO with heteroatoms could be easily applied to carbon nanodots either through *in situ* doping or posttreatment [53]. However, limited by their tiny size in three dimensions and low electrical conductance, carbon nanodots are not preferred to be used as supports for heterogeneous species as GO. Generally, carbon nanodots are often inversely modified onto the surface of other heterogeneous species such as metal or semiconductor nanostructures [62, 63].

2.2.4 Synthesis and Functionalization of Mesoporous Carbon

Mesopores are defined as pores in a narrow pore size range 2–50 nm [14]. Carbon materials with mesoporous morphology are fascinating due to their high surface area. Mesoporous carbon was first reported by Ryoo *et al.* through replicating from a mesoporous silica mould (Figure 2.2) [64]. Ordered mesoporous carbon materials are generally synthesized through a template-assisted

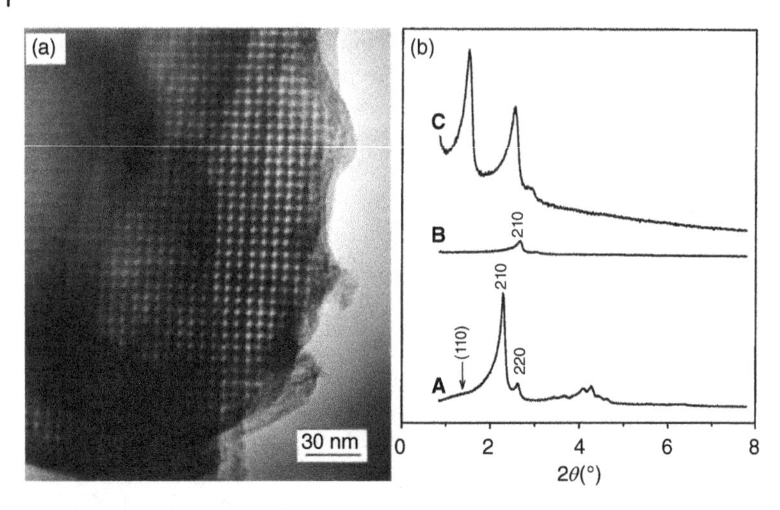

Figure 2.2 (a) TEM image of the ordered carbon molecular sieve CMK-1. (b) Evolution of X-ray diffraction (XRD) patterns during the synthesis of CMK-1 with the silica template MCM-48. A: The silica template MCM-48, B: the carbonized composite, and C: CMK-1 after removal of the template. Ryoo *et al.* 1999 [64]. Reproduced with permission of American Chemical Society.

method. Both hard and soft templates can be employed as the moulds for fabricating of mesoporous carbon. Generally, a hard template method includes three key procedures for synthesizing ordered mesoporous carbon: (i) impregnating the carbon sources into the preformed template; (ii) carbonization of the carbon source; (iii) removing the template [65]. Accessible hard templates for fabricating mesoporous carbon include MCM-48, SBA series, M41S family, MSU-H, hexagonal mesoporous silica, and densely self-assembled colloid silica particles [14, 65]. Mesoporous carbon can also be derived from a soft template, which could simultaneously play the morphology directing role and serve as the carbon source [66]. Four key requirements are generally needed for the successful synthesis of mesoporous carbon materials using soft templates: (i) the ability of the precursors to self-assemble into ordered nanostructures; (ii) the presence of at least one pore-forming component and at least one carbon source; (iii) the pore-forming component can withstand the temperature required for curing the carbon source precursor but can be readily decomposed with low carbon yield during carbonization; and (iv) the ability of the carbon source to form a highly cross-linked polymeric material that can maintain its nanostructure during the decomposition or the extraction of the pore-forming component [65].

The 3-D interconnected and ordered structure endows mesoporous carbon with its high surface area while maintaining good electrical conductance. When being used as the supports for modification with catalytically active species, mesoporous carbon could enhance the volume density of catalytically active sites and interfaces between catalytically active species and reacting substances. Molecules, metal clusters, and ultrasmall nanostructured species have been incorporated into mesoporous carbon to improve their catalytic activities [67–69]. Similarly, doping mesoporous carbon with heteroatoms could also improve its intrinsically catalytic activities [70].

2.3 Applications of Nanocarbon Materials in Electrocatalysis

Due to the intermittence and maldistribution of sustainable energy sources such as solar energy, wind energy, and tide energy, efficient energy conversion and storage are urgently required to satisfy the huge energy demand of future society. Electrocatalysis plays vital roles in highly efficient conversion and storage between electrical and chemical energy. An electrocatalyst is the most hardcore component in an electrocatalytic device, which can lower the activation energy barriers for kinetically inert chemical reactions. An active electrocatalyst can speed up the chemical reaction and reduce the energy exhaustion. In addition to the activity, a catalyst should also be cost-effective, durable, and sustainable for its large-scale and long-term applications.

Precious metal-based materials have been demonstrated to be the most effective electrocatalysts currently for electrochemical reactions such as oxygen reduction reaction (ORR), oxygen evolution reaction (OER), hydrogen evolution reactions, (HER) hydrogen oxidation reaction, CO_2 reduction reaction, and alcohol fuels oxidation reactions [12]. However, the scarcity and high cost of these precious metals make them questionable for large-scale applications. Developing alternatives to replace the precious metal-based catalysts is in demand. However, most of the nonprecious metal catalysts developed still underperform precious metal counterparts. Strategies to develop new materials to improve their electrocatalytic activities and stabilities are highly desirable [28].

Nanocarbon materials have attracted great attentions in the developments of high-performance electrocatalysts due to their excellent characteristics such as high surface area, excellent electrical performances, plentiful surface chemistry, easy processability, low cost, and good stability. Nanocarbon materials can either behave as an intrinsic electrocatalyst by doping heteroatoms into them [28, 33, 46] or be used as a support for anchoring other active materials with improved activities and stability [12, 14, 37, 68]. In either way, applying nanocarbon materials in electrocatalysis is a promising strategy to develop cost-effective electrocatalysts for future large-scale applications. In this section, we focus on some recent progress of the application of nanocarbon materials in several important electrocatalytic reactions.

2.3.1 Oxygen Reduction Reaction

ORR is the cathode reaction for a range of energy-related applications such as fuel cells, metal–air batteries, and chloro-alkali electrolysis. As ORR requires coupled four electron–proton transfer, it is difficult for an electrocatalyst to efficiently lower the activation barriers of all the related elementary reactions. Therefore, ORR is kinetically sluggish, and very few high-performance ORR electrocatalysts have been developed. Currently, carbon-supported Pt-based materials are demonstrated to be the most efficient ORR catalysts with low overpotential and good stability. Other low-cost alternatives, such as complexes with macrocyclic N_4 ligands [71], metal-N complex on carbon matrices (M-N_x/C) [72], metal chalcogenides [73], perovskites [74], and metal oxides [75], have been explored with promising advances. Nevertheless, the catalytic performance of these alternatives still largely lag behind the Pt-based electrocatalysts for ORR. Therefore, developing new catalysts with enhanced activity and stability, as well as low cost, is highly demanding.

Nanocarbon materials have been demonstrated to be excellent supports for loading active ORR electrocatalysts and improving the catalytic activity and stability of the hybridized electrocatalysts through synergistic interactions between the carbon support and the anchored active species [7, 9, 12]. There are several advantages for nanocarbon materials to serve as the catalysts' support. First, nanocarbon materials such as graphene and CNTs could improve the electrical conductance of the hybridized electrocatalyst to facilitate the electron transport between the electrode and the catalytic active sites. Second, the large specific surface area of nanocarbon materials could significantly enhance the volume density of the catalytic active sites. Third, the rich surface chemistry characteristic of defective nanocarbon materials (such as GO or oxidized CNTs) could tune the size and distribution density of catalytic active species growing on them. Fourth, the carbon supports could generate some electronic interactions with the supported catalyst and alter the catalytic activity and stability of the supported species. Fifth, the anchoring effect of the nanocarbon materials could slow the agglomeration of the catalysts during operation to improve the stability.

ORR catalytic active species could be anchored on nanocarbon materials either through π–π interactions or covalent grafting. Recently, Guan *et al.* reported that iron phthalocyanine (FePc) coated on SWCNTs via π–π interactions (Figure 2.3a) exhibited significant enhancement of ORR activity compared to pristine FePc, which usually suffers from severe aggregation and low electron conductivity [76]. Later, Cho *et al.* found that anchored FePc on SWCNTs with an axial ligand linker (Figure 2.3b) could induce significant electronic and geometric interactions, which greatly improved the catalytic performance of the hybridized electrocatalyst for ORR in terms of the onset potential, Tafel slope, and durability [77]. Liu *et al.* also found that covalently

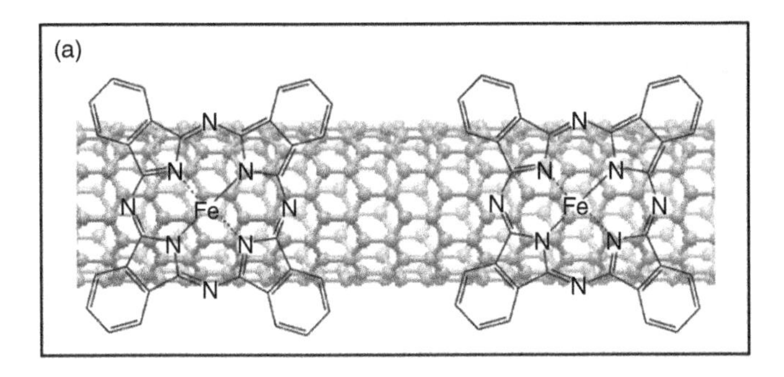

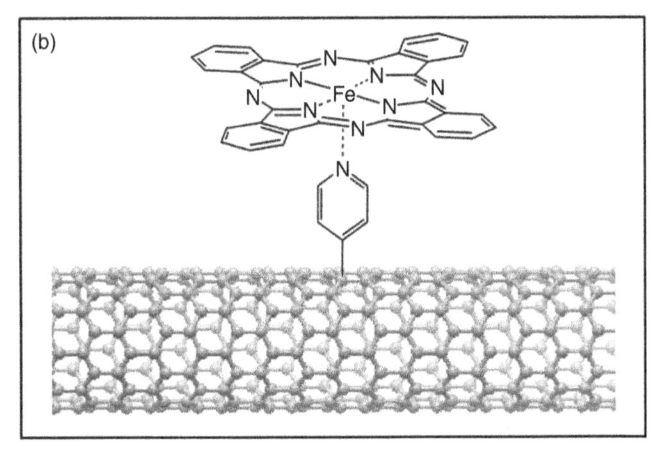

Figure 2.3 Schematic illustration of modifying SWCNTs with FePc via (a) π–π interactions and (b) axial ligands.

grafted iron porphyrin onto multiwalled CNTs (MWCNTs) with an imidazole linker generated obviously positive effects in improving the catalytic activity of the hybridized electrocatalyst for ORR [78].

In situ growth of catalytically active species on nanocarbon materials is also an intriguing method to construct inorganic/nanocarbon hybrids as ORR electrocatalysts. Dai *et al.* reported that inorganic nanomaterials such as manganese oxide, lithium metal phosphate, and cobalt sulfide growing on reduced GO significantly differed from those freely growing in solution without GO in terms of the size, morphology, and distribution [12]. They also developed a Co_3O_4/GO hybrid, which featured small Co_3O_4 nanocrystals that uniformly grew on nitrogen-doped reduced GO (N-RGO). The Co_3O_4/N-RGO hybrids exhibited remarkable enhancement of catalytic activity for ORR compared with the Co_3O_4, N-RGO, or their physical mixtures. They demonstrated that the improvement of catalytic performances of the Co_3O_4/N-RGO hybrids was mainly due to the optimal size and distribution of Co_3O_4 on N-RGO and

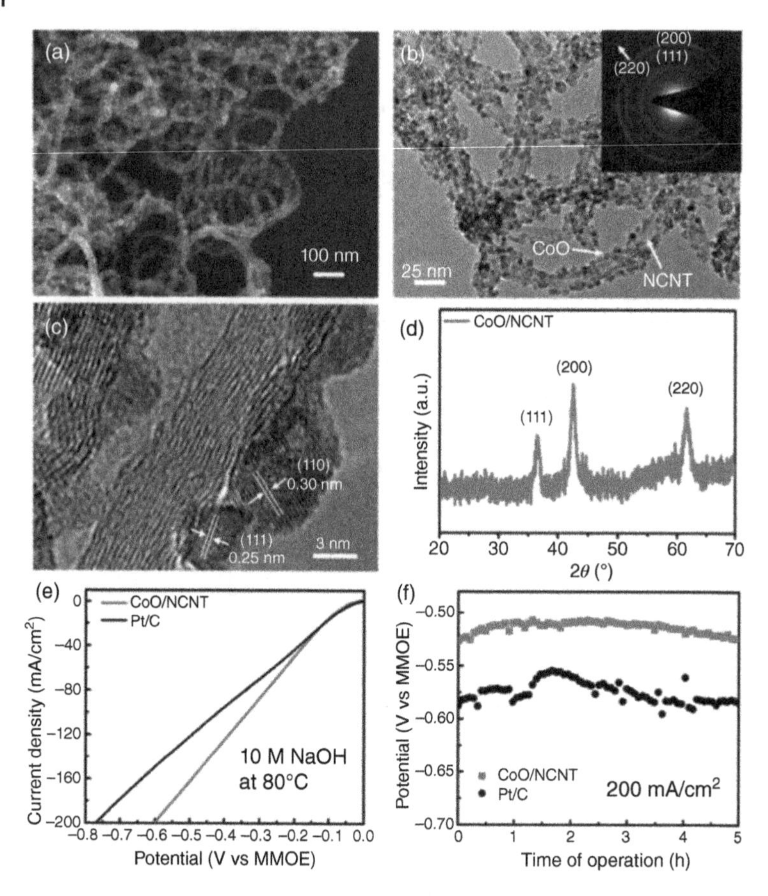

Figure 2.4 (a) Scanning electron microscopy (SEM) image and (b) and (c) TEM images of the CoO/NCNT hybrid. (d) XRD patterns of the CoO/NCNT hybrid. (e) ORR polarization curves of the hybrid and Pt/C. (f) Chronoamperometric curves of the hybrid and Pt/C for ORR at 200 mA/cm^2. Liang *et al.* 2012 [37]. Reproduced with permission of American Chemical Society.

some electronic interactions between Co_3O_4 and N-RGO supports [79]. Later, $Co_{1-x}S$, $MnCo_2O_4$, and CoO nanocrystals were successfully hybridized with either GO or oxidized CNTs as excellent ORR electrocatalysts through *in situ* growing methods [37, 80, 81]. Especially, the hybridized electrocatalysts with CoO nanocrystals that grew on nitrogen-doped oxidized-CNTs (NCNT) outperformed the commercial Pt/C catalysts at intermediate overpotentials in strong alkaline solutions (Figure 2.4).

In addition to serving as supports for heterogeneous ORR catalytically active species, nanocarbon materials could also behave as ORR electrocatalysts after doping with heteroatoms [28, 82]. M-N_x/C catalysts have been regarded as a

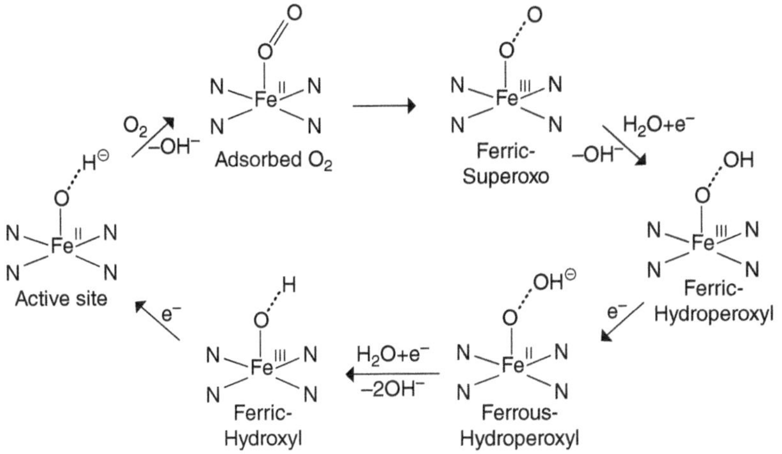

Figure 2.5 Catalytic cycle showing the redox mechanism involved in ORR on pyrolyzed Fe-N_x/C active sites in dilute alkaline medium. Ramaswamy *et al.* 2013 [83]. Reproduced with permission of American Chemical Society.

type of promising nonprecious metal electrocatalysts for replacing Pt-based catalysts for ORR. This type of catalysts can be synthesized either by pyrolyzing carbon-supported N_4-macrocyclic complexes or by pyrolysis of simple mixtures of metal salts, carbon, and nitrogen precursors [83, 84]. Figure 2.5 shows the proposed possible ORR mechanistic pathways on pyrolyzed Fe-N_x/C active sites. Zelenay *et al.* recently designed M-N_x/C catalysts with outstanding activities using polyaniline (PANI) as a carbon–nitrogen template and Co and Fe salts as the metal sources [72]. TEM studies revealed that the catalyst was composed of metal particles encapsulated inside onion-like graphitic carbon nanoshells. Both rotating ring-disk electrode (RRDE) and H_2-fuel cell measurements showed that the M-N_x/C catalysts exhibited activity just slightly inferior to the state-of-the-art Pt/C while owning excellent durability and high selectivity for 4 e^- ORR. More recently, Dai *et al.* reported M-N_x/C catalysts synthesized by controlled oxidation of CNTs to expose buried residual catalysts and subsequent treatment with NH_3 vapor to form M-N_x/C supported on CNT-graphene complex [85]. Compared to benchmark Pt/C, the obtained electrocatalysts exhibited good activity in acid solution and high resistance for methanol crossover effect for ORR. The excellent ORR activities of the catalysts were attributed to the iron and nitrogen species in the complex for forming catalytic active sites and the intact inner walls of the CNTs for highly efficient charge transport.

Noticeably, metal-free and heteroatom-doped nanocarbon materials are also identified as promising candidates for ORR electrocatalysts [28]. In 2009, Dai *et al.* reported that nitrogen-doped vertically aligned CNT arrays could act as metal-free catalysts to catalyze the ORR process with a three-times higher electrocatalytic activity and better long-term operation stability than those of Pt/C [33]. Subsequently, various nonmetal heteroatoms (such as B, N, O, S, P, and Se)-doped nanocarbon materials (such as graphene/GO, CNTs, mesoporous carbon, and carbon dots) had been reported as high-performance electrocatalysts for ORR [32, 34, 35, 48–50, 86–88]. Some of these reports claimed that their metal-free nanocarbon-based catalysts showed better activity and durability than their own referred catalysts prepared from commercial Pt/C. However, some arguments were raised related to the intrinsic active sites in metal-free nanocarbon materials for catalytic ORR because many claimed that metal-free nanocarbon-based catalysts were synthesized with some metal species incorporation. However, metal impurities at levels usually undetectable by techniques such as X-ray photoelectron spectroscopy (XPS), X-ray diffraction (XRD), and energy dispersive X-ray spectroscopy (EDX) could significantly promote the ORR [82]. In addition, strictly metal-free NC catalysts generally performed considerably worse than M-N_x/C and Pt-based catalysts, with H_2O_2 being the main product in acidic electrolytes [82, 89]. Only recently, Dai *et al.* demonstrated that metal-free, nitrogen-doped CNTs and their graphene composites exhibited significantly better long-term operational stabilities and comparable gravimetric power densities with respect to the best nonprecious metal catalysts in acidic proton exchange membrane (PEM) fuel cells [90]. More evidences related to high catalytic activities of truly metal-free nanocarbon-based catalysts in acidic electrolytes are needed to confirm that they are competitive with other nonprecious metal electrocatalysts such as M-N_x/C catalysts.

2.3.2 Oxygen Evolution Reaction

OER is the reverse process of ORR, which is to generate O_2 through electrochemical oxidation of water or hydroxide ions. OER is important for many applications such as water splitting electrolyzers and rechargeable metal–air batteries [12]. Currently, the most active OER catalysts are Ru- or Ir-based electrocatalysts in acidic or alkaline solutions, but these catalysts suffer from the scarcity and high cost of precious metals [91]. Extensive efforts have been taken to develop highly active, durable, and low-cost alternatives, such as first-row transition-metal oxides [91] and perovskites [92]. However, the activities of nonprecious metal electrocatalysts are still worse than noble-metal-based electrocatalysts for OER. It is challenging to develop nonprecious metal catalysts with comparable or even higher performance than the state-of-the-art precious metal-based electrocatalysts.

The first role of nanocarbon materials in improving the activities of OER catalysts was to serve as the support for other catalytic active species. Most of the OER catalysts such as metal oxides/hydroxides and perovskites are not good electrical conductors. Therefore, increasing the electrical conductivity of these catalysts is an effective route to improve the catalytic performances for OER. Extensive efforts have been made to intercalate nanocarbon materials such as CNTs and graphene/GO into the matrices of electrically inert OER active species to facilitate charge transport between the catalytic actives sites and the electrode [79, 81, 93–96]. Recently, Dai *et al.* reported that they synthesized a hybridized catalyst with ultrathin nickel–iron-layered double hydroxide (NiFe-LDH) nanoplates *in situ* grown on mildly oxidized CNTs, which exhibited better activity and durability than benchmark Ir/C catalyst for OER [93]. The excellent OER electrocatalytic activity of the NiFe-LDH/CNTs hybrid was attributed to the high intrinsic activity of the crystalline NiFe-LDH phase and enhanced electron transport by the underlying CNT network. Later, Yang *et al.* developed a high-performance OER electrocatalyst by intercalation of GO between FeNi double hydroxide layers through electrostatic interactions [94]. The catalytic activity of the hybridized electrocatalyst could be further improved by reduction of GO to enhance the whole electrical conductance of the catalyst. Similar to Yang's work, Sasaki *et al.* also synthesized a superlattice hybrid by alternately attaching NiFe hydroxide nanosheets and graphene through direct assembly with the assistance of electrostatic interactions (Figure 2.6) [95]. This hetero-assembled superlattice hybrid electrocatalyst exhibited excellent OER activities with a small overpotential of about 0.23 V and Tafel slope of 42 mV/decade. Other than the hybridization between metal hydroxides and nanocarbon materials, Yu *et al.* also reported an inorganic/nanocarbon hybrid with $CoSe_2$ nanobelts *in situ* grown on nitrogen-doped reduced GO (NG) [96]. In 0.1 M KOH, the hybrid catalyst afforded a current density of 10 mA/cm^2 at a small overpotential of 0.366 V and a small Tafel slope of 40 mV/decade. The excellent catalytic activity of this hybrid catalyst was demonstrated to be attributed to the excellent electrical conductance with the intercalation of NG and beneficial electronic interactions between the NG and $CoSe_2$ nanobelts. More recently, we developed a nickel–iron/nanocarbon hybrid electrocatalyst, which featured *in situ* formed nickel–iron alloy nanoparticles either dispersed on or encapsulated in CNTs. The hybrid electrocatalyst exhibited excellent OER activity superior than the commercial Ir/C benchmark [97].

In addition to these inorganic/nanocarbon hybrids, nonmetal-doped nanocarbon materials have also been reported to exhibit excellent catalytic activities for OER [98–101]. Hashimoto *et al.* reported that nitrogen-doped carbon materials (N/C) derived from pyrolyzing the hybrids of GO and poly(bis-2,6-diaminopyridinesulphoxide) (PDPS) could function as efficient oxygen evolution electrocatalysts, which afforded a current density of 10 mA/cm^2 at the overpotential of 0.38 V [98]. Based on the electrochemical and physical studies, they proposed that the high oxygen evolution activity of the N/C is from the pyridinic nitrogen or/and quaternary nitrogen-related active sites.

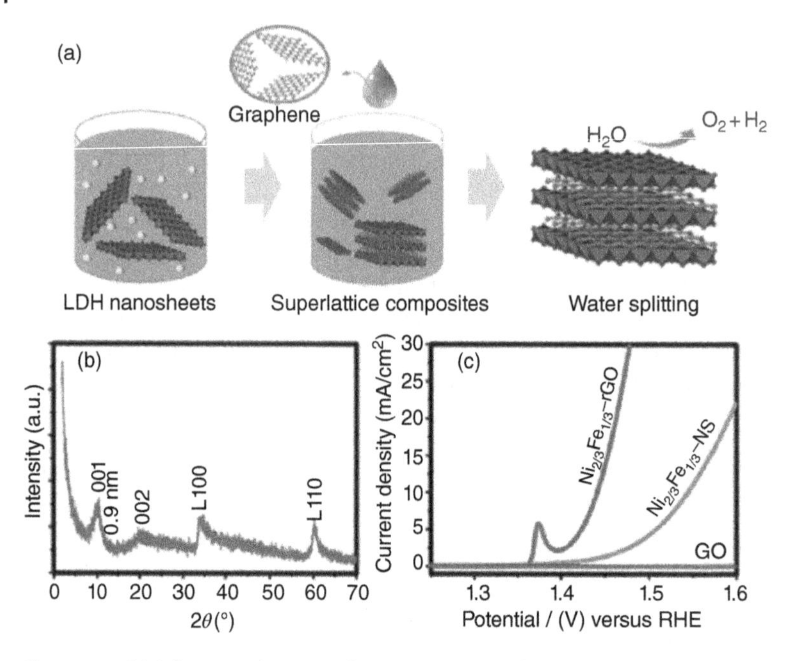

Figure 2.6 (a) Schematic diagram of the preparation of NiFe-LDH nanosheets/graphene for water splitting. (b) XRD pattern of NiFe-LDH nanosheets/graphene. (c) Polarization curves of $Ni_{2/3}Fe_{1/3}$-RGO, $Ni_{2/3}Fe_{1/3}$-NS, and GO in 1 M KOH solution. Ma *et al.* 2015 [95]. Reproduced with permission of American Chemical Society.

Later, Qiao *et al.* first reported that N, O-dual-doped graphene-CNT hydrogel film electrocatalyst exhibited surprisingly high OER activities even outperforming IrO_2 and some transition-metal catalysts [99]. And this metal-free nanocarbon-based electrocatalyst showed excellent stability in both alkaline (0.1 M KOH) and strong acidic (0.5 M H_2SO_4) solutions. More recently, Dai *et al.* developed a metal-free nanocarbon-based catalyst featuring mesoporous carbon foam codoped with nitrogen and phosphorus [101]. This catalyst has a large surface area of $\sim$1663 m^2/g and excellent electrocatalytic properties for both ORR and OER, which was demonstrated feasible in serving as bifunctional cathodic oxygen electrocatalysts for rechargeable Zn–air batteries. Their density functional theory calculations revealed that the N, P codoping and graphene edge effects are essential for the bifunctional electrocatalytic activity of their catalyst. It should also be noted that Zhao *et al.* reported a metal-free OER electrocatalyst (denoted as echo-MWCNTs) consisting of multiwall CNTs, which had been orderly treated in order by mild surface oxidation, hydrothermal annealing, and electrochemical activation (Figure 2.7a) [102]. Their theoretical investigations revealed that the improved catalytic activity for OER of the activated multiwall CNTs was mainly attributed to the

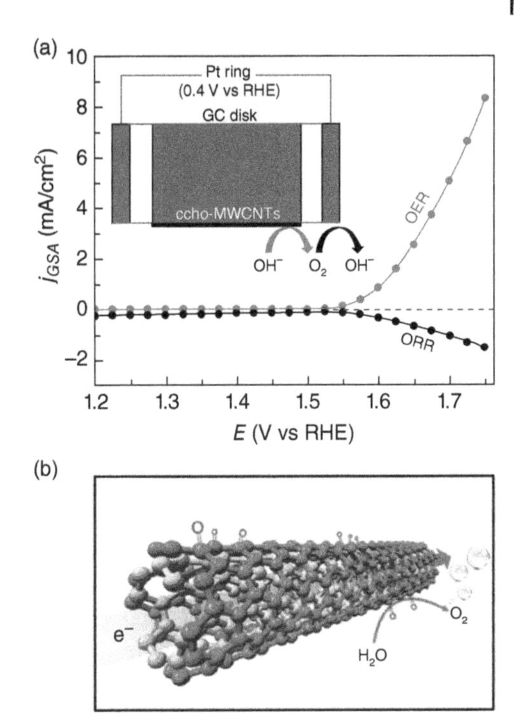

Figure 2.7 (a) Detection of O_2 evolution from the echo-MWCNTs catalysts using rotating ring disk electrode measurements. (b) Schematic diagram for showing the O_2 evolution reactions occurred at the oxygen-containing groups on the surface of MWCNTs. Lu *et al.* 2015 [102]. Reproduced with permission of American Chemical Society.

oxygen-containing functional groups such as ketonic C—O on the outer wall of MWCNTs (Figure 2.7b), which could alter the electronic structures of the adjacent carbon atoms and facilitate the adsorption of OER intermediates.

2.3.3 Hydrogen Evolution Reaction

HER is the process to produce H_2 from water, storing electrical energy in the form of chemical energy. Pt-based materials are still the most efficient electrocatalysts so far for HER with rather small overpotential and Tafel slope. As noted in the earlier sections, Pt-based materials are not good choices for large-scale application due to their high cost and scarcity. More economical electrocatalysts based on nonnoble metal have been identified recently, including transition metal chalcogenides [103], carbides [104], phosphides [105], complexes [106], and metal alloys [107], although they are still less efficient or stable than Pt-based electrocatalysts. Besides searching for other catalytic active species for HER, strategies such as structure engineering, hybridizing, and doping could be developed to improve the performance of these HER catalysts.

Nanocarbon materials have been widely used as the supports to construct inorganic/nanocarbon hybrids as HER electrocatalysts. In addition to the ability to improve the electrical conductance of the hybrid electrocatalysts,

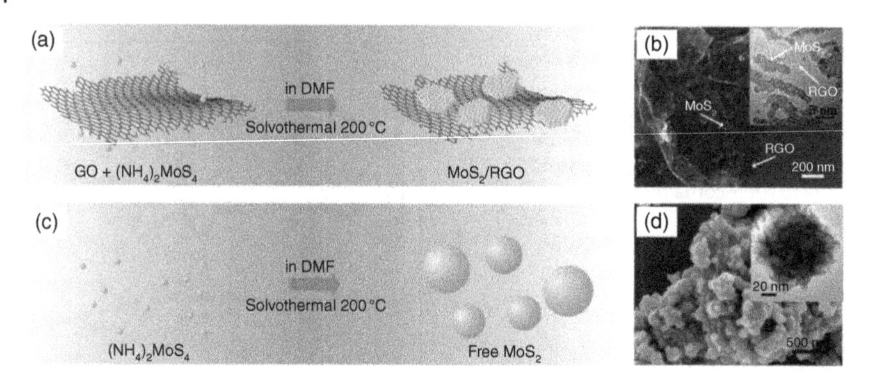

Figure 2.8 (a) Schematic diagram of the preparation of MoS_2/RGO hybrid. (b) SEM and TEM (inset) images of the MoS_2/RGO hybrid. (c) Schematic illustration for the growth of large, free MoS_2 particles without any GO sheets. (d) SEM and TEM (inset) images of the free MoS_2 particles. Li *et al.* 2011 [103]. Reproduced with permission of American Chemical Society.

nanocarbon materials could also efficiently tune the morphology, size, and crystallographic structure of the inorganic catalytic species growing on them in some cases. Dai *et al.* found that MoS_2 grown on RGO afforded a nanoscale few-layer structure with abundant exposed edges stacked on graphene sheets, in strong contrast to large aggregated MoS_2 particles grown freely in solution without RGO (Figure 2.8) [103]. The hybridized MoS_2/RGO catalyst exhibited significantly higher electrocatalytic activity for HER than the free-growing MoS_2 particles. The improved catalytic activity of this hybrid was due to the small size and nanosheet morphology, which was beneficial for exposing the catalytic active edges of MoS_2. As a result, nanocarbon materials could also serve as highly conductive substrates for engineering the morphology of catalysts. Kim *et al.* recently prepared a hybrid HER electrocatalyst composed of amorphous molybdenum sulfide (MoS_x) layers directly bounding to the surface of vertical N-doped CNT forests [108]. The synergistic effects including the dense catalytic sites at the amorphous MoS_x surface, facile charge transport, and mass transport along the NCNT forest conferred the hybrid catalyst with excellent HER activity with a small overpotential of 110 mV at 10 mA/cm^2.

In addition to being supported on the outer surface of nanocarbon materials, catalytic active species encapsulated in graphitic shells of nanocarbon materials could also exhibit excellent HER activity while maintaining good stability even in conditions of strong chemical corrosion [109–111]. Bao *et al.* reported a strategy to encapsulate 3-D transition metals Fe, Co, and the FeCo alloy into NCNTs and investigated their HER activity in acidic electrolytes [109]. They found that the encapsulation could significantly increase the stability of metal species, which might otherwise be easily corroded and dissolved in the acidic electrolytes. Their density functional theory calculations revealed that the introduction of metal and nitrogen dopants could synergistically optimize the

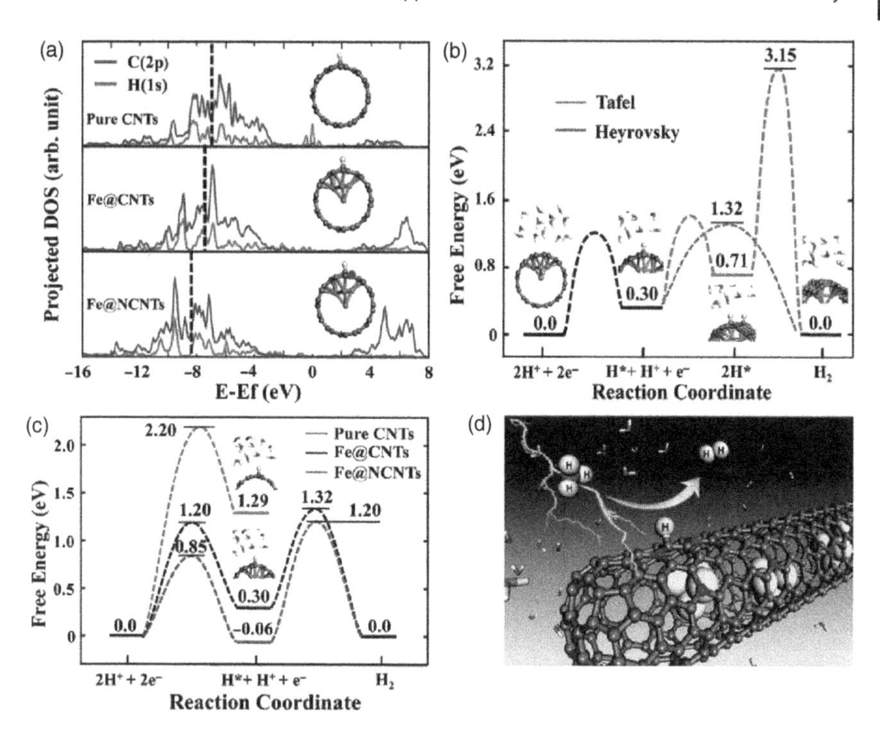

Figure 2.9 (a) Projected density of states (DOS) of H (1 s) and its bonded C (2p) when H is adsorbed on the surface of pristine CNTs, Fe@CNTs, and Fe@NCNTs. The dashed lines present the center of the occupied band. (b) The free energy profiles of Tafel and Heyrovsky routes for Fe@CNTs. (c) The free energy profiles of the Heyrovsky route for pristine CNTs, Fe@CNTs, and Fe@NCNTs. (d) A schematic representation of the HER process on the surface of Fe@NCNTs. Deng *et al.* 2014 [109]. Reproduced with permission of Royal Society of Chemistry.

electronic structure of the CNTs and the adsorption-free energy of H atoms on CNTs, and therefore promoted the HER with a Volmer–Heyrovsky mechanism (Figure 2.9). Later, further theoretical investigations from Bao's group indicated that the HER performance with metal clusters encapsulated in graphene shells originated from the modulation of the electron density and the electronic potential distribution at the graphitic surface by penetrating electrons from the metal core [110]. Their results also suggested that the effect of the enclosed metal clusters on the graphene shells would decline when the graphene shells were more than three layers. Bearing this guidance in mind, they synthesized a high-performance HER electrocatalyst consisted of ultrathin graphene spheres with only one to three graphene layers that encapsulated Co–Ni alloy nanoparticles. This hybrid catalyst showed excellent catalytic activity and durability in acidic electrolyte with an onset overpotential of almost 0 V versus the reversible hydrogen electrode (RHE) and an overpotential of only 142 mV at a current density of 10 mA/cm^2. Subsequently, Laasonen *et al.* reported that a catalyst featuring single-shell carbon encapsulated iron nanoparticles decorated on SWCNTs, which exhibited excellent HER activity just slightly inferior to benchmark Pt/C in terms of the onset potential and Tafel slope [111].

Recently, a type of M-N$_x$/C catalysts was also identified as a promising HER electrocatalyst operating in both acidic and alkaline electrolytes [112]. In this work, a cobalt–nitrogen/carbon catalyst (CoN$_x$/C) was prepared by the pyrolysis of cobalt–N$_4$ macrocycles or cobalt/o-phenylenediamine composites with silica colloids as a hard template (Figure 2.10). With the metal-free N/C and nitrogen-free Co/C as the controls, the authors deduced that the well-dispersed molecular CoN$_x$ sites on the carbon support were the active sites responsible for the HER. This deduction was further confirmed by comparing the HER activities between CoNPs/CoN$_x$/C and CoN$_x$/C catalysts and checking the poisonous effect of SCN$^-$ on the CoN$_x$ active sites.

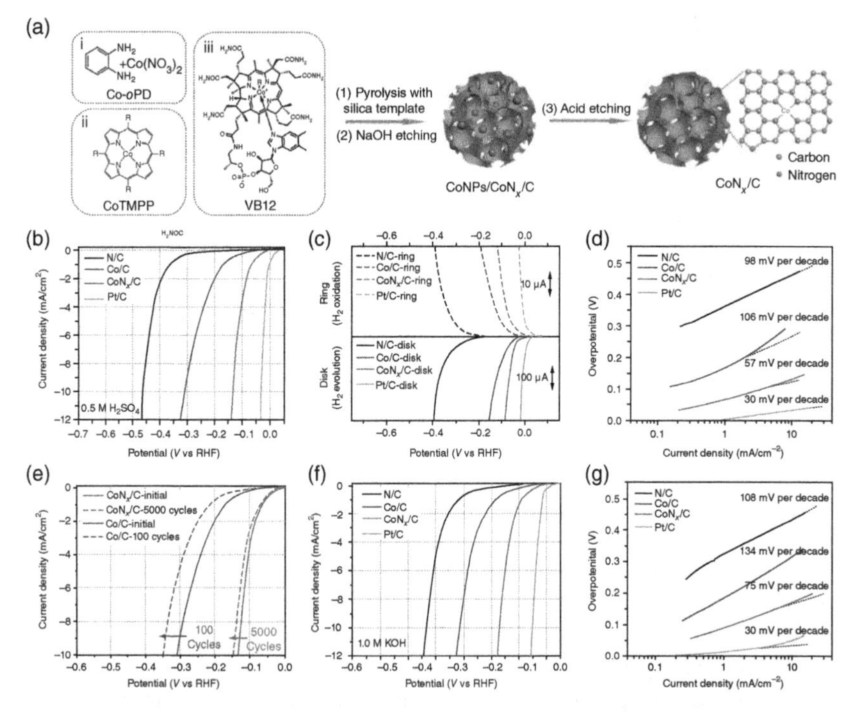

Figure 2.10 (a) Schematic illustration of the synthesis of the CoN$_x$/C electrocatalysts. (b) HER polarization plots, (c) RRDE measurements, and (d) Tafel plots of the CoN$_x$/C, N/C, Co/N, and Pt/C catalysts in 0.5 M H$_2$SO$_4$. (e) Initial and postpotential cyclic voltammograms of CoN$_x$/C and Co/N in 0.5 M H$_2$SO$_4$. (f) HER polarization plots and (g) Tafel plots of CoN$_x$/C, N/C, Co/N, and Pt/C catalysts in 1 M KOH. Liang *et al.* 2015 [112] used under https://creativecommons.org/licenses/by/4.0/

There were also a few intriguing progresses related to metal-free electrocatalysts for HER [113–116]. Qiao *et al.* first reported a metal-free HER electrocatalyst featuring graphitic-carbon nitride hybridized with nitrogen-doped graphene, which exhibited excellent activities in acidic solution and intermediate activities in alkaline solution [113]. Based on their experimental results and theoretical calculations, they proposed that the unusual electrocatalytic properties of the hybrid catalysts originated from an intrinsic chemical and electronic coupling, which synergistically promoted the proton adsorption and reduction kinetics. Later, N, S-dual-doped porous carbon [115] and N, P-dual-doped mesoporous carbon [116] were also been reported as metal-free catalysts with good catalytic activity for HER in acidic electrolytes. However, there were no detailed discussions in these two reports on the specific active sites for the observed HER activity. A more noticeable work reported by Rinzler *et al.* showed that SWCNTs and some graphitic carbons could acquire an activity comparable to Pt, and the materials were activated by cycling the as-prepared electrode in the acid electrolyte in the HER potential region and then leaving it submersed in the same electrolyte for extended time, acquired an activity even comparable to Pt [114]. However, this work was questioned by other researchers who demonstrated that the catalytic performances for HER of these "activated carbons" were greatly affected by the counter electrode employed for the activation. Their work showed that an effective activation was achieved only when platinum wire was used as the counter electrode. Therefore, they argued that the improved HER performances of the graphitic nanocarbons were mainly caused by Pt transfer, rather than the activation of the carbon materials themselves [117]. Even though a high-performance and metal-free HER electrocatalyst are intriguing, more rigorous evidence from both experiment and theory are needed to enable the metal-free HER electrocatalysts to be really competitive to other nonprecious metal HER electrocatalysts.

2.3.4 Roles of Nanocarbon Materials in Catalytic CO_2 Reduction Reaction

Electrocatalytic reduction of CO_2 has been extensively investigated as one of the most promising ways to convert waste CO_2 into useful organic compounds, which can be used as fuels or raw chemicals (Figure 2.11a). The field of electroreduction of CO_2 is far less mature than other electrocatalytic reactions discussed earlier. A main reason is due to the exceptional stability of CO_2, which causes CO_2 reduction in electrolytic devices to require very high electrical energy input. Another reason is that there are generally several concomitant reactions related to cathodic CO_2 reduction and water reduction (Figure 2.11b). Therefore, an effective process for the electrochemical conversion of carbon dioxide to value-added products requires the electrocatalyst to be efficient, selective, and stable [118]. Toward this goal, a

(a)

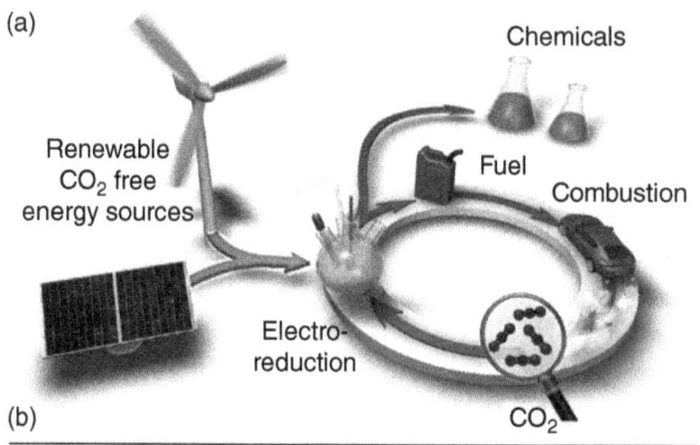

(b)

CO_2 reduction half reactions			[V] vs RHE
$CO_2 + H_2O + 2e^-$	→	$CO + 2OH^-$	−0.10
$CO_2 + H_2O + 2e^-$	→	$HCOO^- + OH^-$	−0.03
$CO_2 + 5H_2O + 6e^-$	→	$CH_3OH + 6OH^-$	0.03
$CO_2 + 6H_2O + 8e^-$	→	$CH_4 + 8OH^-$	0.17
$2CO_2 + 8H_2O + 12e^-$	→	$C_2H_4 + 12OH^-$	0.08
H_2O reduction half reaction			
$2H_2O + 2e^-$	→	$2H_2 + 2OH^-$	0.0

Figure 2.11 (a) Schematic diagram of the electrochemical reduction of CO_2 coupled to renewable electricity sources in carbon cycle. (b) Several incidental reactions related to carbon dioxide reduction and water reduction with relative reduction potentials. Kuhl *et al.* 2014 [118]. Reproduced with permission of American Chemical Society.

number of materials have been studied as cathodic electrocatalysts for CO_2 reduction, including metal-based catalysts (such as Au, Ag, Cu, Pd, Sn, and Zn) [118–120], conducting polymers (such as polyaniline and polypyrrole) [121, 122], pyridinium derivatives [123], complexes of metal (such as ruthenium, iridium, and copper) [124], macrocyclic N_4 ligands (such as cobalt phthalocyanine and cobalt protoporphyrin) [125, 126], and heteroatom-doped nanocarbons [127–131].

Nanocarbon materials have also been explored as the catalysts' support for improving the catalytic activities of the hybrid catalysts for CO_2 reduction. For example, Meyer *et al.* reported that the reduced SnO_2 nanoparticles dispersed on graphene (nano-SnO_2/graphene) exhibited larger cathodic reduction current density at specific overpotentials and higher Faradaic efficiencies for formate production than nano-SnO_2 dispersed on carbon black [119]. They deduced that the higher activity and selectivity for CO_2 reduction of nano-SnO_2/graphene was mainly due to electronic interaction between the graphene supports and reduced SnO_2, which led to enhanced electronic donation, promoting adsorption of CO_2 and $CO_2^{\bullet-}$ and facilitating CO_2 reduction at the Sn surface. Chhowalla *et al.* also found that Cu nanoparticles anchored on GO could obtain higher current density, lower overpotential, and better stability than Cu films and support-free Cu nanoparticles for electrocatalytic CO_2 reduction [132].

An intriguing finding was that heteroatom-doped nanocarbons were also identified as promising electrocatalysts for CO_2 reduction. Salehi-Khojin *et al.* reported that carbon nanofibers with N and O dopants derived from pyrolysis of polyacrylonitrile in Ar atmosphere exhibited better electrocatalytic activity for reducing CO_2 to CO compared with their reference Ag catalyst [127]. Based on their control experiments, they deduced that the catalytic active sites for CO_2 reduction were on the reduced carbons rather than on electronegative nitrogen atoms. The CO_2 reduction reaction was described with three successive steps: (i) intermediate (EMIM–CO_2 complex) formation, (ii) adsorption of EMIM–CO_2 complex on the reduced carbon atoms, and (iii) CO formation. They claimed that the nanofibrillar structure and high binding energy of the key intermediates to the carbon nanofiber surfaces contributed positively to the superior catalytic performances. Subsequently, Meyer *et al.* found that NCNTs were also selective and robust electrocatalysts for CO_2 reduction to formate in aqueous media [128]. In their work, they also demonstrated that a significantly lower overpotential, higher current density, and efficiency could be achieved when the NCNTs were modified with polyethylenimine (PEI) as a cocatalyst, which might help in stabilizing the singly reduced intermediate $CO_2^{\bullet -}$ and concentrating CO_2 in the PEI overlayer. Recently, Ajayan *et al.* reported that NCNTs arrays could behave as an excellent electrocatalyst for reduction CO_2 to CO with a selectivity of 80% at a low overpotential of −0.18 V [129]. Their theoretical calculations proposed that the most selective site toward CO production is pyridinic N and the high selectivity for CO_2 reduction to CO was attributed to the low free energy barrier for the potential-limiting step to form key intermediate COOH as well as strong binding energy of adsorbed COOH and weak binding energy for the adsorbed CO. Later, further theoretical investigations from Ajayan's group showed that pyridinic defects retained a lone pair of electrons that were capable of binding CO_2 while graphitic-like nitrogen electrons were located in the π^* antibonding orbital, which made them less accessible for CO_2 binding [130]. However, their experimental results showed that only the coexistence of graphitic and pyridinic nitrogen in NCNTs could significantly decrease the overpotential (ca. 0.18 V) and increase the selectivity (ca. 80%) toward the formation of CO, opposite to the theoretical predictions (Figure 2.12). The catalytic activity of pyridinic or graphitic nitrogen sites alone could not be distinguished in their experimental data. The proper roles of graphitic nitrogen on the activity and selectivity for reduction CO_2 to CO were not determined in their work. Their theoretical calculations also showed that although pyrrolic defects had a nitrogen-based lone pair of electrons, the geometry of the pyrrolic-like defect moved the nitrogen atom toward the center of the tube, making these electrons harder to access for CO_2 binding. This was consistent with their experimental observation that an increase in pyrrolic N content from 0.78 to 1.3 at.%

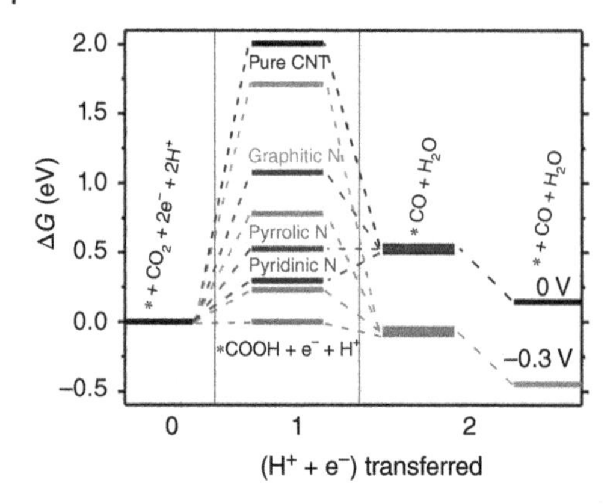

Figure 2.12 Calculated free energy diagram for CO_2 electroreduction to CO on pristine CNTs and NCNTs at 0 V and 0.3 V versus RHE, respectively. Wu *et al.* 2015 [129]. Reproduced with permission of American Chemical Society.

resulted in more cathodic onset potentials from 0.78 to 1.05 V accompanied in company with a decrease in maximum achievable FE for CO from 40% to 25%.

Another noticeable work reported recently was on an M-N_x/C-like catalyst, which exhibited good electrocatalytic activity and selectivity for reduction of CO_2 to CO [131]. Based on the control experiments, it was suggested that the reduction of CO_2 to CO mainly occurred on nitrogen functionalities in competition with hydrogen evolution but was not closely related to the metal's nature. However, the metal's nature was related to the binding intensity between the metal and adsorbed CO, which had an influence on the protonation of CO to form higher CO_2 reduction products such as CH_4 (Figure 2.13).

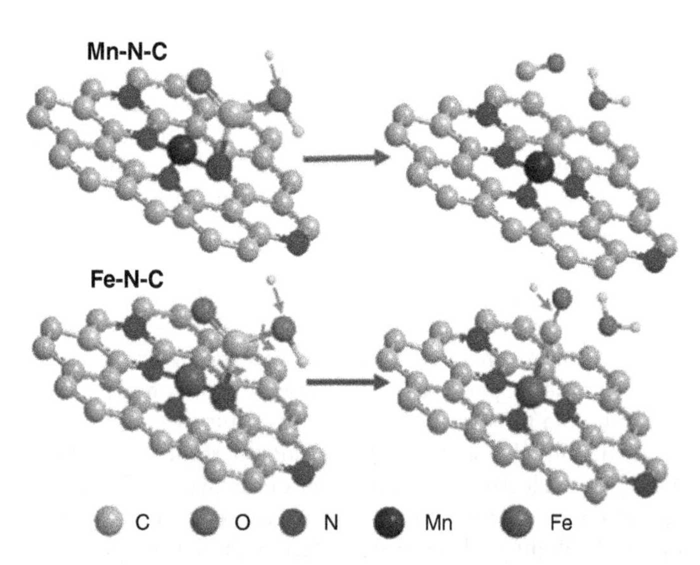

Figure 2.13 Proposed mechanisms for the CO_2 reduction reaction on Mn—N—C and Fe—N—C. Varela *et al.* 2015 [131]. Reproduced with permission of John Wiley & Sons.

2.4 Applications of Nanocarbon Materials in Photocatalysis

In a photocatalysis process, photoactive substance first absorbs the energy of the incident photon and then excites the electrons in the valence band (or highest occupied molecule orbital, HOMO) to the conduction band (or lowest unoccupied molecule orbital, LUMO), leaving an equal number of holes in the valence band (or HOMO). The activated conduction band electrons and the valence band holes could recombine and release their energies through radiation processes or other nonradioactive routes. It is also possible to utilize the activated conduction band electrons and valence band holes before they recombine together to facilitate reduction and oxidation half reactions, respectively. Therefore, a photocatalytic reaction is an ultrafast and energy-uphill process. Photocatalysis processes have been widely used for many applications including water splitting, degradation or oxidization of organic contaminants, photocatalytic conversion of CO_2 to renewable fuels, toxic elimination of heavy metal ions, and antibacterial applications [133]. Nanocarbon materials have been demonstrated to own good compatibilities with most kinds of inorganic semiconductors or organic semiconductors [134]. Therefore, it is promising to hybridize nanocarbon materials with semiconductors to synergistically improve the photocatalytic performances by intentional tuning of the interactions between them. Figure 2.14 shows the schematic diagrams

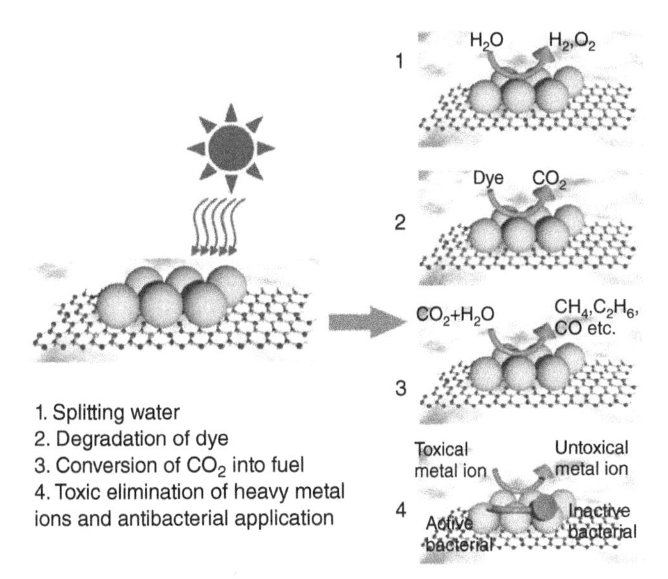

Figure 2.14 Various applications of graphene-based photocatalysts. Tu *et al.* 2013 [133]. Reproduced with permission of John Wiley & Sons.

for various photocatalysis applications of a typical nanocarbon material. In this section, we focus on the roles nanocarbon materials play on hybridized photocatalysts.

2.4.1 Application of Nanocarbon Materials as Photogenerated Charge Acceptors

As nanocarbon materials generally feature high specific surface area, excellent electroconductivity, good mechanical flexibility, and appropriate redox energy levels, they have been extensively used as charge acceptors for efficient separation of photogenerated carriers in contacted semiconductors [135]. Gong *et al.* recently demonstrated that highly efficient photocatalytic H_2 production could be achieved by using graphene nanosheets decorated with CdS clusters as visible-light-driven photocatalysts [136]. The high photocatalytic activity of this hybrid photocatalyst was proposed to be attributed to the presence of graphene, which serves as an electron collector and transporter to efficiently increase the lifetime of the photogenerated charge carriers from CdS nanoparticles. Later, their further work showed that the intercalated RGO in $Zn_xCd_{1-x}S$/RGO hybrid photocatalyst could also facilitate hydrogen evolution during the photocatalytic process, in addition to behaving as a charges acceptor (Figure 2.15) [135]. Beyond graphene, other nanocarbon materials such as CNTs and carbon nanodots have also been extensively used as charge acceptors for efficient separation of photogenerated charge carriers in semiconductor/nanocarbon hybrid photocatalysts [137, 138].

2.4.2 Application of Nanocarbon Materials as Electron Shuttle Mediator

Visible light-driven one-step water splitting into H_2 and O_2 is intriguing for realizing sustainable solar-to-fuel conversion. For a single-phase photocatalyst, there are two basic thermodynamic rules limiting the efficient utilization of visible and even near-infrared photons in the solar spectrum. First, the band gap of the photocatalyst should be at least larger than 1.229 eV. Second, the bottom of conduction band and the top of valence band should straddle over the redox levels of hydrogen evolution and oxygen evolution in a specific electrolyte solution. However, there were very few semiconductors found to simultaneously satisfying these two requirements. And even for an appropriate semiconductor, the limit of band gap also prevents it from fully exploiting the solar spectrum. Z-scheme water splitting, originally introduced by Bard *et al.* in 1979, is an intriguing alternative approach to realize one-step water splitting with visible and even near-infrared light [139].

Figure 2.16 shows the operation principle of a Z-scheme photocatalysis system. This can break through the limit for the band gap of the semiconductor and also release the limit of energy band positions. So far, various semiconductors have been successfully utilized in Z-scheme photocatalytic water

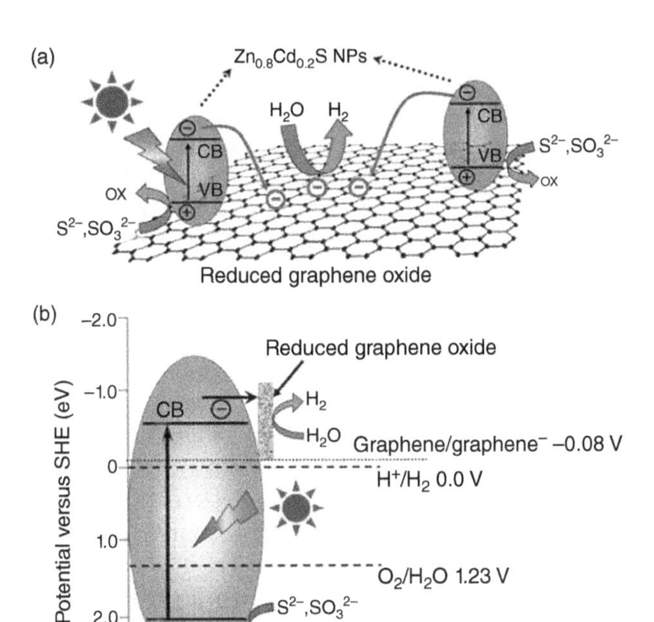

Figure 2.15 (a) Schematic illustration for the charge transfer and separation in the $Zn_{0.8}Cd_{0.2}S$/RGO hybrid photocatalyst; (b) proposed mechanism for photocatalytic H_2-production under simulated solar irradiation. Zhang *et al.* 2012 [135]. Reproduced with permission of American Chemical Society.

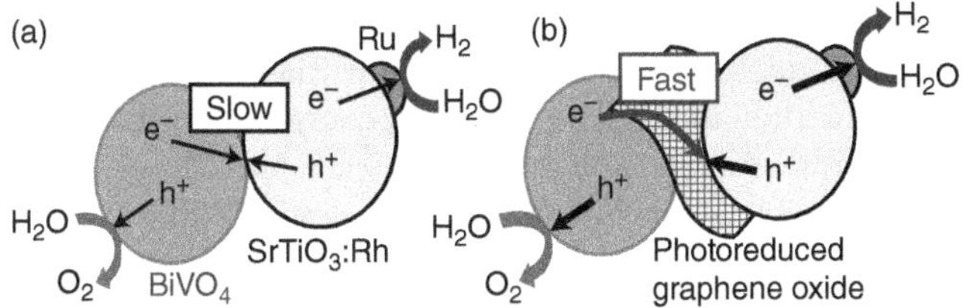

Figure 2.16 Schematic diagrams for a Z-scheme photocatalysis system operating without (a) and with (b) GO as electron shuttle mediator at the interfaces between the n- and p-type semiconductors. Iwashina *et al.* 2015 [140]. Reproduced with permission of American Chemical Society.

splitting systems [140–142]. A Z-scheme system employs two photocatalysts, one producing H_2 (referred as H_2-photocatalyst) and the other producing O_2 (referred as O_2-photocatalyst), and the excess photoexcited holes in the H_2-photocatalyst and electrons in the O_2-photocatalyst recombine together at the interfaces between the two semiconductors [141]. Recent developments of the Z-scheme system have suggested that electron transfer between the two photocatalysts is the rate-determining process [143]. Therefore, the employment of an electron transporter is critical to boost electron relay for efficient consumption of the excess electrons and holes [141]. Currently, Fe^{3+}/Fe^{2+}, IO_3^-/I^- redox couples are the most commonly employed electron mediators for shuttling electrons from the O_2 photocatalyst to the H_2 photocatalyst [144]. However, efficient adsorbing and desorbing of ionic electron mediators on photocatalysts and the associated backward reactions of ionic electron mediators limit the usable semiconductors and solar-to-fuel efficiency [140]. Kudo *et al.* found that a suitable solid electron mediator such as reduced GO, which owned a good dynamic equilibrium between the electron-accepting and -donating abilities, could efficiently shuttle photogenerated electrons from the O_2-photocatalyst to recombine with the generated holes in the H_2-photocatalyst [142]. Their later work demonstrated that using reduced GO as a solid electron shuttle mediator could also get rid of the problems that ionic electron mediators suffered from [140].

2.4.3 Application of Nanocarbon Materials as Cocatalyst for Photocatalysts

In addition to serving as charge acceptors or electron shuttle mediators, nanocarbon materials could also be used as a cocatalyst to facilitate photochemical reaction processes. In a photocatalytic water splitting process, there are general multiple successive intermediate chemical reactions and some by-products may be generated beyond O_2 and H_2 [145]. For example, Chen *et al.* demonstrated that the photocatalytic water oxidation product of a polymeric semiconductor C_3N_4 was mainly hydrogen peroxide (H_2O_2) but not O_2 in a pure water solution without hole scavengers [146]. Therefore, speeding the rate of intermediate reaction and suppressing the generation of unwanted products are also important aspects to improve the solar-to-hydrogen efficiency of a photocatalysis system. Recently, Kang *et al.* reported that carbon nanodots (referred to as CDots) could efficiently catalytically decompose H_2O_2 generated on the surface of C_3N_4 during the photocatalytic process [147]. By hybridizing CDots and C_3N_4 at an appropriate ratio, a solar-to-hydrogen efficiency of 2.0% with a single-phase semiconductor was achieved. It was proposed that the achieved high efficiency was due to the rapid exhaustion of intermediate product H_2O_2 by CDots and thus quick refreshment of the photocatalytic active sites on C_3N_4. And also, the timely removing of H_2O_2

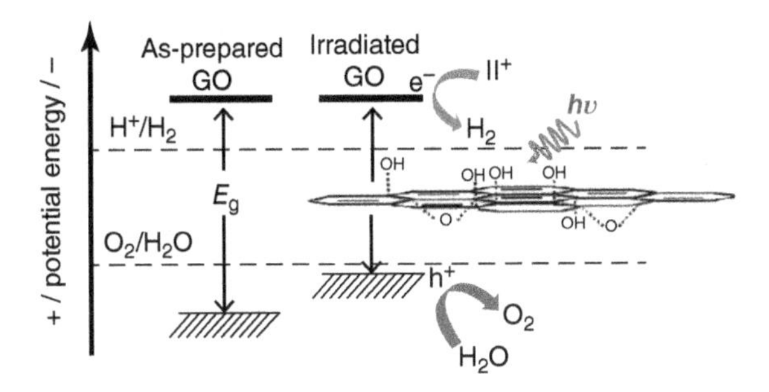

Figure 2.17 Schematic energy level diagrams of the as-prepared GO and irradiated GO and the mechanism for photocatalytic water reduction and oxidation on irradiated GO. Yeh *et al.* 2011 [149]. Reproduced with permission of American Chemical Society.Yeh *et al.* 2011 [149]. Reproduced with permission of American Chemical Society.

efficiently repressed the oxidative degradation of C_3N_4 and enabled the hybrid photocatalyst to be continually operated over 200 days.

2.4.4 Application of Nanocarbon Materials as Active Photocatalyst

Recently, nanocarbon materials such as GO and carbon nanodots have also been identified as photocatalysts for water splitting [62, 148–150]. Teng *et al.* reported that GO with different oxygenated levels showed potential as a photocatalyst for H_2 and O_2 evolution from water with the presence of sacrificial reagents [148, 149]. They performed Mott–Schottky analysis and demonstrated that the conduction and valence band edge levels of GO with appropriate oxidation degree were suitable for both the reduction and the oxidation of water (Figure 2.17). Later, carbon nanodots were also reported to own the ability to inject photogenerated charges into TiO_2 and ZnO and to recover to electroneutrality by exhausting the photogenerated holes through oxygenating hole scavengers [62, 151]. Another interesting work showed that selectively doping nitrogen into specific regions in carbon nanodots could create a p–n-type photochemical diode, which enabled the complete photocatalytic water splitting into H_2 and O_2 under visible light [150].

2.5 Summary

This chapter summarizes some representative progress on applications of nanocarbon materials in electrocatalysis and photocatalysis. The large specific surface area, mechanical flexibility, and easy surface functionalization make the nanocarbon materials intriguing candidates for anchoring catalytically active species with excellent dispersivity. Also, the electroconductivity and charge donating/accepting abilities of nanocarbon materials endow the hybrid catalysts with significantly improved activities in both electrocatalysis and photocatalysis. On the other hand, nanocarbon materials can also exhibit electrocatalytic activities through doping with heteroatoms. In addition, some special nanocarbon materials can also be used as photocatalysts due to their intrinsic optical and catalytic performance induced by nanostructuring and functionality. Nanocarbon materials can pave new ways to develop nonprecious metal or even metal-free electrocatalysts or photocatalysts for large-scale applications. Although recent progresses indicate a bright future for the nanocarbon-based catalysts, further work is needed, including feasible synthetic chemistry, in-depth structure characterization, and real device testing.

Acknowledgments

The authors acknowledge the financial supports from "The Recruitment Program of Global Youth Experts of China," the Shenzhen Key Lab funding (ZDSYS201505291525382), and the Shenzhen Peacock Program (No. KQTD20140630160825828).

References

1 Georgakilas, V., Perman, J.A., Tucek, J., and Zboril, R. (2015) Broad family of carbon nanoallotropes: classification, chemistry, and applications of fullerenes, carbon dots, nanotubes, graphene, nanodiamonds, and combined superstructures. *Chemical Reviews*, **115**, 4744–4822.
2 Li, Z., Liu, Z., Sun, H., and Gao, C. (2015) Superstructured assembly of nanocarbons: fullerenes, nanotubes, and graphene. *Chemical Reviews*, **115**, 7046–7117.
3 Kroto, H.W., Heath, J.R., O'Brien, S.C. *et al.* (1985) C60: buckminsterfullerene. *Nature*, **318**, 162–163.
4 Li, G., Li, Y., Liu, H. *et al.* (2010) Architecture of graphdiyne nanoscale films. *Chemical Communications*, **46**, 3256–3258.
5 Robertson, J. and O'reilly, E. (1987) Electronic and atomic structure of amorphous carbon. *Physical Review B*, **35**, 2946–2957.
6 Dillon, A. (2010) Carbon nanotubes for photoconversion and electrical energy storage. *Chemical Reviews*, **110**, 6856–6872.
7 Wang, Y.-J., Zhao, N., Fang, B. *et al.* (2015) Carbon-supported Pt-based alloy electrocatalysts for the oxygen reduction reaction in polymer electrolyte membrane fuel cells: particle size, shape, and composition manipulation and their impact to activity. *Chemical Reviews*, **115**, 3433–3467.

8 Bao, Q. and Loh, K.P. (2012) Graphene photonics, plasmonics, and broadband optoelectronic devices. *ACS Nano*, **6**, 3677–3694.
9 Liu, M., Zhang, R., and Chen, W. (2014) Graphene-supported nanoelectrocatalysts for fuel cells: synthesis, properties, and applications. *Chemical Reviews*, **114**, 5117–5160.
10 Liu, Z., Robinson, J.T., Sun, X., and Dai, H. (2008) Pegylated nanographene oxide for delivery of water-insoluble cancer drugs. *Journal of the American Chemical Society*, **130**, 10876–10877.
11 Yang, K., Feng, L., Shi, X., and Liu, Z. (2013) Nano-graphene in biomedicine: theranostic applications. *Chemical Society Reviews*, **42**, 530–547.
12 Liang, Y., Li, Y., Wang, H., and Dai, H. (2013) Strongly coupled inorganic/nanocarbon hybrid materials for advanced electrocatalysis. *Journal of the American Chemical Society*, **135**, 2013–2036.
13 Chen, D., Feng, H., and Li, J. (2012) Graphene oxide: preparation, functionalization, and electrochemical applications. *Chemical Reviews*, **112**, 6027–6053.
14 Yang, Z., Ren, J., Zhang, Z. *et al.* (2015) Recent advancement of nanostructured carbon for energy applications. *Chemical Reviews*, **115**, 5159–5223.
15 Eder, D. (2010) Carbon nanotube – inorganic hybrids. *Chemical Reviews*, **110**, 1348–1385.
16 Iijima, S. (1991) Helical microtubules of graphitic carbon. *Nature*, **354**, 56–58.
17 Guo, T., Nikolaev, P., Thess, A. *et al.* (1995) Catalytic growth of single-walled nanotubes by laser vaporization. *Chemical Physics Letters*, **243**, 49–54.
18 Nessim, G.D. (2010) Properties, synthesis, and growth mechanisms of carbon nanotubes with special focus on thermal chemical vapor deposition. *Nanoscale*, **2**, 1306–1323.
19 Seah, C.-M., Chai, S.-P., and Mohamed, A.R. (2011) Synthesis of aligned carbon nanotubes. *Carbon*, **49**, 4613–4635.
20 Kumar, M. and Ando, Y. (2010) Chemical vapor deposition of carbon nanotubes: a review on growth mechanism and mass production. *Journal of Nanoscience and Nanotechnology*, **10**, 3739–3758.
21 Kong, J., Cassell, A.M., and Dai, H. (1998) Chemical vapor deposition of methane for single-walled carbon nanotubes. *Chemical Physics Letters*, **292**, 567–574.
22 Li, Q., Yan, H., Zhang, J., and Liu, Z. (2004) Effect of hydrocarbons precursors on the formation of carbon nanotubes in chemical vapor deposition. *Carbon*, **42**, 829–835.
23 Huang, L., Cui, X., White, B., and O'Brien, S.P. (2004) Long and oriented single-walled carbon nanotubes grown by ethanol chemical vapor deposition. *The Journal of Physical Chemistry B*, **108**, 16451–16456.

24 Zhang, R., Zhang, Y., Zhang, Q. *et al.* (2013) Growth of half-meter long carbon nanotubes based on Schulz–Flory distribution. *ACS Nano*, **7**, 6156–6161.
25 Zhang, Q., Huang, J.Q., Zhao, M.Q. *et al.* (2011) Carbon nanotube mass production: principles and processes. *ChemSusChem*, **4**, 864–889.
26 Li, Y., Zhang, X., Tao, X. *et al.* (2005) Mass production of high-quality multi-walled carbon nanotube bundles on a Ni/Mo/MgO catalyst. *Carbon*, **43**, 295–301.
27 Shi, Z., Lian, Y., Zhou, X. *et al.* (1999) Mass-production of single-wall carbon nanotubes by arc discharge method. *Carbon*, **37**, 1449–1453.
28 Dai, L., Xue, Y., Qu, L. *et al.* (2015) Metal-free catalysts for oxygen reduction reaction. *Chemical Reviews*, **115**, 4823–4892.
29 Dai, H. (2002) Carbon nanotubes: synthesis, integration, and properties. *Accounts of Chemical Research*, **35**, 1035–1044.
30 Chen, C., Tsai, C., and Lin, C. (2003) The characterization of boron-doped carbon nanotube arrays. *Diamond and Related Materials*, **12**, 1500–1504.
31 Denis, P.A., Faccio, R., and Mombru, A.W. (2009) Is it possible to dope single-walled carbon nanotubes and graphene with sulfur? *ChemPhysChem*, **10**, 715–722.
32 Jin, Z., Nie, H., Yang, Z. *et al.* (2012) Metal-free selenium doped carbon nanotube/graphene networks as a synergistically improved cathode catalyst for oxygen reduction reaction. *Nanoscale*, **4**, 6455–6460.
33 Gong, K., Du, F., Xia, Z. *et al.* (2009) Nitrogen-doped carbon nanotube arrays with high electrocatalytic activity for oxygen reduction. *Science*, **323**, 760–764.
34 Shi, Q., Peng, F., Liao, S. *et al.* (2013) Sulfur and nitrogen Co-doped carbon nanotubes for enhancing electrochemical oxygen reduction activity in acidic and alkaline media. *Journal of Materials Chemistry A*, **1**, 14853–14857.
35 Yu, D., Xue, Y., and Dai, L. (2012) Vertically aligned carbon nanotube arrays Co-doped with phosphorus and nitrogen as efficient metal-free electrocatalysts for oxygen reduction. *Journal of Physical Chemistry Letters*, **3**, 2863–2870.
36 Wang, H. and Dai, H. (2013) Strongly coupled inorganic-nano-carbon hybrid materials for energy storage. *Chemical Society Reviews*, **42**, 3088–3113.
37 Liang, Y., Wang, H., Diao, P. *et al.* (2012) Oxygen reduction electrocatalyst based on strongly coupled cobalt oxide nanocrystals and carbon nanotubes. *Journal of the American Chemical Society*, **134**, 15849–15857.
38 Wang, S., Yu, D., and Dai, L. (2011) Polyelectrolyte functionalized carbon nanotubes as efficient metal-free electrocatalysts for oxygen reduction. *Journal of the American Chemical Society*, **133**, 5182–5185.

39 Novoselov, K.S., Geim, A.K., Morozov, S. *et al.* (2004) Electric field effect in atomically thin carbon films. *Science*, **306**, 666–669.
40 Berger, C., Song, Z., Li, X. *et al.* (2006) Electronic confinement and coherence in patterned epitaxial graphene. *Science*, **312**, 1191–1196.
41 Kim, K.S., Zhao, Y., Jang, H. *et al.* (2009) Large-scale pattern growth of graphene films for stretchable transparent electrodes. *Nature*, **457**, 706–710.
42 Bae, S., Kim, H., Lee, Y. *et al.* (2010) Roll-to-roll production of 30-inch graphene films for transparent electrodes. *Nature Nanotechnology*, **5**, 574–578.
43 Dikin, D.A., Stankovich, S., Zimney, E.J. *et al.* (2007) Preparation and characterization of graphene oxide paper. *Nature*, **448**, 457–460.
44 Parvez, K., Li, R., Puniredd, S.R. *et al.* (2013) Electrochemically exfoliated graphene as solution-processable, highly conductive electrodes for organic electronics. *ACS Nano*, **7**, 3598–3606.
45 Wang, X., Tabakman, S.M., and Dai, H. (2008) Atomic layer deposition of metal oxides on pristine and functionalized graphene. *Journal of the American Chemical Society*, **130**, 8152–8153.
46 Qu, L., Liu, Y., Baek, J.-B., and Dai, L. (2010) Nitrogen-doped graphene as efficient metal-free electrocatalyst for oxygen reduction in fuel cells. *ACS Nano*, **4**, 1321–1326.
47 Robinson, J.T., Burgess, J.S., Junkermeier, C.E. *et al.* (2010) Properties of fluorinated graphene films. *Nano Letters*, **10**, 3001–3005.
48 Yang, Z., Yao, Z., Li, G. *et al.* (2011) Sulfur-doped graphene as an efficient metal-free cathode catalyst for oxygen reduction. *ACS Nano*, **6**, 205–211.
49 Sheng, Z.-H., Gao, H.-L., Bao, W.-J. *et al.* (2012) Synthesis of boron doped graphene for oxygen reduction reaction in fuel cells. *Journal of Materials Chemistry*, **22**, 390–395.
50 Zhang, C., Mahmood, N., Yin, H. *et al.* (2013) Synthesis of phosphorus-doped graphene and its multifunctional applications for oxygen reduction reaction and lithium Ion batteries. *Advanced Materials*, **25**, 4932–4937.
51 Wang, X., Li, X., Zhang, L. *et al.* (2009) N-doping of graphene through electrothermal reactions with ammonia. *Science*, **324**, 768–771.
52 Baker, S.N. and Baker, G.A. (2010) Luminescent carbon nanodots: emergent nanolights. *Angewandte Chemie International Edition*, **49**, 6726–6744.
53 Li, H., Kang, Z., Liu, Y., and Lee, S.-T. (2012) Carbon nanodots: synthesis, properties and applications. *Journal of Materials Chemistry*, **22**, 24230–24253.
54 Li, H., He, X., Kang, Z. *et al.* (2010) Water-soluble fluorescent carbon quantum dots and photocatalyst design. *Angewandte Chemie International Edition*, **49**, 4430–4434.

55 Xu, X., Ray, R., Gu, Y. *et al.* (2004) Electrophoretic analysis and purification of fluorescent single-walled carbon nanotube fragments. *Journal of the American Chemical Society,* **126**, 12736–12737.
56 Sun, Y.-P., Zhou, B., Lin, Y. *et al.* (2006) Quantum-sized carbon dots for bright and colorful photoluminescence. *Journal of the American Chemical Society,* **128**, 7756–7757.
57 Ming, H., Ma, Z., Liu, Y. *et al.* (2012) Large scale electrochemical synthesis of high quality carbon nanodots and their photocatalytic property. *Dalton Transactions,* **41**, 9526–9531.
58 Yang, Y., Cui, J., Zheng, M. *et al.* (2012) One-step synthesis of amino-functionalized fluorescent carbon nanoparticles by hydrothermal carbonization of chitosan. *Chemical Communications,* **48**, 380–382.
59 Li, H., He, X., Liu, Y. *et al.* (2011) One-step ultrasonic synthesis of water-soluble carbon nanoparticles with excellent photoluminescent properties. *Carbon,* **49**, 605–609.
60 Zhuo, S., Shao, M., and Lee, S.-T. (2012) Upconversion and downconversion fluorescent graphene quantum dots: ultrasonic preparation and photocatalysis. *ACS Nano,* **6**, 1059–1064.
61 Wang, X., Qu, K., Xu, B. *et al.* (2011) Microwave assisted One-step green synthesis of cell-permeable multicolor photoluminescent carbon dots without surface passivation reagents. *Journal of Materials Chemistry,* **21**, 2445–2450.
62 Zhang, X., Wang, F., Huang, H. *et al.* (2013) Carbon quantum Dot sensitized TiO_2 nanotube arrays for photoelectrochemical hydrogen generation under visible light. *Nanoscale,* **5**, 2274–2278.
63 Choi, H., Ko, S.-J., Choi, Y. *et al.* (2013) Versatile surface plasmon resonance of carbon-Dot-supported silver nanoparticles in polymer optoelectronic devices. *Nature Photonics,* **7**, 732–738.
64 Ryoo, R., Joo, S.H., and Jun, S. (1999) Synthesis of highly ordered carbon molecular sieves via template-mediated structural transformation. *The Journal of Physical Chemistry B,* **103**, 7743–7746.
65 Liang, C., Li, Z., and Dai, S. (2008) Mesoporous carbon materials: synthesis and modification. *Angewandte Chemie International Edition,* **47**, 3696–3717.
66 Liang, C., Hong, K., Guiochon, G.A. *et al.* (2004) Synthesis of a large-scale highly ordered porous carbon film by self-assembly of block copolymers. *Angewandte Chemie International Edition,* **43**, 5785–5789.
67 Li, M., Bo, X., Zhang, Y. *et al.* (2014) Comparative study on the oxygen reduction reaction electrocatalytic activities of iron phthalocyanines supported on reduced graphene oxide, mesoporous carbon vesicle, and ordered mesoporous carbon. *Journal of Power Sources,* **264**, 114–122.
68 Lu, A.-H., Li, W.-C., Hou, Z., and Schüth, F. (2007) Molecular level dispersed pd clusters in the carbon walls of ordered mesoporous carbon as

a highly selective alcohol oxidation catalyst. *Chemical Communications*, **2007**, 1038–1040.
69 Joo, S.H., Choi, S.J., Oh, I. *et al.* (2001) Ordered nanoporous arrays of carbon supporting high dispersions of platinum nanoparticles. *Nature*, **412**, 169–172.
70 Wang, X., Liu, R., Waje, M.M. *et al.* (2007) Sulfonated ordered mesoporous carbon as a stable and highly active protonic acid catalyst. *Chemistry of Materials*, **19**, 2395–2397.
71 Jasinski, R. (1964) A new fuel cell cathode catalyst. *Nature*, **201**, 1212–1213.
72 Wu, G., More, K.L., Johnston, C.M., and Zelenay, P. (2011) High-performance electrocatalysts for oxygen reduction derived from polyaniline, iron, and cobalt. *Science*, **332**, 443–447.
73 Gao, M.R., Jiang, J., and Yu, S.H. (2012) Solution-based synthesis and design of late transition metal chalcogenide materials for oxygen reduction reaction (ORR). *Small*, **8**, 13–27.
74 Suntivich, J., Gasteiger, H.A., Yabuuchi, N. *et al.* (2011) Design principles for oxygen-reduction activity on perovskite oxide catalysts for fuel cells and metal–air batteries. *Nature Chemistry*, **3**, 546–550.
75 Cheng, F., Shen, J., Peng, B. *et al.* (2011) Rapid room-temperature synthesis of nanocrystalline spinels as oxygen reduction and evolution electrocatalysts. *Nature Chemistry*, **3**, 79–84.
76 Dong, G., Huang, M., and Guan, L. (2012) Iron phthalocyanine coated on single-walled carbon nanotubes composite for the oxygen reduction reaction in alkaline media. *Physical Chemistry Chemical Physics*, **14**, 2557–2559.
77 Cao, R., Thapa, R., Kim, H. *et al.* (2013) Promotion of oxygen reduction by a bio-inspired tethered iron phthalocyanine carbon nanotube-based catalyst. *Nature Communications*, **4**, 2076.
78 Wei, P.J., Yu, G.Q., Naruta, Y., and Liu, J.G. (2014) Covalent grafting of carbon nanotubes with a biomimetic heme model compound to enhance oxygen reduction reactions. *Angewandte Chemie International Edition*, **53**, 6659–6663.
79 Liang, Y., Li, Y., Wang, H. *et al.* (2011) Co_3O_4 nanocrystals on graphene as a synergistic catalyst for oxygen reduction reaction. *Nature Materials*, **10**, 780–786.
80 Wang, H., Liang, Y., Li, Y., and Dai, H. (2011) $Co_{1-x}S$–graphene hybrid: a high-performance metal chalcogenide electrocatalyst for oxygen reduction. *Angewandte Chemie International Edition*, **50**, 10969–10972.
81 Liang, Y., Wang, H., Zhou, J. *et al.* (2012) Covalent hybrid of spinel manganese-cobalt oxide and graphene as advanced oxygen reduction electrocatalysts. *Journal of the American Chemical Society*, **134**, 3517–3523.

82 Masa, J., Xia, W., Muhler, M., and Schuhmann, W. (2015) On the role of metals in nitrogen-doped carbon electrocatalysts for oxygen reduction. *Angewandte Chemie International Edition*, **54**, 10102–10120.

83 Ramaswamy, N., Tylus, U., Jia, Q., and Mukerjee, S. (2013) Activity descriptor identification for oxygen reduction on nonprecious electrocatalysts: linking surface science to coordination chemistry. *Journal of the American Chemical Society*, **135**, 15443–15449.

84 Gupta, S., Tryk, D., Bae, I. *et al.* (1989) Heat-treated polyacrylonitrile-based catalysts for oxygen electroreduction. *Journal of Applied Electrochemistry*, **19**, 19–27.

85 Li, Y., Zhou, W., Wang, H. *et al.* (2012) An oxygen reduction electrocatalyst based on carbon nanotube-graphene complexes. *Nature Nanotechnology*, **7**, 394–400.

86 Jin, H., Huang, H., He, Y. *et al.* (2015) Graphene quantum dots supported by graphene nanoribbons with ultrahigh electrocatalytic performance for oxygen reduction. *Journal of the American Chemical Society*, **137**, 7588–7591.

87 Liang, H.-W., Zhuang, X., Brüller, S. *et al.* (2014) Hierarchically porous carbons with optimized nitrogen doping as highly active electrocatalysts for oxygen reduction. *Nature Communications*, **5**, 4973.

88 Li, Y., Zhao, Y., Cheng, H. *et al.* (2011) Nitrogen-doped graphene quantum dots with oxygen-rich functional groups. *Journal of the American Chemical Society*, **134**, 15–18.

89 Fellinger, T.-P., Hasché, F.D.R., Strasser, P., and Antonietti, M. (2012) Mesoporous nitrogen-doped carbon for the electrocatalytic synthesis of hydrogen peroxide. *Journal of the American Chemical Society*, **134**, 4072–4075.

90 Shui, J., Wang, M., Du, F., and Dai, L. (2015) N-doped carbon nanomaterials are durable catalysts for oxygen reduction reaction in acidic fuel cells. *Science Advances*, **1**, e1400129.

91 McCrory, C.C., Jung, S., Ferrer, I.M. *et al.* (2015) Benchmarking hydrogen evolving reaction and oxygen evolving reaction electrocatalysts for solar water splitting devices. *Journal of the American Chemical Society*, **137**, 4347–4357.

92 Suntivich, J., May, K.J., Gasteiger, H.A. *et al.* (2011) A perovskite oxide optimized for oxygen evolution catalysis from molecular orbital principles. *Science*, **334**, 1383–1385.

93 Gong, M., Li, Y., Wang, H. *et al.* (2013) An advanced Ni–Fe layered double hydroxide electrocatalyst for water oxidation. *Journal of the American Chemical Society*, **135**, 8452–8455.

94 Long, X., Li, J., Xiao, S. *et al.* (2014) A strongly coupled graphene and feni double hydroxide hybrid as an excellent electrocatalyst for the oxygen evolution reaction. *Angewandte Chemie*, **126**, 7714–7718.

95 Ma, W., Ma, R., Wang, C. *et al.* (2015) A superlattice of alternately stacked Ni–Fe hydroxide nanosheets and graphene for efficient splitting of water. *ACS Nano*, **9**, 1977–1984.

96 Gao, M.-R., Cao, X., Gao, Q. *et al.* (2014) Nitrogen-doped graphene supported $CoSe_2$ nanobelt composite catalyst for efficient water oxidation. *ACS Nano*, **8**, 3970–3978.

97 Zhang, X., Xu, H., Li, X. *et al.* (2015) Facile synthesis of nickel-iron/nanocarbon hybrids as advanced electrocatalysts for efficient water splitting. *ACS Catalysis*, **6**, 580–588.

98 Zhao, Y., Nakamura, R., Kamiya, K. *et al.* (2013) Nitrogen-doped carbon nanomaterials as Non-metal electrocatalysts for water oxidation. *Nature Communications*, **4**, 2390.

99 Chen, S., Duan, J., Jaroniec, M., and Qiao, S.Z. (2014) Nitrogen and oxygen dual-doped carbon hydrogel film as a substrate-free electrode for highly efficient oxygen evolution reaction. *Advanced Materials*, **26**, 2925–2930.

100 Chen, S., Duan, J., Ran, J., and Qiao, S.Z. (2015) Paper-based N-doped carbon films for enhanced oxygen evolution electrocatalysis. *Advancement of Science*, **2**, 1400015.

101 Zhang, J., Zhao, Z., Xia, Z., and Dai, L. (2015) A metal-free bifunctional electrocatalyst for oxygen reduction and oxygen evolution reactions. *Nature Nanotechnology*, **10**, 44–452.

102 Lu, X., Yim, W.-L., Suryanto, B.H., and Zhao, C. (2015) Electrocatalytic oxygen evolution at surface-oxidized multiwall carbon nanotubes. *Journal of the American Chemical Society*, **137**, 2901–2907.

103 Li, Y., Wang, H., Xie, L. *et al.* (2011) Mos2 nanoparticles grown on graphene: an advanced catalyst for the hydrogen evolution reaction. *Journal of the American Chemical Society*, **133**, 7296–7299.

104 Liu, Y., Yu, G., Li, G.D. *et al.* (2015) Coupling Mo2c with nitrogen-rich nanocarbon leads to efficient hydrogen-evolution electrocatalytic sites. *Angewandte Chemie International Edition*, **54**, 10752–10757.

105 Liu, Q., Tian, J., Cui, W. *et al.* (2014) Carbon nanotubes decorated with CoP nanocrystals: a highly active non-noble-metal nanohybrid electrocatalyst for hydrogen evolution. *Angewandte Chemie International Edition*, **53**, 6710–6714.

106 Helm, M.L., Stewart, M.P., Bullock, R.M. *et al.* (2011) A synthetic nickel electrocatalyst with a turnover frequency above 100,000 S− 1 for H_2 production. *Science*, **333**, 863–866.

107 Lu, Q., Hutchings, G.S., Yu, W. *et al.* (2015) Highly porous Non-precious bimetallic electrocatalysts for efficient hydrogen evolution. *Nature Communications*, **6**, 6567.

108 Li, D.J., Maiti, U.N., Lim, J. *et al.* (2014) Molybdenum sulfide/N-doped Cnt forest hybrid catalysts for high-performance hydrogen evolution reaction. *Nano Letters*, **14**, 1228–1233.

109 Deng, J., Ren, P., Deng, D. *et al.* (2014) Highly active and durable non-precious-metal catalysts encapsulated in carbon nanotubes for hydrogen evolution reaction. *Energy & Environmental Science*, 7, 1919–1923.

110 Deng, J., Ren, P., Deng, D., and Bao, X. (2015) Enhanced electron penetration through an ultrathin graphene layer for highly efficient catalysis of the hydrogen evolution reaction. *Angewandte Chemie International Edition*, **54**, 2100–2104.

111 Tavakkoli, M., Kallio, T., Reynaud, O. *et al.* (2015) Single-shell carbon-encapsulated iron nanoparticles: synthesis and high electrocatalytic activity for hydrogen evolution reaction. *Angewandte Chemie*, **127**, 4618–4621.

112 Liang, H.-W., Brüller, S., Dong, R. *et al.* (2015) Molecular metal-N_x centres in porous carbon for electrocatalytic hydrogen evolution. *Nature Communications*, **6**, 7992.

113 Zheng, Y., Jiao, Y., Zhu, Y. *et al.* (2014) Hydrogen evolution by a metal-free electrocatalyst. *Nature Communications*, **5**, 3783.

114 Das, R.K., Wang, Y., Vasilyeva, S.V. *et al.* (2014) Extraordinary hydrogen evolution and oxidation reaction activity from carbon nanotubes and graphitic carbons. *ACS Nano*, **8**, 8447–8456.

115 Liu, X., Zhou, W., Yang, L. *et al.* (2015) Nitrogen and sulfur Co-doped porous carbon derived from human hair as highly efficient metal-free electrocatalysts for hydrogen evolution reactions. *Journal of Materials Chemistry A*, **3**, 8840–8846.

116 Wei, L., Karahan, H.E., Goh, K. *et al.* (2015) A high-performance metal-free hydrogen-evolution reaction electrocatalyst from bacterium derived carbon. *Journal of Materials Chemistry A*, **3**, 7210–7214.

117 Dong, G., Fang, M., Wang, H. *et al.* (2015) Insight into the electrochemical activation of carbon-based cathodes for hydrogen evolution reaction. *Journal of Materials Chemistry A*, **3**, 13080–13086.

118 Kuhl, K.P., Hatsukade, T., Cave, E.R. *et al.* (2014) Electrocatalytic conversion of carbon dioxide to methane and methanol on transition metal surfaces. *Journal of the American Chemical Society*, **136**, 14107–14113.

119 Zhang, S., Kang, P., and Meyer, T.J. (2014) Nanostructured Tin catalysts for selective electrochemical reduction of carbon dioxide to formate. *Journal of the American Chemical Society*, **136**, 1734–1737.

120 Gao, D., Zhou, H., Wang, J. *et al.* (2015) Size-dependent electrocatalytic reduction of CO_2 over Pd nanoparticles. *Journal of the American Chemical Society*, **137**, 4288–4291.

121 Aydin, R. and Köleli, F. (2004) Electrocatalytic conversion of CO_2 on a polypyrrole electrode under high pressure in methanol. *Synthetic Metals*, **144**, 75–80.

122 Köleli, F., Röpke, T., and Hamann, C.H. (2004) The reduction of CO_2 on polyaniline electrode in a membrane cell. *Synthetic Metals*, **140**, 65–68.

123 Mao, X. and Hatton, T.A. (2015) Recent advances in electrocatalytic reduction of carbon dioxide using metal-free catalysts. *Industrial & Engineering Chemistry Research*, **54**, 4033–4042.

124 Zall, C.M., Linehan, J.C., and Appel, A.M. (2015) A molecular copper catalyst for hydrogenation of CO_2 to formate. *ACS Catalysis*, **5**, 5301–5305.

125 Lieber, C.M. and Lewis, N.S. (1984) Catalytic reduction of carbon dioxide at carbon electrodes modified with cobalt phthalocyanine. *Journal of the American Chemical Society*, **106**, 5033–5034.

126 Shen, J., Kortlever, R., Kas, R. *et al.* (2015) Electrocatalytic reduction of carbon dioxide to carbon monoxide and methane at an immobilized cobalt protoporphyrin. *Nature Communications*, **6**, 8177.

127 Kumar, B., Asadi, M., Pisasale, D. *et al.* (2013) Renewable and metal-free carbon nanofibre catalysts for carbon dioxide reduction. *Nature Communications*, **4**, 2819.

128 Zhang, S., Kang, P., Ubnoske, S. *et al.* (2014) Polyethylenimine-enhanced electrocatalytic reduction of CO_2 to formate at nitrogen-doped carbon nanomaterials. *Journal of the American Chemical Society*, **136**, 7845–7848.

129 Wu, J., Yadav, R.M., Liu, M. *et al.* (2015) Achieving highly efficient, selective and stable CO_2 reduction on nitrogen doped carbon nanotubes. *ACS Nano*, **9**, 5364–5371.

130 Sharma, P.P., Wu, J., Yadav, R.M. *et al.* (2015) Nitrogen-doped carbon nanotube arrays for high-efficiency electrochemical reduction of CO_2: on the understanding of defects, defect density, and selectivity. *Angewandte Chemie*, **127**, 13905–13909.

131 Varela, A.S., Ranjbar Sahraie, N., Steinberg, J. *et al.* (2015) Metal-doped nitrogenated carbon as an efficient catalyst for direct CO_2 electroreduction to CO and hydrocarbons. *Angewandte Chemie International Edition*, **54**, 10758–10762.

132 Alves, D.C., Silva, R., Voiry, D. *et al.* (2015) Copper nanoparticles stabilized by reduced graphene oxide for CO_2 reduction reaction. *Materials for Renewable and Sustainable Energy*, **4**, 2.

133 Tu, W., Zhou, Y., and Zou, Z. (2013) Versatile graphene-promoting photocatalytic performance of semiconductors: basic principles, synthesis, solar energy conversion, and environmental applications. *Advanced Functional Materials*, **23**, 4996–5008.

134 Xiang, Q., Cheng, B., and Yu, J. (2015) Graphene-based photocatalysts for solar-fuel generation. *Angewandte Chemie International Edition*, **54**, 11350–11366.

135 Zhang, J., Yu, J., Jaroniec, M., and Gong, J.R. (2012) Noble metal-free reduced graphene oxide-$Zn_xCd_{1-x}S$ nanocomposite with enhanced solar photocatalytic H2-production performance. *Nano Letters*, **12**, 4584–4589.

136 Li, Q., Guo, B., Yu, J. *et al.* (2011) Highly efficient visible-light-driven photocatalytic hydrogen production of CdS-cluster-decorated graphene nanosheets. *Journal of the American Chemical Society*, **133**, 10878–10884.

137 Silva, C.G., Sampaio, M.J., Marques, R.R. *et al.* (2018) Photocatalytic production of hydrogen from methanol and saccharides using carbon nanotube-TiO_2 catalysts. *Applied Catalysis B: Environmental*, **178**, 82–90.

138 Zhang, H., Huang, H., Ming, H. *et al.* (2012) Carbon quantum dots/Ag 3 Po 4 complex photocatalysts with enhanced photocatalytic activity and stability under visible light. *Journal of Materials Chemistry*, **22**, 10501–10506.

139 Leland, J.K. and Bard, A.J. (1987) Photochemistry of colloidal semiconducting iron oxide polymorphs. *Journal of Physical Chemistry*, **91**, 5076–5083.

140 Iwashina, K., Iwase, A., Ng, Y.H. *et al.* (2015) Z-schematic water splitting into H_2 and O_2 using metal sulfide as a hydrogen-evolving photocatalyst and reduced graphene oxide as a solid-state electron mediator. *Journal of the American Chemical Society*, **137**, 604–607.

141 Maeda, K. (2013) Z-scheme water splitting using Two different semiconductor photocatalysts. *ACS Catalysis*, **3**, 1486–1503.

142 Iwase, A., Ng, Y.H., Ishiguro, Y. *et al.* (2011) Reduced graphene oxide as a solid-state electron mediator in Z-scheme photocatalytic water splitting under visible light. *Journal of the American Chemical Society*, **133**, 11054–11057.

143 Sasaki, Y., Nemoto, H., Saito, K., and Kudo, A. (2009) Solar water splitting using powdered photocatalysts driven by Z-schematic interparticle electron transfer without an electron mediator. *Journal of Physical Chemistry C*, **113**, 17536–17542.

144 Martin, D.J., Reardon, P.J.T., Moniz, S.J., and Tang, J. (2014) Visible light-driven pure water splitting by a nature-inspired organic semiconductor-based system. *Journal of the American Chemical Society*, **136**, 12568–12571.

145 Kormann, C., Bahnemann, D.W., and Hoffmann, M.R. (1988) Photocatalytic production of hydrogen peroxides and organic peroxides in aqueous suspensions of titanium dioxide, zinc oxide, and desert sand. *Environmental Science & Technology*, **22**, 798–806.

146 Sui, Y., Liu, J., Zhang, Y. *et al.* (2013) Dispersed conductive polymer nanoparticles on graphitic carbon nitride for enhanced solar-driven hydrogen evolution from pure water. *Nanoscale*, **5**, 9150–9155.

147 Liu, J., Liu, Y., Liu, N. *et al.* (2015) Metal-free efficient photocatalyst for stable visible water splitting via a two-electron pathway. *Science*, **347**, 970–974.

148 Yeh, T.F., Syu, J.M., Cheng, C. *et al.* (2010) Graphite oxide as a photocatalyst for hydrogen production from water. *Advanced Functional Materials*, **20**, 2255–2262.

149 Yeh, T.-F., Chan, F.-F., Hsieh, C.-T., and Teng, H. (2011) Graphite oxide with different oxygenated levels for hydrogen and oxygen production from water under illumination: the band positions of graphite oxide. *Journal of Physical Chemistry C*, **115**, 22587–22597.

150 Yeh, T.F., Teng, C.Y., Chen, S.J., and Teng, H. (2014) Nitrogen-doped graphene oxide quantum dots as photocatalysts for overall water-splitting under visible light illumination. *Advanced Materials*, **26**, 3297–3303.

151 Guo, C.X., Dong, Y., Yang, H.B., and Li, C.M. (2013) Graphene quantum dots as a green sensitizer to functionalize Zno nanowire arrays on F-doped SnO_2 glass for enhanced photoelectrochemical water splitting. *Advanced Energy Materials*, **3**, 997–1003.

3

Controlling and Characterizing Anisotropic Nanomaterial Dispersion

Virginia A. Davis[1] *and Micah J. Green*[2]

[1] *Department of Chemical Engineering, Auburn University, Auburn, AL, USA*
[2] *Artie McFerrin Department of Chemical Engineering, Texas A&M University, College Station, TX, USA*

3.1 Introduction

In this chapter, we aim to address the critical issue of anisotropic nanomaterial dispersion in fluids. The term fluid is used broadly to include Newtonian solvents, multicomponent solutions, resins, and thermoplastic polymer melts. Dispersion is particularly important in cases where bulk quantities of nanomaterials are needed, as opposed to thin-film applications, where very small masses of nanomaterials are required and can be directly grown using vapor deposition. Due to their broad commercial applications, the availability of industrial quantities, and the exceptionally high forces required for separation of nanomaterial aggregates, this chapter primarily focuses on anisotropic carbon nanomaterials, including single- and multiwalled carbon nanotubes (SWNT and MWNT) and graphene-family nanosheets. In addition, we draw parallels with other cylindrical and sheet-like nanomaterials including inorganic nanocylinders, boron nitride nanosheets, and cellulose nanocrystals [1, 2]. The discussion of nanoclays is limited, given their long history of industrial use and prior thorough reviews [3, 4].

The development and optimization of scalable processes for production, dispersion, and assembly of these nanomaterials falls, in large measure, to the chemical engineering community. Controlling dispersion requires tools that are well known to chemical engineers: thermodynamics, kinetics, fluid dynamics, separations, and quality control, which are correlated with common chemical engineering unit operations (Figure 3.1). Defining, achieving, and maintaining dispersions have long been important and challenging problems in multicomponent materials, including complex fluids and polymer composites. The high level of industrial and scientific interest in nanomaterials in the past two decades is due to both (i) the unique material properties of nanomaterials

Nanotechnology Commercialization: Manufacturing Processes and Products, First Edition.
Edited by Thomas O. Mensah, Ben Wang, Geoffrey Bothun, Jessica Winter, and Virginia Davis.

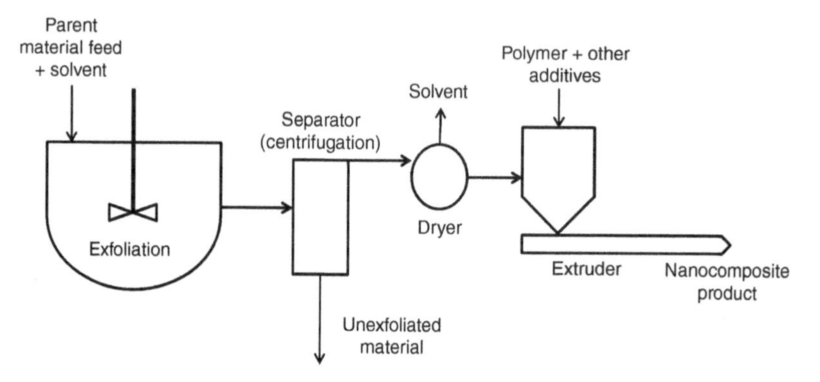

Figure 3.1 An example process flow diagram is shown for nanomaterial exfoliation in liquid, drying, and incorporation into a polymer nanocomposite process. There are clear parallels between these steps and classical chemical engineering technology.

(electrical, thermal, optical, and/or mechanical) and (ii) the high specific surface area of nanomaterials. However, nanomaterials tend to agglomerate into larger, often disordered, structures. This aggregation results in low specific surface area, undermines the intrinsic anisotropic properties that result from nanomaterial structure, and dramatically reduces interfacial interactions with surrounding materials. Dispersion state is therefore a critical consideration at all stages of nanomaterial processing. This includes dispersion in fluids to produce bulk materials consisting solely of the nanomaterial, dispersion in liquids to facilitate processing into a multicomponent material, and direct dispersion into uncured thermoset resins and thermoplastic polymers. In this chapter, we address dispersion in low viscosity liquid solvents, low viscosity uncured resins and composite precursors, as well as dispersion in high viscosity polymer melts. While the emphasis is on graphene and carbon nanotubes (CNTs), much of the dispersion science can easily be extended to other anisotropic nanomaterials, including nanoclays, boron nitride nanosheets, metal dichalcogenides, inorganic nanorods, and cellulose nanocrystals. Note that we do not address the creation of nanomaterial-based films, papers, monoliths, and foams that are later impregnated with a solvent or resin. **Our overall goal is to provide a guide regarding methods to disperse nanomaterials in liquids and melts and to provide techniques for characterizing dispersion quality.**

3.2 What Is Dispersion and Why Is It Important?

Dispersion is the act of disassembling the minor component in a mixture into smaller and smaller pieces [5]. In the complex fluids and nanomaterials communities, dispersion, distribution, alignment, flocculation, and thermodynamic interactions are often described under the general category of dispersion state or dispersion microstructure. More specifically, dispersion state refers to the size of the material as it exists in the system (e.g., polymer melt, liquid dispersion, and final part). Dispersion microstructure refers to not only dispersion state but also spatial characteristics such as alignment, fractal or floc shape, and distribution.

The dispersion state X of a nanomaterial signifies the ratio of the size of the dispersed entities existing in the mixture to the size of the individual species [6, 7]. In the limit of perfect dispersion, X tends toward 1, while in practice, X may become quite large. For example, in the CNT literature, CNTs may exist as millimeter or micron-sized aggregates (large random assemblies), small bundles (assemblies of tens of nanotubes), or individuals. As-received nanomaterials, particularly anisotropic nanomaterials, are often bundled or stacked into micron- or larger scale aggregates. These aggregates must be separated to take advantage of the nanomaterials' geometry. For example, naturally occurring nanoclays, such as montmorillonite, exist as stacked sheets incapable of imparting enhanced mechanical, thermal, and flame retardancy that exfoliated sheets can provide in a polymer matrix [8–10]. Similarly, graphene's mechanical and electrical properties tend toward those of graphite as the number of layers increases [11]. Many CNT synthesis and purification schemes result in bundles and aggregates of CNTs [6, 7]; if not separated, these aggregates can mask individual nanotubes' intrinsic optical, thermal, and electrical properties and form stress risers, which deteriorate mechanical properties in composites [12, 13]. Even materials such as cellulose nanocrystals and inorganic nanocylinders that are extracted or synthesized as individuals may later form aggregates due to Ostwald ripening, sedimentation, or drying. In this regard, colloid science provides a suitable framework for understanding nanomaterial dispersion. By the IUPAC definition, a colloidal system is one in which "molecules or polymolecular particles dispersed in a medium have dimensions between 1 and 1000 nm (1 μm) or in which discontinuities in the system occur at distances on that length scale." [14] Therefore, many dispersed nanomaterials are by definition also colloids and vice versa. However, in the case of anisotropic nanomaterials, the long dimension can be much greater than 1000 nm. For example, MWNTs over 2 cm in length with diameters on the order of 10 nm have been synthesized [15].

In all complex mixtures, including liquid dispersions, polymer melts, and solidified multicomponent materials, both dispersion and distribution affect properties. Dispersion is how fine the individual particles are, and distribution designates their spatial placement within a sample. For example, Figure 3.2 shows two aggregates of nine rodlike nanomaterials; mixing could be used to reduce the number of rods in each aggregate to three and subsequently individual rods. This measure of particles/aggregate is an indicator of dispersion quality [7]. However, distribution, that is, where the materials are located and how they are oriented, is also important. Typically, a uniform

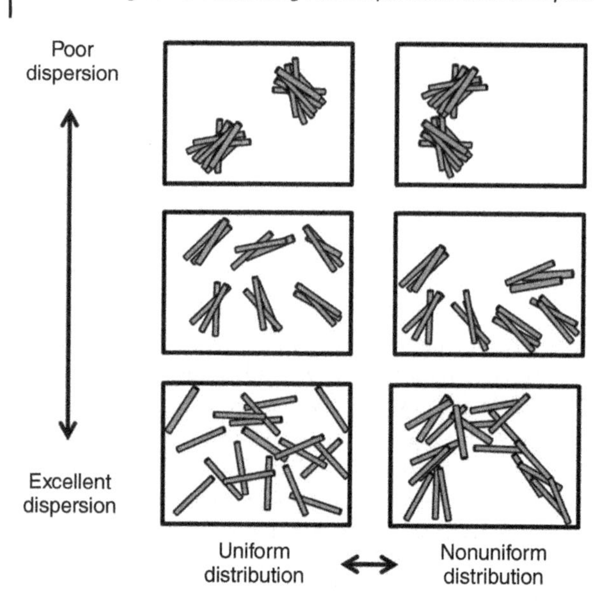

Figure 3.2 Schematic distinguishing degrees of dispersion and distribution.

distribution throughout the volume is desirable, but for a graded material, a nonhomogenous distribution is desired. In some cases, the goal with anisotropic nanomaterials is to achieve not only uniform distribution but also orientation. Orientation can be achieved through liquid crystalline phase formation, and/or the application of shear, electrical, or magnetic forces. There are very few reports of true thermodynamic solutions of nanomaterials, with one notable exception being SWNT dispersed in chlorosulfonic acid [16, 17]. Fortunately, from a practical engineering point of view, the key factor is that the dispersion stability exceeds the timescale of processing, not whether a true thermodynamic solubility is achieved.

Dispersion state is of greater importance for anisotropic nanomaterials than for classical colloidal dispersions or polymer composites. Agglomerates whose small dimension is greater than 100 nm are, by definition, no longer nanomaterials. The specific surface area of nanomaterials in a matrix is a key consequence of the level of dispersion. Whether the goal is interfacial adhesion, sensor activity, or antimicrobial activity, the available surface area determines the performance of the material. Nanomaterials that are poorly dispersed will have "wasted" surface area inside agglomerates that do not contribute to material performance. In other words, the effective aspect ratio of the material, as it exists in the fluid or solidified structure is what controls the interfacial volume between the nanomaterial and its surroundings (e.g., other nanomaterials, polymer, and resin). Vaia and Wagner illustrated this concept by plotting logarithmic isolines of interfacial area per volume of filler ($\mu m^{-1} = m^2/ml$) on a plot of aspect ratio $\alpha = H/R$ against the long dimension, H, and short dimension, radius R (Figure 3.3) [18]. Values of $\alpha > 1$ indicate rods, while $\alpha < 1$ represents disks or sheets. Classical mineral fillers have low aspect ratios and cluster sizes between 0.1 and 10 μm; these low aspect ratios result in an interfacial area per particle volume of 1–1000 μm^{-1}. Traditional carbon and glass fibrous materials used in composites result in much lower interfacial areas per particle volume on the order of 0.1–10 μm^{-1}.

In the fiber composite community, it is well recognized that differences in dispersion state can dramatically alter composite properties and that small differences in surface chemistry can dramatically affect dispersion state. In the case of nanomaterials, the interfacial areas per particle are orders of magnitude greater than in traditional composites. For example, a 5-nm-diameter SWNT rope (bundle) has an interfacial area of 100–1000 μm^{-1}, while an individual SWNT with similar length has an area of 10,000 μm^{-1}. In addition to affecting the interfacial area, dispersion state can have a significant effect on properties. Even small stacks or bundles of nanomaterials can prevent manifestation of the materials' interesting optical, electronic, or thermal properties. For example, individualized SWNT exhibit characteristic UV–vis spectra due to van Hove singularities, but these features are quenched when SWNT bundle [19, 20]. Similarly, as graphene sheets are stacked together, the electronic properties begin to change, with 10 layers resulting in properties nearly indistinguishable from graphite [21, 22].

The differences in dimensionality between nanomaterials and traditional fillers are also the reason that loadings required for nanocomposites are orders of magnitude lower than those required for traditional composites. To the traditional composites community, where using tens of percent of fillers is typical, adding a material at concentrations less than 1 wt% is unheard of. This misconception that nanocomposites need loadings of 20–50 wt% was part of the reason for economic concerns that have slowed applications development. The reason that the required loadings are so much lower for nanocomposites is that the volume fraction, not the weight fraction, is the relevant metric for evaluating potential changes in properties. The conversion between weight and volume percent is simply achieved by implementing the densities of the nanomaterial (ρ_n) and solvent (ρ_s)

$$\phi = \frac{m_n/\rho_n}{m_n/\rho_n + m_s/\rho_s}$$

where m_n and m_s are the masses of the nanomaterial and solvent, respectively. For many systems, the difference between mass and volume fraction is relatively minor. However, for inorganic high density nanomaterial systems in low density fluids, the difference is dramatic. For example, for polymer-coated silver nanowires dispersed in water, 1 wt% is approximately 0.1 vol%.

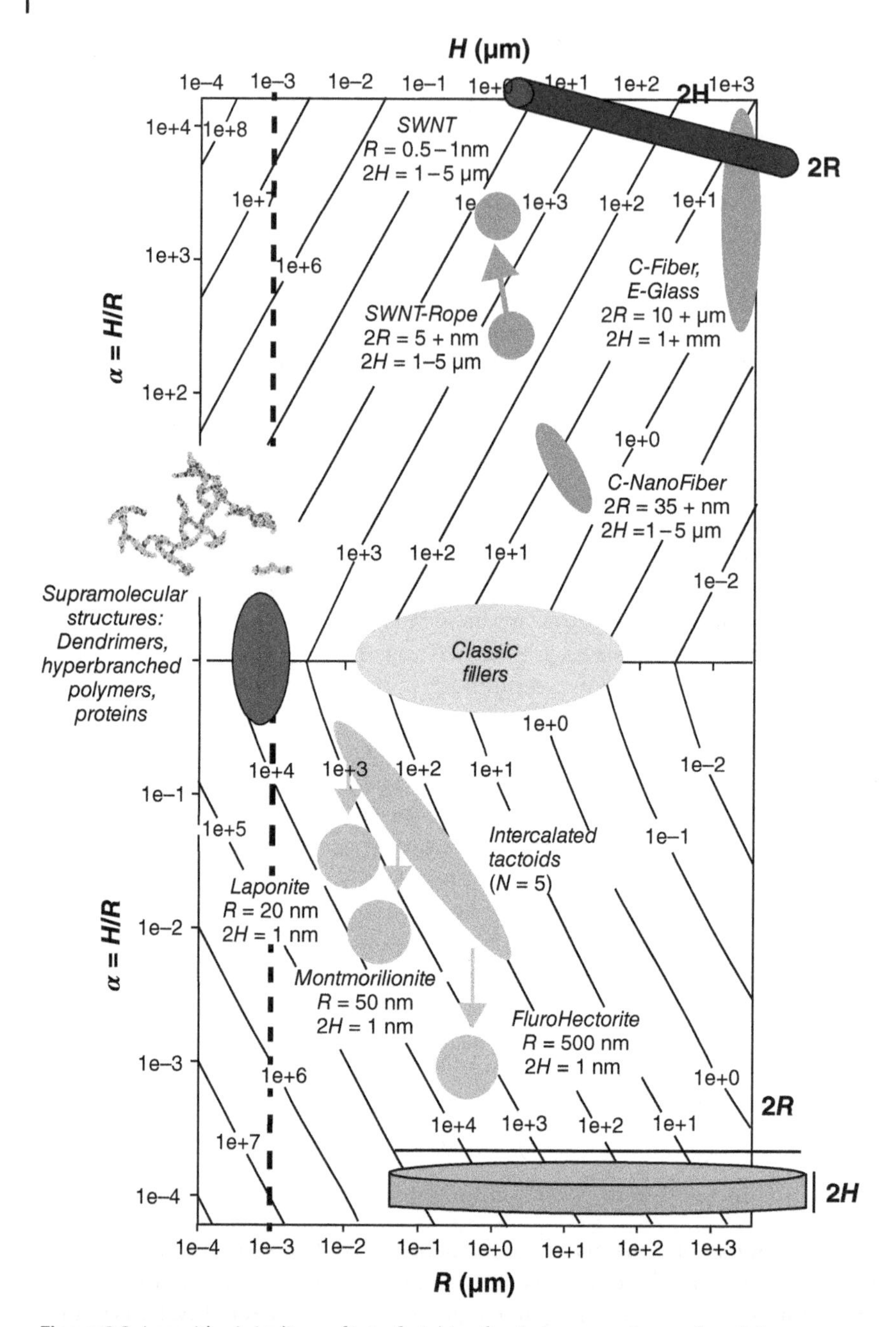

Figure 3.3 Logarithmic isolines of interfacial (surface) area per volume of particles ($\mu m^{-1} = m^2/ml$ with respect to aspect ratio $\alpha = H/R$, where H is the height or length and R is the radius based on approximating particles as cylinders (area/volume $= 1/H + 1/R$). Vaia and Wagner 2004 [18]. Reproduced with permission of Elsevier.

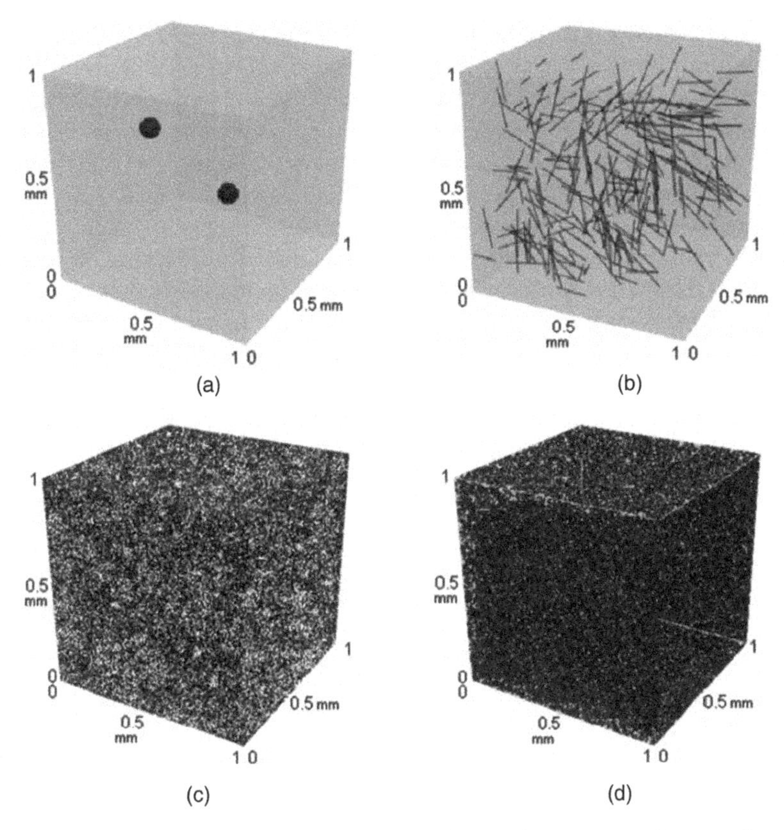

Figure 3.4 Distribution of micro- and nanoscale fillers at the same 0.1 vol% in a reference volume (a): alumina (Al_2O_3) particle; (b): carbon fiber; (c): graphene nanoplatelet (GNP); (d): carbon nanotubes (CNT). Ma *et al.* 2010 [23]. Reproduced with permission of Elsevier.

The number and size of entities that constitute a given volume fraction is particularly important. Figure 3.4 shows the differences in the space filling properties of 0.1 vol% of uniformly distributed particles with different sizes in a 1 mm^3 reference volume. For 100 μm alumina particles, 0.1 vol% is only two particles and there are significant voids in the structure. In the case of a composite, this would mean a large fraction of the polymer matrix was unaffected by the material. In contrast, for a 5-μm-diameter, 200-μm-long carbon fiber system, there are a total of 255 fibers, with fewer voids in the structure. For 45-μm long, 7.5-nm-thick square carbon sheets (referred to as graphene nanoplatelets in the reference), there are over 65,000 sheets and significantly less void volume. Finally, in the case of 12-nm diameter, 20-μm-long MWNTs, 0.1 vol% of a 1 mm^3 volume is over 4.4 billion nanotubes, and the structure is almost completely filled by a network of CNTs [23].

One of the chief considerations in characterizing dispersion microstructure is determining whether a percolated network exists and/or what concentration is required to achieve this percolated network. Percolation occurs when a connected network of nanofillers spans the entire system and can be detected using a number of bulk measurement techniques. In cases where the nanomaterial is conductive and the surrounding liquid or matrix is not, electrical conductivity can be used to determine the percolation threshold. For systems where the nanomaterial network becomes the chief load-bearing structure, rather than the matrix, percolation can also be detected using rheological or mechanical measurements. Rheological percolation is important to understand because it provides significant insight into the system's behavior, results in significant changes in rheological properties, and provides an indication of the potential for mechanical reinforcement. It should be noted that increasing concentration above percolation results in a reduction in nanomaterial separation. This reduction may manifest as jamming or aggregation, both of which are correlated with potential processing difficulties. Therefore, as shown by Dondero and Gorga [13], higher concentrations are not always better. In their work on PP/MWNT, composites fibers with 0.25 wt% MWNT had higher mechanical properties than those with 1 wt% MWNT, and some properties further deteriorated at 3 wt% [13].

The volume fraction required for percolated network formation ϕ_c varies with aspect ratio as

$$\phi_c \sim \frac{R}{L}$$

where R is the effective radius of the dispersed element and L is the effective length of the dispersed element. The high aspect ratios associated with nanomaterials demonstrates that such networks may be attained with relatively low nanomaterial content. Similar percolation scaling relationships (with a different prefactor) exists for disk-like or sheet-like nanomaterials. (Note that these expressions only hold in the limit of infinite rigidity.) Dispersion quality directly impacts this threshold; if rodlike nanomaterials are bundled together, then the bundle rather than the individual is the connecting element. The aspect ratio of the bundle is significantly lower than that of the individual rods (i.e., $X \gg 1$), so the percolation ratio is higher for the poorly dispersed case.

However, note that perfect dispersion is not always the sole goal. Even if nanomaterials can be perfectly dispersed in a matrix, two additional factors come into play. First, matrix–nanofiller interactions are critically important. The interfacial load transfer between the matrix and nanomaterial ultimately determines the degree of mechanical reinforcement; load transfer is typically dictated by the interactions between the chemical functionalities on the nanomaterial surface with the surrounding matrix. Second, nanofiller–nanofiller interactions are also important for network formation; recent studies on nanomaterial-based 3-D structures (such as networks, gels, and foams) demonstrate a strong dependence on the quality of mechanical and electrical point contacts.

One key practical question is this: How does aggregation influence the desired properties of the final product? The answer is problem specific. In the case of colorants, well-established fillers, and even nanoclays, decades of research and production have enabled well-defined relationships and quality control specifications. However, for materials at an earlier stage of research and development, different perceptions of "acceptable levels of dispersion" from one research problem to another have led to miscommunication across fields and journals. This is particularly odious in the case of nanosheets such as graphene where miscommunication on the level of dispersion has actually blurred the lines between graphene and graphite. Many researchers have been frustrated by the ambiguity of terms such as "graphite nanoplatelets," "graphene nanoplatelets," "nanographite platelets," and "nanographene platelets." [24–27] Similar problems persist for other nanosheet types. Our hope is that the current chapter will shed light and improve characterization, communication, and processing of nanomaterials, particularly in regard to classic chemical engineering problems.

3.2.1 Factors Affecting Dispersion

We must distinguish between the dissolution of nanomaterials as a solute in a solvent (similar to polymer dissolution) in contrast to dispersion, where an energetic potential prevents the free energy minimum (aggregation) from being obtained. Both of these realms lie well within the chemical engineering-centric fields of colloid and polymer physics.

3.2.2 Thermodynamic Dissolution of Pristine Nanomaterials

In general, the thermodynamic driving force for dissolution of a solute into a solvent can be defined by the Gibbs free energy of mixing ΔG_{mix}. Alternatively, the partial molar derivative of the Gibbs free energy, the intensive chemical potential, can be used. At constant temperature and pressure,

$$\Delta G_{\text{mix}} = \Delta H_{\text{mix}} - T\Delta S_{\text{mix}}$$

where T is the temperature, and ΔH_{mix} and ΔS_{mix} are the respective changes in enthalpy and entropy occurring with mixing. Spontaneous dissolution occurs only if the change in free energy upon mixing ΔG_{mix} decreases, leading to the following criteria

$$\Delta H_{\text{mix}} < T\Delta S_{\text{mix}}$$

Thus, it can be deduced that at constant temperature and pressure, minimizing the enthalpy of mixing ΔH_{mix} and maximizing the entropy of mixing

ΔS_{mix} will drive solute solubility. Classical theories for the mixing entropy of rodlike molecules (Flory-Huggins, Onsager, DiMarzio) [28–31], in conjunction with the high molecular stiffness and molecular weight, predicts a small positive ΔS_{mix} [32, 33]. Therefore, the free energy of mixing is largely dependent on the magnitude of ΔH_{mix}.

A number of studies (chiefly out of Trinity College, Dublin) have focused on the use of organic solvents (with no dispersant) and extensive sonication to form dispersions of SWNTs and graphene. The rationale is that by matching interfacial energy parameters between the nanomaterial and the surrounding solvent, the nanomaterial should remain stably dispersed [34–38]. Substantial efforts have been made on this front, even leading to multiple corporate efforts aiming to take advantage of this approach [39]. However, significant debate remains over whether these nanomaterial dispersions in these organic solvents can be classified as solutions. True thermodynamic dissolution should (in principle) be spontaneous and not require sonication. Also, the obtained concentrations are quite small and the nanomaterials often suffer from sonication-induced reduction in their longest dimension [40].

In contrast, substantial work (chiefly out of Rice University) has been devoted to the use of superacids as a means to directly dissolve CNTs, graphene, and other nanomaterials [1, 16, 40–45]. The original impetus behind this effort was the natural analogy between SWNTs and other rigid-rod polymers; rigid-rod polymers such as PPTA (used in Kevlar fibers) have been processed in strong acids for some time [40]. The strong acid protonates the polymer (or SWNT, or graphene), allowing for delocalized charge and repulsion between the nanomaterials. Using this approach, the spontaneous formation of nanomaterial solutions has been enabled, even allowing for high-concentration liquid crystalline phases [16, 41, 44]. Due to the corrosiveness of superacids, they are predominately suitable for processing neat films, foams, and fibers, rather than polymer composites. Therefore, additional dispersion routes are still needed.

3.2.3 Intermolecular Potential in Dispersions

In contrast to true solutions, most nanomaterial dispersion involves an attempt to exfoliate the nanomaterial from an aggregate parent material (covered in more detail as follows) and processing to achieve a repulsive interparticle potential between individual dispersed entities such that the dispersion may remain as a stable colloid.

The overall intermolecular potential controlling stability, flocculation, and/or aggregation of a colloidal (or nanomaterial) dispersion is a result of the balance between repulsive and attractive forces. This can generally be described by Derjaguin, Landau, Verwey, and Overbeek (DLVO) theory and subsequent refinements [46, 47], but the application of this theory to experimental nanomaterial dispersions needs additional development. Nonetheless, attempts to predict or control dispersion should consider the interparticle potential as a complex function of van der Waals attractions, electrostatic repulsions, depletion interactions, and sedimentation forces.

Although van der Waals are often viewed as weak forces, in the case of nanomaterials they can be stronger than covalent bonds. This is perhaps best highlighted by the case of SWNTs where continuous fibers with Young's modulus of 120 GPa and tensile strength of 1.0 GPa can be produced and held together solely by van der Waals forces [48]. These same forces coupled with their amphiphobic characteristics are what make CNTs so difficult to disperse. Due to the smooth and polarizable sp^2 hybridized carbon interface, SWNT readily form undesirable aggregates and ropes, intimately held together by van der Waals forces [49, 50]. Continuum-based methods have been employed for modeling the pairwise interaction potential of parallel nanotubes showing a deep potential energy well at equilibrium separation on the order of 20–40 k_BT/nm [50–52]. Since typical CNT aspect ratios can be on the order of 100–1000, the work required to separate individual tubes is substantial [53]. For example, to separate a 1-μm-long nanotube from a rope, the energy required is on the order of 10^4 k_BT.

3.2.4 Functionalization of Nanomaterials

There are two approaches for tailoring nanomaterial surface chemistry to improve the thermodynamics of nanomaterial dispersion: covalent and noncovalent functionalization. In covalent functionalization, a chemical reaction is used to covalently attach a functional group to the nanomaterial surface. Noncovalent functionalization relies on interactions between the nanomaterial and other species such as polymers, biological molecules, or surfactants. In the case of carbon nanomaterials, noncovalent functionalization often relies on π–π interactions [54, 55]. For both types of functionalization, if the nanomaterial is not debundled/exfoliated during the functionalization process, it is only the surface of the aggregate that will be functionalized, and the impact on dispersion will be limited. For noncovalent functionalization, this is often achieved with sonication in the presence of the dispersant. For covalent functionalization, chemical modifications are typically used to facilitate intercalation and attachment of functional groups on the surface of individual dispersed entities. For example, the Billups reaction for SWNT functionalization relies on alkyl halides, which dissociate in ammonia [56]. The halide, typically lithium, intercalates the nanotube bundles creating a charge on, and access to, the nanotube surface, which is then functionalized with the alkyl. This and similar approaches can also be used to achieve *in situ* polymerization or selective functionalization based on diameter or electronic character.

The interplay between the quality of the dispersion state and the interfacial interaction between the CNTs and solvent/polymer is a delicate balance; too little functionalization results in poor dispersion quality while too much results in adverse effects on material structure and performance. An excellent illustration was seen in the work of Gojny *et al.* who studied the performance of nanocomposites fabricated from DWNTs and amine functionalized DWNTs in an epoxy resin matrix [57]. For the purpose of benchmarking performance, the group was able to compute the expected maximum theoretical Young's modulus for mixtures of pure components. Excellent agreement between predicted and experimental Young's modulus were found at low nanotube loadings of 0.1 wt% using the amine functionalized DWNTs, whereas the pristine DWNTs underperformed. This result is due to favorable interactions between the amino groups and the epoxy resin/amine hardener. However, deviations were found when the loading was increased by an order of magnitude to 1.0 wt%; this departure is explained by the nonideal dispersion state unaccounted for in the model. Essentially, decreased separation distance between DWNTs resulted in aggregation. For an exhaustive review on CNT functionalization routes, the reader is referred to Dyke and Tour [58], Banerjee *et al.* [59], and Tasis *et al.* [60]

The question of graphene functionalization is easily divided between the pristine graphene and graphene oxide (GO) routes [61, 62]. For covalent functionalization, the most common route involves the production of graphite oxide (using strong oxidizers such as potassium permanganate, sulfuric acid, and nitric acid [63]) and subsequent exfoliation of individual graphene oxide nanosheets [11, 64, 65]. The oxidation results in the creation of functional groups such as hydroxyl, carboxyl, and epoxide groups [66]. These ionic groups allow for interlayer electrostatic repulsion (as measured by ζ-potential) between nanosheets, such that graphene oxide is soluble in water and other solvents [67]. Subsequent chemical alteration of these functional groups allows for more sophisticated chemical decoration of GO in order to allow for favorable interactions with polymer matrices. However, the oxidation process compromises the electrical properties of graphene, so chemical and thermal reduction methods are often used to restore graphene oxide to "reduced graphene oxide" (rGO) [61, 68].

In contrast, pristine (i.e., non-GO-derived) graphene must be exfoliated in solution in the presence of dispersants such as surfactants, polyaromatic hydrocarbons, or adsorbing polymers [69–84]. These dispersants noncovalently functionalize the nanomaterial and impart the necessary steric or electrostatic repulsion, allowing for the stabilized pristine graphene nanosheets to remain as stable colloids. (Note that high graphene content in composites may indeed result in decreasing toughness, similar to the CNT case discussed earlier [85].) Similar dispersant schemas have been noted for other nanosheets such as transition metal dichalcogenides and boron nitride nanosheets [82, 86], which is somewhat surprising as it implies that the dispersant interactions with the nanosheet surface is materials-agnostic.

Considerations for choosing a functionalization scheme involve the balance of structural damage and possible property degradation resulting from covalent functionalization versus the retention of dispersants or biological materials used for noncovalent functionalization in the final product. Surfactants and other dispersants may not only prevent aggregation in solvents or resins, they can also impart additional properties. For example, a number of researchers have added flexible polymers to dispersions to enhance both dispersion and mechanical toughness in the final materials. Many biological materials not only enhance dispersion through π–π stacking (and other interactions), they can also enable sensing or antimicrobial properties. It should be noted, however, that dispersants can also have detrimental effects including depletion-induced aggregation and phase separation with the final target polymer matrix. Researchers typically use a "like interacts with/dissolves like" or solubility parameter approach when choosing a functionalization scheme, but the complexities of the system can yield unexpected results. For instance, in noncovalent schemes in particular, the strength of the interaction and/or the possibility of depletion-induced aggregation need to be considered. Both types of functionalization schemes can be used to introduce additional functionality or viscosity modification to tailor the properties of the final nanocomposite.

3.2.5 Physical Mixing

In addition to the dispersion thermodynamics, the process parameters involved in mixing the nanomaterial and solvent or matrix must be considered. As for other chemical processes, order of material addition, energy input, mixer design, viscosity changes, and temperature effects all must be considered. Within nanomaterial dispersion research, many methods to date have been small scale and/or used long mixing times. Successful applications development and reduction of production costs will require more efficient and scalable methodologies. Current methods include sonication, ball mixing, paddle mixing, extrusion, and solvent intercalation methods.

3.2.5.1 Sonication

One of the most commonly used laboratory mixing methods is sonication of nanomaterial dispersions in low viscosity solvents. In sonication, sound waves are used to induce cavitation in the solvent and the localized high energy results in dispersion [87]. The two main methods are bath sonication and the higher energy tip sonication; however, sonication has also been used in conjunction with melt extrusion [88]. Due to the potential for increased temperatures and solvent volatilization, caution must be used when sonicating flammable and volatile materials. Other important considerations are the sample volume, the frequency and amplitude of the sonicating waves, time, and temperature control. Heating and solvent evaporation during sonication can also change the dynamics of the system and affect the final dispersion state. Due to the high energy input, sonication can be very effective at separating CNTs and other materials held together by large thermodynamic attractive forces. However, this same high energy can damage the nanomaterials and other components of the dispersion. For example, tip sonication can dramatically shorten CNTs [89]; this shortening decreases the potential for mechanical reinforcement and increases the concentration required for percolation [90]. Similarly, sonication can damage the dispersant; it can shorten the length of polymers and biological molecules such as DNA and denature proteins [91, 92]. In addition, if sonication is performed in the absence of dispersion aides that associate with the nanomaterial and retard aggregation, the nanomaterial will simply reaggregate after the cessation of sonication. Therefore, successful sonication requires conditions that do not overly damage the materials and result in a stabilized system. For example, the sonication of SWNT in aqueous lysozyme causes individual nanotubes to be peeled away from nanotube bundles. At the same time, the sonication energy partially opens the lysozyme and exposes its tryptophan residue, which strongly interacts with SWNT through π–π stacking. After cessation of sonication, the lysozyme refolds engulfing the nanotube and stabilizing the dispersion. Remaining aggregates can be removed by centrifugation. If an ice bath is used to minimize the temperature increase caused by sonication, the antimicrobial activity of the lysozyme can be retained and the dispersions can be used to make antimicrobial films and fibers [93–97].

3.2.5.2 Solvent Intercalation Methods

The introduction of high aspect ratio materials into a mixture can result in significant increases in viscosity. This is particularly true in the case of rigid rods such as SWNT. Dispersion of 100 ppm vol of individual SWNT can result in an order of magnitude increase in viscosity. Increased viscosity makes uniform mixing and further dispersion more challenging. Solvent intercalation methods are used to circumvent this problem, particularly in composite processing. Typically, the nanomaterial is dispersed in a low viscosity solvent. The uncured resin or polymer matrix is also often dissolved in a solvent. The two dispersions are then combined and mixed and the solvent removed. Volatile solvents are preferred because of the ease of removal, but for materials such as polyolefins only a few solvents can dissolve the polymer, and these often have to be removed via filtration. For example, Haggenmueller *et al.* used a hot coagulation method where CNTs were subjected to sonication in hot dichlorobenzene and then mixed with polypropylene dissolved in hot dichlorobenzene. The mixtures were combined and the nanocomposite was allowed to crystallize. The results showed individual dispersion that resulted in a significant increase in thermal stability [98, 99].

3.2.5.3 Shear Mixing Methods

Shear mixing methods common to the polymer and chemical industry are also employed and are advantageous due to the existence of large-scale production equipment and mixer design knowledge. The only fundamental requirement for successful shear dispersion is that the shear energy density be sufficiently high to overcome attractive forces. However, in Huang *et al.*'s systematic study of the shear dispersion state of MWNTs in viscous poly(dimethylsiloxane) (PDMS) in a simple paddle mixer, the results revealed that a long characteristic mixing time, on the order of days, was required to achieve a uniform reproducible dispersion state [100]. The authors argued that this characteristic time was so extensive that many previous shear-based dispersions studies should be revisited. The long mixing time was a function of the time required to impart the shear stress needed to peel the CNT bundles. As the authors noted, this will vary with polymer viscosity and vessel geometry, but the result highlights a key challenge of scaling up CNT dispersion and the necessity of optimizing the mixing process.

For shear mixing in low viscosity liquids, recent reports have demonstrated that high shear mixing can exfoliate nanosheets from the parent material with a power law relationship written as

$$P_R \sim V^{\alpha}$$

where P_R is the production rate, V is the volume of the exfoliation vessel, and α depends on the choice of nanosheet, solvent, and dispersant. The key issue is this: if α is greater than or equal to 1, then scaling up to large volumes becomes feasible from a manufacturing standpoint. If α is less than 1, then scaling up results in diminishing returns and this nanomanufacturing approach will remain confined to the labscale. Paton *et al.* determined values of $\alpha > 1$ for several dispersant/solvent combinations for exfoliation of pristine graphene, boron nitride nanosheets, and transition metal dichalcogenides [37]. This shows excellent promise for scalable exfoliation of nanosheets from their parent materials.

In spite of the effect of mixing time, a number of groups have used high shear strategies to disperse CNT in polymers. Sandler *et al.* recognized the relationship between high matrix viscosity and high shear stresses [101]. To exploit this effect, the group mixed MWNTs at high shear (1000 rpm; 2 h) by using dry ice to lower the viscosity of the epoxy resin. Likewise, Rahatekar *et al.* dispersed aligned MWNT carpets using a high shear protocol (1000 rpm, 2 h) but noted that the initial aligned state of their tubes made dispersion much easier as compared to highly entangled MWNTs [102]. Gojny *et al.* first used a calendaring (3-roll mill) process to homogeneously disperse DWNTs into an epoxy resin with very short mixing times on the order of minutes [57]. The authors note that this method is an attractive technique that can be readily scaled up in industry since calendaring is a common industrial process. Shortly after, Gojny *et al.* used this method to compare the dispersion of various MWNTs and SWNTs into an epoxy resin matrix [103]. From this work, two important conclusions were drawn. The first was that while SWNTs possess the highest potential for mechanical property enhancement, the resulting nanocomposites could not outperform the amine-functionalized MWNTs due to the poorer SWNT dispersion state. The second conclusion was that the three-roll mill process could not produce composite materials with an MWNT loading above 0.3 wt%. This limitation was later addressed by Wichmann *et al.* who developed a postcalendaring method by means of a vacuum dissolver to construct MWNT epoxy composites at loading up to 2 wt% [104]. Seyhan *et al.* achieved dispersion of MWNTs and amine-functionalized MWNTs in a vinyl ester/unsaturated polyester resin (VE/UPR) at short mixing times by means of a three-roll milling process [105]. Similarly, the technique was later used for processing MWNT-UPR dispersions [106]. It should be noted that evaporation of components, such as the styrene in UPR, can result in undesirable polymerization during mixing.

Effective melt processing relies on typical extrusion process considerations such as the use of a single or twin screw, the type and placement of distributive and dispersive mixing elements, temperature profile, and residence time [88, 107]. Typically, twin screw extruders with a combination of dispersive and distribution elements are most effective. However, research is often conducted on less effective systems with smaller volumes such as conical screw extruders or batch mixers. For all melt processes, the selection of processing conditions needs to balance the competing needs. High shear, high temperatures, and long residence times are needed to facilitate nanomaterial dispersion while lower shear, temperatures, and residence times are needed to preserve the polymer backbone. In addition, the effect of stabilizers, colorants, and other additives on processing and the final dispersion state need to be considered [90]. Due to the number of competing and interacting effects, design of experiments is typically the most effective way to optimize processing for a given property (e.g., Young's modulus, tensile strength, thermal stability, and flame retardancy).

To reduce the shear required in the melt state, nanomaterials should be combined with the polymer before melt extrusion. Radhakrishnan *et al.* compared the effects of three premixing methods on the dispersion of both SWNT and dodecyl functionalized SWNT (C12SWNTs) on dispersion in 12 MF PP [108]. They found that the mixing enabled by the hot coagulation method described in Haggenmueller *et al.* gave the most uniform dispersion while simply dry mixing the nanotubes and PP powder resulted in numerous large aggregates [98]. The third method, rotary evaporation, consisted of dispersing the nanotubes as bundles in isopropanol and using a rotary evaporator to coat polymer powder with the nanotubes. This method resulted in an intermediate level of dispersion that was nearly equivalent to that from hot coagulation but avoided the large quantities of dichlorobenzene and polymer degradation associated with the hot coagulation method [108].

3.3 Characterizing Dispersion State in Fluids

Here we summarize the various techniques for characterizing liquid dispersion in Table 3.1 (adapted from Ref. [7]). There are several important issues to note: (i) The distinction between measurements that focus on very small samples of the liquid dispersion (hundreds of nanoparticles) as opposed to bulk measurements (> trillions of nanoparticles) [109, 110]; (ii) the statistics associated with small-scale measurements are far less reliable, and (iii) sample preparation. Liquids often have to be dried to produce films appropriate for characterization; this drying process can induce agglomeration. Since high-resolution imaging methods such as atomic force microscopy (AFM), transmission electron microscopy (TEM), and scanning electron microscope (SEM) typically require solidification/drying and only sample a limited area, they may not always provide an accurate depiction of the dispersion state in a liquid or melt.

3.3.1 Visualization

The first method for characterizing dispersion state is visual inspection. While individual nanomaterials are not visible to the eye, aggregates are. In the case of carbon materials, the black color of aggregates facilitates this initial check. Large visible black particles surrounded by clear fluid or polymer matrix are an indication of poor dispersion. However, since low levels of dispersion of even fairly large bundles results in a dark color, it is difficult to distinguish between mediocre and excellent dispersions. Images of vials of gray or black fluids provide little information about the dispersion state; they simply show that some amount of material was distributed throughout the fluid.

For dispersions with micron-sized or larger entities, optical microscopy is a standard technique for characterizing dispersion and distribution. As for higher magnification images, one must consider whether the area viewed is representative. This challenge can be partially resolved by using a scanning stage to stitch together images over a larger area than can be visualized in a single field of view. After determining how large a sampling area is required to obtain representative data, visualization can be used as the basis for quality control standards for size, number or distribution of the minor component in a mixture. Statistical methods for assessing gross uniformity, texture, and the scale of segregation are well defined [5]. Anisotropic nanomaterials may or may not be visible with an optical microscope, depending on the individual nanomaterial size, shape, and optical properties. Nonetheless, simple visual inspection and optical microscopy can readily detect large aggregates and textural features such as alignment and connectivity. Higher magnification imaging techniques are often used to identify physical features such as whether nanosheets are aggregated, intercalated, or exfoliated. However, as noted in

Table 3.1 Summary of analytical techniques for dispersion characterization.

Name	What does it characterize?	Advantages and disadvantages
Visualization/optical microscopy	Distribution, size of agglomerates	No information on nano-dimensions
Transmission electron microscopy (TEM)	Number of individuals within agglomerates, nanoscale details	Most detailed technique; can even discern number of layers. Nonbulk. Drying effects can alter sample. (Cryo-TEM can capture *in situ* details.) Small sample size can be misleading
Atomic force microscopy (AFM)	Addresses both lateral size and nano-dimensions	Nonbulk. Carried out on dried samples
Light scattering	Hydrodynamic radius (can be correlated with lateral size)	Bulk. No information on nano-dimensions
Rheology (dilute)	Aspect ratio distribution	Bulk, rapid
Rheology (concentrated)	Transition from isotropic to liquid crystalline (scales inversely with aspect ratio). Fractal network formation, scaling behavior	Bulk
UV–vis/fluorescence spectroscopy	Concentration (using Beer–Lambert law), particular CNT species can be distinguished as individuals	Often cannot distinguish between nanomaterial and bulk
Raman spectroscopy	Degree of functionalization, number of layers in nanosheets. Load transfer in composites	Typically carried out on solid materials. Spatial mapping is possible to provide quantitative information on nanomaterial distribution. Probes near the surface. Most applicable to carbon nanotubes but also widely used in industry for other materials
TGA/DSC	Stabilization and/or degradation of polymer changes	Difficult to sort out nanomaterial effects on properties versus nanomaterial effects on the surrounding polymer
FTIR	Functional groups (associated with peaks in the spectra)	Vague, can be difficult to see at low degrees of functionalization

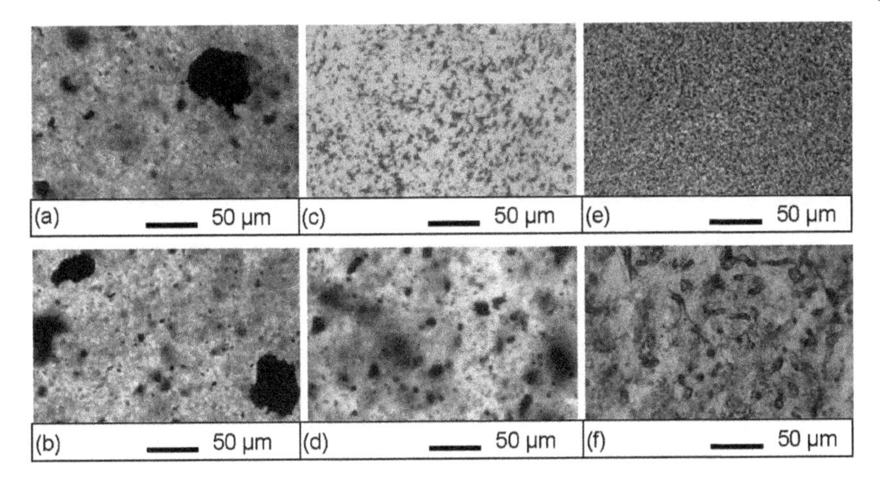

Figure 3.5 Optical microscopy images of melt extruded nanocomposites (a) and (b) PP/SWNT and PP/C12SWNT by dry mixing, (c) and (d) PP/SWNT and PP/C12SWNT by rotary evaporation, and (e) and (f) PP/SWNT and PP/C12SWNT by hot coagulation. Radhakrishnan *et al.* 2010 [108]. Reproduced with permission of John Wiley & Sons.

Table 3.1, caution must be used when interpreting such images. Micrographs may provide information about the dispersion, but lack information regarding the distribution or uniformity of the dispersion state.

Figure 3.5 compares optical microscopy images for the previously described mixing methods compared in Radhakrishnan *et al.* [108] Even after extrusion, nanotube aggregates, tens of microns in diameter, were visible in the dry mixed nanocomposites. Nanotube aggregates were also visible in the extruded rotary evaporator mixed nanocomposites, but the aggregate size was much smaller. The hot-coagulated SWNT nanocomposites appeared very uniform. However, for this method, as well as the others, the functionalized C12SWNT nanocomposites appeared less uniform than the pristine SWNT nanocomposites.

3.3.2 Spectroscopy

A second method useful for numerous nanomaterials is spectroscopic characterization. For dilute dispersions, UV-visible-near-infrared (UV-vis-NIR) spectroscopy can be used to verify the presence of spectral features that are characteristic of individual or small bundles of nanomaterials. For example, the presence of van Hove singularities in SWNT samples provides a first indication of dispersion. The size-dependent plasmon resonance of inorganic particles such as gold can also be used to validate the presence of nanomaterials with the expected aspect ratio. UV–vis-NIR is often used in conjunction with the Beer–Lambert law to establish calibration curves so that concentrations, particularly in supernatants, can be determined directly from

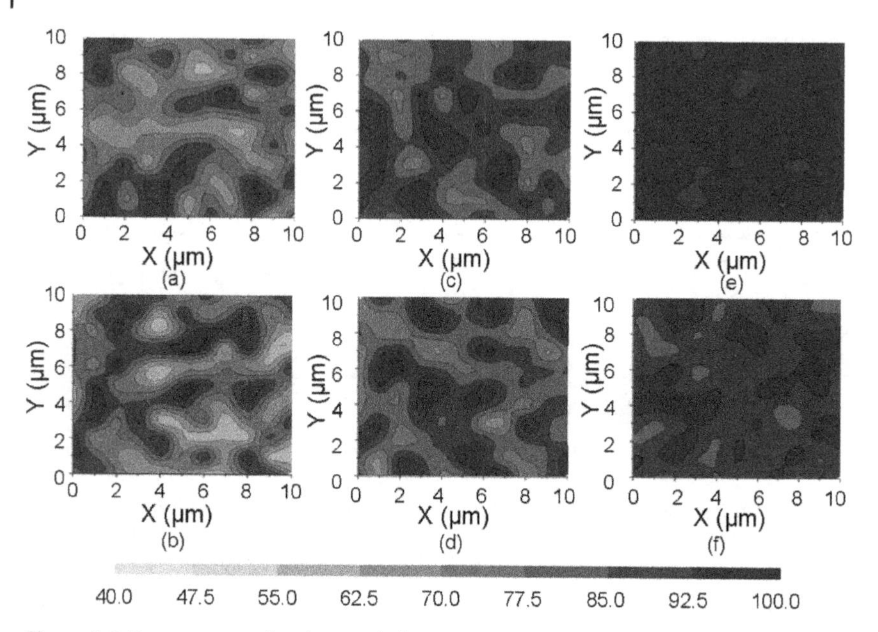

Figure 3.6 Raman maps of melt extruded nanocomposites (a) and (b) PP/SWNT and PP/C12SWNT by dry mixing, (c) and (d) PP/SWNT and PP/C12SWNT by rotary evaporation, and (e) and (f) PP/SWNT and PP/C12SWNT by hot coagulation. Radhakrishnan *et al.* 2010 [108]. Reproduced with permission of John Wiley & Sons.

the absorbance at a specific wavelength. The key limitation of UV–vis-NIR lies within the necessity of using dilute dispersions, particularly in the case of carbon nanomaterials. Other spectroscopic methods include FTIR for confirming interaction between the nanomaterial and dispersant/matrix.

Microspectroscopy methods, particularly micro-Raman, can be used to map the distribution of the nanomaterial in a larger sample. This is a much more quantitative analog to the information provided by microscopy alone. Figure 3.6 shows Raman maps corresponding to the samples in Figure 3.5; the maps were prepared using a scanning stage to measure changes in the intensity of the CNT G peak with a 514-nm laser. Within a given sample, the relative G peak intensity is simply a function of nanotube concentration; higher intensity indicates a higher localized SWNT concentration. Raman intensity maps provided a clear visualization of the differences between composite dispersion from sample to sample, and the data was quantified in terms of the standard deviation of the intensities [99] in the 100 μm^2 map (Table 3.2). Since Raman mapping is quantitative and not affected by sample opacity, it provided a superior assessment of composite uniformity when compared with SEM or optical microscopy. The Raman mapping results confirmed that hot coagulation provided the most uniform composites and that in all cases the

Table 3.2 Standard deviation of G band intensities over the entire map for melt extruded samples obtained by Raman mapping, giving a quantitative measure of distribution and uniformity.

	Dry mixed	Rotary evaporation	Hot coagulation
SWNT	11.5	7.9	2.7
C12SWNT	13.9	9.5	5.1

Source: Radhakrishnan *et al.* 2010 [108]. Reproduced with permission of John Wiley & Sons.

C12SWNTs were more poorly distributed than SWNTs. The poorer distribution of the C12SWNT composites is contrary to the expectation that surface modification, even below the length required for entanglement with polymer chains, should result in lower aggregation tendencies due to a combination of increased steric hindrance and improved interfacial compatibility.

3.3.3 TEM

One of the most common techniques used for inferring dispersion quality in fluids is TEM; this technique allows for determination of *both* dimensions (the nano-dimension and the larger dimension) of anisotropic nanomaterials such as nanosheets, nanocrystals, nanowires, and nanotubes. This is somewhat subtle in the case of nanosheets, where the nano-dimension is in the z-direction of the image. The number of atomic layers within the nanosheet can be determined at the edges of the nanosheet by counting the number of distinct edges.

3.3.4 AFM

AFM is also typically performed on dried samples. It has the distinct advantage of mapping both lateral dimensions and thickness (diameter) of anisotropic nanomaterials. In addition, AFM can often access a large sample size (relative to TEM) such that meaningful statistical quantities can be computed; in addition, a number of programs are available that can allow such statistics to be gathered and computed automatically. However, it should be noted that AFM samples are typically prepared by drop casting or spin casting, which can lead to drying and aggregation effects.

3.3.5 Light Scattering

Light scattering is a powerful tool for colloid science and very relevant for dispersion of low aspect ratio nanomaterials in fluids. Even highly anisotropic nanomaterials such as 10-μm-long, 20-nm-diameter germanium nanowires exhibit Brownian behavior in low viscosity solvents [111]. However, light scattering is not effective at measuring the length of such long entities.

Dynamic light scattering (DLS) can distinguish the hydrodynamic radius (R_H) of dispersed colloidal nanomaterials in liquid media and gives a population distribution of the measured sizes. DLS software measures the Brownian diffusion of the particle and backs out a hydrodynamic radius as

$$R_H = \frac{k_B T}{6\pi \eta_s D_t}$$

where k_B is Boltzmann's constant, T is temperature, D_t is the measured translational diffusivity of the particles, and η_s is the viscosity of the solvent. Particle–particle effects are neglected. If the particles are actually spheres, the hydrodynamic radius and actual radius are equal.

Such measurements can generally distinguish the presence of large aggregates. Hydrodynamic theory allows for clear relationships between as-measured R_H and the actual geometry of individual nanomaterials if the nanomaterials conform to prespecified geometries such as rigid rods or rigid disks. For rigid rods and rigid disks [112],

$$D_t = \frac{1}{3} \frac{k_B T(ln(p) + C_t)}{\pi \eta_s L}$$

where p is the aspect ratio and C_t denotes end effects. Deviations from these theories for well-dispersed nanomaterials is typically caused by the flexibility of the materials and by the extreme polydispersity in shape observed for nanosheets. Although the nano-dimensions in anisotropic nanomaterials (diameter of nanowires, number of atomic layers in nanosheets) cannot be probed using DLS, the flexibility of such materials generally decreases with bundle size (for rods) or number of layers (for nanosheets). This may explain recent anomalous scaling data [113] for various nanosheet types; the data indicate $L \sim {R_H}^{2/3}$, whereas the theory indicates $L \sim {R_H}^{1}$.

3.3.6 Rheology

While the aforementioned methods are useful and dispersion characterization should always rely on multiple methods, rheology is one of the most powerful methods for understanding dispersion because it provides quantitative data on bulk fluid (not dried) samples. Rheological measurements on dispersions involve measurements of viscosity as a function of nanomaterial concentration and shear rate.

In the case of rodlike nanomaterial dispersions where the geometry of the particle is well known, comparison against hydrodynamic theory in the dilute limit can prove helpful. For instance, the intrinsic viscosity of rodlike dispersions can be derived as [110]

$$[\eta] = \lim_{\phi \to 0} \frac{\eta_0 - \eta_s}{\phi \eta_s} = \frac{8}{45}\left(\frac{L}{D}\right)^2 \frac{1}{\ln(L/R_s)} f_M$$

where L is the length of the rod, D is the diameter of the rod, R_s is the effective radius of the rod (including any dispersant effects), η_0 is the zero-shear viscosity, η_s is the solvent viscosity, ϕ is the volume fraction of the rods in solution, and f_M is an aspect ratio-dependent correction factor. Thus, measurement of the intrinsic viscosity can provide a direct, bulk means to determine dispersed nanomaterial dimensions and distinguish between individuals and bundles. Extensions of this theory allow for polydispersity in aspect ratio and additional bundling effects. However, for many commercial rheometers, such measurements occur at torque levels that can be close to the noise in the measurement.

For concentrated dispersions, rheological measurements can determine the presence of a percolated network, the nature of fractals and aggregation, and the relative degree of nanomaterial-solvent/polymer versus nanomaterial–nanomaterial interactions. As previously mentioned, for rodlike entities in a fluid, the onset of rheological percolation occurs at $\phi_c \approx R/L$. For CNC, $\phi_c = 1.4\ R/L$ has been used in the literature, but for CNT a common prefactor has not been chosen [114]. There are several methodologies for determining ϕ_c, one of the more common of which is fitting the low-frequency storage moduli of a series of concentrations to $G'_0 \propto (\phi - \phi_c)^\beta$, which also provides qualitative information about the microstructure [115, 116]. The magnitude of β defines the network response to stress-induced deformations; similar values of β between systems indicate similar network structure and relative nanoparticle–polymer attraction. In simulation work on percolated elastic networks, Arbabi and Sahimi categorized elastic networks that undergo attraction-induced flocculation and found that β ~2.10 is indicative of particle bonds that are resistant to stretching but free to rotate [117], and β ~3.75 is indicative of networks where both bond stretching and rotation were resisted. Surve *et al.* extended this to polymer nanocomposites and found β ~1.88 when network formation is assisted by polymer bridging and there are significant polymer–nanoparticle interactions [118]. Filippone *et al.* compared several nanoparticle–polymer systems and concluded that higher values of β result from a higher degree of nanoparticle–nanoparticle interactions in comparison to nanoparticle–polymer interactions. ***In other words, lower values of β indicate better interactions between the nanoparticle and surrounding matrix*** [119]. Stronger nanoparticle–polymer interactions are associated with improved mechanical properties, result in fewer voids and better load transfer at the interphase, decreased likelihood of nanomaterial pullout from the matrix, and possibly a smoother topology. Rewriting the equation as $G'_0 = A(\phi - \phi_c)^\beta$ enables comparison of the relative nanomaterial–nanomaterial attraction in each system. As shown by Prasad *et al.* in their study of model colloid systems, higher values of A indicate greater interparticle attraction [120]. In summary, the average L/D of the as-dispersed nanomaterial can be estimated from ϕ_c, and the values of β and A can be used to assess relative thermodynamic interactions. This enables direct quantitative comparison of the different composite and provides a means to explore the effects these parameters have on performance properties.

In addition, the fractal dimensions of the nanomaterial network can be determined from scaling relations of low-frequency plateau modulus and critical strain using methods developed by Shih *et al.* [121] and Wu and Morbidelli [122]. The equations for the scaling relations are as follows:

$$G' \propto \phi^{\frac{B}{d-d_f}}$$

$$\gamma_c \propto \phi^{\frac{d-B-1}{d-d_f}}$$

where d_f is the fractal dimension and d is the system dimensionality (2-D or 3-D) [123, 124]. It should be noted that comparisons between systems require consistent criteria for the selection of the critical strain γ_c. The value of d_f can be calculated by equating the exponents of the two scaling relations. According to Wu and Morbidelli, $B = (d - 2) + (2 + x)(1 - \alpha)$, where $1 \le x \le 1.3$ is the backbone fractal dimension, $d = 3$ for a three-dimensional network, and α ranges from 0 to 1 and defines the interfloc and intrafloc strength [122, 123]. For $\alpha = 0$, the interfloc links are stronger than intrafloc links, while for $\alpha = 1$ the intrafloc strength dominates [122]. Fractal dimensions of $1.7 < d_f < 1.8$ have been associated with fast aggregating systems that form more open networks, while $2.0 < d_f < 2.2$ have been associated with slower interpenetrating flocs [124]. The radius of gyration of the flocs (ξ) should have a power law relation with shear rate, $\xi \propto \dot{\gamma}^m$. For rigid aggregates, the interparticle potential of particle is noncentral, and $m = 1/\beta$ is in the range 0.23–0.29 [7]. For soft aggregates (or flocs), the interaction is central and depends on the distance between particle centers, and the internal structure does not respond elastically to small deformations.

3.4 Characterization of Dispersion State in Solidified Materials

A poor dispersion in a liquid or melt will not improve during solidification processes (e.g., drying, curing, and cooling a thermoplastic melt). However, dispersion and distribution can be adversely affected as a result of these processes. Therefore, characterization of nanomaterial dispersion in solid matrices is of direct relevance for understanding the properties of final materials.

3.4.1 Microscopy

The presence of large aggregates in polymeric matrices can be detected using lower resolution microscopy such as optical microscopy and SEM. In most cases, the actual nanoscale dimensions of the nanofillers require microtoming and TEM in order to be visible. Such methods suffer from low sample sizes but have been helpful in distinguishing various processing methods in terms of dispersion quality [125–127].

3.4.2 Electrical Percolation

As noted earlier, the bulk conductivity of composites increases if the conductive nanomaterial concentration reaches above the electrical conductivity threshold. This can be measured by a variety of methods; four-point probe measurements are the most common and require care to ensure good contact between the sample and probes. Similar to the rheological percolation behavior noted earlier, percolation behavior can be modeled as $\sigma_o \propto (\phi - \phi_c)^\beta$. In this way, the as-measured percolation threshold can provide nanoscale insight into the level of dispersion. Despite the simplicity of this model, the actual reported percolation behavior in the literature varies to amazing degrees, with a wide range of reported β and ϕ_c values, even for the same systems. These variations occur chiefly because of differences in dispersion quality, which can be related back to processing conditions [90].

3.4.3 Mechanical Property Enhancement

Mechanical (tensile strength and toughness) tests are often used as a probe for the dispersion state of nanomaterials dispersed in solid matrices. Although some qualitative, empirical comparisons can be made, it is difficult to draw a clear correlation between these properties and improved dispersion quality. Halpin–Tsai rules of mixtures have been used to estimate the expected mechanical enhancement for nanosheet fillers [128]:

$$E_c = E_m \left(\frac{1 + \varsigma\eta\phi}{1 - \eta\phi} \right), \eta = \frac{\frac{E_f}{E_m} - 1}{\frac{E_f}{E_m} + \varsigma}$$

where ϕ is filler volume fraction, E is the modulus (of composite, filler, or matrix), and ζ is twice the aspect ratio of the nanosheets. Note, however, that these mechanical enhancements are often undermined by poor interfacial load transfer between the high strength filler and the surrounding matrix. In such cases, the use of either covalent functionalization of the filler surface or the addition of compatibilizers can improve the load transfer. This is a critical point because techniques used for stable dispersions of nanomaterials in a precursor fluid must allow for not only excellent dispersion but also compatible interfacial interactions between the nanofiller and target matrix.

3.4.4 Thermal Property Changes

In polymer composites, changes in thermal properties measured using differential calorimetry (DSC) and thermal gravimetric analysis (TGA) can result from both processing-induced polymer degradation and the effects of nanomaterials on polymer chain crystallization and stability. Changes in melting point, crystallization temperature, crystallization kinetics, and glass transition temperature are a complex function of both dispersion state and processing-induced polymer degradation. Deconvoluting these effects typically requires additional analysis using other methods such as rheology, X-ray diffraction, and X-ray scattering. On the other hand, increases in thermal decomposition temperature are often directly attributed to nanomaterials stabilizing nearby polymer chains and the nanomaterial network slowing degradation by decreasing the diffusion rate of intermediates in polymer degradation reactions.

3.5 Conclusion

On the timescale of materials research, development, and commercialization, most anisotropic nanomaterials are still in their infancy. New formulations, processing routes, and applications are still being developed for polymers that were first synthesized in the mid-twentieth century. While research and development for materials often takes longer for other fields, the results have greater longevity than in industries such as microelectronics. Although anisotropic nanomaterial research has thus far engaged scientists from a wide range of disciplines, the difficult choices and innovations involved in economic scale-up will be handled by the chemical engineering community. To the degree that chemical engineers can cross-apply concepts from the polymer processing community and colloidal physics community, these novel materials will become a regular part of the broader chemical processing field and the global economy.

Acknowledgments

We acknowledge the helpful insight and research efforts of Sriya Das, Dorsa Parviz, Fahmida Irin, Smit Shah, Matthew Kayatin, Joyanta Goswami, Esteban Urena-Benavides, Amber Hubbard, and Geyou Ao. The authors acknowledge funding from the National Science Foundation under Grant Numbers EPS-115886 and CBET-1565490 (Davis) and CMMI-200489 and CBET-1437073 (Green). Any opinions, findings, and conclusions or recommendations expressed in this material are those of the authors and do not necessarily reflect the views of the National Science Foundation.

References

1 Davis, V.A. (2011) Liquid crystalline assembly of nanocylinders. *Journal of Materials Research*, **26**, 140–153.
2 Urena-Benavides, E.E., Ao, G., Davis, V.A., and Kitchens, C.L. (2011) Rheology and phase behavior of lyotropic cellulose nanocrystal suspensions. *Macromolecules*, **44**, 8990–8998.
3 Pavlidou, S. and Papaspyrides, C.D. (2008) A review on polymer-layered silicate nanocomposites. *Progress in Polymer Science*, **33**, 1119–1198.
4 Choudalakis, G. and Gotsis, A.D. (2009) Permeability of polymer/clay nanocomposites: a review. *European Polymer Journal*, **45**, 967–984.
5 Tadmor, Z. and Gogos, C.G. (2006) *Principles of Polymer Processing*, Wiley.
6 Peigney, A., Laurent, C., Flahaut, E. *et al.* (2001) Specific surface area of carbon nanotubes and bundles of carbon nanotubes. *Carbon*, **39**, 507–514.
7 Green, M.J. (2010) Analysis and measurement of carbon nanotube dispersions: nanodispersion versus macrodispersion. *Polymer International*, **59**, 1319–1322.
8 Brown, J.M., Curliss, D., and Vaia, R.A. (2000) Thermoset-layered silicate nanocomposites. Quaternary ammonium montmorillonite with primary diamine cured epoxies. *Chemistry of Materials*, **12**, 3376–3384.
9 Kashiwagi, T., Harris, R.H., Zhang, X. *et al.* (2004) Flame retardant mechanism of polyamide 6–clay nanocomposites. *Polymer*, **45**, 881–891.
10 Qin, H., Zhang, S., Zhao, C. *et al.* (2005) Flame retardant mechanism of polymer/clay nanocomposites based on polypropylene. *Polymer*, **46**, 8386–8395.
11 Zhu, Y., Murali, S., Cai, W. *et al.* (2010) Graphene and graphene oxide: synthesis, properties, and applications. *Advanced Materials*, **22**, 3906–3924.
12 Muthu, J. and Dendere, C. (2014) Functionalized multiwall carbon nanotubes strengthened GRP hybrid composites: improved properties with optimum fiber content. *Composites Part B: Engineering*, **67**, 84–94.
13 Dondero, W.E. and Gorga, R.E. (2006) Morphological and mechanical properties of carbon nanotube/polymer composites via melt compounding. *Journal of Polymer Science Part B: Polymer Physics*, **44**, 864–878.
14 Slomkowski, S., Alemán, J.V., Gilbert, R.G. *et al.* (2011) Terminology of polymers and polymerization processes in dispersed systems (IUPAC Recommendations 2011). *Pure Applied Chemistry*, **83**, 2229–2259.
15 Cho, W., Schulz, M., and Shanov, V. (2014) Growth and characterization of vertically aligned centimeter long CNT arrays. *Carbon*, **72**, 264–273.
16 Davis, V.A., Parra-Vasquez, A.N.G., Green, M.J. *et al.* (2009) True solutions of single-walled carbon nanotubes for assembly into macroscopic materials. *Nature Nanotechnology*, **4**, 830–834.

17 Green, M.J., Parra-Vasquez, A.N.G., Behabtu, N., and Pasquali, M. (2009) Modeling the phase behavior of polydisperse rigid rods with attractive interactions with applications to single-walled carbon nanotubes in superacids. *The Journal of Chemical Physics*, **131**, 041401.

18 Vaia, R.A. and Wagner, H.D. (2004) Framework for nanocomposites. *Materials Today*, **7**, 32–37.

19 Tabakman, S.M., Welsher, K., Hong, G., and Dai, H. (2010) Optical properties of single-walled carbon nanotubes separated in a density gradient: length, bundling, and aromatic stacking effects. *The Journal of Physical Chemistry C*, **114**, 19569–19575.

20 Crochet, J.J., Sau, J.D., Duque, J.G. *et al.* (2011) Electrodynamic and excitonic intertube interactions in semiconducting carbon nanotube aggregates. *ACS Nano*, **5**, 2611–2618.

21 Choi, W. and Lee, J. (2011) *Graphene: Synthesis and Applications (Nanomaterials and their Applications)*, CRC Press.

22 Metzger, R.M. (2012) *Unimolecular and Supramolecular Electronics I: Chemistry and Physics Meet at Metal-Molecule Interfaces*, vol. **1**, Springer Science & Business Media.

23 Ma, P.-C., Siddiqui, N.A., Marom, G., and Kim, J.-K. (2010) Dispersion and functionalization of carbon nanotubes for polymer-based nanocomposites: a review. *Composites Part A: Applied Science and Manufacturing*, **41**, 1345–1367.

24 Yu, A., Ramesh, P., Sun, X. *et al.* (2008) Enhanced thermal conductivity in a hybrid graphite nanoplatelet – carbon nanotube filler for epoxy composites. *Advanced Materials*, **20**, 4740–4744.

25 Shen, J., Hu, Y., Li, C. *et al.* (2009) Synthesis of amphiphilic graphene nanoplatelets. *Small*, **5**, 82–85.

26 Jang, B.Z. and Zhamu, A. (2008) Processing of nanographene platelets (NGPs) and NGP nanocomposites: a review. *Journal of Materials Science*, **43**, 5092–5101.

27 Narimissa, E., Gupta, R.K., Choi, H.J. *et al.* (2012) Morphological, mechanical, and thermal characterization of biopolymer composites based on polylactide and nanographite platelets. *Polymer Composites*, **33**, 1505–1515.

28 Flory, P.J. (1953) *Principles of Polymer Chemistry*, Cornell University Press, http://www.cornellpress.cornell.edu/book/?GCOI=80140100145700.

29 Onsager, L. (1949) The effects of shape on the interaction of colloidal particles. *Annals of the New York Academy of Sciences*, **51**, 627–659.

30 Usrey, M.L., Chaffee, A., Jeng, E.S., and Strano, M.S. (2009) Application of polymer solubility theory to solution phase dispersion of single-walled carbon nanotubes. *Journal of Physical Chemistry C*, **113**, 9532–9540.

31 DiMarzio, E. and Gibbs, J. (1958) Chain stiffness and the lattice theory of polymer phases. *The Journal of Chemical Physics*, **28**, 807–813.

32 Romanko, W.R. and Carr, S.H. (2002) Comparing the Flory approach with the DiMarzio theory of the statistical mechanics of rodlike particles. *Macromolecules*, **21**, 2243–2249.
33 Kayatin, M.J. (2008) *Rheology, Structure, and Stability of Carbon nanotube-unsaturated polyester resin dispersions*, Auburn University.
34 Hernandez, Y., Nicolosi, V., Lotya, M. *et al.* (2008) High-yield production of graphene by liquid-phase exfoliation of graphite. *Nature Nanotechnology*, **3**, 563–568.
35 Hernandez, Y., Lotya, M., Rickard, D. *et al.* (2010) Measurement of multicomponent solubility parameters for graphene facilitates solvent discovery. *Langmuir*, **26**, 3208–3213.
36 Ciesielski, A. and Samori, P. (2014) Graphene via sonication assisted liquid-phase exfoliation. *Chemical Society Reviews*, **43**, 381–398.
37 Paton, K.R., Varrla, E., Backes, C. *et al.* (2014) Scalable production of large quantities of defect-free few-layer graphene by shear exfoliation in liquids. *Nature Materials*, **13**, 624–630.
38 Bergin, S.D., Sun, Z., Rickard, D. *et al.* (2009) Multicomponent solubility parameters for single-walled carbon nanotube–solvent mixtures. *ACS Nano*, **3**, 2340–2350.
39 Thomas, G. (2013) Improving Composites Using Graphene: An Interview with Allen Clauss. http://www.azom.com/article.aspx?ArticleID=8151.
40 Green, M.J., Behabtu, N., Pasquali, M., and Adams, W.W. (2009) Nanotubes as polymers. *Polymer*, **50**, 4979–4997.
41 Davis, V.A., Ericson, L.M., Parra-Vasquez, A.N.G. *et al.* (2004) Phase behavior and rheology of SWNTs in superacids. *Macromolecules*, **37**, 154–160.
42 Ericson, L.M., Fan, H., Peng, H.Q. *et al.* (2004) Macroscopic, neat, single-walled carbon nanotube fibers. *Science*, **305**, 1447–1450.
43 Behabtu, N., Green, M.J., and Pasquali, M. (2008) Carbon nanotube-based neat fibers. *Nano Today*, **3**, 24–34.
44 Behabtu, N., Lomeda, J.R., Green, M.J. *et al.* (2010) Spontaneous high-concentration dispersions and liquid crystals of graphene. *Nature Nano*, **5**, 406–411.
45 Mirri, F., Ma, A.W.K., Hsu, T.T. *et al.* (2012) High-performance carbon nanotube transparent conductive films by scalable Dip coating. *ACS Nano*, **6**, 9737–9744.
46 Hiemenz, P. and Rajagopalan, R. (1997) *Principles of Colloid and Surface Chemistry*, Marcel Dekker Inc., New York.
47 Israelachvili, J. (1992) *Intermolecular and Surface Forces*, Second edn, Academic Press, London.
48 Behabtu, N., Young, C.C., Tsentalovich, D.E. *et al.* (2013) Strong, light, multifunctional fibers of carbon nanotubes with ultrahigh conductivity. *Science*, **339**, 182–186.

49 Thess, A., Lee, R., Nikolaev, P. *et al.* (1996) Crystalline ropes of metallic carbon nanotubes. *Science*, **273**, 483–487.
50 O'Connell, M.J., Bachilo, S.M., Huffman, C.B. *et al.* (2002) Band gap fluorescence from individual single-walled carbon nanotubes. *Science*, **297**, 593–596.
51 Girifalco, L.A., Hodak, M., and Lee, R.S. (2000) Carbon nanotubes, buckyballs, ropes, and a universal graphitic potential. *Physical Review B*, **62**, 13104–13110.
52 Sun, C.H., Yin, L.C., Li, F. *et al.* (2005) Van der Waals interactions between two parallel infinitely long single-walled nanotubes. *Chemical Physics Letters*, **403**, 343–346.
53 Dresselhaus, M.S., Dresselhaus, G., and Eklund, P.C. (1996) *Science of Fullerenes and Carbon Nanotubes: Their Properties and Applications*, Academic Press.
54 Arnold, M.S., Green, A.A., Hulvat, J.F. *et al.* (2006) Sorting carbon nanotubes by electronic structure using density differentiation. *Nature Nanotechnology*, **1**, 60–65.
55 Arnold, M.S., Stupp, S.I., and Hersam, M.C. (2005) Enrichment of single-walled carbon nanotubes by diameter in density gradients. *Nano Letters*, **5**, 713–718.
56 Liang, F., Sadana, A.K., Peera, A. *et al.* (2004) A convenient route to functionalized carbon nanotubes. *Nano Letters*, **4**, 1257–1260.
57 Gojny, F.H., Wichmann, M.H.G., Köpke, U. *et al.* (2004) Carbon nanotube-reinforced epoxy-composites: enhanced stiffness and fracture toughness at low nanotube content. *Composites Science and Technology*, **64**, 2363–2371.
58 Dyke, C.A. and Tour, J.M. (2004) Covalent functionalization of single-walled carbon nanotubes for materials applications. *Journal of Physical Chemistry A*, **108**, 11151–11159.
59 Banerjee, S., Hemraj-Benny, T., and Wong, S.S. (2005) Covalent surface chemistry of single-walled carbon nanotubes. *Advanced Materials*, **17**, 17–29.
60 Tasis, D., Tagmatarchis, N., Bianco, A., and Prato, M. (2006) Chemistry of carbon nanotubes. *Chemical Reviews*, **106**, 1105–1136.
61 Stankovich, S., Dikin, D.A., Piner, R.D. *et al.* (2007) Synthesis of graphene-based nanosheets via chemical reduction of exfoliated graphite oxide. *Carbon*, **45**, 1558–1565.
62 Park, S., An, J.H., Jung, I.W. *et al.* (2009) Colloidal suspensions of highly reduced graphene oxide in a wide variety of organic solvents. *Nano Letters*, **9**, 1593–1597.
63 Hummers, W.S. and Offeman, R.E. (1958) Preparation of graphitic oxide. *Journal of the American Chemical Society*, **80**, 1339.

64 Dreyer, D.R., Park, S., Bielawski, C.W., and Ruoff, R.S. (2010) The chemistry of graphene oxide. *Chemical Society Reviews*, **39**, 228–240.
65 Pei, S. and Cheng, H.-M. (2012) The reduction of graphene oxide. *Carbon*, **50**, 3210–3228.
66 Gao, W., Alemany, L.B., Ci, L., and Ajayan, P.M. (2009) New insights into the structure and reduction of graphite oxide. *Nature Chemistry*, **1**, 403–408.
67 Li, D., Muller, M.B., Gilje, S. *et al.* (2008) Processable aqueous dispersions of graphene nanosheets. *Nature Nanotechnology*, **3**, 101–105.
68 Shin, H.-J., Kim, K.K., Benayad, A. *et al.* (2009) Efficient reduction of graphite oxide by sodium borohydride and its effect on electrical conductance. *Advanced Functional Materials*, **19**, 1987–1992.
69 Das, S., Irin, F., Tanvir Ahmed, H.S. *et al.* (2012) Non-covalent functionalization of pristine few-layer graphene using triphenylene derivatives for conductive poly (vinyl alcohol) composites. *Polymer*, **53**, 2485–2494.
70 Parviz, D., Yu, Z., Hedden, R.C., and Green, M.J. (2014) Designer stabilizer for preparation of pristine graphene/polysiloxane films and networks. *Nanoscale*, **6**, 11722–11731.
71 Bourlinos, A.B., Georgakilas, V., Zboril, R. *et al.* (2009) Aqueous-phase exfoliation of graphite in the presence of polyvinylpyrrolidone for the production of water-soluble graphenes. *Solid State Communications*, **149**, 2172–2176.
72 Guardia, L., Fernández-Merino, M., Paredes, J. *et al.* (2011) High-throughput production of pristine graphene in an aqueous dispersion assisted by non-ionic surfactants. *Carbon*, **49**, 1653–1662.
73 Green, A.A. and Hersam, M.C. (2009) Solution phase production of graphene with controlled thickness via density differentiation. *Nano Letters*, **9**, 4031–4036.
74 Liang, Y.T. and Hersam, M.C. (2010) Highly concentrated graphene solutions via polymer enhanced solvent exfoliation and iterative solvent exchange. *Journal of the American Chemical Society*, **132**, 17661–17663.
75 Secor, E.B., Prabhumirashi, P.L., Puntambekar, K. *et al.* (2013) Inkjet printing of high conductivity, flexible graphene patterns. *The Journal of Physical Chemistry Letters*, **4**, 1347–1351.
76 Marsh, K.L., Souliman, M., and Kaner, R.B. (2015) Co-solvent exfoliation and suspension of hexagonal boron nitride. *Chemical Communications*, **51**, 187–190.
77 Stankovich, S., Piner, R.D., Chen, X. *et al.* (2006) Stable aqueous dispersions of graphitic nanoplatelets via the reduction of exfoliated graphite oxide in the presence of poly(sodium 4-styrenesulfonate). *Journal of Materials Chemistry*, **16**, 155–158.

78 Bari, R., Tamas, G., Irin, F. *et al.* (2014) Direct exfoliation of graphene in ionic liquids with aromatic groups. *Colloids and Surfaces A: Physicochemical and Engineering Aspects*, **463**, 63–69.

79 Lu, F., Wang, F., Gao, W. *et al.* (2013) Aqueous soluble boron nitride nanosheets via anionic compound-assisted exfoliation. *Materials Express*, **3**, 144–150.

80 Parviz, D., Das, S., Ahmed, H.S.T. *et al.* (2012) Dispersions of Non-covalently functionalized graphene with minimal stabilizer. *ACS Nano*, **6**, 8857–8867.

81 Wajid, A.S., Das, S., Irin, F. *et al.* (2012) Polymer-stabilized graphene dispersions at high concentrations in organic solvents for composite production. *Carbon*, **50**, 526–534.

82 Bari, R., Parviz, D., Khabaz, F. *et al.* (2015) Liquid phase exfoliation and crumpling of inorganic nanosheets. *Physical Chemistry Chemical Physics: PCCP*, **17**, 9383–9393.

83 Das, S., Wajid, A.S., Shelburne, J.L. *et al.* (2011) Localized in situ polymerization on graphene surfaces for stabilized graphene dispersions. *ACS Applied Materials & Interfaces*, **3**, 1844–1851.

84 Irin, F., Hansen, M.J., Bari, R. *et al.* (2015) Adsorption and removal of graphene dispersants. *Journal of Colloid and Interface Science*, **446**, 282–289.

85 May, P., Khan, U., O'Neill, A., and Coleman, J.N. (2012) Approaching the theoretical limit for reinforcing polymers with graphene. *Journal of Materials Chemistry*, **22**, 1278–1282.

86 Yang, H., Withers, F., Gebremedhn, E. *et al.* (2014) Dielectric nanosheets made by liquid-phase exfoliation in water and their use in graphene-based electronics. *2D Materials*, **1**, 011012.

87 Hennrich, F., Krupke, R., Arnold, K. *et al.* (2007) The mechanism of cavitation-induced scission of single-walled carbon nanotubes. *The Journal of Physical Chemistry B*, **111**, 1932–1937.

88 Isayev, A.I., Kumar, R., and Lewis, T.M. (2009) Ultrasound assisted twin screw extrusion of polymer–nanocomposites containing carbon nanotubes. *Polymer*, **50**, 250–260.

89 Pagani, G., Green, M.J., Poulin, P., and Pasquali, M. (2012) Competing mechanisms and scaling laws for carbon nanotube scission by ultrasonication. *Proceedings of the National Academy of Sciences*, **109**, 11599–11604.

90 Grady, B.P. (2011) *Carbon nanotube – polymer composites*, Wiley, Hoboken, N.J.

91 Ao, G., Nepal, D., Aono, M., and Davis, V.A. (2011) Cholesteric and nematic liquid crystalline phase behavior of double-stranded DNA stabilized single-walled carbon nanotube dispersions. *ACS Nano*, **5**, 1450–1458.

92 Nepal, D. and Geckeler, K.E. (2007) Proteins and carbon nanotubes: close encounter in water. *Small*, **3**, 1259–1265.

93 Nepal, D., Balasubramanian, S., Simonian, A.L., and Davis, V.A. (2008) Strong antimicrobial coatings: single-walled carbon nanotubes armored with biopolymers. *Nano Letters*, **8**, 1896–1901.
94 Nepal, D., Minus, M.L., and Kumar, S. (2011) Lysozyme coated DNA and DNA/SWNT fibers by solution spinning. *Macromolecular Bioscience*, **11**, 875–881.
95 Horn, D.W., Ao, G., Maugey, M. *et al.* (2013) Dispersion state and fiber toughness: antibacterial lysozyme-single walled carbon nanotubes. *Advanced Functional Materials*, **23**, 6082–6090.
96 Horn, D.W., Tracy, K., Easley, C.J., and Davis, V.A. (2012) Lysozyme dispersed single-walled carbon nanotubes: interaction and activity. *The Journal of Physical Chemistry C*, **116**, 10341–10348.
97 Nyankima, A.G., Horn, D.W., and Davis, V.A. (2013) Free-standing films from aqueous dispersions of lysozyme, single-walled carbon nanotubes, and polyvinyl alcohol. *ACS Macro Letters*, **3**, 77–79.
98 Haggenmueller, R., Fischer, J.E., and Winey, K.I. (2006) Single wall carbon nanotube/polyethylene nanocomposites: nucleating and templating polyethylene crystallites. *Macromolecules*, **39**, 2964–2971.
99 Du, F., Scogna, R.C., Zhou, W. *et al.* (2004) Nanotube networks in polymer nanocomposites: rheology and electrical conductivity. *Macromolecules*, **37**, 9048–9055.
100 Huang, Y.Y., Ahir, S.V., and Terentjev, E.M. (2006) Dispersion rheology of carbon nanotubes in a polymer matrix. *Physical Review B*, **73**, 125422.
101 Sandler, J.K.W., Kirk, J.E., Kinloch, I.A. *et al.* (2003) Ultra-low electrical percolation threshold in carbon-nanotube-epoxy composites. *Polymer*, **44**, 5893–5899.
102 Rahatekar, S.S., Koziol, K.K.K., Butler, S.A. *et al.* (2006) Optical microstructure and viscosity enhancement for an epoxy resin matrix containing multiwall carbon nanotubes. *Journal of Rheology*, **50**, 599–610.
103 Gojny, F.H., Wichmann, M.H.G., Fiedler, B., and Schulte, K. (2005) Influence of different carbon nanotubes on the mechanical properties of epoxy matrix composites: a comparative study. *Composites Science and Technology*, **65**, 2300–2313.
104 Wichmann, M., Sumfleth, J., Fiedler, B. *et al.* (2006) Multiwall carbon nanotube/epoxy composites produced by a masterbatch process. *Mechanics of Composite Materials*, **42**, 395–406.
105 Seyhan, A.T., Gojny, F.H., Tanoglu, M., and Schulte, K. (2007) Rheological and dynamic-mechanical behavior of carbon nanotube/vinyl ester-polyester suspensions and their nanocomposites. *European Polymer Journal*, **43**, 2836–2847.
106 Seyhan, A.T., Gojny, F.H., Tanoglu, M., and Schulte, K. (2007) Critical aspects related to processing of carbon nanotube/unsaturated thermoset polyester nanocomposites. *European Polymer Journal*, **43**, 374–379.

107 Erik, T.T. and Tsu-Wei, C. (2002) Aligned multi-walled carbon nanotube-reinforced composites: processing and mechanical characterization. *Journal of Physics D: Applied Physics*, **35**, L77.

108 Radhakrishnan, V.K., Davis, E.W., and Davis, V.A. (2010) Influence of initial mixing methods on melt-extruded single-walled carbon nanotube–polypropylene nanocomposites. *Polymer Engineering & Science*, **50**, 1831–1842.

109 Parra-Vasquez, A.N.G., Stepanek, I., Davis, V.A. *et al.* (2007) Simple length determination of single-walled carbon nanotubes by viscosity measurements in dilute suspensions. *Macromolecules*, **40**, 4043–4047.

110 Parra-Vasquez, A.N.G., Duque, J.G., Green, M.J., and Pasquali, M. (2014) Assessment of length and bundle distribution of dilute single-walled carbon nanotubes by viscosity measurements. *AIChE Journal*, **60**, 1499–1508.

111 Marshall, B.D., Davis, V.A., Lee, D.C., and Korgel, B.A. (2009) Rotational and translational diffusivities of germanium nanowires. *Rheologica Acta*, **48**, 589–596.

112 Ortega, A. and Garcı´ a de la Torre, J. (2003) Hydrodynamic properties of rodlike and disklike particles in dilute solution. *The Journal of Chemical Physics*, **119**, 9914–9919.

113 Lotya, M., Rakovich, A., Donegan, J.F., and Coleman, J.N. (2013) Measuring the lateral size of liquid-exfoliated nanosheets with dynamic light scattering. *Nanotechnology*, **24**, 265703.

114 Khoshkava, V. and Kamal, M.R. (2014) Effect of cellulose nanocrystals (CNC) particle morphology on dispersion and rheological and mechanical properties of polypropylene/CNC nanocomposites. *ACS Applied Materials & Interfaces*, **6**, 8146–8157.

115 Goswami, J. and Davis, V.A. (2015) Viscoelasticity of single-walled carbon nanotubes in unsaturated polyester resin: effects of purity and chirality distribution. *Macromolecules*, **48**, 8641–8650.

116 Kayatin, M.J. and Davis, V.A. (2009) Viscoelasticity and shear stability of single-walled carbon nanotube/unsaturated polyester resin dispersions. *Macromolecules*, **42**, 6624–6632.

117 Arbabi, S. and Sahimi, M. (1993) Mechanics of disordered solids. I. Percolation on elastic networks with central forces. *Physical Review B*, **47**, 695.

118 Surve, M., Pryamitsyn, V., and Ganesan, V. (2006) Universality in structure and elasticity of polymer-nanoparticle gels. *Physical Review Letters*, **96**, 177805.

119 Filippone, G. and Salzano de Luna, M.A. (2012) Unifying approach for the linear viscoelasticity of polymer nanocomposites. *Macromolecules*, **45**, 8853–8860.

120 Prasad, V., Trappe, V., Dinsmore, A. *et al.* (2003) Rideal Lecture Universal features of the fluid to solid transition for attractive colloidal particles. *Faraday Discussions*, **123**, 1–12.

121 Shih, W.-H., Shih, W.Y., Kim, S.-I. *et al.* (1990) Scaling behavior of the elastic properties of colloidal gels. *Physical Review A*, **42**, 4772.
122 Wu, H. and Morbidelli, M. (2001) A model relating structure of colloidal gels to their elastic properties. *Langmuir*, **17**, 1030–1036.
123 Khalkhal, F. and Carreau, P.J. (2011) Scaling behavior of the elastic properties of non-dilute MWCNT–epoxy suspensions. *Rheologica Acta*, **50**, 717–728.
124 Ureña-Benavides, E.E., Kayatin, M.J., and Davis, V.A. (2013) Dispersion and rheology of multiwalled carbon nanotubes in unsaturated polyester resin. *Macromolecules*, **46**, 1642–1650.
125 Kim, H., Abdala, A.A., and Macosko, C.W. (2010) Graphene/polymer nanocomposites. *Macromolecules*, **43**, 6515–6530.
126 Kim, H., Miura, Y., and Macosko, C.W. (2010) Graphene/polyurethane nanocomposites for improved gas barrier and electrical conductivity. *Chemistry of Materials*, **22**, 3441–3450.
127 Kim, H. and Macosko, C.W. (2009) Processing-property relationships of polycarbonate/graphene composites. *Polymer*, **50**, 3797–3809.
128 Kim, S.-K., Wie, J.J., Mahmood, Q., and Park, H.S. (2014) Anomalous nanoinclusion effects of 2D MoS2 and WS2 nanosheets on the mechanical stiffness of polymer nanocomposites. *Nanoscale*, **6**, 7430–7435.

4

High-Throughput Nanomanufacturing via Spray Processes

Gauri Nabar[1], *Matthew Souva*[1], *Kil Ho Lee*[1], *Souvik De*[2], *Jodie Lutkenhaus*[2], *Barbara Wyslouzil*[1,3], *and Jessica O. Winter*[1,4]

[1] *William G. Lowrie Department of Chemical and Biomolecular Engineering, The Ohio State University, Columbus, OH, USA*
[2] *Artie McFerrin Department of Chemical Engineering, Texas A&M University, College Station, TX, USA*
[3] *Department of Chemistry and Biochemistry, The Ohio State University, Columbus, OH, USA*
[4] *Department of Biomedical Engineering, The Ohio State University, Columbus, OH, USA*

4.1 Introduction

Over the past two decades, significant advances in synthesis procedures have enabled the creation of nanocomposites that comprised metals, semiconductors, polymers, and ceramics. While many of these methods are effective, they cannot yield industrially relevant quantities of material. The need for ***scalable nanomanufacturing*** has therefore been cited by the National Science and Technology Committee as a critical barrier in the commercialization of nanomaterials [1]. Unfortunately, many routes to forming nanocomposites, such as sequential growth approaches [2, 3], rely almost exclusively on small-scale batch processing. One possible path forward is the use of spray-based methods since these can be performed in semibatch or continuous configurations. Such approaches have the potential to dramatically increase throughput compared to batch synthesis, enabling breakthroughs in scalable nanomanufacturing.

The most traditional method for creating nanocomposites is to simply mix the desired components together; however, this approach provides little control over the organization of the components with respect to each other. Nonetheless, the majority of nanocomposites presently commercialized rely on synthesis techniques that do not control spatial organization [4, 5]. The challenge is to identify methods that permit production of industrial scale quantities of nanocomposites with defined structures. This review will focus specifically on nanocomposites that combine polymers with each other and/or inorganic nanoparticles.

Nanotechnology Commercialization: Manufacturing Processes and Products, First Edition.
Edited by Thomas O. Mensah, Ben Wang, Geoffrey Bothun, Jessica Winter, and Virginia Davis.

Polymer-based nanocomposites with structural organization are generally formed via self-assembly, and include thin films, micelles, liposomes, and polymersomes. Polymers with different structural characteristics, including hydrophilicity, hydrogen bonding capability, and charge, can be used to create composites that segregate into different phases based on polymer structural characteristics (Figure 4.1). If nanoparticle additives preferentially segregate in one of these phases, the result is an ordered system exhibiting both polymer and nanoparticle patterns with defined orientation and spacing (Glass *et al.* [7]).

At the bench scale, these composites are formed in a number of ways. Short surfactant or block copolymer (BCP) systems can be directly dispersed in water to yield micelles; however, this approach is only viable for polymers with small hydrophobic segments [1]. For longer polymer chains, more sophisticated dispersion techniques utilizing multiple solvents or mixed phases are required. Typically, the polymer, nanoparticles, and sometimes a surfactant are added to organic and aqueous phases based on their respective miscibility. These phases are mixed, primarily via sonication or agitation, and the organic phase is then removed using evaporation or dialysis. Reducing the organic concentration eventually leads to nanocomposite formation; and, depending on the polymer and operating conditions, possible outcomes include solid particles, micelles, liposomes, and polymersomes. If the rate of solvent removal is high, the polymer chains may not reach thermodynamic equilibrium and the structure can be kinetically trapped in a metastable conformation.

For example, one well-established approach for self-assembly of polymers with a long hydrophobic component chain is "cosolvent addition/solution-phase assembly." In this approach, the polymer is dispersed in a high-affinity organic solvent or solvent–water mixture. Altering the solvent quality by water addition or organic solvent extraction induces microphase separation [2]. Eisenberg and coworkers demonstrated the feasibility of this approach for producing glassy as well as soft-core micelles with varying block lengths [2, 3]. Micelles with different shapes and drug loading capacities have been synthesized by employing different cosolvents [4], cosolvent ratios [5], and antisolvent addition rates [7]. However, a number of protocols require dialysis

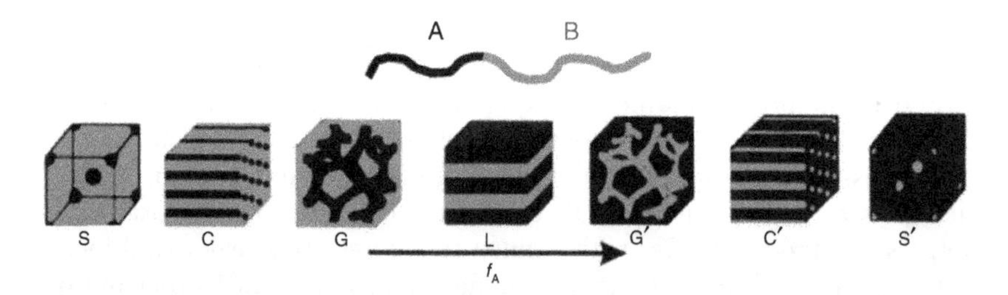

Figure 4.1 Block copolymers can adopt several phases depending on their concentration and temperature. Mai *et al.* 2012 [6]. Reproduced with Royal Society of Chemistry.

spanning multiple days to extract the cosolvent [4], thereby limiting the throughput of this otherwise versatile technique.

More recently, Hayward and colleagues introduced the interfacial instability process [8–10] to produce micelles from high-molecular-weight BCPs. Here, the organic and aqueous phases are largely immiscible. A surfactant is added to disperse the organic-BCP solution in the aqueous phase and create an emulsion with a large droplet surface area. As the organic evaporates, the high surfactant concentrations at the aqueous-organic interface can lead to transiently negative surface tension, resulting in droplet fission. The emulsion droplets continue to decline in size until the final self-assembled structures are formed (Figure 4.2). In contrast to cosolvent approaches, interfacial instability appears to be more suitable for the processing of bulkier polymers and is the only method that has yielded highly loaded, long, worm-like micelles [9]. Finally, for high-molecular-weight polymers with limited solubility in the aqueous phase, the final structures that form are not necessarily the most stable from a thermodynamic standpoint. Rather the structures may correspond to kinetically trapped conformations, and, therefore, the final structures are more sensitive to processing conditions than structures produced from more mobile components.

A second common approach to forming nanocomposites focuses on thin-film morphology manipulation. In this case, a polymer is deposited on a surface and additional polymer or nanoparticles are then introduced. For example, BCPs may orient on a solid substrate to yield patterns, which preferentially solvate nanoparticles in one of the blocks. Alternatively, a method known as LbL assembly may be used to sequentially deposit materials based on their chemical properties [11]. Complementary species are alternately deposited onto a surface. A combination of electrostatic, hydrogen bonding, hydrophobic interactions, and entropic considerations are responsible for the stability of the film, its growth behavior, and morphology. One of the most unique features of this technique is that polymers and nanomaterials may be arranged selectively in certain layers, giving fine control of materials placement vertically within a film

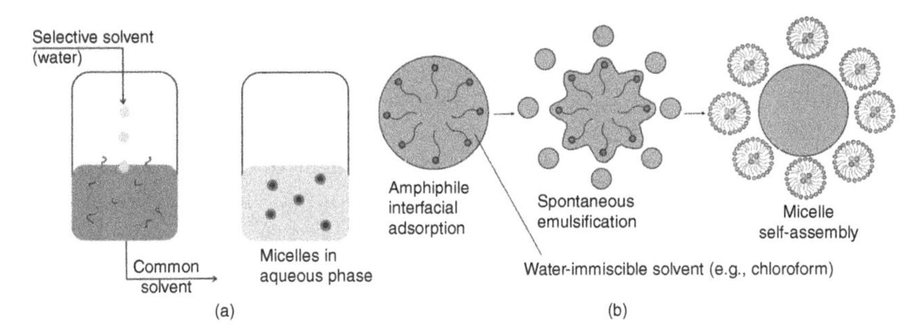

Figure 4.2 Schematics of the (a) cosolvent addition and (b) interfacial instability processes.

or coating. Because the assembly method is based on noncovalent interactions, polar or charged materials are most suitable for the LbL assembly process. Within the umbrella of LbL assembly, several processing methods are available, such as dipping, spraying, spin-coating. Dip-assisted LbL assembly, the most popular of these methods on the lab scale, is typically a slow batch process.

All of these approaches focus on manufacturing small quantities of material, far below commercial scale. Self-assembly processes that require thermodynamic equilibrium can require hours or more to complete. Sequential processes, such as dip-assisted LbL, are inherently slow and difficult to perform at high throughput. Approaches based on emulsification require energy input to control initial droplet size. If the energy input is not uniform, initial droplet size distribution may be large; final products may in turn exhibit variation in size. Polymer and nanoparticle orientation within the composite may also vary as different energy states may be favored based on processing conditions, such as droplet size and evaporation rate. These fundamental barriers must be overcome to consistently produce self-assembled polymer-based nanocomposites at useful scale.

4.2 Flash Nanoprecipitation

4.2.1 Overview

FNP is one of the most advanced techniques for scalable nanomanufacturing of polymer/nanoparticle nanocomposites and a model for aspiring spray-based approaches. This technique, pioneered by Johnson and Prud'homme [12], is a continuous flow adaptation of the cosolvent addition/solution phase assembly approach developed by Eisenberg. FNP is a bottom-up technology that uses rapid mixing to quickly produce stable, water-dispersible nanocomposites at high yield with narrow size distributions. In FNP, a water-miscible, organic stream containing BCPs and hydrophobic components, that is, nanoparticles or other hydrophobic molecules, undergoes rapid mixing with water in a specialized microreactor (Figure 4.3). The solvent quality, with respect to the hydrophobic components and BCPs, deteriorates on a timescale of milliseconds and generates supersaturations as high as 1000:1, that is, the hydrophobic component/BCP concentrations are as much as 1000 times higher than the equilibrium concentrations supported by the solution. The rapid addition of antisolvent induces precipitation of both the hydrophobic additives and the BCPs. If the hydrophobic components precipitate first, the nuclei that form continue to grow by fusion with other aggregates until further growth is arrested by BCP adsorption, or heterogeneous precipitation, onto their surfaces.

The final size of the nanocomposites depends on the precipitation/aggregation kinetics of the hydrophobic components and the BCPs for a

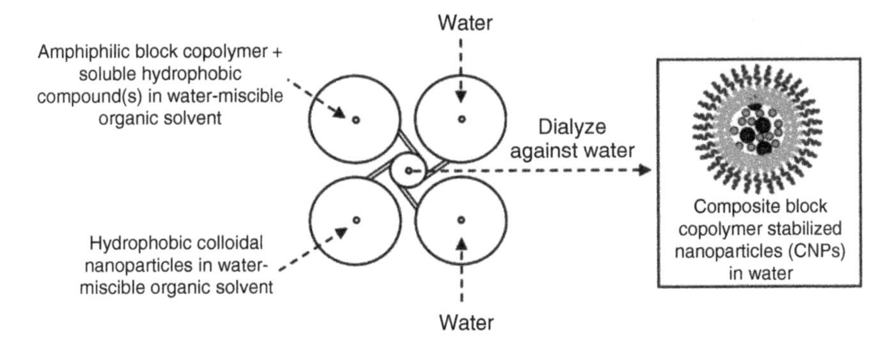

Figure 4.3 Schematic of flash nanoprecipitation process (four inlet stream multi-inlet vortex mixer shown). Gindy *et al.* 2008 [13]. Reproduced with permission of American Chemical Society.

given mixing condition [14]. Solvents with high diffusivities in water enhance this process, because in addition to mechanical mixing, micromixing is also achieved via molecular diffusion, and rapid diffusion yields smaller, more monodisperse nanocomposites. A range of water-miscible solvents can be used, including tetrahydrofuran (THF), methanol, ethanol, dimethyl sulfoxide (DMSO), dimethylformamide (DMF), and *N*-methyl pyrrolidone (NMP) [15]. Although a wide variety of polymers can serve as stabilizers, BCPs are most commonly employed because their rate of exchange in water is low – increasing the stability of the nanocomposite – and the hydrophilic blocks prevent fusion [12]. In addition to these considerations, extensive research over the past decade has examined the effect of mixing time, mixer geometry, hydrodynamics, solute–polymer interactions, and concentration on the self-assembly process on the FNP process [16].

4.2.2 Importance of Rapid Mixing

A key element of the FNP process is rapid mixing of the phases to drive the solvent quality from organic-rich to aqueous-rich and induce coprecipitation of the hydrophobic materials. In multicomponent systems, differences in component hydrophobicity can lead to one component precipitating more rapidly the other. If the mixing time is longer than the characteristic time for nucleation and growth of the more hydrophobic component, aggregates of this component may grow before precipitation of the other component is initiated. This can result in poorly stabilized, large particles [14]. Rapid mixing (e.g., less than ~100 ms) ensures that supersaturation of all components is rapidly achieved, potentially inhomogeneous nucleation and growth are limited, and the particle size is uniform. Further decrease in mixing time does not yield smaller nanoparticles. This regime is characterized by homogeneous kinetics and is the preferred mode of operation of FNP [12].

Another important element in minimizing inhomogeneity is to ensure that the nucleation and growth rates for BCPs and additive components are similar. If the BCP assembles too quickly, a population of unloaded micelles may result, in addition to the desired nanocomposites. On the other hand, delayed BCP assembly allows the formation of large clusters of the hydrophobic components before encapsulation by BCPs occurs, potentially yielding nonuniform products [17]. Fortunately, the concentration and architecture of the BCP can usually be manipulated to match the aggregation kinetics of the additive. The composition of the final particles depends on the initial stoichiometry [18]; and FNP can yield up to 100% encapsulation of additives. Further, as the precipitation of hydrophobic components is governed solely by solubility, incorporation of one hydrophobic component in the nanocomposite is largely unaffected by the presence of other coencapsulants [13].

4.2.3 Mixers Employed in FNP

4.2.3.1 Confined Impinging Jet Mixers (CIJMs)

Johnson and Prud'homme first described the use of confined impinging jet mixers (CIJMs) for FNP in 2003 [12]. In the CIJM configuration, two linear turbulent jets impinge inside a small chamber. Typical dimensions of the CIJM are in millimeters, and typical flow rates range between 3 and 120 ml/min [19]. High feed flow rates generate and control turbulence inside the chamber, ensuring thorough mixing. The product exits the reactor and is then subjected to postprecipitation treatment, such as dialysis or drying to remove the remaining solvent.

Experimental and modeling investigations of CIJM performance show that inlet jet velocity, mixer dimensions, antisolvent:solvent ratio, and fluid properties can all impact final nanoparticle size distribution. Numerous studies confirm that mixing efficiency is improved by maintaining high flow rates and high antisolvent:solvent ratios [20]. An important constraint for CIJM operation is the requirement that the colliding streams have equal flow rates. Unequal momentum of the impinging streams can displace the impingement plane from its central location, resulting in fluid bypass and inefficient mixing [21]. This stringent requirement limits the supersaturation attainable inside the chamber, enhancing the solubility of the precipitating species in the final mixture and, thereby, making the products susceptible to Ostwald ripening [22]. A secondary dilution tank can increase supersaturation levels, increasing stability [23]. Scale-up criteria for CIJM are based on mixing times of the various hydrophobic components [19].

4.2.3.2 Multi-Inlet Vortex Mixers (MIVMs)

To overcome the constraints of the CIJM, Liu and coworkers [24] developed the four-stream multi-inlet vortex mixer (MIVM). In the MIVM, four turbulent jets with 90° offset enter tangentially into a chamber. The collision of the streams results in a swirling vortex pattern. All streams contribute to micromixing in the chamber and may be introduced into the reactor at unequal flow rates. However, the mixing efficiency is highest when opposing streams have equal concentrations and flow rates [24]. This flexibility in operation allows the selection of feed streams with different concentrations and flow rates, and the product stream can be antisolvent rich, suppressing Ostwald ripening. The MIVM configuration can be scaled up to increase throughput; however, experiments and modeling suggest that even at increased Reynolds number (*Re*), mixing is inferior and the mixing time is longer in the larger MIVM than in the small-scale versions of this device [25].

4.2.3.3 Mixer Selection

Selection of the mixer is typically based on the desired properties of the final nanocomposite. In a head-to-head comparison of the CIJM and MIVM configurations operated under the same conditions, nanocomposites produced in the MIVM were roughly spherical, whereas those produced in the CIJM (with a dilution tank) were smaller and more irregularly shaped, but less polydisperse [26]. In addition, the MIVM configuration permits independent manipulation of component concentrations. For example, separate streams can be devoted to BCPs, nanoparticles, and antisolvents (Figure 4.3) [13].

4.2.4 FNP Product Structure

The final structure of products produced via FNP reflects a careful balance between initial supersaturation, concentrations, and mixer configuration for a given interaction strength between the BCP and the hydrophobic components to be encapsulated [27]. Despite the fact that FNP depends most strongly on precipitation kinetics of the individual participating components, interactions between the BCPs and the other hydrophobic components can drastically influence the assembly process [27]. Unfavorable interactions between the hydrophobic components and the BCPs increase the probability of aggregate fusion and, therefore, broaden the particle size distribution. For example, weak hydrophobic component-BCP interactions can drive clustering of the BCP on the surface of the hydrophobic component aggregate, reducing protection against fusion of the aggregates [28]. It can even result in a population of empty BCP micelles accompanied by large, poorly stabilized solute aggregates. This effect is further compounded at higher concentrations where the probability of aggregate collision is increased.

BCP architecture can also influence final particle size [29]. Long hydrophilic blocks in the BCP can shield a larger surface area, reducing aggregate fusion. This, in turn, makes the final particles smaller. Increasing the BCP concentration for the same amount of hydrophobic components has a similar effect because the larger number of small aggregates have higher surface area, which the BCP can stabilize against fusion.

4.2.5 Applications of FNP Nanocomposites

FNP provides a simple route to generate monodisperse polymer nanoparticles with tunable size and composition at high yield [30] and has, therefore, led to a wide range of applications. Drug loading remains at the forefront because very high payloads with nearly 100% solute encapsulation are attained via this technique [16], much higher than what traditional emulsion-based approaches can achieve. Multifunctional nanocomposites with prespecified solute concentrations are also being developed for sunscreen formulations [31], and FNP is being employed to produce nanocomposites for pesticide applications because of their improved dispersion, lower toxicity, and greater colloidal stability [32]. The ability to design nanocomposites with morphology and composition specified a priori, in a safe, reliable, and scalable manner, has made FNP a promising process for scalable nanomanufacturing.

4.3 Electrospray

4.3.1 Overview

Another approach to scale-up of small, batch nanomanufacturing processes is electrospray – also called electrohydrodynamic (EHD) spraying, or EHD atomization – a technique that uses electrical forces to atomize liquids [33]. Perhaps the most well-known application of electrospray is in mass spectrometry, where electrospray ionization is used to generate molecular ions, particularly of large biomolecules, for analysis via mass spectrometry [34]. However, electrospray is also widely used in aerosol science to generate fine particles via solvent evaporation [35] or to generate an aerosol from preexisting particles [36]. In this section, we focus on particle formation via electrospray of liquids into a gas.

In the context of nanocomposite manufacturing, electrospray can be used to generate distinct micro- and nanoscale particles [37, 38] and to produce thin films [39]. Recently, coaxial electrospray has been used to demonstrate scale-up of micelle and nanocomposite production via the interfacial instability process [40] and liposome formation via cosolvent addition [41]. Compared to FNP, electrospray has some advantages. In particular, it can process smaller volumes and is therefore better suited to exploratory studies, as well as the production of high-value materials that require only intermediate levels of production.

In the electrospray process, an electric field is applied to a charged liquid flowing from a capillary. In most implementations of electrospray, the voltage is increased until the fluid exiting the capillary is deformed into a conical shape that emits a jet of liquid from the tip [42]. Kink or varicose instabilities break the jet into droplets [43], creating a fine mist of charged droplets that disperse because of Coulombic repulsion. Depending on the properties of the fluid, evaporation may decrease the surface area of the original droplets to the point that the surface charge on the droplets reaches the Rayleigh limit and the original droplets undergo further fissioning [42].

4.3.2 Single Nozzle Electrospray

4.3.2.1 Forces and Modes of Electrospray

Electrospray was first described by John Zeleny in 1917 [44], but it was Taylor, in 1964, who enhanced our understanding of electrospray physics by deriving equations that govern the basic shape of the "Taylor cone" in the cone-jet mode of electrospray [45]. As illustrated in Figure 4.4, the shape of the fluid stream exiting the nozzle results from the competition between the body forces acting on the fluid and the normal and tangential stresses acting at the fluid's interface with the continuous medium [46].

Although electrospray is most often associated with the cone-jet mode illustrated in Figure 4.4, other modes exist including dripping, spindle, oscillating jet, and precession modes (Figure 4.5) [47, 48]. The choice of the spray mode determines the resulting aerosol droplet size and charge, and influences final particle size and distribution. For example, dripping mode is used to produce large monodisperse alginate particles (~200–600 μm) for cell encapsulation [49], and cone-jet mode, a mode characterized by small (micro- and nanoscale droplets) [47, 48], can be used to produce emulsions with uniform droplet sizes by collecting the spray in a liquid. The particular spray mode depends

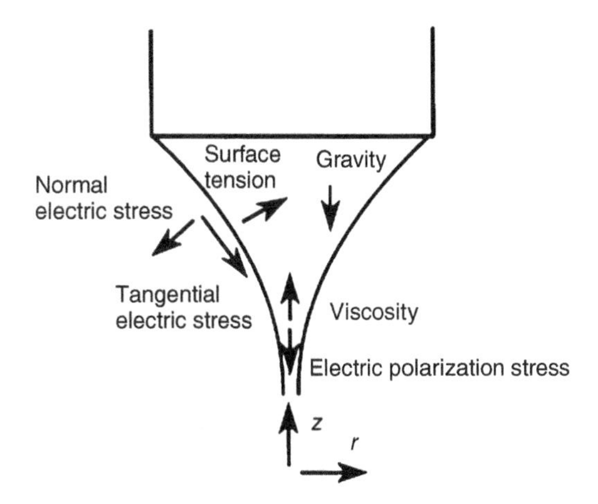

Figure 4.4 Force and stress diagram for an electrospray in cone-jet mode. The variables indicated can impact the size and shape of the liquid cone. Hartman *et al.* 1999 [46]. Reproduced with permission of Elsevier.

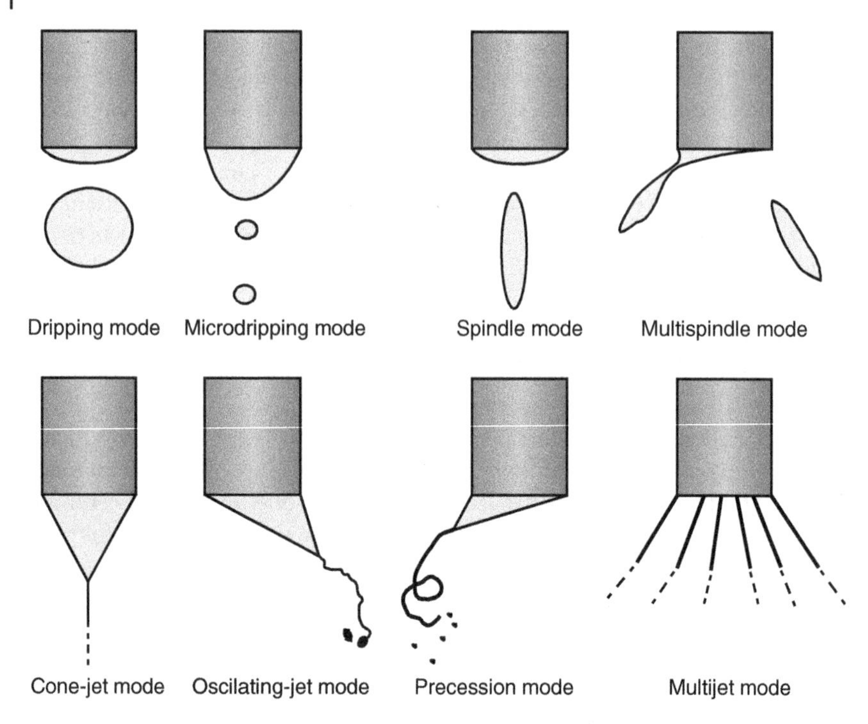

Figure 4.5 Modes of electrospray. Jaworek and Sobczyk 2008 [33]. Reproduced with permission of Elsevier.

in a complex manner on both the fluid characteristics (e.g., viscosity, surface tension, and electrical conductivity) and the electrospray operating parameters (e.g., flow rate and applied voltage). Very generally, for a given fluid and flow rate, the modes progress from dripping, through spindle, cone-jet, precessing jet, to multijet as the voltage increases. This progression is illustrated in Figure 4.5. Transitions between the modes can be rapid, and hysteresis is also commonly observed [47, 50]. When the viscosity becomes too high, electrospray transitions to electrospinning [51].

4.3.2.2 Applications of Single Nozzle Electrospray

In the context of nanoscale material production, electrospray processes can be used to generate thin films for applications in solar cells [52], fuel cells [53], lithium batteries [54], and catalysis [55]. To generate a film via electrospray, the material to be deposited is first dissolved or suspended in an appropriate solvent. The solution is then electrosprayed, and the resultant particles are collected on a substrate to form films. Various film thickness have been reported for different materials [39], including sub-100 nm films of poly(vinylidene fluoride) [56]. In addition, electrospray film deposition has been suggested as a way to generate multilayer films [57].

In the context of nanocomposite particle generation, electrospray can be used to generate either pure or composite solid particles with sizes ranging from as small as 4 nm [38] to hundreds of micrometers [49]. One particularly important application is using electrospray to manufacture drug-loaded polymeric microparticles [37]. Particles manufactured this way are generally more monodisperse than those produced via conventional emulsion techniques and drug encapsulation efficiency is usually higher [58, 59]. The latter stems from the fact that even small drug molecules have very low vapor pressures and cannot partition into the continuous gas phase as easily as they can into a continuous liquid phase. Electrospray has also been used to encapsulate preexisting micro- and nanoparticles by spraying colloidal suspensions in a polymer solution [60]. As the solvent evaporates, the polymer hardens to generate the microencapsulated material.

A less common application of electrospray is to generate emulsions in a process that can be referred to as electroemulsification [61]. Here, droplets produced via electrospray are collected in a continuous liquid or solution phase to produce an emulsion with well-defined droplet sizes. The spray can be initially generated in the gaseous phase or directly in the continuous phase. We refer to the latter case as liquid–liquid electrospray, and discuss it in detail in the following section. In either case, electroemulsification is one way to continuously create the initial emulsion required to scale up the interfacial instability technique. To make an emulsion by spraying a polymer dissolved in a volatile organic solvent is, however, challenging because the particles are likely to harden before reaching the collection solution. One way to overcome this problem is to use a coaxial or compound electrospray setup and slow evaporation of the volatile organic by surrounding it with a less volatile species.

4.3.3 Coaxial Electrospray

4.3.3.1 Configuration

Coaxial electrospray is a more complex electrospray configuration first reported by Loscertales and coworkers [62]. In the simplest configuration (Figure 4.6), two concentric needles are used to combine different fluids before forming the spray in order to produce more complex, structured particles [63]. The liquids may be miscible or immiscible, although immiscible liquids are most commonly reported [41, 64]. An electric potential is applied to the needle assembly, deforming the fluid and forming a compound jet. In coaxial electrospray, the liquid most responsive to changes in the electrical current acts as the driving liquid and controls the composite flow rate [65]. Similar to single nozzle systems, the fluid properties (including flow rate, viscosity, and

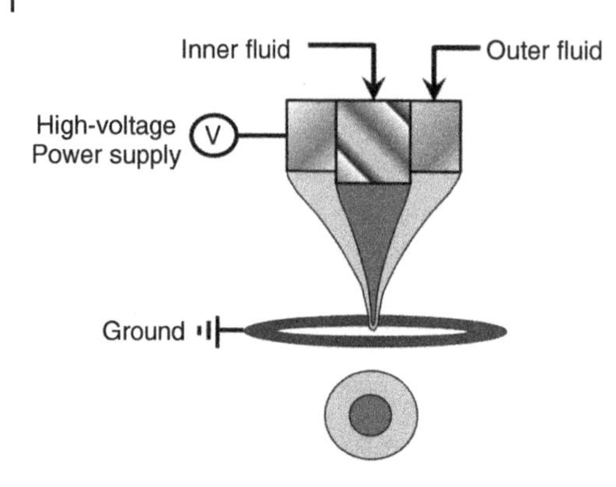

Figure 4.6 Schematic of coaxial electrospray for immiscible fluids. Duong *et al.* 2014 [40]. Reproduced with permission of American Chemical Society.

electrical conductivity) influence the operating modes, the transitions between them, and the final droplet size [65].

4.3.3.2 Applications

As in single nozzle electrospray, coaxial electrospray can also produce monodisperse micron-sized particles [62], with the added advantage that the particles can now exhibit structure. Thus, coaxial electrospray has been used as an alternative to water/oil/water emulsion techniques to encapsulate biological materials, initially suspended in an aqueous solution, in a polymeric shell. In one application, Lee and coworkers used both dual- and tri-capillary coaxial electrospray to generate PLGA particles [66] containing two different drugs. By varying the production methods, they were able to significantly change the release kinetics of the drugs. In another example, Nie and coworkers used coaxial electrospray to generate core-shell microspheres and demonstrated parallel release of two different drugs from the resultant particles [67]. Thus, coaxial electrospray is a highly flexible method to produce particles with tuned diffusion rates and to achieve parallel encapsulation.

In an alternative application, Wu and coworkers used coaxial electrospray to generate lipoplex nanoparticles containing short pieces of DNA [41]. Here, the inner (aqueous) and outer (ethanol/lipid mixture) fluids were miscible; thus, the electrospray setup functioned as a scaled-up version of the cosolvent addition method. In another novel application, Duong and coworkers flowed an organic solution containing BCP and nanoparticle additives through the inner needle and an aqueous poly(vinyl) alcohol (PVA) through the outer needle [40]. The presence of the outer aqueous layer slowed evaporation of the volatile solvent, and the droplets collected in a water dish below the nozzle generated an emulsion. The emulsion then underwent the interfacial instability process to yield micelles. This approach greatly increased the throughput compared to previous batch processing methods based on generating the emulsion via probe sonication. The nanoparticle-loaded micelles have been used in imaging and cell separations [40].

4.3.4 Future Directions

The utility of electrospray in nanomanufacturing depends on scalability, and several steps have been taken to improve throughput. The development of sophisticated models to correctly predict the effect of important geometric, fluid, and operating parameters on the structure and stability of an electrospray jet is an ongoing area of research [46, 68]. Preliminary work has focused on developing physical scaling laws that relate the key variables, that is, electrical current, fluid flow rate, and spray cone diameter, to particle size. These laws exist for both single nozzle [69, 70] and coaxial electrospray [65] configurations. Recent electrospray modeling has expanded to include detailed computational approaches [71]. The goal of these efforts is to improve fundamental understanding of electrospray, to better predict the resulting particles produced in electrospray, and to reduce the often lengthy process of optimizing spray conditions.

Another important advance is the development of multiplexed nozzles. For example, Almeria and coworkers used a multiplexed single nozzle array to increase electrospray throughput in a drug delivery application [72] by an order of magnitude. Conceptually, multiplexing is straightforward. The desired fluid from a single reservoir is delivered to multiple nozzles in a single assembly, and the same high voltage is then applied to each nozzle enabling simultaneous spray from each needle. Extending the concept to coaxial assemblies appears feasible and is important to scaling up the production of more complex particle types.

4.4 Liquid-in-Liquid Electrospray

4.4.1 Overview

In most applications of electrospray, a liquid phase is sprayed into a gaseous medium before droplet/particle collection. Stable atomization in a liquid-in-gas system requires the liquid to have some conductivity, and a surfactant may be needed to reduce the surface tension of the liquid so that it can more easily deform and disperse [73]. Although less common, directly spraying one liquid into a second liquid is also possible. In this configuration, known as liquid-in-liquid electrospray (LLE) (Figure 4.7a), a semiconducting dispersed phase moves through a nozzle submerged in a continuous phase that contains a grounded electrode. A high electrical potential gradient between the nozzle tip and the grounded electrode is established by applying a high voltage directly to the nozzle. The resultant electrical field in the continuous phase interacts with the charges on the dispersed phase: Coulombic forces overcome the interfacial tension at the liquid–liquid interface, resulting in electroemulsification [61, 75]. LLE droplet size and uniformity are strongly influenced by the applied

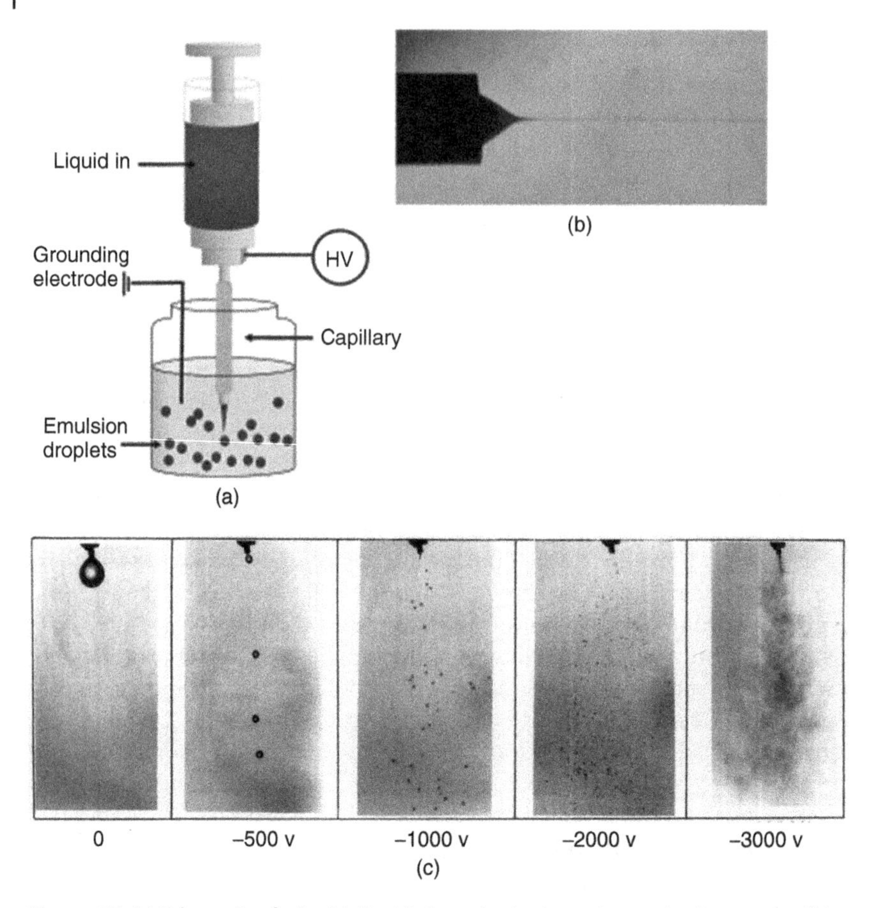

Figure 4.7 (a) Schematic of a liquid–liquid electrohydrodynamic atomization system. (b) Cone-jet mode spraying of glycerin in hexane. Barrero *et al.* 2004 [74]. Reproduced with permission of Elsevier. (c) Spraying carbon tetrachloride in distilled water at various voltages applied using high-voltage power supply. Sato *et al.* 1993 [73]. Reproduced with permission of Elsevier.

voltage and dispersed phase solution viscosity [73]. Increased voltage increases the driving force for dispersion, decreasing droplet size. Similarly, low viscosity of the dispersed phase favors stream breakup, reducing the droplet size.

4.4.2 Importance of Relative Conductivities of the Dispersed and Continuous Phases

The electrical conductivity and the dynamic viscosity of the dispersed phase relative to that of the continuous phase are crucial determinants of spray characteristics and mode of electroemulsification. In the liquid-in-gas systems described earlier, steady-state cone-jet spray is achieved by spraying a conducting fluid into an insulating fluid (i.e., air) [74]. Similarly, stable cone-jet spray in LLE can be achieved by spraying conducting fluids into a highly insulating continuous phase. For example, glycerin sprayed into hexane yields a stable cone-jet (Figure 4.7b). This is possible because the electrical conductivity of glycerin is $\sim 2.87 \times 10^{-6}$ S/m, within the range of dispersed phase conductivity requirements for a liquid-in-gas system, whereas hexane is highly insulating with an electrical conductivity of less than 10^{-12} S/m [74]. Using this approach, it has been shown that LLE operating in steady cone-jet mode follows the same scaling laws as liquid-in-gas systems [74].

Unfortunately, these results highlight the difficulties inherent in successfully spraying an organic into aqueous media that stem from the high conductivity of water (4×10^{-6} S/m). Sato *et al.* first addressed this issue by insulating the metal nozzle from the aqueous environment using a glass tube [73], and then successfully sprayed carbon tetrachloride into distilled water by applying a DC pulsed potential (Figure 4.7c). In addition, Tsouris *et al.* demonstrated that an electrostatic field sufficient to spray a nonconductive liquid into a conductive liquid could be created with the metal capillary encased in a ceramic insulation tube [76]. Although electrospray does not occur via cone-jet mode, the fundamental mechanism of atomization is the same and emulsification of the dispersed phase is possible. However, emulsification can only proceed if the presence of ions in the continuous phase is low enough. If the continuous phase becomes too conductive, emulsification stops because the ions cause a short circuit. Thus, it is crucial that the conductivity of dispersed phase be higher than that of the continuous phase to achieve successful electroemulsification [75].

An important advantage of LLE relative to gas-in-liquid electrospray is that highly insulating phases can be dispersed into relatively conductive aqueous phases without using a surfactant. Compared to the coaxial liquid–gas electrospray process described by Duong *et al.* [40], in LLE, the surfactant (i.e., poly(vinyl) alcohol, or PVA) is not only unnecessary, but it cannot be present. The presence of the surfactant (or impurities in the surfactant) increases the conductivity of the continuous phase, increasing the applied current required to maintain the voltage and the same level of dispersion. If the current is too high, dispersion becomes impractical and unsafe.

4.4.3 Modified Liquid-in-Liquid Electrospray Designs

LLE approaches can be combined with existing technologies for highly controlled nanoparticle generation and nanofluid flow, such as microfluidic chips. Although electrospray-incorporated microfluidic devices have been widely used for analytical purposes, such as mass spectrometry; by applying the same fundamental principles described earlier, emulsification can be accomplished to produce nano-particles or -composites (Figure 4.8a). As with

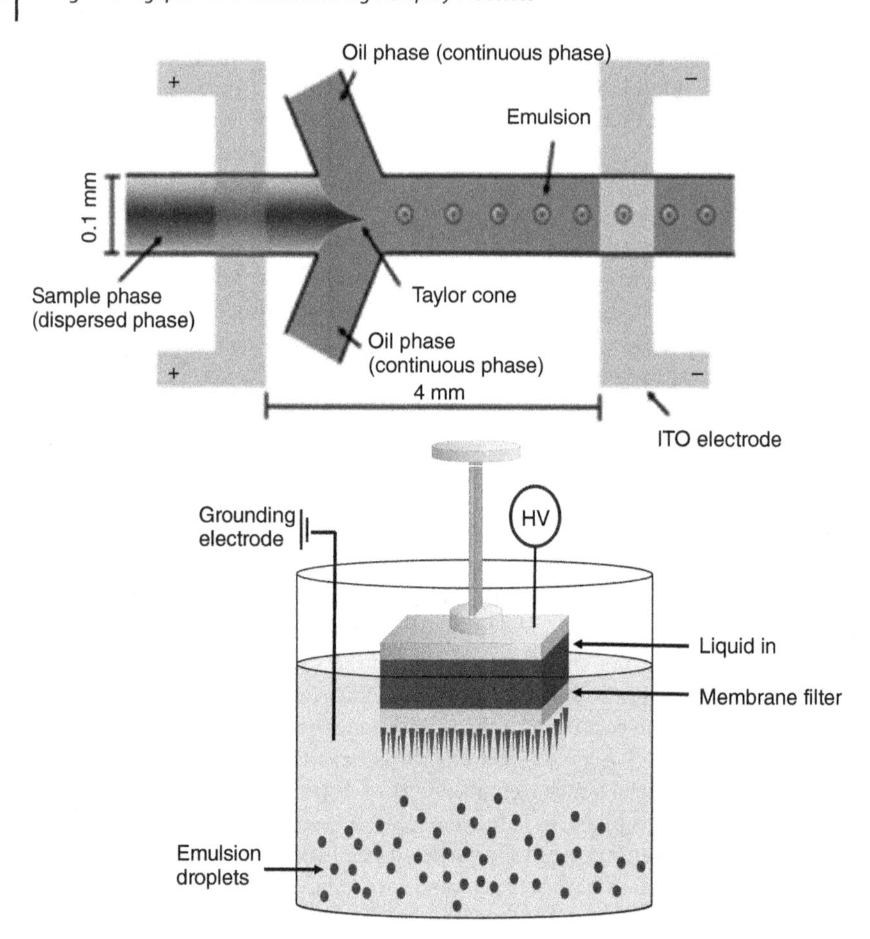

Figure 4.8 (a) Electrospray-incorporated microfluidic device, showing emulsification of water in oil. Yeh *et al.* 2012 [77]. Reproduced with permission of Springer. (b) Schematic of membrane filter-incorporated liquid-in-liquid electrospray.

single nozzle LLE, the size of the resulting emulsion droplets is dependent on fluid flow rate, applied voltage, and the emulsifier concentration, if necessary. As described earlier, a Taylor cone can be achieved, and it has been shown that the angle of the Taylor cone can be altered to control the size of emulsion droplets ejected [77, 78]. This approach has been utilized for manufacturing of poly(lactic-*co*-glycolic acid) (PLGA) microparticles [77].

In addition, LLE approaches can be operated in a massively parallel configuration using a nozzle-less, membrane filter system (Figure 4.8b) [79]. This approach offers a number of advantages over the gas-in-liquid approach, including the lack of corrosion-prone metal capillaries, an ability to easily scale from one to multiple nozzles, and a dramatically increased throughput. For example, significantly higher production rates of droplets by volume have been reported for both dripping (53 times) and spray (6.4 times) modes relative to single nozzle LLE [79]. This approach is versatile and has been demonstrated with mixed cellulose ester, hydrophilic polytetrafluoroethylene (PTFE), and polycarbonate membranes with pore sizes of ~0.4 μm.

4.4.4 Applications and Future Directions

LLE methods are relatively new, and the nanomanufacturing applications space is still emerging. The approach appears to offer several distinct advantages in the synthesis of nanomaterials and composites. LLE is suitable for the emulsification of oil/water or water/oil systems and, therefore, has the ability to replace emulsion batch processes, such as those employing dialysis and sonication, for nanocomposite synthesis. For example, LLE has been used to produce solid nanoparticles (i.e., nano-silica particles) [80, 81]. Similar to gas-in-liquid electrospray, energy is applied primarily to the dispersed phase, reducing energy costs and increasing scale-up potential. LLE can be operated in semibatch or continuous modes (e.g., microfluidic chip incorporation). In contrast to the gas-in-liquid approach, surfactants are not required, reducing downstream purification challenges. Thus, it is anticipated that use of LLE in nanomanufacturing will greatly increase in the future.

4.5 Spray-Assisted Layer-by-Layer Assembly

4.5.1 Overview

Spray-assisted LbL assembly is the alternating spray deposition of oppositely charged species onto a substrate or surface. The adsorbing species are dissolved or suspended in a solvent, usually water. An optional rinsing step is often included between adsorption steps (Figure 4.9a–c). Each cycle deposits nanometers to tens of nanometers of material [11]. Although electrostatically interacting materials are the most commonly used, other molecular interactions such as hydrophobic, hydrogen bonding, host–guest, stereochemical, and covalent interactions are also well established. Similar to other variants of LbL assembly (dip- and spin-assisted), the key aspect of spray-assisted LbL assembly is that the newly adsorbed layer should have multiple interactions with the layer underneath. Spin- and spray-assisted LbL assembly is another variant, in which the substrate spins during the spraying process, but is beyond the scope of this review.

In general, LbL films deposited by spraying are similar to films deposited by traditional immersion or dip-assisted LbL assembly (Figure 4.9d) [82, 83]. The only major difference in terms of the resulting film is that the thickness increment per cycle is higher for dip-assisted LbL assembly than that of

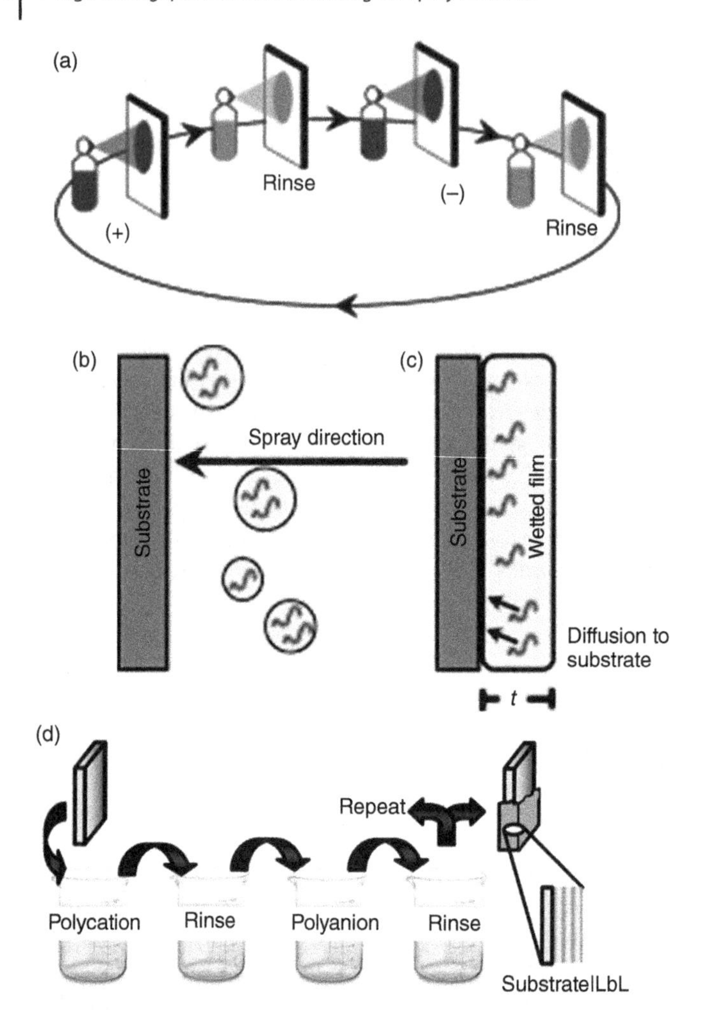

Figure 4.9 (a) Rapid spray-assisted LbL assembly. The completion of each cycle indicates of deposition of a layer pair. (b) A substrate is sprayed with a solution (or dispersion) containing the adsorbing species. (c) The result is a liquid film that wets the substrate, and species diffuse and adsorb to the substrate's surface. (d) Dip-assisted LbL assembly. Lutkenhaus and Hammond 2007 [82]. Reproduced with permission of Royal Society of Chemistry.

spray [11]. The reason for this is that spraying is a faster process, and thus the newly adsorbed layer does not reach an equilibrium thickness within the short contact time with the substrate.

Moreover, as a result of short contact time, several operational factors become even more critical in controlling the thickness, roughness, and quality of the resultant film, especially for nanoparticles. The reason being, unlike charged polymers, nanoparticles have a tendency to aggregate during the

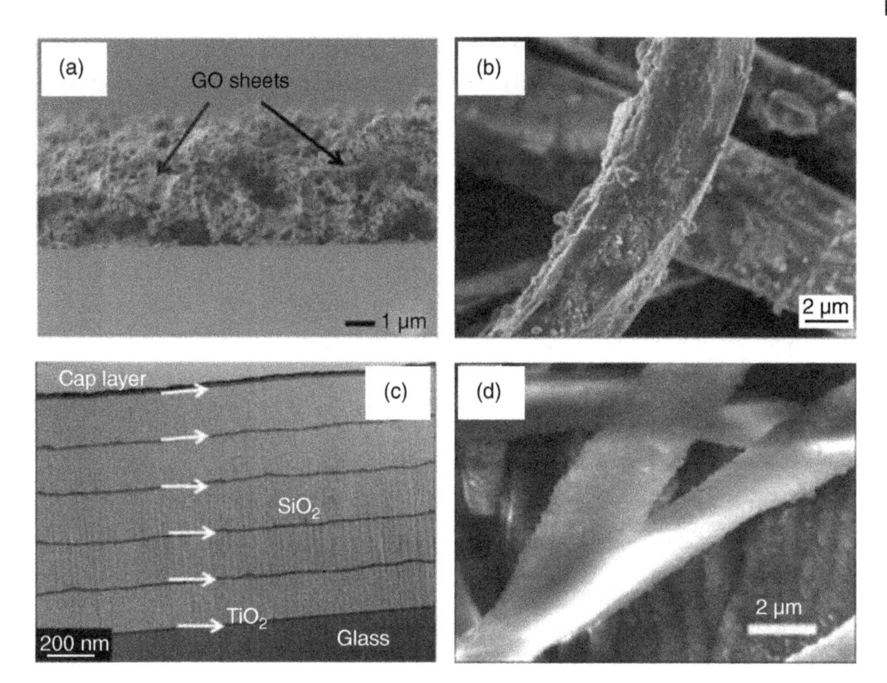

Figure 4.10 (a) Cross-sectional scanning electron microscopy (SEM) of polyaniline nanofiber/graphene oxide sheet spray-assisted LbL films. Kwon *et al.* 2015 [88]. Reproduced with permission from The Royal Society of Chemistry. (b) SEM of spray-assisted LbL deposition of positively and negatively charged silica nanoparticles on fabrics. Carosio *et al.* 2013 [89]. Reproduced with permission of American Chemical Society. (c) Cross-sectional transmission electron microscope image of an alternating layers of TiO_2 (dark gray) and SiO_2 (light gray) nanoparticles assembled via spray-assisted LbL assembly. Nogueira *et al.* 2011 [84]. Reproduced with permission of American Chemical Society. (d) SEM of conformal coatings made by spray-assisted LbL assembly on Tyvek®. Krogman *et al.* 2007 [85]. Reproduced with permission of American Chemical Society.

spraying process. Several publications describe the spraying of nanoparticles for spray-assisted LbL assembly [84–87], but their success is system specific. No set of universal guidelines yet exists for spray-assisted LbL assembly of nanoparticles (Figure 4.10).

4.5.2 Influence of Processing Parameters on Film Quality

One of the major advancements in the domain of spray-assisted LbL assembly is the computer-aided automation of the spraying process [85], which was later commercialized by Svaya Nanotechnology Inc. This allowed independent and precise examination of the effect of several parameters, such as concentration, spray volume, spraying time, spraying distance, and air pressure on the final film properties. As mentioned earlier, electrostatic interactions are the most common interactions in the fabrication of LbL films; thus, our discussion will be mainly focused on electrostatically assembled LbL films if not stated otherwise.

4.5.2.1 Effect of Concentration

Among different parameters that can have direct impact on film properties, concentration of the depositing species is one of the most vital. When a particular species is sprayed onto the substrate, it is adsorbed on the surface until a point at which surface-charge reversal takes place. Beyond a critical concentration, the charge reversal becomes strong enough so that no further adsorption of like-charged species takes place, and no further increase in incremental film thickness occurs even at increased concentrations. Film roughness also increases with increasing concentration, possibly because of the fact that at higher concentration, the sprayed species interact with each other before they are adsorbed to the substrate. It is a common practice to use concentrations of adsorbing species of 20–50 mM to obtain a smooth film with steady growth behavior.

4.5.2.2 Effect of Spraying Time

Spraying time is another parameter that can be tuned to engineer film thickness and roughness [90–100]. For successful adsorption of the adsorbing species, the contact time should be comparable to the time required for diffusion through the wetted film (Figure 4.9c). When this condition is met, the species can diffuse through the wetted film and subsequently adsorb to the substrate. Therefore, increasing the spraying time leads to larger contact time and higher incremental thickness up to a point. However, longer spray time may give rise to enhanced roughness as in the case of high concentration. Hence a smaller spray time should be used. In general, a spraying time between 10 and 15 s is a common practice and produces smooth and steady growth of LbL films.

4.5.2.3 Effect of Spraying Distance

Another parameter that can be tuned to adjust film thickness is the distance of the spray nozzle from the substrate, which affects the film thickness in several ways. With increasing distance between the nozzle and the adsorbing surface, the diameter of the spray cone expands. Therefore, if the substrate is too close to the spray nozzle, the spray cone will not cover the entire substrate. On the other hand, if the distance is too far, most of the impinging droplets will not have a chance to impact and interact with the substrate strongly. Therefore, an optimal distance is important for the uniform coating of the surface. The effect of spraying distance on film thickness is a combination of two parameters. First, as the distance between the spray nozzle and the substrate is reduced, a higher number of droplets impact the substrate at a particular time leading to the availability of more species to adsorb on the surface, which gives rise to a thicker incremental thickness. The other parameter that varies with nozzle-substrate distance is the impinging droplet velocity. The impinging droplets form a liquid film on the substrate, the thickness of which varies inversely with droplet velocity (Figure 4.9b, c). As the nozzle-substrate distance is reduced, the droplets impact the substrate with a higher velocity, resulting in the formation of thinner wetted film, which in turn reduces the diffusion length of the adsorbing species. This allows more species to be adsorbed at a particular time giving rise to a larger incremental thickness.

Film roughness is also affected by the spraying distance. If the substrate is too close to the nozzle, the droplets impact the surface with high velocity and potentially disturb the film, resulting in high roughness. On the other hand, if the distance between the substrate and the nozzle is too far, the number of droplets impacting the film is low, which may result in island formation during initial deposition steps and may lead to a roughened morphology. It is, therefore, a common practice to adjust the spraying distance between 19 and 28 cm for consistent impact velocity as well as homogeneous wetting and uniform coverage.

4.5.2.4 Effect of Air Pressure

Air pressure affects the thickness and roughness of the resulting LbL film as well. Increasing air pressure leads to higher droplet velocity and, similar to spraying distance, leads to a larger incremental thickness. However, at higher air pressure droplets impacting the surface with higher velocity may cause deformation of the film and higher roughness. Therefore, for smoother films, lower air (25–30 psi) pressures are better suited.

4.5.2.5 Effect of Charge Density

The charge density of the adsorbing species also plays a pivotal role in tuning the thickness of the resulting film. This effect is most prominent in the case of charged polymers. Charge density of the sprayed species can be tuned by either varying the degree of ionization through adjusting the pH of the spraying solution, as in case of species bearing weakly ionizing groups, or by charge screening via salt. At higher charge density, more charged groups are able to interact and bind strongly with the surface, and consequently, fewer species are required for charge reversal, producing a smaller incremental thickness. On the other hand, at lower charge density, charge reversal requires more species to be adsorbed, leading to a larger incremental thickness. In addition, in the case of polyelectrolytes, variation of charge density also affects the polymer conformation. For example, at higher charge density, the polymer chains remain in a more stretched conformation because of Coulombic repulsion between like charges, producing a smoother and thinner film. On the contrary, at low ionization levels, the polymer chains attain a more coiled conformation and fewer charged groups are able to interact with the surface because of conformational constraints, producing a larger incremental thickness. It is worth mentioning that the effect of charge density is much more pronounced on thickness than roughness values.

4.5.2.6 Effect of Rinsing and Blow-Drying

Rinsing is an important parameter, especially for dip-assisted LbL, to eliminate loosely bound species from the surface. However, in the case of spraying, it is not clear that this remains the case. Because of shear force on the film surface and continuous draining of excess material from the surface, one might opt to remove the rinsing step to further speed the manufacturing process. However, it has been demonstrated that films deposited without rinsing are thicker films than films prepared with the rinsing step. Whether to rinse or not becomes important in the case of anisotropic nanomaterials. Anisotropic nanomaterials often require longer times to orient and diffuse through the wetted films and adsorb to the surface at a favorable orientation. If the timescale for spraying and rinsing is shorter than that required for the diffusion–adsorption process, then rinsing will partially destroy the wetted film via removal of non-adsorbed anisotropic materials, which leads to poor film growth. In these cases, blow-drying reduces the thickness of the wetted film, resulting in a shorter diffusion length and faster adsorption [88]. Should rinsing be used, rinsing with deionized water for 10 s is a common practice. Moreover, rotation of the substrate at a low speed may also be performed in order to avoid drip patterns in the cascading film [85].

4.5.2.7 Effect of Rinsing Solution

Variation of ionic strength of the rinsing solution does not influence film thickness [11]. However, addition of a cosolvent changes the solvent quality and wetting behavior of the surface [101–103]. Therefore, it might affect the thickness as well as roughness of the resulting film, or, in the worst case, precipitation, especially for fast processes such as spray-assisted LbL assembly in which wetting of the surface plays a crucial role on the diffusion–adsorption process. Consequently, for spray-assisted LbL assembly, the use of a cosolvent to fine-tune film properties is not a popular practice and is rarely performed.

4.5.3 Applications

Spray-assisted LbL assembly has the advantages of fast processing (relative to traditional, immersive dip-assisted LbL assembly), versatility in materials and substrate selection, and conformal deposition. Materials ranging from polyelectrolytes [11], nanomaterials [84, 88, 104], metal oxides [105], therapeutic compounds [106], and DNA [107] have been deposited by spray-assisted LbL assembly. The unprecedented choice of materials that can be rapidly and sequentially assembled into multilayered films with nanoscale precession has made spray-LbL assembly an extremely attractive tool for applications such as energy storage [88], membrane separations [104], biocompatibility [106], tissue engineering [108], biocatalysis [109], microreactors [109], drug delivery [106, 110], fire resistance [111], and corrosion protection [112], among others.

4.5.4 Future Directions

With the advantage of rapid processing, spray-assisted LbL assembly has become a versatile tool for fabrication of conformal coatings with a wide range of materials. The development of automated spray-assisted LbL assembly allows one to examine the effect of several operating parameters, and can be fine-tuned to adjust the thickness and roughness as well as the final film properties. The effects and trends of operational parameters such as concentration, spraying time, spraying distance, and air pressure, as well as the charge density on the final film properties apply over wide range of material combinations. On the other hand, parameters such as composition of the rinsing solution, rinsing, or blow drying are system specific and are only exploited for special cases in which state-of-art fabrication protocols fail or do not provide satisfactory growth. Finally, spray-assisted LbL deposition of nanoobjects as well as anisotropic materials is far more challenging compared to traditional polyelectrolytes; therefore, special care must be taken during assembly of those materials.

4.6 Conclusion and Future Directions

Improvements in nanomanufacturing technology are a current bottleneck in nanocomposite commercialization. Most synthesis approaches developed in the laboratory rely on batch processes that can be difficult to scale because of mass or energy transfer considerations. Conversion of these processes to continuous approaches with improved mixing can alleviate these challenges, increasing throughput by orders of magnitude. Spray processes in particular offer promise because of their broad range of fluids and solutes that can be employed and because of their ability to disperse phases with minimal energy input. Further, spray technologies can be integrated with existing nanomanufacturing approaches such as microfluidics and can be performed in high-throughput, multiplexed configurations. Thus, spray-based nanomanufacturing strategies offer promise to overcoming limitations of batch scale processes.

References

1 NIST, NSF, DOE *et al.* (2010) *Sustainable Nanomanufacturing – Creating the Industries of the Future*, NSTC Committee on Technology and E. Subcommittee on Nanoscale Science, and Technology.

2 Hines, M.A. and Guyot-Sionnest, P. (1996) Synthesis and characterization of strongly luminescing ZnS-Capped CdSe nanocrystals. *Journal of Physical Chemistry*, **100** (**2**), 468–471.

3 Selvan, S.T., Patra, P.K., Ang, C.Y., and Ying, J.Y. (2007) Synthesis of silica-coated semiconductor and magnetic quantum dots and their Use in the imaging of live cells. *Angewandte Chemie International Edition*, **46** (**14**), 2448–2452.

4 Hussain, F., Hojjati, M., Okamoto, M., and Gorga, R.E. (2006) Review article: polymer-matrix nanocomposites, processing, manufacturing, and application: an overview. *Journal of Composite Materials*, **40** (**17**), 1511–1575.

5 Hanemann, T. and Szabo, D.V. (2010) Polymer-nanoparticle composites: from synthesis to modern applications. *Materials*, **3** (**6**), 3468–3517.

6 Mai, Y. and Eisenberg, A. (2012) Self-assembly of block copolymers. *Chemical Society Reviews*, **41** (**18**), 5969–5985.

7 Glass, R., Moller, M., and Spatz, J.P. (2003) Block copolymer micelle nanolithography. *Nanotechnology*, **14** (**10**), 1153–1160.

8 Zhu, J.T. and Hayward, R.C. (2008) Spontaneous generation of amphiphilic block copolymer micelles with multiple morphologies through interfacial instabilities. *Journal of the American Chemical Society*, **130** (**23**), 7496–7502.

9 Bae, J., Lawrence, J., Miesch, C. *et al.* (2012) Multifunctional nanoparticle-loaded spherical and wormlike micelles formed by interfacial instabilities. *Advanced Materials*, **24** (**20**), 2735–2741.

10 Granek, R., Ball, R.C., and Cates, M.E. (1993) Dynamics of spontaneous emulsification. *Journal De Physique Ii*, **3** (**6**), 829–849.

11 Izquierdo, A., Ono, S., Voegel, J.-C. *et al.* (2005) Dipping versus spraying: exploring the deposition conditions for speeding up layer-by-layer assembly. *Langmuir*, **21** (**16**), 7558–7567.

12 Johnson, B.K. and Prud'homme, R.K. (2003) *Flash nanoprecipitation of organic actives and block copolymers using a confined impinging jets mixer. Australian Journal of Chemistry*, **56** (**10**), 1021–1024.

13 Gindy, M.E., Panagiotopoulos, A.Z., and Prud'homme, R.K. (2008) Composite block copolymer stabilized nanoparticles: simultaneous encapsulation of organic actives and inorganic nanostructures. *Langmuir*, **24** (**1**), 83–90.

14 Johnson, B.K., Saad, W., and Prud'homme, R.K. (2006) Nanoprecipitation of pharmaceuticals using mixing and block copolymer stabilization. *Polymeric Drug Delivery II: Polymeric Matrices and Drug Particle Engineering*, **924**, 278–291.

15 Cheng, J.C., Vigil, R.D., and Fox, R.O. (2010) A competitive aggregation model for flash nanoprecipitation. *Journal of Colloid and Interface Science*, **351** (**2**), 330–342.

16 D'Addio, S.M. and Prud'homme, R.K. (2011) Controlling drug nanoparticle formation by rapid precipitation. *Advanced Drug Delivery Reviews*, **63** (**6**), 417–426.

17 Akbulut, M., Ginart, P., Gindy, M.E. *et al.* (2009) Generic method of preparing multifunctional fluorescent nanoparticles using flash NanoPrecipitation. *Advanced Functional Materials*, **19** (**5**), 718–725.

18 Gindy, M.E. and Prud'homme, R.K. (2009) Multifunctional nanoparticles for imaging, delivery and targeting in cancer therapy. *Expert Opinion on Drug Delivery*, **6** (**8**), 865–878.

19 Valente, I., Celasco, E., Marchisio, D.L., and Barresi, A.A. (2012) Nanoprecipitation in confined impinging jets mixers: production, characterization and scale-up of Pegylated nanospheres and nanocapsules for pharmaceutical use. *Chemical Engineering Science*, **77**, 217–227.

20 Somashekar, V., Liu, Y., Fox, R.O., and Olsen, M.G. (2012) Turbulence measurements in a rectangular mesoscale confined impinging jets reactor. *Experiments in Fluids*, **53** (**6**), 1929–1941.

21 Lince, F., Bolognesi, S., Marchisio, D.L. *et al.* (2011) Preparation of poly(MePEGCA-co-HDCA) nanoparticles with confined impinging jets reactor: experimental and modeling study. *Journal of Pharmaceutical Sciences*, **100** (**6**), 2391–2405.

22 Liu, Y., Kathan, K., Saad, W., and Prud'homme, R.K. (2007) Ostwald ripening of beta-carotene nanoparticles. *Physical Review Letters*, **98** (**3**), 036102.

23 Han, J., Zhu, Z., Qian, H. *et al.* (2012) A simple confined impingement jets mixer for flash nanoprecipitation. *Journal of Pharmaceutical Sciences*, **101** (**10**), 4018–4023.

24 Liu, Y., Cheng, C., Prud'homme, R.K., and Fox, R.O. (2008) Mixing in a multi-inlet vortex mixer (MIVM) for flash nano-precipitation. *Chemical Engineering Science*, **63** (**11**), 2829–2842.

25 Liu, Z.P., Ramezani, M., Fox, R.O. *et al.* (2015) Flow characteristics in a scaled-up multi-inlet vortex nanoprecipitation reactor. *Industrial & Engineering Chemistry Research*, **54** (**16**), 4512–4525.

26 Chow, S.F., Sun, C.C., and Chow, A.H.L. (2014) Assessment of the relative performance of a confined impinging jets mixer and a multi-inlet vortex mixer for curcumin nanoparticle production. *European Journal of Pharmaceutics and Biopharmaceutics*, **88** (**2**), 462–471.

27 Spaeth, J.R., Kevrekidis, I.G., and Panagiotopoulos, A.Z. (2011) Dissipative particle dynamics simulations of polymer-protected nanoparticle self-assembly. *Journal of Chemical Physics*, **135** (**18**), 10.

28 Chen, T., D'Addio, S.M., Kennedy, M.T. *et al.* (2009) Protected peptide nanoparticles: experiments and Brownian dynamics simulations of the energetics of assembly. *Nano Letters*, **9** (**6**), 2218–22.

29 D'Addio, S.M., Saad, W., Ansell, S.M. *et al.* (2012) Effects of block copolymer properties on nanocarrier protection from in vivo clearance. *Journal of Controlled Release*, **162** (**1**), 208–17.

30 Budijono, S.J., Russ, B., Saad, W. *et al.* (2010) Block copolymer surface coverage on nanoparticles. *Colloids and Surfaces A: Physicochemical and Engineering Aspects,* **360** (**1-3**), 105–110.

31 Shi, L., Shan, J.N., Ju, Y.G. *et al.* (2012) Nanoparticles as delivery vehicles for sunscreen agents. *Colloids and Surfaces A:Physicochemical and Engineering Aspects,* **396**, 122–129.

32 Liu, Y., Tong, Z., and Prud'homme, R.K. (2008) Stabilized polymeric nanoparticles for controlled and efficient release of bifenthrin. *Pest Management Science,* **64** (**8**), 808–812.

33 Jaworek, A. and Sobczyk, A.T. (2008) Electrospraying route to nanotechnology: an overview. *Journal of Electrostatics,* **66** (**3-4**), 197–219.

34 Yamashita, M. and Fenn, J.B. (1984) Electrospray ion-source - another variation on the free-jet theme. *Journal of Physical Chemistry,* **88** (**20**), 4451–4459.

35 Okuyama, K. and Lenggoro, I.W. (2003) Preparation of nanoparticles via spray route. *Chemical Engineering Science,* **58** (**3-6**), 537–547.

36 Jennerjohn, N., Eiguren-Fernandez, A., Prikhodko, S. *et al.* (2010) Design, demonstration and performance of a versatile electrospray aerosol generator for nanomaterial research and applications. *Nanotechnology,* **21** (**25**), 255603.

37 Ciach, T. (2006) Microencapsulation of drugs by electro-hydro-dynamic atomization. *International Journal of Pharmaceutics,* **324** (**1**), 51–55.

38 Chen, D.R., Pui, D.Y.H., and Kaufman, S.L. (1995) Electrospraying of conducting liquids for monodisperse aerosol generation in the 4 nm to 1.8 mu-m diameter range. *Journal of Aerosol Science,* **26** (**6**), 963–977.

39 Jaworek, A. (2007) Electrospray droplet sources for thin film deposition. *Journal of Materials Science,* **42** (**1**), 266–297.

40 Duong, A.D., Ruan, G., Mahajan, K. *et al.* (2014) Scalable, semicontinuous production of micelles encapsulating nanoparticles via electrospray. *Langmuir,* **30** (**14**), 3939–3948.

41 Wu, Y., Yu, B., Jackson, A. *et al.* (2009) Coaxial electrohydrodynamic spraying: a novel one-step technique to prepare oligodeoxynucleotide encapsulated lipoplex nanoparticles. *Molecular Pharmaceutics,* **6** (**5**), 1371–1379.

42 de la Mora, J.F. (2007) The fluid dynamics of Taylor cones. *Annual Review of Fluid Mechanics,* **39**, 217–243.

43 Hartman, R.P.A., Brunner, D.J., Camelot, D.M.A. *et al.* (2000) Jet break-up in electrohydrodynamic atomization in the cone-jet mode. *Journal of Aerosol Science,* **31** (**1**), 65–95.

44 Zeleny, J. (1917) Instability of electrified liquid surfaces. *Physical Review,* **10** (**1**), 1–6.

45 Taylor, G. (1964) Disintegration of water drops in an electric field. *Proceedings of the Royal Society of London. Series A, Mathematical and Physical Sciences*, **280** (**1382**), 383–397.
46 Hartman, R.P.A., Brunner, D.J., Camelot, D.M.A. *et al.* (1999) Electrohydrodynamic atomization in the cone-jet mode physical modeling of the liquid cone and jet. *Journal of Aerosol Science*, **30** (7), 823–849.
47 Cloupeau, M. and Prunetfoch, B. (1994) Electrohydrodynamic spraying functioning modes - a critical-review. *Journal of Aerosol Science*, **25** (**6**), 1021–1036.
48 Cloupeau, M. and Prunetfoch, B. (1989) Electrostatic spraying of liquids in cone-jet mode. *Journal of Electrostatics*, **22** (**2**), 135–159.
49 Xie, J.W. and Wang, C.H. (2007) Electrospray in the dripping mode for cell microencapsulation. *Journal of Colloid and Interface Science*, **312** (**2**), 247–255.
50 Cloupeau, M. and Prunetfoch, B. (1990) Electrostatic spraying of liquids - main functioning modes. *Journal of Electrostatics*, **25** (**2**), 165–184.
51 Fong, H., Chun, I., and Reneker, D.H. (1999) Beaded nanofibers formed during electrospinning. *Polymer*, **40** (**16**), 4585–4592.
52 Chandrasekhar, R. and Choy, K.L. (2001) Electrostatic spray assisted vapour deposition of fluorine doped tin oxide. *Journal of Crystal Growth*, **231** (**1-2**), 215–221.
53 Benitez, R., Soler, J., and Daza, L. (2005) Novel method for preparation of PEMFC electrodes by the electrospray technique. *Journal of Power Sources*, **151**, 108–113.
54 Cao, F. and Prakash, J. (2002) A comparative electrochemical study of $LiMn_2O_4$ spinel thin-film and porous laminate. *Electrochimica Acta*, **47** (**10**), 1607–1613.
55 Lapham, D.P., Colbeck, I., Schoonman, J., and Kamlag, Y. (2001) The preparation of $NiCo_2O_4$ films by electrostatic spray deposition. *Thin Solid Films*, **391** (**1**), 17–20.
56 Rietveld, I.B., Kobayashi, K., Yamada, H., and Matsushige, K. (2006) Morphology control of poly(vinylidene fluoride) thin film made with electrospray. *Journal of Colloid and Interface Science*, **298** (**2**), 639–651.
57 Ju, J., Yamagata, Y., and Higuchi, T. (2009) Thin-film fabrication method for organic light-emitting diodes using electrospray deposition. *Advanced Materials*, **21** (**43**), 4343.
58 Almeria, B., Deng, W.W., Fahmy, T.M., and Gomez, A. (2010) Controlling the morphology of electrospray-generated PLGA microparticles for drug delivery. *Journal of Colloid and Interface Science*, **343** (**1**), 125–133.
59 Duong, A.D., Sharma, S., Peine, K.J. *et al.* (2013) Electrospray encapsulation of toll-like receptor agonist resiquimod in polymer microparticles for the treatment of Visceral leishmaniasis. *Molecular Pharmaceutics*, **10** (**3**), 1045–1055.

60 Ding, L., Lee, T., and Wang, C.H. (2005) Fabrication of monodispersed Taxol-loaded particles using electrohydrodynamic atomization. *Journal of Controlled Release*, **102** (**2**), 395–413.

61 Jaworek, A. (2008) Electrostatic micro- and nanoencapsulation and electroemulsification: a brief review. *Journal of Microencapsulation*, **25** (7), 443–468.

62 Loscertales, I.G., Barrero, A., Guerrero, I. *et al.* (2002) Micro/nano encapsulation via electrified coaxial liquid jets. *Science*, **295** (**5560**), 1695–1698.

63 Zhang, L.L., Huang, J.W., Si, T., and Xu, R.X. (2012) Coaxial electrospray of microparticles and nanoparticles for biomedical applications. *Expert Review of Medical Devices*, **9** (**6**), 595–612.

64 Si, T., Zhang, L.L., Li, G.B. *et al.* (2013) Experimental design and instability analysis of coaxial electrospray process for microencapsulation of drugs and imaging agents. *Journal of Biomedical Optics*, **18** (7), 075003.

65 López-Herrera, J.M., Barrero, A., López, A. *et al.* (2003) Coaxial jets generated from electrified Taylor cones. Scaling laws. *Journal of Aerosol Science*, **34** (**5**), 535–552.

66 Lee, Y.H., Bai, M.Y., and Chen, D.R. (2011) Multidrug encapsulation by coaxial tri-capillary electrospray. *Colloids and Surfaces B-Biointerfaces*, **82** (**1**), 104–110.

67 Nie, H.M., Dong, Z.G., Arifin, D.Y. *et al.* (2010) Core/shell microspheres via coaxial electrohydrodynamic atomization for sequential and parallel release of drugs. *Journal of Biomedical Materials Research Part A*, **95A** (**3**), 709–716.

68 Hartman, R.P.A., Borra, J.P., Brunner, D.J. *et al.* (1999) The evolution of electrohydrodynamic sprays produced in the cone-jet mode, a physical model. *Journal of Electrostatics*, **47** (**3**), 143–170.

69 Chen, D.R. and Pui, D.Y.H. (1997) Experimental investigation of scaling laws for electrospraying: dielectric constant effect. *Aerosol Science and Technology*, **27** (**3**), 367–380.

70 Delamora, J.F. and Loscertales, I.G. (1994) The current emitted by highly conducting Taylor cones. *Journal of Fluid Mechanics*, **260**, 155–184.

71 Xie, J.W., Jiang, J., Davoodi, P. *et al.* (2015) Electrohydrodynamic atomization: a two-decade effort to produce and process micro-/nanoparticulate materials. *Chemical Engineering Science*, **125**, 32–57.

72 Almeria, B., Fahmy, T.M., and Gomez, A. (2011) A multiplexed electrospray process for single-step synthesis of stabilized polymer particles for drug delivery. *Journal of Controlled Release*, **154** (**2**), 203–210.

73 Sato, M., Saito, M., and Hatori, T. (1993) Emulsification and size control of Insulating and/or viscous-liquids in liquid liquid-system by electrostatic dispersion. *Journal of Colloid and Interface Science*, **156** (**2**), 504–507.

74 Barrero, A., Lopez-Herrera, J.M., Boucard, A. *et al.* (2004) Steady cone-jet electrosprays in liquid insulator baths. *Journal of Colloid and Interface Science*, **272** (**1**), 104–108.
75 Watanabe, A., Higashitsuji, K., and Nishizawa, K. (1978) Studies on electrocapillary emulsification. *Journal of Colloid and Interface Science*, **64** (**2**), 278–289.
76 Tsouris, C., Depaoli, D.W., Feng, J.Q. *et al.* (1994) Electrostatic spraying of nonconductive fluids into conductive fluids. *AIChE Journal*, **40** (**11**), 1920–1923.
77 Yeh, C.H., Lee, M.H., and Lin, Y.C. (2012) Using an electro-spraying microfluidic chip to produce uniform emulsions under a direct-current electric field. *Microfluidics and Nanofluidics*, **12** (**1-4**), 475–484.
78 Kim, H., Luo, D.W., Link, D. *et al.* (2007) Controlled production of emulsion drops using an electric field in a flow-focusing microfluidic device. *Applied Physics Letters*, **91** (**13**), 133106.
79 Sato, M., Okubo, N., Nakane, T. *et al.* (2010) Nozzleless EHD spraying for fine droplet production in liquid-in-liquid system. *IEEE Transactions on Industry Applications*, **46** (**6**), 2190–2195.
80 Sato, M., Kuroiwa, I., Ohshima, T., and Urashima, K. (2009) Production of nano-silica particles in liquid–liquid system by pulsed voltage application. *IEEE Transactions on Dielectrics and Electrical Insulation*, **16** (**2**), 320–324.
81 Sato, M., Morita, N., Kuroiwa, I. *et al.* (2009) Dielectric liquid-in-liquid dispersion by applying pulsed voltage. *IEEE Transactions on Dielectrics and Electrical Insulation*, **16** (**2**), 391–395.
82 Lutkenhaus, J.L. and Hammond, P.T. (2007) Electrochemically enabled polyelectrolyte multilayer devices: from fuel cells to sensors. *Soft Matter*, **3** (**7**), 804–816.
83 Sung, C., Hearn, K., Reid, D.K. *et al.* (2013) A comparison of thermal transitions in dip-and spray-assisted layer-by-layer assemblies. *Langmuir*, **29** (**28**), 8907–8913.
84 Nogueira, G.M., Banerjee, D., Cohen, R.E., and Rubner, M.F. (2011) Spray-layer-by-layer assembly can more rapidly produce optical-quality multistack heterostructures. *Langmuir*, **27** (**12**), 7860–7867.
85 Krogman, K., Zacharia, N., Schroeder, S., and Hammond, P. (2007) Automated process for improved uniformity and versatility of layer-by-layer deposition. *Langmuir*, **23** (**6**), 3137–3141.
86 Lu, C., Dönch, I., Nolte, M., and Fery, A. (2006) Au nanoparticle-based multilayer ultrathin films with covalently linked nanostructures: spraying layer-by-layer assembly and mechanical property characterization. *Chemistry of Materials*, **18** (**26**), 6204–6210.
87 Hong, J. and Kang, S.W. (2011) Carbon decorative coatings by dip-, spin-, and spray-assisted layer-by-layer assembly deposition. *Journal of nanoscience and nanotechnology*, **11** (**9**), 7771–7776.

88 Kwon, S.R., Jeon, J.-W., and Lutkenhaus, J.L. (2015) Sprayable, paintable layer-by-layer polyaniline nanofiber/graphene electrodes. *RSC Advances*, **5** (**20**), 14994–15001.
89 Carosio, F., Di Blasio, A., Cuttica, F. *et al.* (2013) Flame retardancy of polyester fabrics treated by spray-assisted layer-by-layer silica architectures. *Industrial & Engineering Chemistry Research*, **52** (**28**), 9544–9550.
90 Shin, Y., Roberts, J.E., and Santore, M.M. (2002) The relationship between polymer/substrate charge density and charge overcompensation by adsorbed polyelectrolyte layers. *Journal of Colloid and Interface Science*, **247** (**1**), 220–230.
91 Qiao, B., Cerdà, J.J., and Holm, C. (2011) Atomistic study of surface effects on polyelectrolyte adsorption: case study of a poly (styrenesulfonate) monolayer. *Macromolecules*, **44** (**6**), 1707–1718.
92 Cheng, H. and de la Cruz, M.O. (2003) Adsorption of rod-like polyelectrolytes onto weakly charged surfaces. *The Journal of Chemical Physics*, **119** (**23**), 12635–12644.
93 Shin, Y., Roberts, J.E., and Santore, M.M. (2002) Influence of charge density and coverage on bound fraction for a weakly cationic polyelectrolyte adsorbing onto silica. *Macromolecules*, **35** (**10**), 4090–4095.
94 Lavalle, P., Picart, C., Mutterer, J. *et al.* (2004) Modeling the buildup of polyelectrolyte multilayer films having exponential growth x. *The Journal of Physical Chemistry B*, **108** (**2**), 635–648.
95 Muthukumar, M. (2004) Theory of counter-ion condensation on flexible polyelectrolytes: adsorption mechanism. *The Journal of Chemical Physics*, **120** (**19**), 9343–9350.
96 Carrillo, J.-M.Y. and Dobrynin, A.V. (2007) Molecular dynamics simulations of polyelectrolyte adsorption. *Langmuir*, **23** (**5**), 2472–2482.
97 Carrillo, J.-M.Y. and Dobrynin, A.V. (2011) Layer-by-layer assembly of polyelectrolyte chains and nanoparticles on nanoporous substrates: molecular dynamics simulations. *Langmuir*, **28** (**2**), 1531–1538.
98 Dobrynin, A.V., Deshkovski, A., and Rubinstein, M. (2001) Adsorption of polyelectrolytes at oppositely charged surfaces. *Macromolecules*, **34** (**10**), 3421–3436.
99 Dobrynin, A.V. and Rubinstein, M. (2003) Effect of short-range interactions on polyelectrolyte adsorption at charged surfaces. *The Journal of Physical Chemistry B*, **107** (**32**), 8260–8269.
100 Dobrynin, A.V. and Rubinstein, M. (2005) Theory of polyelectrolytes in solutions and at surfaces. *Progress in Polymer Science*, **30** (**11**), 1049–1118.
101 Dubas, S.T. and Schlenoff, J.B. (1999) Factors controlling the growth of polyelectrolyte multilayers. *Macromolecules*, **32** (**24**), 8153–8160.
102 Poptoshev, E., Schoeler, B., and Caruso, F. (2004) Influence of solvent quality on the growth of polyelectrolyte multilayers. *Langmuir*, **20** (**3**), 829–834.

103 Zhang, H., Wang, Z., Zhang, Y., and Zhang, X. (2004) Hydrogen-bonding-directed layer-by-layer assembly of poly (4-vinylpyridine) and poly (4-vinylphenol): effect of solvent composition on multilayer buildup. *Langmuir*, **20** (**21**), 9366–9370.
104 Liu, L., Son, M., Chakraborty, S. *et al.* (2013) Fabrication of ultra-thin polyelectrolyte/carbon nanotube membrane by spray-assisted layer-by-layer technique: characterization and its anti-protein fouling properties for water treatment. *Desalination and Water Treatment*, **51** (**31-33**), 6194–6200.
105 Krogman, K.C., Lowery, J.L., Zacharia, N.S. *et al.* (2009) Spraying asymmetry into functional membranes layer-by-layer. *Nature Materials*, **8** (**6**), 512–518.
106 Kim, B.-S., Smith, R.C., Poon, Z., and Hammond, P.T. (2009) MAD (multiagent delivery) nanolayer: delivering multiple therapeutics from hierarchically assembled surface coatings†. *Langmuir*, **25** (**24**), 14086–14092.
107 Qi, A., Chan, P., Ho, J. *et al.* (2011) Template-free synthesis and encapsulation technique for layer-by-layer polymer nanocarrier fabrication. *ACS Nano*, **5** (**12**), 9583–9591.
108 Mironov, V., Boland, T., Trusk, T. *et al.* (2003) Organ printing: computer-aided jet-based 3D tissue engineering. *Trends in Biotechnology*, **21** (**4**), 157–161.
109 Michel, M., Arntz, Y., Fleith, G. *et al.* (2006) Layer-by-layer self-assembled polyelectrolyte multilayers with embedded liposomes: immobilized submicronic reactors for mineralization. *Langmuir*, **22** (**5**), 2358–2364.
110 Chong, S.-F., Sexton, A., De Rose, R. *et al.* (2009) A paradigm for peptide vaccine delivery using viral epitopes encapsulated in degradable polymer hydrogel capsules. *Biomaterials*, **30** (**28**), 5178–5186.
111 Alongi, J., Carosio, F., Frache, A., and Malucelli, G. (2013) Layer by layer coatings assembled through dipping, vertical or horizontal spray for cotton flame retardancy. *Carbohydrate Polymers*, **92** (**1**), 114–119.
112 Shchukin, D.G., Zheludkevich, M., Yasakau, K. *et al.* (2006) Layer-by-layer assembled nanocontainers for self-healing corrosion protection. *Advanced Materials*, **18** (**13**), 1672–1678.

5

Overview of Nanotechnology in Military and Aerospace Applications

Eugene Edwards[1], Christina Brantley[1], and Paul B. Ruffin[2]

[1] *U. S. Army Research, Development, and Engineering Command, Redstone Arsenal, Alabama, USA*
[2] *Alabama A&M University, Normal, Alabama, USA*

5.1 Introduction

There has been tremendous progress made in the field of nanotechnology since the renown physicist, Richard Feynman, suggested that it should be possible to build machines small enough to manufacture objects with atomic precision in his talk, "There's Plenty of Room at the Bottom," in 1959 [1]. During the past two decades, nanotechnology has been widely studied in academia, industrial sector, and the government, during which time it has progressed from the concept and laboratory research stages to use in a plethora of commercial applications. Carbon nanotubes (CNTs) are currently used in the automobile industry to improve safety of fuel lines in passenger vehicles and in the electronics industry for packaging material to better protect the goods and aid in the removal of electrical charges. In the household, nano-ceramic particles are used to improve the smoothness, and heat resistance of common household equipment such as the flat iron and engineered nanofibers are used to make clothes water/stain-repellent or wrinkle-free. In the nano-medicine area, nanoparticles are used in drug delivery [2, 3].

The primary advantages of nanotechnology for use in military and aerospace systems are in the areas of weaponry, protection, and communication. Nanotechnology can be used to produce military devices, structures, and materials that are smaller, lighter, smarter, stronger, cheaper, cleaner, and more precise. Nanotechnology can also be used in military systems to minimize weaponry or used as weaponry (e.g., microfusion missiles), thus increasing the efficiency of military weaponry and reducing production and transportation costs [4]. In the area of protection, nanotechnology can be used to produce body armor that are light and thin and yet dense and strong enough to resist the impact of a high-speed bullet.

Nanotechnology Commercialization: Manufacturing Processes and Products, First Edition.
Edited by Thomas O. Mensah, Ben Wang, Geoffrey Bothun, Jessica Winter, and Virginia Davis.

The major challenges in nanotechnology for use in military and aerospace systems are in the areas of (i) advanced materials for coatings, including thin-film optical coatings, light-weight, strong armor and missile structural components, embedded computing, and "smart" structures; (ii) nanoparticles for explosives, warheads, turbine engine systems, and propellants to enhance missile propulsion; (iii) nanosensors for autonomous chemical detection; and (iv) nanotube arrays for fuel storage and power generation [5]. Some of these challenges have been addressed by researchers at component organizations within DoD, to include Air Force Office of Scientific Research (AFOSR); Army Engineering R&D Center (ERDC); Army Research Laboratory (ARL); Army Research Office (ARO); Defense Advanced Research Projects Agency (DARPA); Office of the Director, Defense Research & Engineering (ODDR&E); Defense Threat Reduction Agency (DTRA); and Office of Naval Research (ONR) [6]. During the past decade, a research and development program was conducted at the Army Aviation and Missile Research, Development, and Engineering Center (AMRDEC) to develop composites for missile motor casings and structural components, nanosensors for the detection of chemical agents, nanoparticles for energetics, nanoelectronics, nano-thermal batteries, nanoplasmonics, and nanoelectromechanical systems (NEMS) for insertion in missile, aviation, and unmanned aerial systems [7]. The developmental progress of these technologies is the thrust of this chapter.

Implications of nanotechnology in military and aerospace systems applications are presented in the following section. Nano-based materials designed for (i) microsensors, (ii) structural health monitoring, (iii) missile solid propellants, and (iv) composites and structural components are presented in Sections 5.3–5.6, respectively. Nanoplasmonics is presented in Section 5.7. Nanothermal batteries and fuel cells are presented in Section 5.8. A conclusion is provided in Section 5.9.

5.2 Implications of Nanotechnology in Military and Aerospace Systems Applications

Due to its desirable material properties such as high tensile strength, large surface area to volume ratio, light weight, high thermal conductivity, and the tremendous developmental progress, nanotechnology has the possibility of being fully implemented in the military within the next 10–20 years. Since 2001, the Department of Defense has invested more than $0.5B per year under the National Nanotechnology Initiative (NNI) in the areas of (i) fundamental phenomena and processes, (ii) nanomaterials, (iii) nanoscale devices and systems, (iv) instrument research, metrology, and standards, (v) nanomanufacturing, (vi) major research facilities, (vii) environment, health, and safety, and (viii) education and societal dimensions [8]. The US Department of the

Army's primary nanotechnology investment (~\$10 M per annum) has been in the Institute of Soldier Nanotechnologies (ISN) at Massachusetts Institute of Technology (MIT), which is a research initiative that aims to advance soldier survivability through advanced nanotechnology [9].

Near-term uses of nanotechnology in the military range from improved communications to smart armor for the soldier, as well as very advanced weaponry. In the area of communications, the focus is on less bulky communication devices that are able to be inserted in the ear, and permits the soldier to be able to hear sound on the battlefield on whatever side the device is on. Although much progress has been made, the technologies are far from maturity. Nanotechnology is currently being used in very small roles for uniforms (hence nanoscale, very small) in the military [10].

In the same manner as nuclear devices presented a potential threat to society during the 1950s, it appears that nanotechnology for military use may present an alleged concern to society. Although any chemical warfare is banned by the United Nations, it is not beyond reason to believe that it may be explored by some adversarial country, which is willing to cause maximum damage. There is a degree of caution when using nanotechnology in the military due to possible destructive devices such as nanochemicals. The threat of nuclear weapons led to the cold war. The same trend is foreseen with nanotechnology, which may lead to the so-called nanowars, a new age of destruction [10].

5.3 Nano-Based Microsensor Technology for the Detection of Chemical Agents

Nano-based sensor technology was one of the first areas initiated to develop nano-based structures and components for insertion into sophisticated missile, aviation, and ground-based weapon systems. The objective of the research was to exploit unique phenomena for the development of novel technology to enhance warfighter capabilities and monitor the longevity of precision weapons. By integrating nano-based sensors into weaponry, the Army can develop health-wise smart weaponry with significantly reduced size, weight, and cost. One of the major objectives was to develop nano-based chemical sensors to detect rocket motor off-gassing and general energetic/toxic chemicals.

5.3.1 Surface-Enhanced Raman Spectroscopy

Generally, the amplitude of the enhancement associated with laser beam-generated surface plasmons is influenced by the texture of the surface that the laser beam strikes. In Figure 5.1a, the majority of surface plasmons from absorbed particles of the incident laser beam become reflected heat loss; in Figure 5.1b, surface plasmons enhancement is generated as a product of

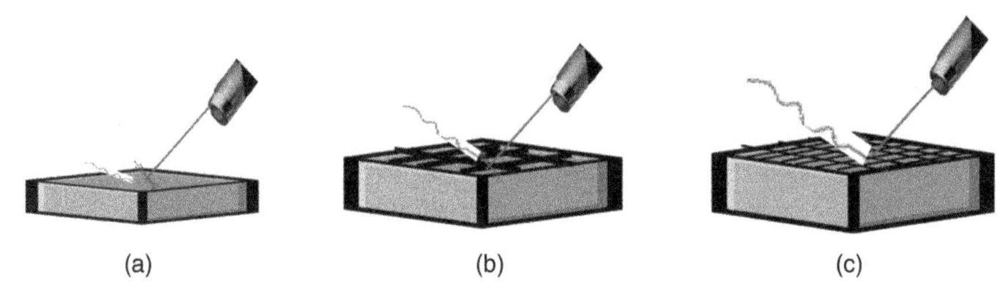

Figure 5.1 (a) Plasmons become reflected heat loss; (b) plasmons enhanced via roughed surface; (c) plasmons greater enhancement via nano-roughed surfaces.

the rough surface; and in Figure 5.1c, greater enhancement is evident on a nano-roughed surface. The greatest enhancement is realized when the individual components of the nano-roughed surface are significantly less in size than the wavelength of the incident beam, which is resonant with occurring Raman scattering. Nano-roughed surfaces can come in all textures, shapes, and configurations (including squares, bars, spheres, bowties, and antennas generated from an array of a chosen shape).

As indicated and described in Ref. [11], various surface-enhanced Raman spectroscopy (SERS) substrates were studied in order to achieve the best enhancement of the Raman spectrum for residual amounts of materials [11]. New substrates produced by nanospheres imprint technique were investigated. Two different-size polystyrene nanospheres, 625 and 992 nm, are used to produce nanopatterns and nanocavities on the surface of a glass slide, which is coated with sputtered gold. Results from laboratory-version substrates were compared to a commercial gold-coated substrate (as shown in Figure 5.2, which consists of an array of resonant cavities that give the SERS effect). Sample concentration, starting from 1000 ppm was gradually diluted to the smallest detectable amount. The Raman spectrum was obtained using a portable spectrometer operating at a wavelength of 780 nm.

Although there are many benefits to the use of SERS for sensors, the design challenge was generally centered around the inability to easily and consistently produce uniform substrates, whether using electrodes, colloids (small particles of gold or silver dispersed in an aqueous solution-after now referred to as nanoparticles), or metal surfaces [11]. Some SERS experts contend that the difficulty in manufacturing uniform colloids and producing consistent substrate surfaces has been a major limiting factor in the commercialization of SERS. Others, however, say the primary limitation of SERS for chemical agents' detection has been finding applications where selectivity is not an issue [11].

5.3.1.1 Design Approach

In agreement with the well-known elementary models, the basis for the AMRDEC/Alabama A&M University (AAMU) team concept was previously

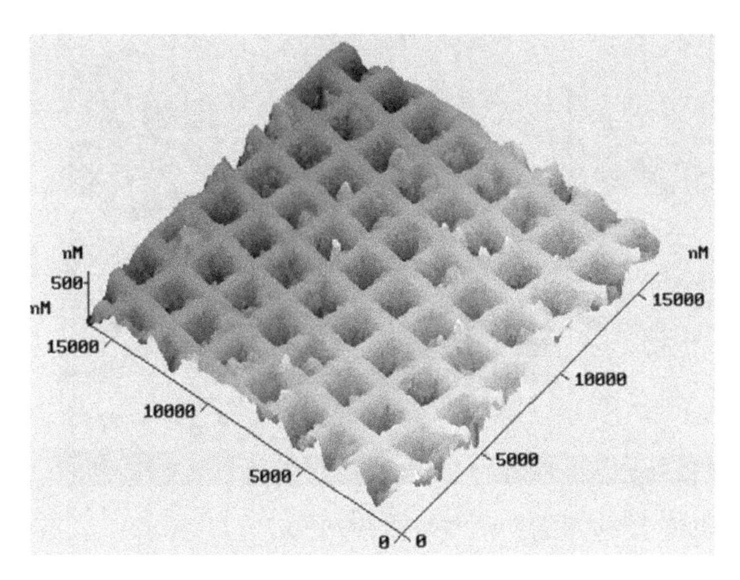

Figure 5.2 Commercial gold-coated substrate with an array of resonant cavities for SERS effect.

depicted in Figure 5.1. Novel research efforts were made to enhance the development of chemical and biological sensors to detect the presence of explosive devices, which may threaten national security, as well as a host of military applications. Collaborative research efforts between the US Army AMRDEC and AAMU focused on direct detection/identification of residual traces of explosive materials (in various forms) and the detection of undesired outgassing from weaponry rocket motors. Detailed results from this effort were published in Ref. [11]. The AMRDEC/AAMU team successfully developed highly sensitive sensors/detectors that are capable of detecting extremely small traces of substance in order to determine when a targeted location and/or an individual has explosives on his/her person or has been exposed to explosive materials. The team made use of concurrent research associated with SERS techniques in order to analyze residual traces of explosives in highly diluted solutions. An array of explosives (including TNT, HMX, RDX, and nitroglycerin) was successfully analyzed (evaluated at different concentrations) while simultaneously evaluating the detection sensitivity of the chosen SERS approach.

5.3.1.2 Experiment

The SERS measurements were implemented using a commercially available Raman Spectroscopy Instrumentation System (EZRaman) purchased from Enwave Optronics Inc. as shown in Figure 5.3. The system was equipped with a 785 nm excitation wavelength from a diode laser source yielding 300–400 mW

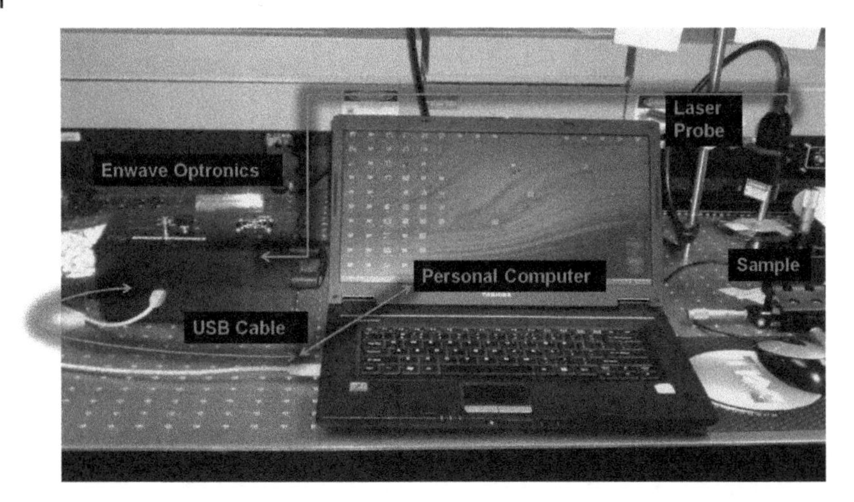

Figure 5.3 Experimental setup of SERS instrumentation system.

power at the laser output port. The light was coupled to the SERS substrate via a permanently aligned two single fiber combination fiber-optic probe with a 100 μm excitation fiber and a 200 μm diameter collection fiber with a 0.22 numerical aperture. The standard spectral resolution and spectral coverage of this instrument were 6 and ~250–2350 cm^{-1}, respectively. The SERS substrates were placed at an optimum working distance of 7 mm from the Raman probe. The SERS spectra presented in this chapter were recorded for a typical integration time of 20 s [12].

5.3.1.3 Results

The experimental setup in Figure 5.3 was used to provide the data in Figures 5.4 and 5.5. Laboratory samples of TNT and RDX were deposited on the SERS substrate and exposed to the laser source to examine the performance of the Raman spectroscopy system. Various concentrations of the TNT (10, 50, 125, and 250 ppm) were measured. The results clearly show the increased intensity as the concentration of the TNT increased. Also, the results can be compared to the standard SERS spectrum [13]. The dominant wavenumbers of TNT are 792, 825, and 1351, which are also experimentally observed (as shown in Figure 5.4). As indicated from the results in Figure 5.5, RDX was also deposited on a gold substrate in various concentrations (125, 250, and 1000 ppm). The experimentally observed Raman spectrum was compared to the standard dominant wavenumbers of RDX, which are 874, 930, 1258, 1312, and 1560 cm^{-1}, with 874 cm^{-1} being the most dominant [14]. Results clearly show that by using a nanostructured gold surface, the signals of the explosives adsorbed at the surface are enhanced due to the SERS effect by several orders of magnitude.

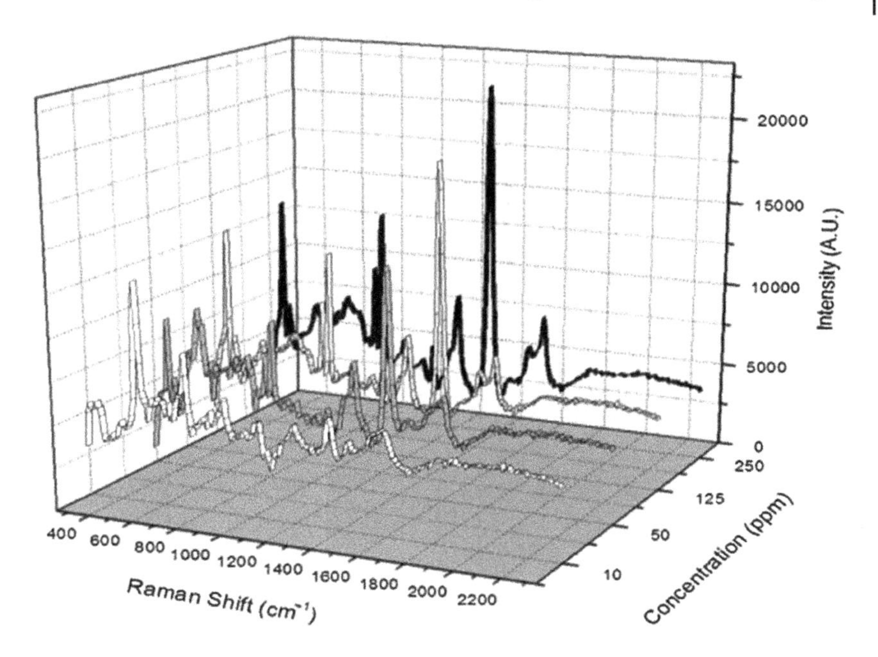

Figure 5.4 Experimental Raman spectrum for TNT (1 pg).

5.3.2 Voltammetric Techniques

The thick-film voltammetric sensor technology, originally developed by Argonne National Laboratory and licensed by Streamline Automation for use in the detection of gases, was adopted as the baseline for the AMRDEC research and development project [15, 16]. The thick-film sensor, as shown in Figure 5.6, has a solid electrolyte with multiple electrodes for simultaneously sensing three different analytes or targeted chemical agents. Test results for the subsequent thick-film sensor design indicate that the sensor can successfully detect hypergolic propellants off-gassing.

The primary objective of the AMRDEC research and development project was to develop a miniaturized chemical sensor array package that could be mounted onto an uninhabited vehicle to detect and identify multiple analytes from explosive devices, as well as from chemical warfare agents. The sensor had to be ruggedized for military environments and miniaturized for minimal payload additions on uninhabited vehicles. The size, power consumption, and surface temperature dependence of the sensor were minimized by redesigning the sensor substrate of the thick-film device to a thin-film configuration. Useful lessons learned from the thick-film design were utilized for the thin-film devices, which display the same electrochemical behavior as the thick film, especially when adequate attention is given to fabrication steps (such as

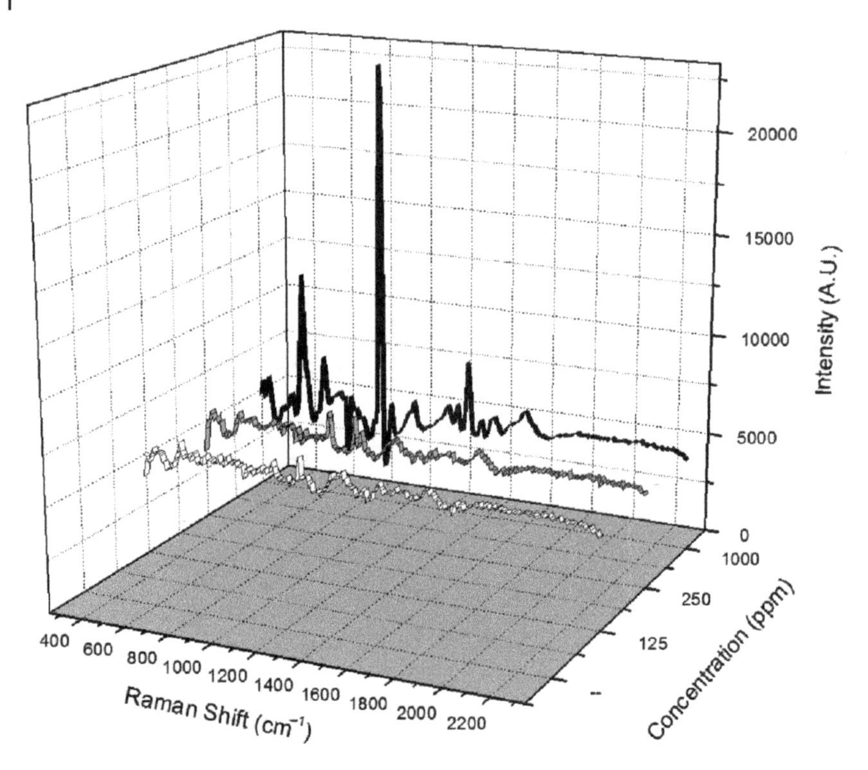

Figure 5.5 Experimental Raman spectrum for RDX (1 pg).

cleaning, surface preparation, material purity, sputtering, target preparation, and control of process parameters during reactive magnetron sputtering) [17]. It should likewise be noted that the response and recovery speeds of thin-film devices were faster than similarly configured thick-film devices.

Research on microelectrodes for gas sensors on bulk or thin-film substrates is extremely rare for voltammetric methods; however, all electrochemical methods share similar physics, differing mostly by the method of interrogation (passive or active) and variable measured (current or potential) [18].

5.3.2.1 Design Approach

A study of various geometrical configurations for the electrode deposition was conducted in order to generate a design layout for a thin-film sensor. Three configurations were investigated in order to determine the relative sensitivity of different sensing electrode designs and the response to the atomic structures of various chemical species. The resulting designs were the combed (a), solid (b), and grid (c) patterns, as shown in Figure 5.7.

The combed design was selected for the thin-film sensor electrode deposition. The preliminary sensor design, as shown in Figure 5.8, demonstrates

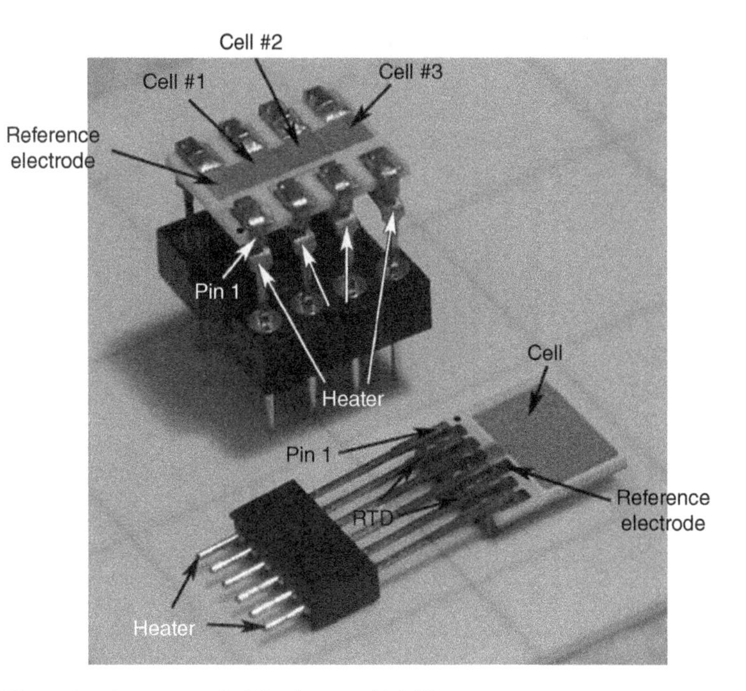

Figure 5.6 Prototype of triple element thick-film sensor.

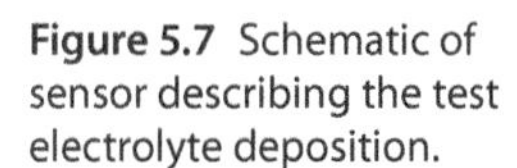

Figure 5.7 Schematic of sensor describing the test electrolyte deposition.

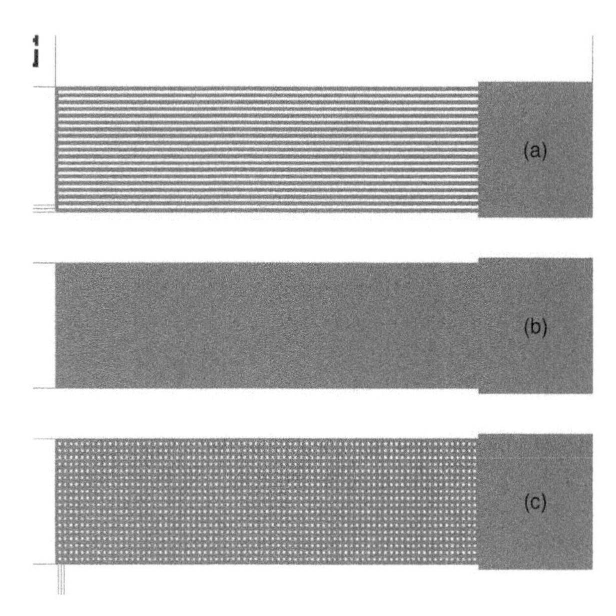

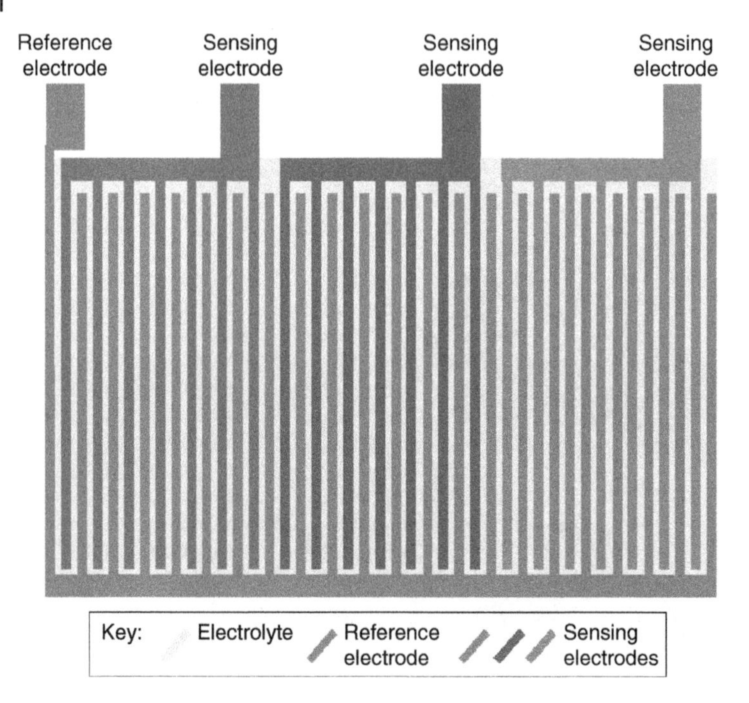

Figure 5.8 Schematic of triple element thin-film sensor.

the combed configuration. Each sensor has a reference electrode intertwined in a combed design with three different sensing electrodes, allowing for the detection of at least three different gas contaminants. The design parameters include response speed, sensitivity, dynamic range, short- and long-term stability, resolution, selectivity, operating ambient conditions, operating lifetime, output format, size, weight, and cost. Material selection for the microsensor is dependent on sensor performance characteristics.

5.3.2.2 Experimental/Test Setup

The sensor system was tested to measure the sensitivity to undiluted trace amounts of ammonia, mustard gas stimulant (CEES), venomous (V) series of a chosen nerve agent/insecticide (X) VX stimulant (diazonin), nitrogen tetraoxide (NTO), monomethyl hydrazine (MMH), and hydrazine. The sensors were enclosed in a Pyrex glass enclosure, 3-l three-neck flask. Gases from the stock cylinders were injected into the septum and directed through precision flow meters into the glass chamber, as shown in Figure 5.9. The chamber was flushed with air, and the sensor readings were measured and recorded. The response/output of the voltammetry sensor system was complex waveforms containing a large amount of data, referred to as voltammograms.

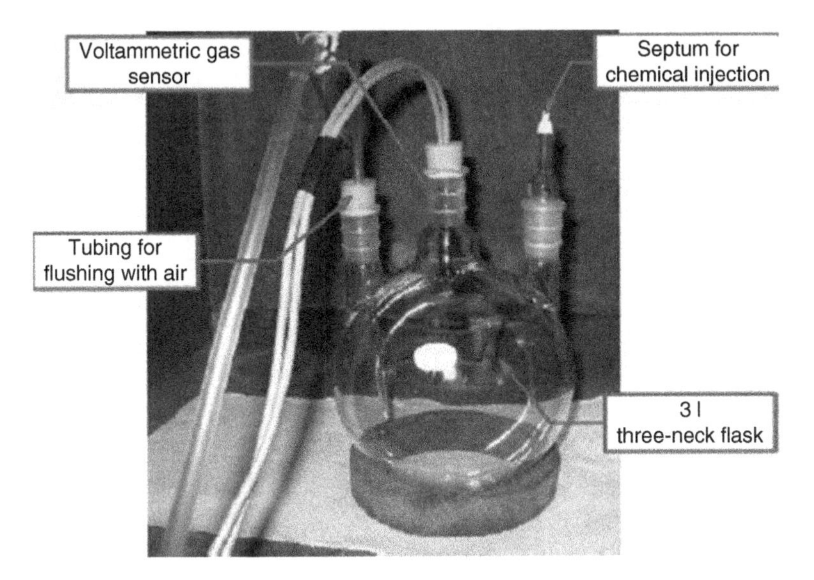

Figure 5.9 Voltammetry test setup.

The voltammogram depends on the type of electroactive gaseous components present in the target analyte. The chemical signature resulting from the voltammogram produced the identity and quantity of the present chemical species that interacted with the sensor's electrochemical cell. The advantage to this type of sensing method over single sample point responses is the significant amount of information on the rate-of-change and intervariable relationships produced by time- and parameter-based features on the voltammograms. These response features also aid in the identification of signal trends. The occurrence of chemical reactivity was determined from the calculation and comparison of the target analyte chemical signature to the background air signature. Pattern recognition tools and neural network analysis were used to extract the analog values representing the chemical composition of the target analytes.

5.3.2.3 Results

Test results were obtained when a single element sensor was exposed to ammonia from lowest to highest concentration, followed by a background air sample. The data from the sensor's reaction to ammonia is presented in Figure 5.10. The voltammogram (Figure 5.10) shows ±2 V sweep limits, and focuses on the difference between the ammonia responses and the background air responses. Each sweep occurred over a 1-s interval. Measurements were taken from 0 to 109 ppm of ammonia. As the measurement cycle was repeated, each successive peak narrowed, representing a function of less and less other species present to react. An increase in concentration of the ammonia is depicted on

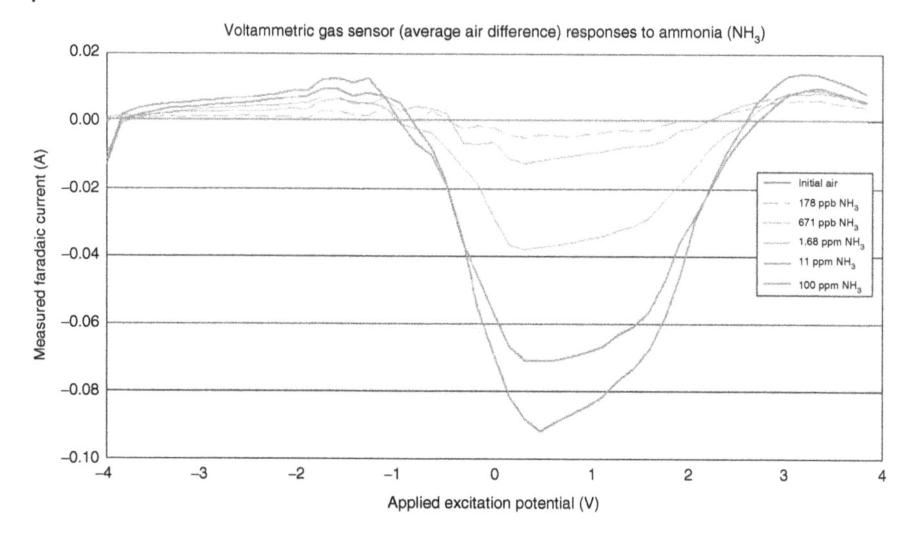

Figure 5.10 Raw data of sensor response to ammonia.

the graph where each successive peak rose in amplitude. Each successive peak in the chemical signature also shifts toward a stable dissociation potential.

The sensing element was also exposed to common rocket motor outgassing toxic compounds [19]. The spectral response of the sensing element was analyzed using common spectral analysis techniques to identify the chemical at various concentrations in the gas sample. Curves typifying the sensitivities of the selected gas samples [0.05 ppm of hydrazine, 1.5 ppm of monomethyl hydrazine (MMH), and 6 ppm of nitrogen tetroxide (NTO)] are shown in Figure 5.11. The raw voltage sensor relates the applied voltage to the measured

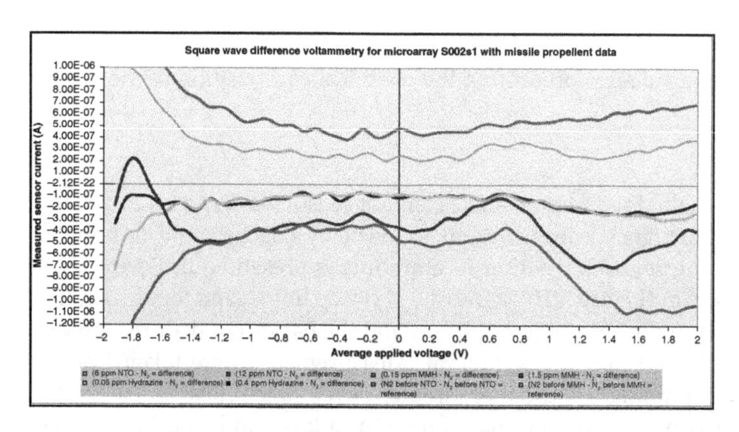

Figure 5.11 Graph of thick-film gas sensor sensitivities to rocket outgassing chemicals.

current passing through the electrochemical cell. As the voltage sensor output was analyzed, the location of the peaks along the voltage axis indicates which species are present, and the magnitude of peaks reflects the concentrations of each component. This information was used to identify and quantify chemical species in the environment.

5.3.3 Functionalized Nanowires – Zinc Oxide

Research was accomplished by AMRDEC and Dr Bruce Kim (the University of Alabama) to design, develop, and embed functionalized Zinc Oxide (ZnO) sensors onto remote-controlled vehicles [20]. The goal of the research was to provide an unmanned ZnO nanowire-based sensing platform for sensing explosive vapors under field deployment conditions. The crucial features of the sensor are its high sensitivity and selectivity. It was envisioned that the sensor would be capable of cost-effectively and accurately ascertaining complex explosives and variants of explosive vapors that are challenging to detect with conventional techniques. In order to achieve the aforementioned objective, ZnO nanowire-based sensor designs were implemented that incorporate the current state-of-the-art fluorescent explosive vapor detection methodologies. Results were to be semiconducting ZnO nanowire backbone to achieve nanomolar sensitivity.

5.3.3.1 Design Approach

In the laboratory, ZnO nanowires were successfully synthesized on sapphire substrates for the purpose of device fabrication. Sapphire substrates were chosen because of several inspiring advantages. First of all, sapphire substrates are insulating in nature, which makes them suitable for fabricating devices and helpful in eliminating current leakage through substrate. Secondly, sapphire substrates' lattice mismatch with ZnO is negligible that aids in controlled vertical growth of ZnO nanowires. An interlinked vertical morphology of ZnO nanowires is crucial for analyte permeation and helps to enhance the sensitivity of the device (as compared against thin-film structures).

Based on electrical testing, it was clear that the most suitable overall design concept for integrating into conventional electronics was a ZnO nanowire array device, as opposed to a single nanowire device, which would be expensive and time consuming to fabricate and integrate. Hence, the nanowire array device was utilized for further experimentation and sensor development. Figure 5.12 shows the schematic CAD design of the ZnO nanowire sensor.

It should be noted that silver paste was utilized to form electrodes on these nanowire chips (as the work function of silver and electron affinity of ZnO bulk is extremely close). This is crucial as it aids in formation of ohmic contact on the nanowire device in contrast to Schottky contacts. Ohmic contact helps to prevent loss of signal and facilitates smooth transport of electrons from the

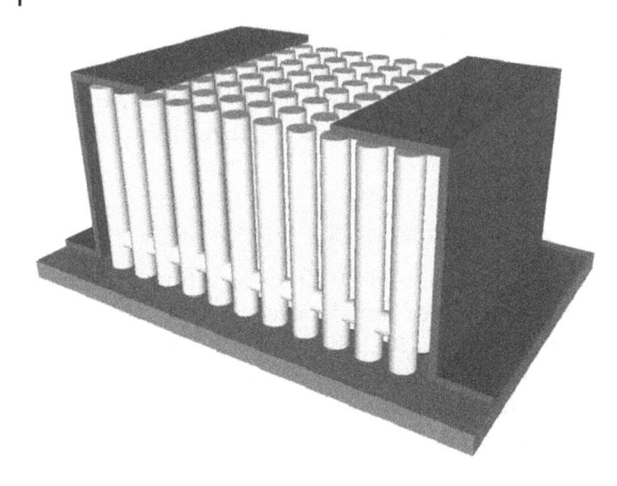

Figure 5.12 CAD illustration of the ZnO nanowire array device on sapphire substrate. Flat S-shaped electrodes consist of silver to form ohmic contact on ZnO nanowires (tube shaped).

semiconducting nanowire backbone to the external circuitry. This is crucial as the change in electrical current upon detection of p-nitrophenol analyte is in microamperes and must be detected reliably without any interference from the device itself.

5.3.3.2 Experimental/Test Setup

Figure 5.13 shows the actual prototype ZnO nanowire device based on the CAD design previously shown in Figure 5.12.

5.3.3.3 Results

The AMRDEC research team successfully demonstrated p-nitrophenol vapor detection with the functionalized ZnO nanowire sensing device through the

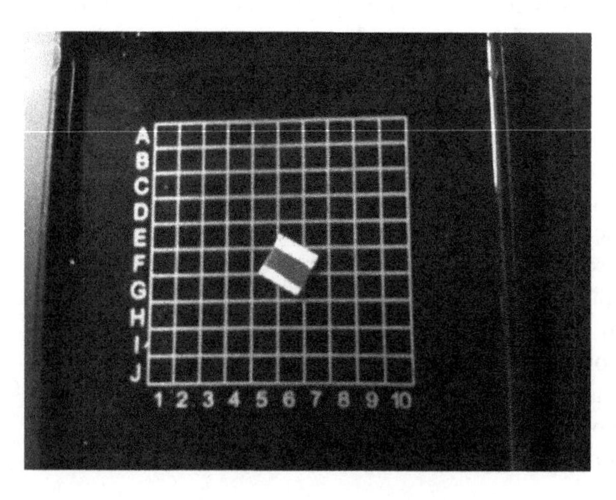

Figure 5.13 Actual prototype ZnO nanowire array device.

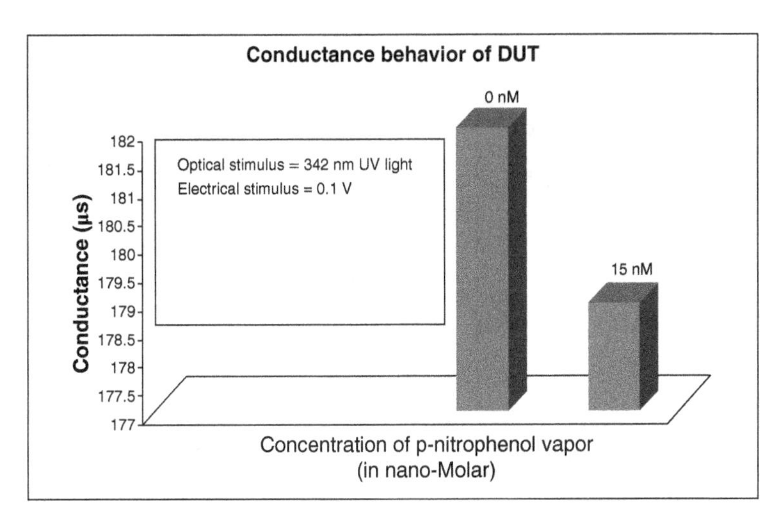

Figure 5.14 Amperometric detection of p-nitrophenol vapor using ZnO nanowire device.

optoelectronic route. When the phenylboronic acid (PBA)-functionalized sensing chip was subjected to a concurrent optical and electrical stimulus, it was observed that under the influence of p-nitrophenol vapor the sensing device was able to detect resistivity change under the trace concentration of 15 nM that transforms to parts per billion (ppb) level sensitivity under ambient conditions. These highly desirable results provide empirical evidence that using a concurrent approach, relying on both optical and electrical responses to provide sensitivity and selectivity, one can successfully determine vapor of p-nitrophenol under ambient and field conditions. The remaining task was just to extend this for the actual explosive vapor in a test facility. Figure 5.14 represents the results from amperometric detection of p-nitrophenol vapors using the functionalized PBA receptor/ZnO nanowire device.

5.3.4 Functionalized Nanowires – Tin Oxide

Researchers from AMDREC, in collaborations with Professor V. K. Varadan (University of Arkansas), developed a wireless prototype system for detecting/monitoring chemical analytes [19]. The chemical sensor portion of this technology (with resistance/ohmic output) is capable of detecting and identifying bioterrorism agents and hazardous gases. The sensing techniques utilize tin oxide (SnO)-coated nanowires as the sensing material. Vertically aligned gold (Au) nanowires were grown uniformly on the enhanced substrate. When the selectively grown nanowires reached the desired length and density, the sample was prepared from the deposition of the spin-coated tin oxide. The coating was achieved using a sol-gel technique in optimum conditions, preparing the SnO-coated nanowires for device-level fabrication.

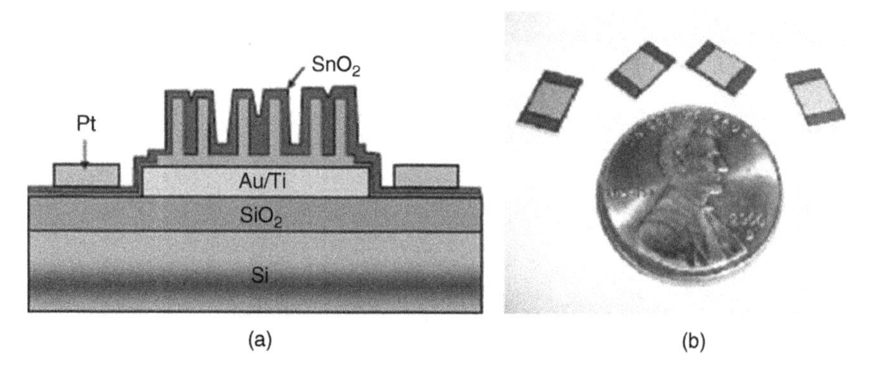

Figure 5.15 (a) Design concept schematic for the nanowire sensor. (b) Photo images of the nanowire sensor-fabricated devices.

5.3.4.1 Design Approach

The focus of the design approach is the development of vertically aligned nanostructures and the integration of nanosensors within wireless sensor networks. Device fabrication utilizes low temperature processes to facilitate the nanostructured tin oxide thin film on a silicon substrate. The conductance of the thin film is dependent on the concentration of those gases, which can potentially act as reducing agents for any adsorbed oxygen on the surface. Based on this design, the sensor is capable of detecting isopropyl alcohol, ethanol, water vapor, carbon monoxide (CO), methane, and other gases. The schematic configuration allowing the most acceptable sensitivity and response to chemical analytes is shown in Figure 5.15a. Figure 5.15b depicts the fabricated devices in comparison to the size of a penny [19].

5.3.4.2 Prototype Configuration/Testing

The nanowire sensors were integrated into a wireless system. The wireless system consisted of the wireless module and data acquisition board as shown in Figure 5.16. The operation of the wireless gas sensing system was initiated by the resistance change of the nanowire sensor resulting from the change of gas concentration. Afterward, the preamplifier converted resistances to voltage levels and amplifies the signals for conversion to a digital signal interpreted by analog to digital converters (A/DC) on the sensor node. The digitized signals were transmitted to the base station (laptop) through wireless transmission. Finally, the base station received the signals and displayed the data on the laptop monitor [19].

5.3.4.3 Results

The resistance changes of the wireless nanowire sensor were amplified via the signal processing and conditioning circuitry and sent wirelessly using a MICA2 wireless module. The data was recorded on a laptop, which is connected to another wireless node acting as a receiver. Figure 5.17 shows a set of data,

Figure 5.16 Image of the wireless sensing system, (a) wireless module and (b) data acquisition board.

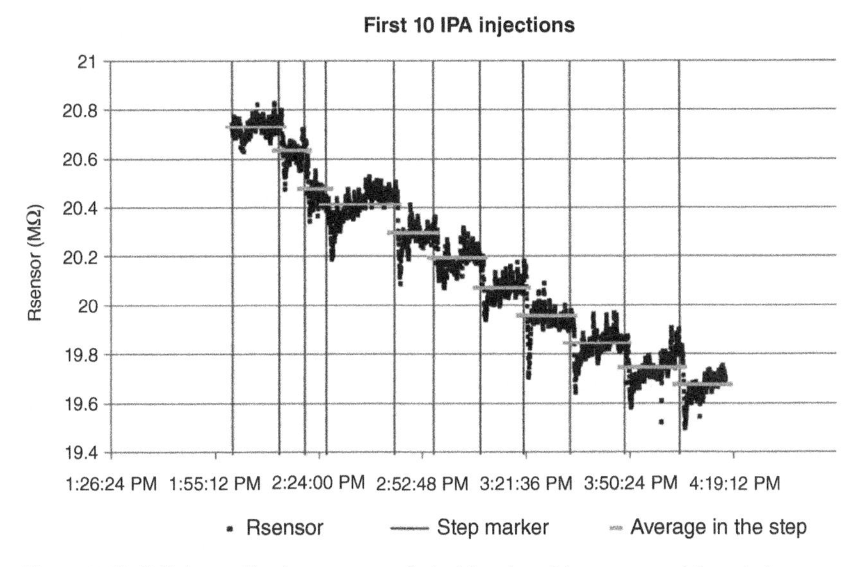

Figure 5.17 Full data collection run recorded with a tin oxide sensor and the wireless system.

which were transformed from raw voltage data to the resistance value of the tin oxide thin-film sensor. The graph clearly depicts the relationship of the sensor resistance to the concentration of isopropyl alcohol (IPA) [19].

5.4 Nanotechnology for Missile Health Monitoring

Nanotechnology for missile health monitoring became another of the nano-sensing technologies that was somewhat of a companion to the nano-based sensor technology for monitoring chemical ingredients associated with weaponry propulsion systems. The use of nanotechnology to monitor the health of missiles was initiated to develop nano-based elements to specifically insert in missiles to monitor degradation in propellant, warheads, and the overall weaponry structure. The research targeted reduced-size technology demonstrations that allowed for the monitoring of the durability of weapons. The utilization of missile health monitoring technology supplements the Army's ability to obtain embedded condition-based maintenance capabilities.

5.4.1 Nanoporous Membrane Sensors

While working with Dr Stuart Yin (Research Professor at Pennsylvania State University and founder of General Opto Solutions), the AMRDEC scientists developed and patented a technique that utilized a nanoporous alumina membrane as the filtering component of a sensing device that detected gases generated during the degradation of solid rocket propellant. The nanoporous alumina membrane not only filtered the gas molecules but also accumulated them such that the shelf life of propellant's stabilizer could be estimated. During the propellant's degradation process, gases such as CO, CO_2, NO, NO_2, and N_2O were released; the sensing method must be able to distinguish between the several molecules that were present. Furthermore, only some of the molecules released gases as indicators of the propellant's health. The rate of evolution of N_2O, for example, is a direct indicator of the available amount of stabilizer that remained in the propellant. Figure 5.18 shows the production of gas before and after the stabilizer, methyl-*p*-nitroaniline (MNA), is depleted. The timescale is on the order of years [19, 21]. As the N_2O gas is generated by the degrading rocket propellant, the stabilizer binds to the N_2O such that the N_2O is neutralized; however, once the stabilizer is depleted, the amount of N_2O increases exponentially.

5.4.1.1 Design Approach

The gas sensor developed used a nanoporous membrane to trap particles from rocket motor propellants. The average pore size in the nanoporous alumina is ~250 nm, which was verified by SEM (Figure 5.19).[22] The nanoporous membrane was able to trap particles that were larger than the pores while smaller gas particles could pass through.

5.4.1.2 Experimental Setup and Prototype Configuration

The experimental setup shown in Figure 5.20a simulates the gases passing from the motor through the nanoporous membrane. The mechanism for filling the gas collecting chamber is the micro pump connected to the gas sensor such that the ambient air is pumped out of the gas collecting chamber, which forcibly brings gas through the nanoporous membrane (Figure 5.20b). The nanoporous

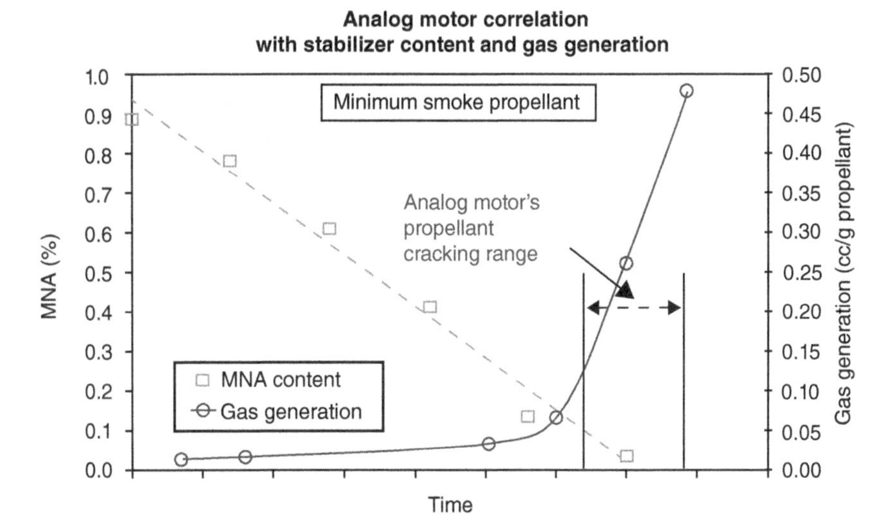

Figure 5.18 Plot of production of gas from degrading propellant shown in relationship to depletion of a stabilizer (MNA); time frame is on the order of years. After the graphical crossover point, gas generation increases exponentially (because stabilizer is near depletion).

Figure 5.19 SEM image of nanoporous alumina membrane; the average pore size is ~250 nm.

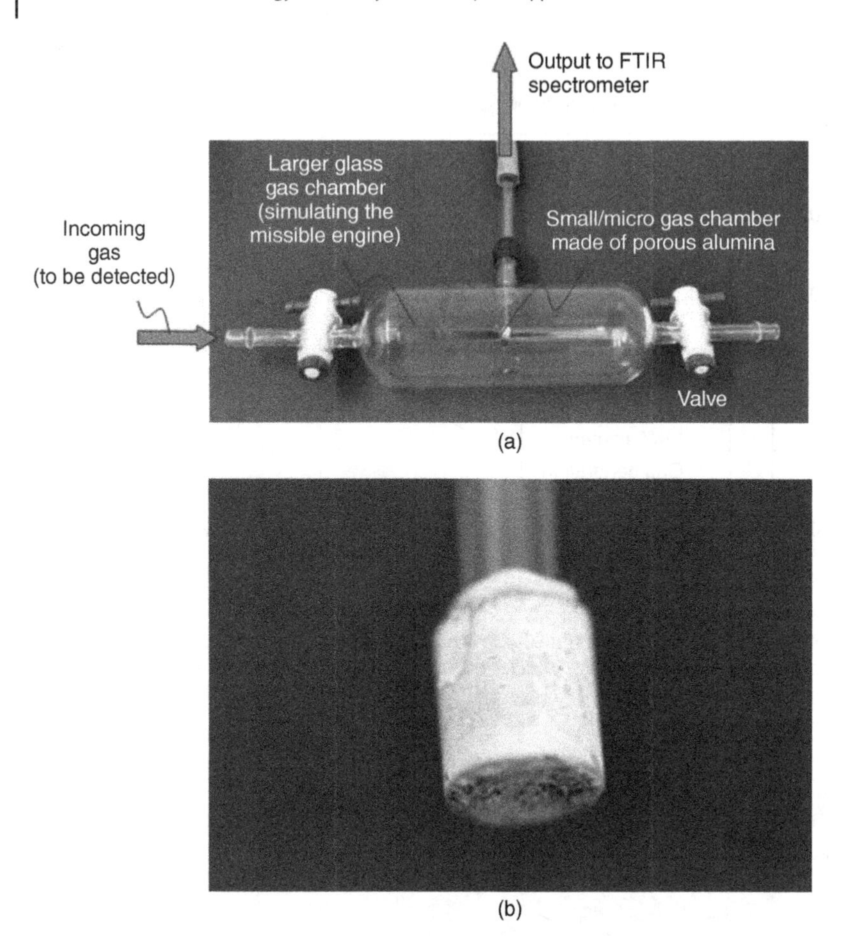

Figure 5.20 (a) The experimental setup for testing the nanoporous alumina membrane; (b) the prototype nanoporous membrane.

membrane filters the gases traveling from the tubular chamber into the gas collecting chamber when the ambient air is pumped from the gas collecting chamber by means of a micro pump (Figure 5.20a).

5.4.1.3 Results

In order to test the gas sensor concept, a gas mixture was allowed to flow into the chamber and was collected by the nanoporous membrane for analysis. Two types of related, less toxic gases (CO_2 and N_2O) were tested. Figure 5.21a shows the experimentally measured IR absorption spectrum displayed by the Fourier transform infrared (FTIR) spectrometer, which clearly shows major IR absorption peaks for both gases. These experimental results demonstrated the feasibility of the nanoporous alumina membrane sensor approach. In order

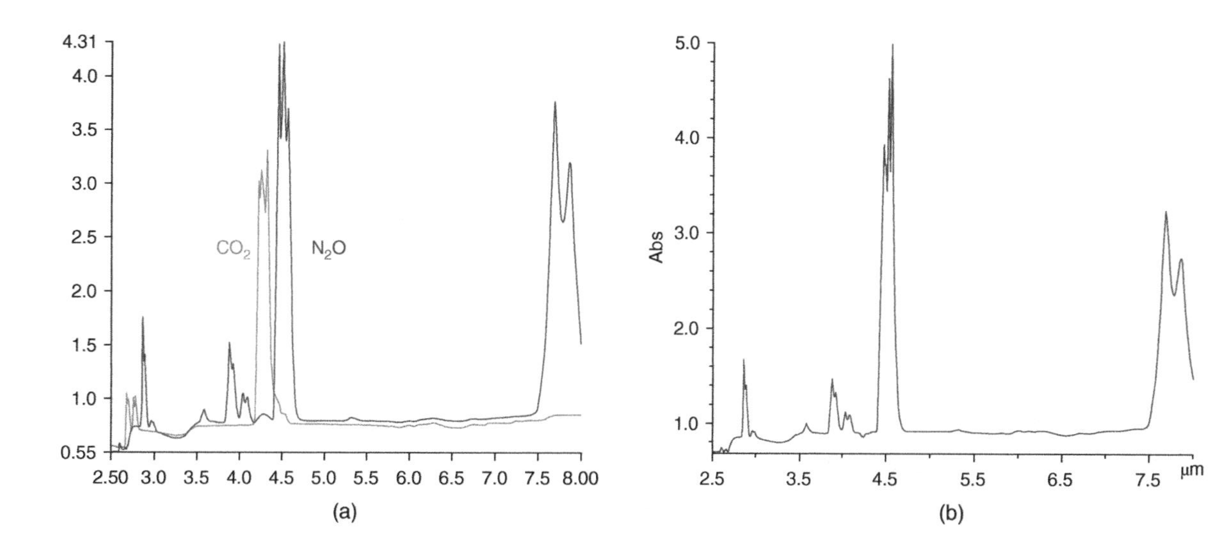

Figure 5.21 (a) IR absorption spectrum of CO_2 and N_2O collected by nanoporous membrane; (b) IR absorption spectrum of 1 ppm of N_2O.

to verify that detection can be achieved at the required level of sensitivity, the concentration of N_2O gas of the larger chamber was gradually reduced to 1 ppm. A Bacharach's trace gas analyzer was used to monitor the concentration of N_2O in the larger gas chamber. Afterward, the IR absorption level associated with the small gas chamber was measured, as shown in Figure 5.21b. Absorption peaks for N_2O could still be clearly observed, which confirmed the parts-per-million level of sensitivity and selectivity.

5.4.2 Multichannel Chip with Single-Walled Carbon Nanotubes Sensor Arrays

The sensing technique described in this section utilizes the electrical characteristics of CNTs to detect the presence of gas molecules. CNTs have been studied for use in sensor applications [23–25]. The gas particles that interact with the CNT cause a charge transfer, which changes the electrical resistance [23]. The viability of propellant can be detected based on changes in electrical resistance in the sensor.

5.4.2.1 Design Concept

Multichannel components, utilizing single-walled carbon nanotubes (SWCNT) sensor arrays (Figure 5.22), originally developed by researchers from the National Aeronautics and Space Administration (NASA) Ames Research Center (ARC), were used as the baseline sensor for AMRDEC's propellant degradation research [26, 27]. The SWCNT sensor array has approximately 32 individually functionalized CNT sensor elements for simultaneously sensing of up to 32 different analytes or targeted chemical agents (including propellant off-gassing).

5.4.2.2 Experimental Configuration

The CNTs are depicted in Figure 5.23 as the randomly shaped particle dispersed across gold electrodes (mainly the dark gray area of the figure). The CNT-based chemical sensor concept consisted of the interdigitated electrode (usually fabricated with gaps using photolithography) with purified SWCNT-based materials for chemical sensing (acting as a chemical resistor).

Figure 5.22 Multichannel sensor array (NASA-AMES).

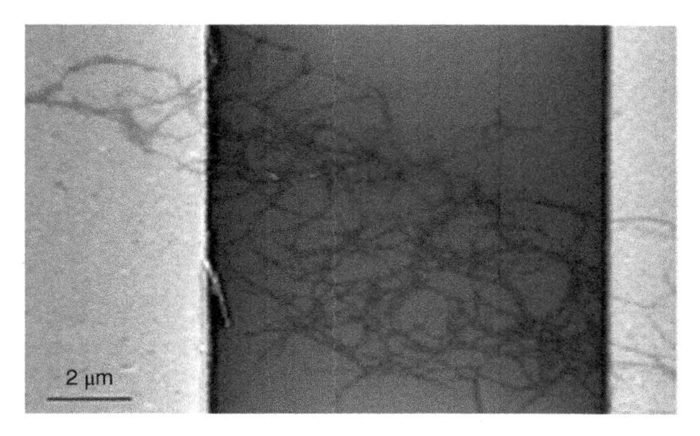

Figure 5.23 Single sensor element with carbon nanotubes across gold electrodes.

Table 5.1 Resulting CNT sensors and supplementary characteristics.

	Sensor 1	**Sensor 2**	**Sensor 3**	**Sensor 4**
Substrate	Steel	Steel	Steel	Foam
CNT mass	1.53 g	1.8 g	0.5 g	1.8 g
Epoxy mass	10.00 g	10.16 g	1.20 g	10.16 g
Hardener mass	9.975 g	10.010 g	1.200 g	10.010 g
Baseline resistance	1.550 kΩ	0.890 MΩ	2.600 kΩ	0.298 MΩ

5.4.2.3 Results

The developed sensors detect gases based on changes in the electrical properties of CNTs. Four sensors were fabricated for testing (Table 5.1). The sensors were a polymer composite using CNTs as the filler material. The substrates used for the sensors were steel and foam; the baseline resistances were measured for each sensor. The differences in resistance are a combination of resistance at the interfaces of the CNTs and within the polymer, which is an electrically insulating material.

5.4.3 Optical Spectroscopic Configured Sensing Techniques – Fiber Optics

The sensing technique, described in this section, uses noninvasive fiber optic spectroscopy to monitor the status of double-base propellants. It is a patented concept/collaborative initiative created by AMRDEC scientists and Dr Stuart Yin (Research Professor at Pennsylvania State University and founder of General Opto Solutions). In order to noninvasively determine (in real time) the concentration level of stabilizer in the double-base propellant, a novel noninvasive

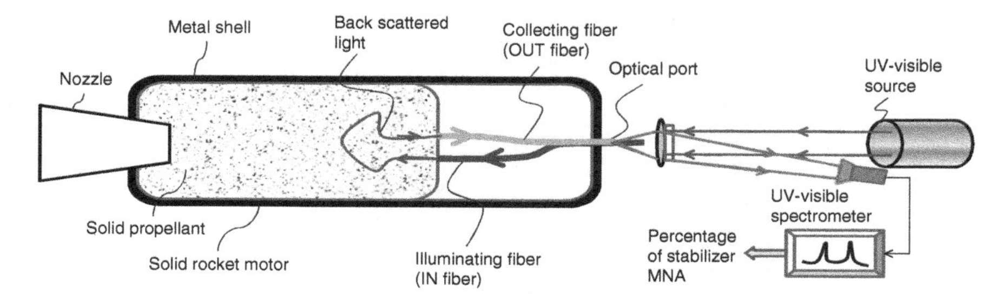

Figure 5.24 Conceptual sketch of fiber optic backscattering spectroscopic system used for measuring the concentration level of propellant stabilizer (MNA) and other ingredients inside a rocket motor.

inspection method (based on volume backscattering spectroscopy) was developed. The technique is based on the fact that the absorption property of propellant is directly related to its chemical composition. As the propellant ages, the chemical composition is constantly changing due to the chemical reaction between the stabilizers and outgas NO_x. By measuring the volume backscattering spectrum, the concentration level of the stabilizer can be detected; this in turn determines the status of the double-base propellant.

5.4.3.1 Design Concept Spectroscopic Sensing

The conceptual sketch of the fiber optic backscattering spectroscopic system is shown in Figure 5.24. The system can be used for measuring the concentration level of propellant stabilizer (2-methyl-4-nitroaniline, i.e., MNA) and other ingredients inside a rocket motor. The sensor head includes two optical fibers (an illuminating fiber and a collecting fiber), which are close to each other with a separation around a couple of hundreds of microns. Both fibers are in contact with the propellant surface. Thus, the collecting fiber can be used to collect the back scattered light from the illuminating fiber; the collected back scattered light is related to the absorption coefficient of the propellant material.

5.4.3.2 Experimental Approach/Aged Propellant Samples

Since nitrate ester propellants are widely used as low emission signature solid-state propellants, they were chosen for experimental analysis. Their stabilizers can be consumed during the time that the propellant is stored as ingredient of the missile's motor. A typical double-base solid-state propellant is composed of nitrocellulose (NC) and nitroglycerin (NG). The propellant studied was the M9 grain that is the minimum smoke and low signature brand produced by the ATK motor manufacturer. The stabilizer used was the typically preferred 2-methyl-4-nitroaniline (MNA). Figure 5.25 shows fresh and aged sample propellant.

As indicated in Figure 5.25, sample #0 is fresh propellant that was initially manufactured with the normal amount of 2% 2-methyl-4-nitroaniline (MNA).

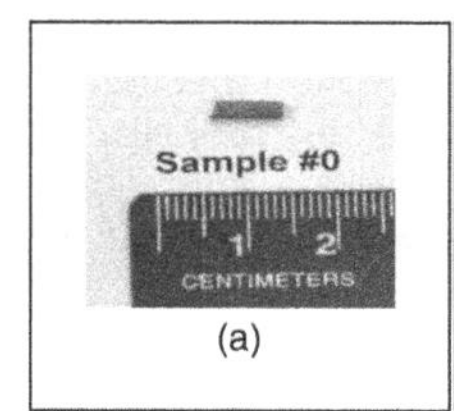

(a)

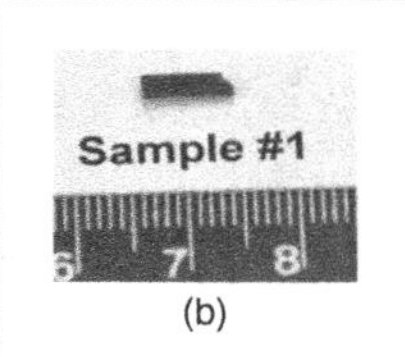

(b)

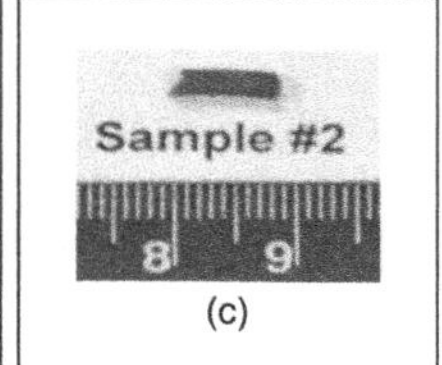

(c)

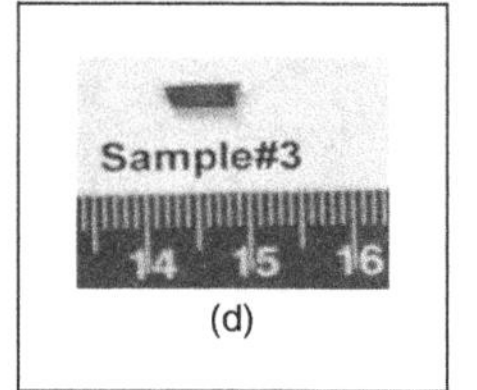

(d)

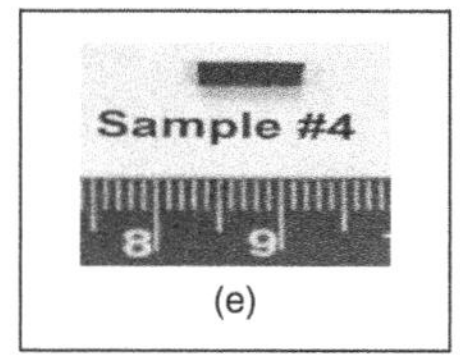

(e)

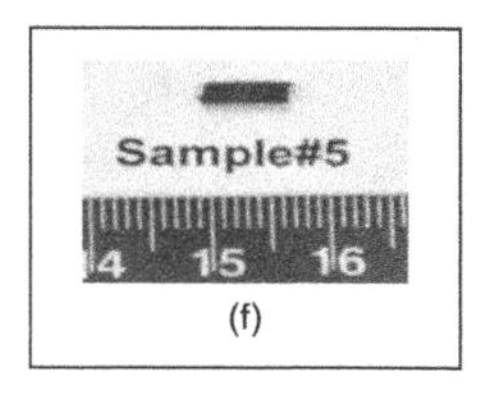

(f)

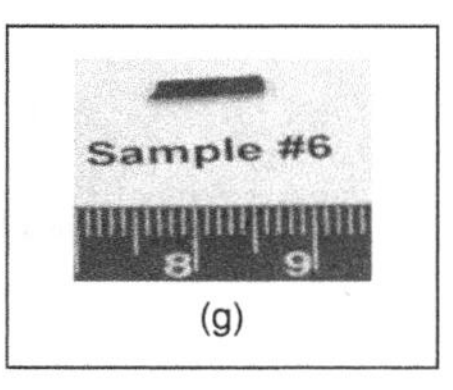

(g)

Figure 5.25 Photos of sample propellant that is fresh/un-aged (sample #0) and accelerated aged for up to 25 years. (a) Propellant un-aged with 2% MNA. (b) Propellant aged 4.2 years, MNA depleting MNA. (c) Propellant aged 8.2 years, MNA depleting. (d) Propellant aged 12.6 years, MNA depleting. (e) Propellant aged 16.8 years, MNA depleting. (f) Propellant aged 21.0 years, MNA depleting. (g) Propellant aged 25.0 years, MNA depleting.

After undergoing accelerated aging (to simulate long-term storage), the other samples (including sample #6) show the resulting discoloration of the same sampled lot of propellant after 25 years of replicated aging. Resulting aging is near equivalent to the normal shelf life of the propellant.

5.4.3.3 Results from Absorption Measurements

The absorption coefficient of MNA has strong absorption in the UV/blue spectral range [28]. Determining the concentration of MNA by measuring the reflection, back scatter, or transmission spectrum within the UV/visible/IR region of the propellant is possible [22]. The higher the concentration of MNA, the larger the absorption within this spectral range will be. On the other hand, the absorption spectrum at other wavelengths (e.g., green or red), the absorption of MNA will be compatible to other compositions, including NC, NG, and carbon. Thus, one can determine the concentration of MNA by measuring the spectrum at the UV/visible/IR spectral range. Figure 5.26 shows measured results indicating absorption spectrums for samples of fresh propellant (un-aged, sample #0) and the varying sample propellants accelerated aged up to 25 years.

In Figure 5.26, the decrease in magnitude of scattered light is due to the depletion of stabilizer. After approximately 25 years of storage (i.e., 6 weeks of accelerated aging at 75 °C), the stabilizer is depleted from 2.0% concentration level (fresh sample) to 0.5% concentration level. The propellant can become unstable/unusable after stabilizer depletes to less than 0.5%.

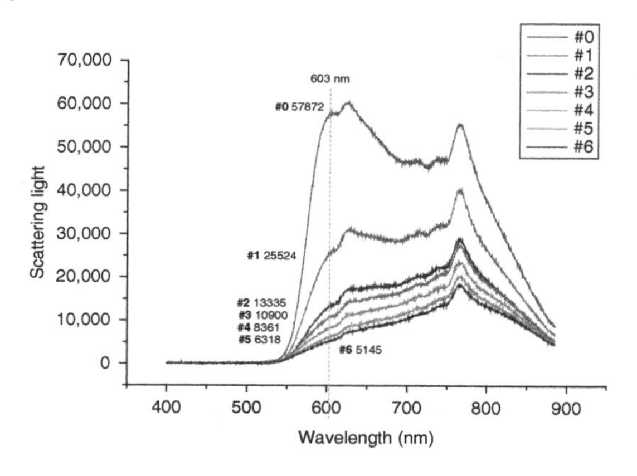

Figure 5.26 Measured results showing absorption spectrums for samples of fresh propellant (un-aged, sample #0) and sample propellants accelerated aged up to 25 years.

5.5 Nanoenergetics – Missile Propellants

Missile propellants emit gases that are harmful to humans, can damage sensitive missile components, and can potentially cause deterioration to the missile structure. Nanoenergetic-based propellants offer a solution to a safer and more environmentally friendly missile propellant. The development of a minimum signature solid propellant with increased ballistic and physical properties to meet insensitive munition requirements is an area of research at AMRDEC led by scientist Larry Warren.

Propellants doped with CNTs were investigated to meet the Army's insensitive munitions (IM) requirement. Research involved experimenting with both multiwalled carbon nanotubes (MWCNTs) and SWCNTs to test the similarity in performance to rebar when added to a propellant formulation. The burning rate of the propellant was also measured by conducting heat below the burning surface of the propellant grain. These two properties are important and potentially beneficial reasons to evaluate CNTs in propellant formulations. Increasing the burning rate of minimum signature propellants is of great interest to the propellant industry, with a goal of doubling the burning rate.

5.5.1 Multiwall Carbon Nanotubes

MWCNTs were analyzed to determine optimum chemical and mechanical properties for insertion into missile solid propellants. The MWCNTs were substituted for the carbon powder in the propellant grains. The stress and strain, as well as the burn rate, were among the properties characterized in the research.

5.5.1.1 Design Approach

CNTs are available with either the hydroxyl (—OH) or carboxyl (—COOH) functional group attached to the carbon structure. The propellant mechanical

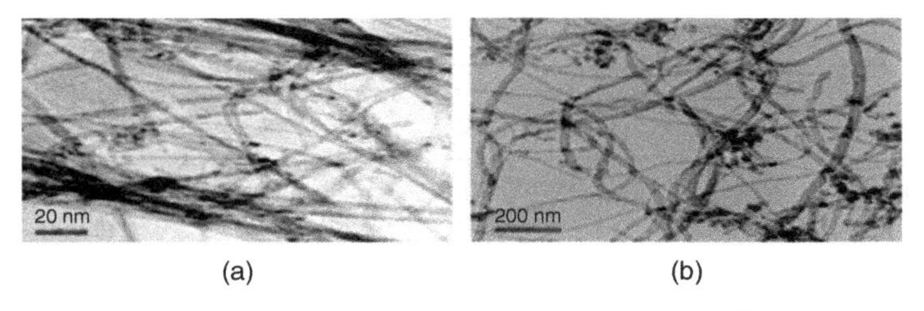

Figure 5.27 SEMs showing the structure of CNTs, multiwalled (a); and single-walled (b).

properties were further enhanced by the addition of the —OH or —COOH functional group, and the bonding of the CNTs to the propellant binder network. The polymer binder network of double-base propellants was primarily formed via hydrogen bonding occurring during the plasticization of nitrocellulose with nitrate esters. The most common nitrate esters used in double-base propellants are nitroglycerin (NG) and butanetriol-trinitrate (BTTN). Figure 5.27 is an SEM image of the structure of the CNTs.

5.5.1.2 Experiment

Researchers from AMRDEC exploited the properties of CNTs to improve the physical and performance properties of a potential castable double-base minimum signature propellant formulation. Castable double-base propellant technology was evaluated extensively in a Joint Insensitive Munitions Technology Program (JIMTP) for potential next-generation tactical missile applications with an emphasis on achieving insensitive munitions objectives [29–31]. The general double-base propellant formulation is a combination of nanoparticle carbon powders, bismuth and aluminum compounds utilized to control burning rate and burning rate stability. CNTs were substituted for the carbon powder and the properties of the propellants were compared. Propellant burning rate and rate exponent were determined. A burning rate of at least one inch per second at 1000 psi for minimum signature propellants is greatly desired by the military. Achieving the performance objectives and meeting the insensitive munitions requirement is a technical challenge.

In addition to evaluating the burning rate properties of CNTs, the propellant physical properties were assessed. The potential to enhance propellant physical properties (such as stress, strain, modulus, and cross-link density) is an added benefit of the —OH or —COOH functional group attached to the carbon structure of the CNTs.

Several CNTs were evaluated for their effect on burning rate and rate exponent. It is well known that carbon powder particle size greatly influences propellant burning rate and rate exponent. The most commonly used carbon powder has an average particle size of 70 nm. Carbon powders of even smaller average particle size were evaluated for future propellant applications.

Table 5.2 Evaluation of multiwalled CNT results.

CNT	Rate (in./s) 1000 psi	Rate (in./s) 1500 psi	Rate (in./s) 2000 psi	Rate exponent
A. Baseline: 70 nm carbon powder	0.40	0.49	0.55	0.45
B. CNT 30–50 nm dia, 0.5–2 μm	0.32	0.39	0.51	0.7
C. CNT, 10–20 nm dia, 0.5–2 μm	0.30	0.40	0.54	0.85
D. CNT 10–20 nm dia/10–30 μm	0.33	0.42	0.57	0.80
E. CNT 8 nm dia, 10–30 μm	0.35	0.42	0.54	0.65
F. –COOH 95% MWCNT, 50–80 nm dia/10–20 μm	0.34	0.41	0.46	0.43
G. –OH 90% MWCNT, 8–15 mm dia	0.30	0.37	0.43	0.50

5.5.1.3 Results

The results from the evaluation of the multiwalled CNTs are outlined in Table 5.2. The MWCNTs (items B, C, D, and E) all reduced the propellant's burning rate and significantly increased burning rate exponent to performance levels lower than the objective values. MWCNTs, with either the —OH or —COOH functional group (F, G), also lowered burning rate, but the burning rate exponent is comparable to the baseline formulation containing the 70 nm carbon powder. From these results, the multiwalled CNTs with either the —OH or —COOH group is considered to be more promising for possible further evaluations.

5.5.2 Single-Wall Carbon Nanotubes

The second group of CNTs evaluated was the single-walled with —OH or —COOH functional group. The SWCNTs are expected to give comparable results to the carbon powders with the additional mechanical property enhancement provided by the functional groups.

5.5.2.1 Design Approach

Researchers substituted SWCNTs for the carbon powder and the properties of the propellants were compared. Similar to the approach utilized with the MWCNTs, the propellant burning rate and rate exponent were determined. A burning rate of at least one inch per second at 1000 psi for minimum signature propellants is greatly desired by the military.

The burning rate data of the SWCNTs are outlined in Table 5.3. The propellant burning rate at the 0.7% (B, D) level was slightly reduced from the baseline, but the burning rate pressure exponent was less than the baseline. A reduced rate exponent is always desirable for tactical missile applications. However, at

Table 5.3 Evaluation of single-walled CNT results.

Carbon nanotube (CNT)	Rate (in./s) 1000 psi	Rate (in./s) 1500 psi	Rate (in./s) 2000 psi	Rate exponent
A. Baseline, 70 nm carbon powder	0.42	0.53	0.61	0.43
B. —OH SWCNT 60%, 1–2 μm, od @0.7%	0.35	0.45	0.49	0.35
C. —OH SWCNT 60%, 1–2 μm, od @1.0%	0.41	0.46	0.51	0.30
D. —COOH SWCNT 60%, 1–2 μm, od @0.7%	0.35	0.45	0.49	0.35
E. —COOH SWCNT 60%, 1–2 μm, od @1.0%	0.40	0.47	0.52	0.40

the 1% level (C, E) the propellant burning rate was the same as the baseline with a further significant reduction of the burning rate exponent.

5.5.2.2 Experiment

The results indicated that both types of CNTs could be potential candidates for future propellant applications to replace carbon. The lower burning rate exponents with the CNTs were very encouraging. The results were from small 300-g propellant mixes. Scale-up to larger mixes may yield different results.

In addition to enhancing propellant burning rate properties, CNTs with the —OH or —COOH functional groups were expected to enhance propellant mechanical properties. The preliminary results are outlined in Table 5.4. Both the propellant stress and strain properties were improved, but the propellant modulus property did not show any improvement. Additional experiments, with various mixes, are needed to make a determination on the ability of CNTs with the —OH or —COOH functional groups to improve

Table 5.4 Initial comparison of mechanical properties of CNTs in propellant with the baseline.

Carbon nanotube (CNT)	Stress (psi)	Strain (%)	Modulus (psi)
Baseline, 70 nm carbon powder	141	44	132
OH SWCNT 60%, 1–2 μm, od @0.7%	161	74	140
OH SWCNT 60%, 1–2 μm, od @1.0%	170	77	147
COOH SWCNT 60%, 1–2 μm, od @0.7%	204	80	156
COOH SWCNT 60%, 1–2 μm, od @1.0%	156	66	157

propellant mechanical properties. One theory is that both the —OH and —COOH compete for the curing agent of the formulation, which influences the overall propellant properties, especially the propellant modulus. Future efforts include a thorough investigation of how the CNTs change the cure ratio of the formulation. Small changes to the formulation are common when new ingredients are added and during each major scale-up. Lastly, methods to determine if the small improvements to mechanical property values can be further increased were evaluated.

5.5.2.3 Results

From the preliminary results, both the —OH or —COOH functional SWCNTs may be candidates to replace carbon powder in double-base propellant formulations. There was not a significant drop in propellant burning rate when either was used, but the rate exponent was further reduced. Propellant rate exponents approaching a "plateau" are very desirable for double-base propellants [31]. Further evaluations would have to be performed with scale-up to larger mixes to exploit this property of —OH and —COOH functional CNTs. The largest challenge to overcome for considering CNTs for propellant applications is the cost. CNT cost is several times the cost of carbon powders. However, the cost of CNTs should decrease with the production of larger quantities.

5.6 Nanocomposites for Missile Motor Casings and Structural Components

Nanocomposites integrated into the missile casing can potentially provide a method to reduce the electronics overheating and structural damage of the missile. Thermal and vibrational methods were evaluated to increase the structural integrity of missile platforms.

5.6.1 Thermal Methods

Scientists and engineers, led by J. Keith Roberts, at AMRDEC developed solutions for thermal management within composite missile structures. Conventional tactical missile airframes are constructed using aluminum, which has moderate strength and reasonably low density to ensure electronics survivability, as well as attractive thermal conductivity and high specific heat capacity to meet the thermal requirements. Fiber-reinforced composites (FRCs), which offer many benefits from a structural standpoint due to their high specific strength and stiffness, were considered for the existing and future tactical weapons. FRCs exhibit high specific strength and stiffness that make them extremely attractive in many structural applications. Unfortunately,

Figure 5.28 SEM of multiwall carbon nanotube array used as a thermal interface.

the insulative nature of the polymer matrix makes them unattractive in applications where thermal management is needed. Various material considerations have been explored to improve both through-thickness and in-plane thermal conductivity to enhance the thermal performance of structural composites. AMRDEC's work focused on a number of methods for improving overall thermal management capability in FRC structures. Finite element models were used to explore various material solutions as applied to a representative missile airframe structure. Experimental data was acquired using a test-bed developed at the Composite Structures Lab to compare various fiber-reinforced structures against a baseline aluminum structure [32].

Research was conducted to determine the use of multiwall carbon nanotubes for thermal management and improved electrical conductivity in composite materials used in missile applications (Figure 5.28). Aligned multiwall nanotube mats were integrated into a filament-wound multilayer cylinder that represents a motor case analog. The filament winding was performed at the Pennsylvania State University Applied Research Laboratory. The nanotube mats were continuous vertically aligned nanotubes infiltrated with epoxy. The continuous nanotubes are much more conducive to phonon transport than random dispersions of nanotubes that require phonons to jump across interfaces. Epoxy with dispersed nanotubes was also integrated into the laminate to determine electrical conductivity effects. The motor case analog was tested for large-scale thermal and electrical conductivity and low-energy impact effects. The research effort was a collaboration between AMRDEC, Materials Sciences Corporation (MSC), and The University of Kentucky Center for Applied Energy Research [32, 33].

Tests were performed to demonstrate new approaches to thermal management in fiber-reinforced polymer composites including integrated heat spreader approaches and through thickness thermal pathways. Experimental results were obtained from a series of tests that were conducted to characterize thermal interface behavior. A number of thermal interface materials were integrated into a test setup and interfacial temperatures are recorded. Finite element analysis was used to further understand the thermal transport and its effect on overall system performance.

5.6.2 Vibrational Methods

The development of nanocomposite structures to provide enhanced vibratory protection of the weaponry's inertial measurement unit (IMU – the key component of guidance and navigation systems aboard airborne weapons) were also investigated by AMRDEC engineer, J. Keith Roberts. In order to accomplish this vibratory protection effort, AMRDEC researchers developed advanced composite materials that provided high-frequency damping for the IMU's packaging.

AMRDEC, in collaboration with Materials Sciences Corporation (MSC), developed and validated analytical models to characterize the high-frequency vibration response of nanocomposite materials. Nanofiller-loaded material was characterized via dynamic mechanical analysis (DMA) and vibration testing of material specimens fabricated by MSC/AMRDEC. The target application for this technology is a mounting structure for instrumentation subject to interference from high-frequency (3–20 kHz) mechanical vibration as a result of its operational environment.

5.6.2.1 Design Approach

Two material systems were identified as desirable for construction of the instrument mount: (i) IM7/8552 unidirectional prepreg and (ii) IM7/PSS compression modeled thermoplastic. The IM7/8852 utilizes conventional hand layup fabrication methodologies, unlike the more advanced IM7/PPS system, which requires techniques such as fiber placement or compression molding. Nanotubes were sprayed between layers of the IM7/8852 prepreg during the layup process. Other programs have demonstrated the successful even dispersion of the nanotubes. A similar approach is currently underway for the compression molded IM7/PSS component. This process has not been previously demonstrated; therefore, both approaches to achieve even distribution of the nanotubes in the compression molding process were evaluated.

Material testing under this program was conducted in two stages to evaluate matrix-level effect (doped polymer) and laminate-level effect (doped polymer with continuous carbon fiber reinforcement):

5.6.2.2 Experiment

Testing of three replicates each of nanotube doped and nondoped 8552 epoxy and PPS thermoplastic was conducted to determine if nanotubes provide high-frequency vibration attenuation. During this level of the study, DMA and coupon-level beam testing were conducted. DMA testing requires the materials to behave as thermorheologically simple materials; validating the application of the time–temperature superposition (TTS) principle [34]. This allows a correspondence between tests over a low-frequency bandwidth and temperature range to be used to develop the high-frequency response of the material. Coupon-level testing was conducted using a mechanical shaker over the entire desired frequency bandwidth (20 kHz) to compare with the TTS assumption. The coupon-level testing was coupled with an analytical solution based on higher order plate theory. This approach assumed a displacement field that results in a parabolic through thickness shear distribution in the specimen. The shear was zero on the free surfaces of the specimen. The analytical model is preferable over a finite element (FE)-based approach because at higher order modes of vibration, the FE-based approach requires meshes that become computationally impractical. At high frequencies mode interaction is common and distinction of a single mode of vibration to relate a modal damping ratio to the material damping factor is a technically difficult task; coupling experimental analysis with analytical models will add a measure of justification to bandwidth averaging techniques that will require understanding of the complex modal behavior of the structure at high frequency.

5.6.2.3 Results

Based on the results of the matrix level testing, DMA and coupon testing of doped and nondoped IM7/8552 laminates and IM7/PPS specimens were conducted. The testing was contingent on two factors:

- Does the addition of nanotubes provide increased high-frequency damping for the doped matrix samples?
- Do the DMA TTS results agreed with the coupon tests?

The first factor determines which materials are carried forward as feasible material systems, and the second factor determines if the DMA TTS method is valid. If the DMA TTS method is demonstrated for these material systems, the time and effort required to generate test data is greatly reduced.

Initial trials were performed using nanotube-doped epoxy to determine the effect at the matrix level and the applicability of the DMA TTS method [35]. These coupons were fabricated at Rensselaer Polytechnic Institute (RPI) and are pictured as follows (Figure 5.29). Beam samples for classic modal analysis and DMA samples were provided to MSC.

Shear modulus and loss factor data were collected on the RPI DMA samples from 0.3 to 10 Hz over the temperature range of −50 to 50 °C in increments of 5 °C. These data sets were then shifted in the frequency domain using the Williams, Landel, and Ferry (WLF) equation. The Havriliak–Negami (HN) equation, which describes the complex shear modulus as a function frequency, was fit to the shifted data. The analytical equation results in values of the shear modulus and loss factor of the material at a given temperature across the entire frequency domain.

Figure 5.30 shows the RPI samples loss factor as a function of frequency as a result of a TTS analysis of the experimental DMA data. All of the samples doped with nanotubes exhibited increased damping at high frequency with respect

Panel ID	Date/Time	Vendor ID	Origin	Material
911	9/15/2010 9:55:28 AM	RPI_EPOXY_01	RPI	Fibre Glast System 2000 Epoxy Resin
910	9/15/2010 9:55:10 AM	RPI_0.4%_MWNT_02	RPI	Fibre Glast System 2000 Epoxy Resin
909	9/15/2010 9:54:58 AM	RPI_0.2%_SWNT_01	RPI	Fibre Glast System 2000 Epoxy Resin
908	9/15/2010 9:54:49 AM	RPI_0.4%_SWNT_01	RPI	Fibre Glast System 2000 Epoxy Resin
907	9/15/2010 9:54:32 AM	RPI_0.2%_MWNT_01	RPI	Fibre Glast System 2000 Epoxy Resin

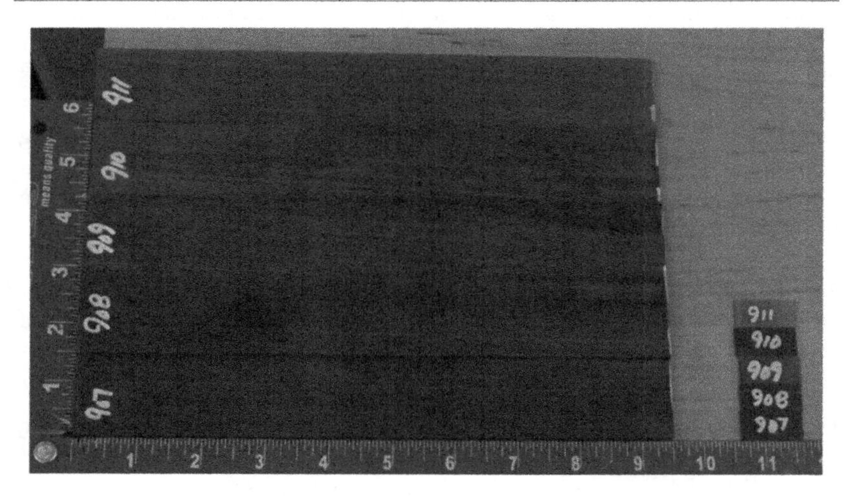

Figure 5.29 Carbon nanotube doped epoxy specimens from RPI.

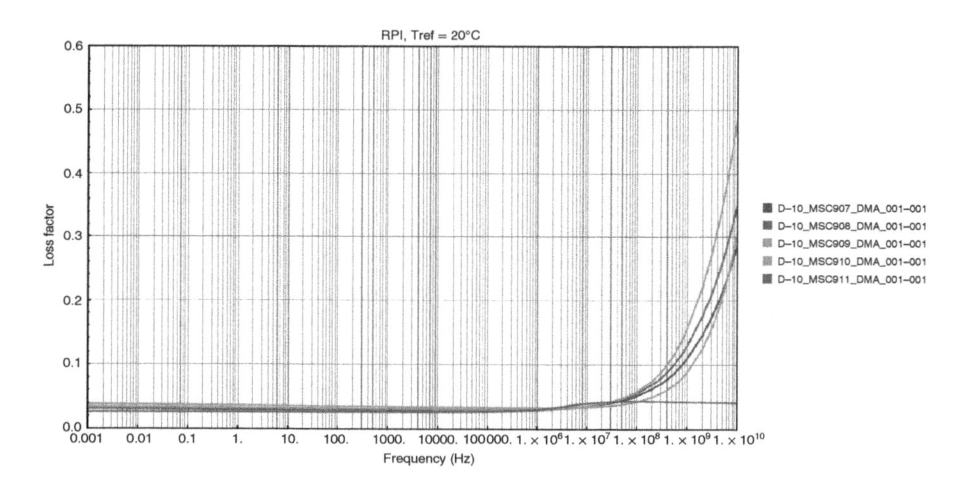

Figure 5.30 RPI samples loss factor as a function of frequency.

to the undoped sample; however, it is not in the bandwidth of interest. These material systems did not result in DMA data sets that are typical of the TTS application making use of the WLF and HN equations. Additional research of interest includes determining if other functional forms are more appropriate for TTS analysis of this material system; in addition high-frequency modal analysis is still required to support these findings.

The technical challenges associated with developing and validating analytical models to characterize the high-frequency vibration response of nanocomposite materials are numerous. Various analytical models based on classical micromechanics are extended via linear viscoelastic theory to evaluate frequency-dependent damping. Understanding of the nanotube and matrix properties and spatial distribution of the nanotubes is of primary importance to the application of these models. Since it is currently not possible to develop this data for nanotubes, test data from doped and undoped samples is used to infer this information. Application of this modeling process is not difficult; generation of the analysis samples and accurate test data to date has been exigent. Dispersion of nanotubes in the material systems during fabrication has been technically challenging, a limitations on matrix viscosity during processing has governed the practical limits of nanotube concentration. In addition, uniformity of dispersion and repeatability are also issues associated with nanotube-doped material systems.

The most challenging aspect of this effort was the generation of test data for the material system in the bandwidth of interest. Classical modal dynamics methods of beams are troubled by system resonances and mixed modes, making determination of damping properties at a specific frequency difficult. DMA data and TTS analysis offered a possible solution to the problems associated with modal testing, but the method is unproven and may require significant research to determine the appropriate functional forms for proper application of the TTS. The combination of a high-frequency modal shaker and the analytical dynamical system models used by the DMA to compute complex modulus data may provide a method damping investigation at high frequency that could be used to validate TTS analysis in the bandwidth of interest.

5.7 Nanoplasmonics

Nanoplasmonics, which is plasmonics involving nanofabricated systems, is the science of the detection, production, and manipulation of nanoscale waves that pass through the field of electrons existing on the surface of metals. Plasmonics can be referred to as information transfer in nanoscale structures by means of surface plasmons, which are coherent electron oscillations that propagate along the surface of metallic and certain dielectric materials. Plasmons are quanta of plasma oscillations that are created when light is incident on the surface of a metal under specific circumstances. They can theoretically encode much more information than what is possible for conventional electronics. Scientists, led by Dr Henry Everitt, at AMRDEC investigated the plasmonic properties of metallic nanoparticles.

5.7.1 Metallic Nanostructures

Metallic nanostructures play an important role in the emerging field of plasmonics. Due to their remarkable plasmonic ability to enhance local electromagnetic fields, the development and exploitation of metallic nanostructures have made great progress during the past decade. Nanofabricated systems offer unprecedented opportunities in sensing applications due to the enhanced sensitivity of light to external parameters. Ultraviolet (UV) plasmonics were considered by the AMRDEC research team to enhance Raman spectroscopy in chemical sensing.

Metallic nanostructures used as plasmonic sensors in biosensing, spectroscopy, and so on are usually made of silver and gold and operate in the visible and near-infrared regions. Since the nano-based chemical detection sensors developed by the research team at AMRDEC respond in the UV part of the spectrum, gallium nanoparticles deposited on sapphire were investigated as alternative substrates to enhance Raman spectral signatures.

Size-controlled gallium nanoparticles deposited on sapphire were explored as alternative substrates to enhance Raman spectral signatures [36]. Ga nanoparticles (NPs) were grown using a simple, molecular beam epitaxy-based fabrication protocol. The first demonstrations of enhanced Raman signals from reproducibly tunable self-assembled Ga nanoparticles were done using visible wavelength lasers. Nonoptimized aggregate enhancement factors of ~80 were observed from the substrate with the smallest Ga nanoparticles for CFV dye solutions down to a dilution of 10 ppm.

Ga NPs can be tuned into the deep ultraviolet (UV), owing to its high plasma frequency, $\omega p = 14\,\text{eV}$, with demonstrated tunability over a broad spectral range from 0.75 to 6.5 eV – a significant advantage over the limited range achievable by both Ag and Au, especially for simultaneous UV Raman/PL spectroscopy [37–39]. Moreover, substrate-supported Ga NPs exhibit no postdeposition aggregation or attendant modification of the plasmon; therefore, the plasmon resonance is stable and reproducible. The Ga SPR remains stable and protected once oxidized even after over a year of air exposure. Conversely, Ag oxidizes excessively and becomes quenched within 36 h of air exposure [40]. In addition, the Ga plasmon mode's remarkable thermal stability foreshadows Ga's advantageous use for applications in thermally harsh and diverse environments.

The Raman enhancement and longevity of Ga NPs substrates were tested using the standard Raman dye cresyl fast violet (CFV) [36]. By adjusting the deposition time at a fixed beam flux, scientists modified the mean NP diameter, and tuned the surface plasmon resonance of three Ga NPs/sapphire substrates to 2.9, 1.96, or 1.58 eV [36, 37, 41].

5.7.2 Gallium-Based UV Plasmonics

Solid gallium, which has Raman-active vibrational modes, exhibits SERS [42]. Everitt and the AMRDEC research team demonstrated that gallium is a highly promising and compelling material for UV nanoplasmonics through synthesis of size-controlled nanoparticle arrays [7].

Gallium nanoparticles (Ga NPs) were deposited on sapphire substrates to enhance Raman spectral signatures [7]. Enhanced Raman signals were observed for these Ga NPs, however, the enhancement factor was smaller than reported for Ag nanostructures. It is important to emphasize that the enhanced Raman signals were observed for nonoptimized NP density and size distributions. More uniformly sized NPs or more controllable NP densities via tremendous precision could produce self-assembled plasmonic Ga NPs for SERS extended into the ultraviolet [7, 36].

5.8 Nanothermal Batteries and Supercapacitors

A thermal battery is a heat-activated battery consisting of a combination of thermal cells. A typical battery contains a stack of cells, each of which having its own anode (lithium), cathode (chromates or sulfides), electrolyte (mixture of lithium chloride and potassium chloride), and a heat pellet. The battery is an excellent power supply for applications needing to perform in harsh environments, such as in military applications.

Thermal batteries, which are predominantly used in guided missile systems, rockets, torpedoes, and so on, have long shelf life, high peak power density, reduced startup time, and require low maintenance. They are activated by thermal energy supplied by a pyroheater. As missiles are required to travel greater distances, the thermal battery designed for missile use needs to provide a significantly higher level of power to the missile's electronics, actuation components, the fuse, and so on, while fitting in a compact volume.

Nanomaterials have the potential of producing batteries with much improved performance characteristics (e. g., higher voltages, increased current densities). Using nano-sized materials for battery components could significantly reduce battery size and increase operating life and power density. The state-of-the-art battery technologies do not meet the Army's requirement for performance (capacity, mission life, and internal resistance) and size/weight constraints. The AMRDEC research team, led by Mark Temmen, conducted a program to develop and demonstrate supercapacitor components for power systems of quick reaction munitions using the latest developments in nanoscale materials and coatings. Another objective of the research was to develop (design, fabricate, test, and demonstrate) novel nanostructured anode and cathode materials to improve thermal battery technology for longer range munitions [43].

The Army goals for miniature power sources are a 10× increase in power density, lower internal resistance for supercapacitors, and a 3× increase in energy density. To provide a substantial increase in thermal battery specific energy, new cathode and/or anode materials and electrolytes must be developed. These innovative materials will potentially provide higher specific capacity at higher operating voltages across the range of discharge rates typically required of thermal batteries. The combination of higher specific capacity and higher operating voltage translates directly to higher specific energy at the battery level [44].

In collaboration with multiple industrial partners, AMRDEC fabricated silicon nanowire (SiNW) anode materials for thermal battery solutions as seen in Figure 5.31. The novel nanostructured anode materials enhanced electronic

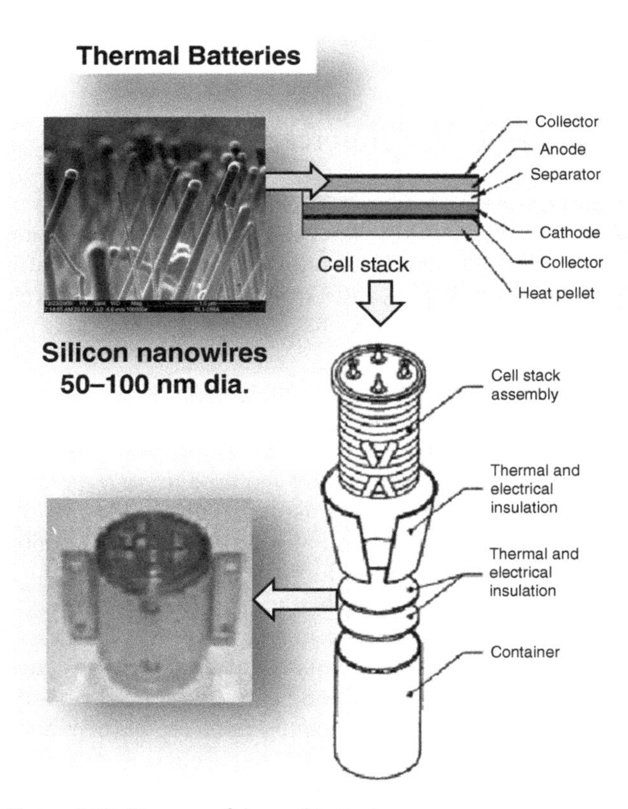

Figure 5.31 Diagram of thermal batteries.

conductivity and storage capacity. The concept is based on the application of nanostructured silicon composite material for the battery anode. Silicon has the highest known capacity for lithium storage of 4.2 Ah/g, which is >10× higher than graphite and other traditional anode materials. Furthermore, Silicon has a low discharge potential. However, its usage has been limited due to its 400% volume change with Li+ intercalation, in which such expansion causes strong mechanical strain in the material resulting in mechanical damage of the electrode material [44].

Half-cell testing of the SiNW samples was performed at room temperature and evaluated for battery cycle life and thermal stability. Research techniques were conducted to develop test methods providing the capability to test samples at temperatures exceeding 500 °C. The high temperature testing addressed issues related to potential internal short of smaller anodes, uniform electrode compression, and provided an estimate of the time constants in the T-cell heated in the furnace.

AMRDEC also developed supercapacitors based on oriented single-walled carbon nanotubes. The aligned CNTs were successfully grown directly on Al foils (typically, this is quite hard since CNT growth recipes were above the melting point of Al, ~666 °C) as shown in Figure 5.32.

The technology to develop a high power and capacity cathode material for thermal batteries is applicable to multiple Army missile systems. The development of other military applications of this technology may include future urban warfare surveillance/reconnaissance unmanned aerial vehicles. Future plans include an enhanced design of the anode for thermal batteries, and an improved synthesis of CNT on metal foil with improved alignment, height, and density, and specific capacitance.

Figure 5.32 Aluminum on CNT.

The AMRDEC research team conducted research to develop long-time storage thermal batteries for missiles using nano-sized anode, cathode, separator, and heat pellet materials to improve battery power capabilities and compactness. Silicon was considered to produce improved batteries with higher energy densities and smaller sizes.

Moreover, the AMRDEC research team investigated "supercapacitors," or "electrochemical capacitors" that exhibited low voltage with high capacity. These high energy devices filled a niche where a battery cannot meet the requirement for some systems that need a high-power dump, that is, enormous energy very fast (a quick wake-up munition). [45]

5.9 Conclusion

Nanotechnology, which is an enabling technology, is not so futuristic anymore. The Army (under the MIT ISN Program) has developed the building blocks (multifunctional fibers, nanoscale coatings, core-shell, nanostructures, carbon nanotubes, fibers, fabrics, and layered and membrane structures) required for a twenty-first-century battle suit that combines high-technology protection and survivability capabilities with low weight and increased comfort. Nanotechnology has the potential to increase the efficiency of military weaponry and reduce production and transportation costs.

The major nanotechnology research thrust areas, considered at AMRDEC, have been discussed in this chapter, including

- nano-based chemical sensors to detect rocket motor off-gassing and toxic industrial chemicals;
- nano-based minimum signature solid propellant with increased ballistic and physical properties that meet insensitive munitions requirements;
- nano-based composite materials that provide high-frequency damping for inertial measurement units' packaging;
- nano-based composite materials for missile motor casings, missile structural components, and strong armor for soldiers;
- metallic nanostructures for ultraviolet surface-enhanced Raman spectroscopy; and
- nanostructured material for higher voltage thermal batteries and higher energy density storage.

The Applied Nanotechnology program at AMRDEC developed prototypes and performed analysis and laboratory testing of nano-based sensors, composites, and propellants. The next step is to transition the technology for integration into military systems as well as product commercialization.

The proposed transition plan for each functional area includes integration and relevant environment testing of the nanomaterials (sensor, composite, and/or propellant) for a nano-enhanced missile. In the near future, the nano-enhanced missile could potentially be integrated and tested in a military environment. The integration of these critical innovative nanotechnologies into current and future weapon systems will provide the solution to missile outgassing, resulting in increased soldier safety, less missile misfires due to malfunctioning of stored missiles, and a decreased load for the soldier. The rapid transition of these technologies will result in more advanced, more powerful, smaller, lighter, cheaper, and safer missiles being rapidly transitioned to the soldier.

References

1 Drexler, E. (2009) There's Plenty of Room at the Bottom. *Metamodern: The Trajectory of Technology*, http://metamodern.com/2009/12/29/theres-plenty-of-room-at-the-bottom%E2%80%9D-feynman-1959/ (accessed 05 May 2017).

2 Azonano (2015) Emerging Products and Applications Using Nanotechnology in the Transport and Automotive Industry, http://www.azonano.com/article.aspx?ArticleID=1655 (accessed 05 May 2017).

3 Azonano (2015) Commercial Applications of Nanotechnology in the Military, Space and Security, http://www.azonano.com/article.aspx?ArticleID=1050 (accessed 05 May 2017).

4 Sanders, D. (2015) Negative Effects of Artificial Intelligence. *eHow.com*, http://www.ehow.com/info_8590188_benefits-nanotechnology-military-devices.html (accessed 05 May 2017).

5 Ruffin, P.B. (2004) Nanotechnology for Missiles, *Proc. SPIE* 5359. *Quantum Sensing and Nanophotonic Devices*, July 6, 2004 p. **177**.

6 Nano.gov. (2015) Department of Defense, http://www.nano.gov/node/144 (accessed 05 May 2017).

7 Ruffin, P.B., Brantley, C.L., Edwards, E. *et al.* (2011) Nanotechnology Research and Development for Military and Industrial Applications, *Proc. SPIE* 7980. *Nanosensors, Biosensors, and Info-Tech Sensors and Systems*, **798002**, April 15, 2011.

8 Nanotechnology Now (2010) National Nanotechnology Initiative at Ten: Nanotechnology Innovation Summit, http://www.nanotech-now.com/news.cgi?story_id=39336 (accessed 05 May 2017).

9 Institute for Soldier Nanotechnologies. (2015). http://isnweb.mit.edu/ (accessed 05 May 2017).

10 Soutter, W. (2015) Nanotechnology in the Military. *Azonano*, http://www.azonano.com/article.aspx?ArticleID=3028 (accessed 05 May 2017).

11 Calzzani, F.A., Silesh, R., Kassu, A.S. *et al.* (2008) Detection of residual traces of explosives by surface enhanced Raman scattering using

gold coated substrates produced by nanospheres imprint technique, *Proc. SPIE* 6945. *Optics and Photonics in Global Homeland Security IV*, **694510**, April 15, 2008.

12 Sadate, S., Kassu, A., Farley, C.W. *et al.* (2011) Standoff Raman measurement of nitrates in water, *Proc. SPIE* 8156. *Remote Sensing and Modeling of Ecosystems for Sustainability VIII*, **81560D**, September 16, 2011.

13 Gundrum, L., Hüttner, W., and Wackerbarth, H. (2010) *Detection of Explosives Based on Surface Enhanced Raman Spectroscopy, Laser-Laboratorium Göttingen.*

14 Hatab, N.A., Gyula, E., Hatzinger, P.B., and Baohua, G. (2010) Detection and analysis of cyclotrimethylenetrinitramine (RDX) in environmental samples by surface-enhanced Raman spectroscopy. Published online January 5, 2010. *Journal of Raman Spectroscopy*, **41** (**10**), 1131–1136.

15 Vogt, M., Shoemaker, E., and Turner, T. (1996) A trainable cermet gas microsensor technology using cyclic voltammetry and neural networks, *Proc. of the Sixth Int'l Meeting on Chemical Sensors. Sensors and Actuators B: Chemical*, **35-36** (**1–3**), 370–376.

16 Kincade, K., et al. (2007) Optoelectronic applications: nanophotonics - an 'old' technique finds new life in the nano world, http://www.laserfocusworld.com/articles/274732, 1 (accessed 05 May 2017).

17 Currie, J., Essalik, A., and Marusic, J. (1999) Micromachined thin film solid state electrochemical CO_2, NO_2 and SO_2 gas sensors. *Sensors and Actuators B: Chemical*, **59** (**2–3**), 235 241.

18 Dubbe, A., Nafe, H., and Aklinger, F. (2002) *Patterned Micro Electrodes on Yttria-Stabilized Zirconia Thin Films*, Book of Abstracts, 4th International Symposium on Electrochemical Micro- and Nanosystem Technology, Dusseldorf: Germany, pp. 15–20.

19 Brantley, C.L., Ruffin, P.B., and Edwards, E. (2008) Innovative smart microsensors for army weaponry applications, *Proc. SPIE* 6931. *Nanosensors and Microsensors for Bio-Systems*, **693102**, March 26, 2008.

20 Kim, B. (University of Alabama) Unmanned Zinc Oxide (ZnO) Nanowire-based Sensing Platform, Annual Report 2012-2013. US Army Aviation Missile Research Development & Engineering Center. 23 p. Task 5.2.2. Contract No.: W31P4Q-09-D-0028.

21 Ruffin, P.B., Brantley, C.L., Edwards, E., and Mensah, T. (2012) *Nano-based Materials for Use in Missile. Proc. of AIAA*, AIAA Strategic and Tactical Missile Systems Conference and AIAA Missile Science Conference, Monterey, CA.

22 Luo, C. and Yin, S. (General Opto Solutions, LLC), Unconventional broadband IR optical system for the advanced multi-functional, low scattering loss, high discrimination capability LADAR seeker and imager, Final Report

25 August 2010. US Army Aviation and Missile Command. 45 p. Topic No.: A083-162. Contract No.: W31P4Q-10-P-0119.
23 Chopra, S., McGuire, K., Gothard, A. *et al.* (2003) Selective gas detection using a carbon nanotube sensor. *Applied Physics Letters*, **83** (**11**), 2280–2282.
24 Kang, I., Schulz, M.J., Kim, J.H. *et al.* (2006) A carbon nanotube strain sensor for structural health monitoring. *Smart Materials and Structures*, **15** (**3**), 737–748.
25 Li, J., Lu, Y., Ye, Q. *et al.* (2003) Carbon nanotube sensors for gas and organic vapor detection. *Nano Letters*, **3** (**7**), 929–933.
26 Li, J. (2009) Gas Sensors Based on Coated and Doped Carbon Nanotubes, NASA Tech Briefs *ARC-15566-1*, http://www.techbriefs.com/component/content/article/2700 (accessed 05 May 2017).
27 Li, J., Carbon nanotube applications: chemical and physical sensors in *Carbon Nanotubes: Science and Applications*, Editor: M. Meyyappan, CRC Press, Boca Raton, FL, USA. (2004).
28 Sugihara, C. *et al.* (1991) Phase-matched second harmonic generation in poled dye/polymer waveguide. *Applied Optics*, **30**, 2957.
29 Warren, L.C. and Neidert, J.B. (2010) *Insensitive Minimum Signature Propellants for Joint Services Applications*, JANNAF, Orlando, FL.
30 Clubb, J.W., Turnbaugh, D.G., and White, D.W. (2010) *Phase II Development of Next Generation Insensitive Minimum Signature Propellants Utilizing Energetic Polymers and Novel Ingredients*, JANNAF, Orlando, FL.
31 Headrick, S.A., Warren, L., and Stiles, S. (2010) Nano-composite samples, W31PQ-08-C-0189, in *Development of Lead-Free Rocket Propellant* (ed. D. Hayduke), JANNAF.
32 Wlodarski, J. F., Roberts, K. (2008) AMRDEC Significant Actions Report.
33 Owens, A.T., Russell, C.D., Weisenberger, M.C. *et al.* (2012) *Thermal Interface Materials in Tactical Missile Airframe Applications, Proc. of AIAA*, AIAA Strategic and Tactical Missile Systems Conference and AIAA Missile Science Conference, Monterey, CA.
34 Christensen, R. (1982) *Theory of Viscoelasticity: Second Edition*, Academic Press, New York.
35 Dlubac, J., *Polymer Test Procedures for Naval Applications*, NSWCCD-72-TR—2010/103, A published technical report by the Naval Surface Warfare Center (Carderock Division), West Bethesda, MD20817, (2010).
36 Wu, P.C., Khoury, C.G., Kim, T.H. *et al.* (2009) Demonstration of surface-enhanced raman scattering by tunable, plasmonic gallium nanoparticles. *Journal of the American Chemical Society*, **131**, 12032–12033.
37 Wu, P.C., Kim, T.H., Brown, A.S. *et al.* (2007) Real-time plasmon resonance tuning of liquid Ga nanoparticles by in situ spectroscopic ellipsometry. *Applied Physics Letters*, **90**, 103119–103123.

38 Haes, A.J., Haynes, C.L., McFarland, A.D. *et al.* (2005) Plasmonic materials for surface-enhanced sensing and spectroscopy. *MRS Bulletin*, **30**, 368–375.

39 Quinten, M. (1996) Optical constants of gold and silver clusters in the spectral range between 1.5 eV and 4.5 eV. *Zeitschrift für Physik B Condensed Matter*, **101**, 211–217.

40 McMahon, M., Lopez, R., Meyer, H. *et al.* (2005) Rapid tarnishing of silver nanoparticles in ambient laboratory air. *Applied Physics B*, **80**, 915–921.

41 Oates, T.W.H. and Mucklich, A. (2005) Evolution of plasmon resonances during plasma deposition of silver nanoparticles. *Nanotechnology*, **16**, 2606–2611.

42 Creighton, J.A. and Withnall, R. (2000) The Raman spectrum of gallium metal. *Chemical Physics Letters*, **326** (**18**), 311–313.

43 Svoboda, V. (2011) Silicon Nanostructure Anode for Thermal Batteries. Presented at *Pacific Power Source Symposium 2011*, Waikoloa Village, HI.

44 CFDRC (2015) Battery Technoloy Development. http://www.cfdrc.com/emt/emerging-energy-technologies/battery-technology-development (accessed 05 May 2017).

45 Gourle, S.R. (2011) The 27th Army Science Conference: wandering through tomorrow's technology. *ARMY Magazine*, **61** (**2**), 43–47.

6

Novel Polymer Nanocomposite Ablative Technologies for Thermal Protection of Propulsion and Reentry Systems for Space Applications

Joseph H. Koo[1] and Thomas O. Mensah[2]

[1] *Department of Mechanical Engineering, The University of Texas at Austin, Austin, TX, USA*
[2] *Georgia Aerospace Systems, Nano technology Division, Georgia Aerospace, Inc., Atlanta, GA, USA*

6.1 Introduction

This chapter deals with the use of thermal protection materials for two key components of a solid rocket motor (SRM), the exhaust nozzle and internal motor insulation. The problems associated with these items are similar to those of thermal protection systems (TPS) for reentry probes or ballistic vehicles. The role of the materials utilized for SRM and TPS thermal protection depends on the engineers involved, who would ideally be using data obtained from full-scale ground or in-flight testing to help guide their choices.

In 1971, D'Alelio and Parker edited a book entitled *Ablative Plastics*, which was the first book solely dedicated to the subject of "ablation" [1]. This book was a collection of 20 papers presented at a 1968 symposium structured to emphasize both the interdisciplinary character of ablation research and individual disciplines concerned with this subject matter. These papers, assembled from late 1968 to early 1969, were provided for the first time in one publication. In this publication, the comprehensive and rational approach required to design and to produce reliable heat shields was discussed for future space missions, with a minimum of redundancy being a key theme. The first four papers in this historical book were based on invited tutorial lectures by individuals, representing the U.S. Air Force, NASA, and the U.S. Navy, who were at the symposium to discuss the fundamental research viewpoints of polymeric material ablation as a means for thermal protection. They deal with the contribution of (i) heat rejection mode as a function of heating rate, (ii) the nature of the heat transfer, both radiative and conductive, (iii) the nature of the degrading polymer, (iv) the molecular weight and geometric structure of the polymer, and (v) the fabrication of the polymer. The remaining papers particularly emphasize the individual disciplines of chemical synthesis, characterization, and polymer processibility

Nanotechnology Commercialization: Manufacturing Processes and Products, First Edition.
Edited by Thomas O. Mensah, Ben Wang, Geoffrey Bothun, Jessica Winter, and Virginia Davis.

required to obtain reliable ablative structures. They also consider the problems of thermochemical kinetics and thermophysics, physical elements that are required to predict the ablative properties of these polymer systems [1].

A recent book on ablation modeling was published in 2013 by Georges Duffa, entitled *Ablative Thermal Protection Systems Modeling* [2]. This book discusses the modeling of the phenomena of ablation. It includes (from microscopic to macroscopic) atomic physics, thermodynamics, gas kinetics, radiative heat transfer, physical and chemical reactions (both homogenous and heterogeneous), fluid mechanics, and turbulence applied to previously studied areas, such as rough walls. The main objective of this book was to develop physical skills in the above cited research areas applied to the modeling of thermal protection. Chapters dedicated to (i) TPS Conception (Chapter 1), (ii) Ablation of Carbon (Chapter 5), (iii) Pyrolysis and Pyrolyzable Materials (Chapter 8), (iv) Testing and Specific Testing Facilities (Chapter 12), and (v) An Example: Apollo (Chapter 13) are closely related to the "Thermal Protection System" materials from the ablation modeling point of view. This book by Duffa is an excellent reference that deals with the modeling of ablation from a historical viewpoint, as well as dealing with the current status of ablation research for reentry spacecraft.

Natali *et al.* recently published a very comprehensive article entitled, "Science and technology of polymeric ablative materials for TPSs and propulsion devices: A review" in *Progress in Materials Science* [3]. This review paper covers all main topics related to the science and technology of ablative materials with current and potential applications in the aerospace industry, after a short, yet comprehensive introduction on nonablative materials. This review paper summarizes 50 years of research efforts on polymeric ablatives, starting from the state-of-the-art solutions currently used as TPS, up to covering the most recent efforts for nanostructured ablative formulations. The contents of this excellent review with 454 references consist of nine different topics: (i) TPS materials, (ii) Nonablative TPS materials, (iii) Ablative TPS materials, (iv) Advanced testing techniques for TPS materials, (v) Erosion rate sensing techniques for ablative TPS, (vi) Polymer ablatives as insulator materials, (vii) A new class of TPS ablatives: lightweight ceramic ablators, (viii) An introduction to modeling of ablation phenomena, (ix) Nanostructured ablative materials.

Thermal protection materials are required to protect structural components of space vehicles during the reentry stage, SRMs, liquid rocket engines, and missile launching systems. A thorough literature survey was conducted by Koo *et al.* [3–9] to review the development of these thermal protection materials for different military and aerospace applications. This series of literature surveys were grouped into (i) numerical modeling [4], (ii) materials' thermophysical properties characterization [5], (iii) experimental testing [6], and (iv) polymer nanocomposite (PNC) ablatives [7, 8]. In this chapter, research materials are drawn from these review articles and from several references [1, 10–16], supplemented by the new research tools that have been developed by Koo *et al.* recently, such as in situ ablation recession and thermal sensors (Section 6.4.1), and char strength sensors (Section 6.4.2) to evaluate PNC ablatives. A brief discussion on important thermophysical properties and their measurement techniques (Section 6.5.1) as well as selective industry ablation modeling codes (Section 6.5.2) are included as the technologies needed to advance PNC ablative research.

6.2 Motor Nozzle and Insulation Materials

SRMs never reach thermal equilibrium during motor firing. The temperatures of all the components exposed to the heat flow increase continuously during operation. In a good thermal design, the critical locations reach a maximum allowable temperature a short time after the rocket motor stops running. The nozzle components rely on their heat-absorption capacity (high specific heat and high energy demand for material decomposition) and slow heat transfer (good insulation with low thermal conductivity) to withstand the stresses and strains imposed by their thermal gradients and loads. The maximum allowable temperature for any of the motor materials is just below the temperature at which excessive degradation occurs. The operating duration is limited by the design and amount of heat-absorbing and insulation material pieces. The objective is to design a nozzle with just enough heat-absorbing material mass and insulation mass at the various locations within the nozzle assembly that its structures and joints will do the job for the duration of the application under all likely operating conditions.

The selection and application of the proper material is key to the successful design of an SRM nozzle. The high-temperature exhaust of an SRM presents an unusually severe environment for the nozzle materials, especially when metalized solid propellants are employed.

About 75 years ago, some solid rocket nozzles were made of molded polycrystalline graphite, and others were supported by metal housing structures. They eroded easily, but were low cost. For more severe conditions, a throat insert or *integral throat entrance* (ITE) was placed into the graphite piece. This insert was a denser, better grade of graphite. Later, pyrolytic graphite washers and fiber-reinforced carbon materials were used. For a period of time, tungsten inserts were used. They had very good erosion-resistant characteristics, but were heavy and often cracked during motor operation. Pyrolytic graphite was introduced and is still being used as washers for the throat insert of small nozzles [10]. The introduction of high-strength carbon fibers and carbon matrix were major advances in high-temperature materials. For small- and medium-sized nozzles, ITE pieces were made of carbon–carbon composite.

The orientation of the fibers can be 2D, 3D, or 4D. Some properties of all these materials can be found in [10]. For large nozzles, previously existing technology did not allow the fabrication of large 3D carbon–carbon ITE pieces, so layups of carbon fiber or silica fiber cloth in a phenolic resin matrix were used.

Several types of materials are potentially suitable for being used in the critical throat region of various SRM nozzles. The materials suitable for use can be divided into two general categories: (i) refractories and (ii) ablatives. The refractories, characterized by high melting temperatures ranging from 2204 to 3316 °C (4000–6000 °F) include materials, such as tungsten, molybdenum, certain oxides, and carbides. These materials maintain their strength at high temperatures, so they are sufficiently tough to withstand the erosive effects of the hot gas stream. During motor operation, the internal surfaces of the throats and nozzles of these materials are heated to temperatures close to that of the gas stream. The heat is absorbed by the material and dissipated by normal thermal radiation and convective cooling by the atmosphere on the outside.

In contrast to the refractories, ablative materials are not resistant to melting, and would not typically be characterized as being tough. They absorb heat from the gas stream by chemical reactions as well as through melting, spallation, and vaporization. One type of ablative material is carbon–phenolic composite. This material consists of a tape woven from carbon fibers and impregnated with a phenolic resin, and compression molded to high density. A trade name is MX-4926, manufactured by Cytec Engineered Materials. The tape is wound on a mandrel to form the nozzle, which is heated under pressure to bond the plies together with the phenolic resin. Under the high temperatures experienced during motor firing, the resin decomposes to form graphite and organic compounds, which melt and vaporize (ablate) to form a carbonaceous charred layer. This compound has a fairly high melting point and is relatively tough, and imparts a certain degree of resistance to mechanical erosion to the nozzle. A ablative nozzles are inexpensive and easily fabricated, and can be used in motors where the operating conditions are not too severe as to require the use of a tougher refractory material. This type of nozzle ablative is the focus of this chapter.

The regions immediately upstream and downstream of the throat have less heat transfer, less erosion, and lower temperatures than the throat region, making them regions where less expensive materials are usually used [10, 14]. This includes various grades of graphite ore ablative materials, and strong high-temperature fibers (carbon or silica) in a matrix of phenolic resin. A movable nozzle can employ multilayer insulators behind the graphite nozzle piece, which is directly exposed to heat. These insulators (between the very hot throat piece and the surrounding housing) limit the heat transfer and prevent excessive housing temperatures [10, 14].

In the diverging exit cone section, the heat transfer and temperature are even lower, thus less capable and less expensive materials can be used in the region. This exit segment can be built as an integral part with the nozzle throat segment, or it can be separate as a one or two-piece subassembly, which is then fastened to the smaller diameter throat segment. Ablative materials, without oriented fibers as in cloth or ribbons, but with short fibers or insulating ceramic particles, can be used in this region. For large area ratios (upper stages and space transfer), the nozzle will often protrude beyond the vehicle's boat tail surface. This allows for radiation cooling, since the exposed exit cone can reject heat by radiation to space. Lightweight, thin, high-temperature metals (niobium, titanium, stainless steel, or a thin carbon–carbon shell) with radiation cooling have been used in a few upper-stage or spacecraft exit cone applications. Since radiation-cooled nozzle sections reach thermal equilibrium, their duration is unlimited.

The housing or structure of the nozzle uses the same material as the metal case, such as steel or aluminum. The housings are never allowed to become very hot. Some of the simpler, smaller nozzles do not have a separate housing structure but use the ITE for the structure. Typical materials used for this purpose are those used for the ITE or nozzle throat insert. They are exposed to the most severe conditions of heat transfer, thermal stresses, and high temperatures. Their physical properties are often anisotropic, that is, their properties vary with the orientation or direction of the crystal structure or the direction of reinforcing fibers. Polycrystalline graphites are extruded or molded. Different grades with different densities and capabilities are available. Pyrolytic graphite is strongly anisotropic and has excellent conductivity in a preferred direction. It is fabricated by depositing graphite crystals on a substratum containing methane gas. Its use is declining, but it is still installed in current rocket motors of older design.

Carbon-carbon materials are made of carefully oriented sets of carbon fibers (woven, knitted, threaded, or laid up in patterns) in a carbon matrix. Two-dimensional (2D) material has fibers in two directions, 3D has fibers oriented in three directions (at right angle to each other), and 4D has an extra set of fibers at approximately a 45° angle to the other three directions. An organic liquid resin is injected into the spaces between the fibers. The assembly is pressurized and the filler is transformed into a carbon char by heating and is compacted by further injection and densification processes. The graphitization is then performed at temperatures higher than 2000 °C (3632 °F). This material is expensive but suited to nozzle applications. Highly densified material is superior in high heat transfer regions, such as the nozzle throat. The multidirectional fiber reinforcements allow them to better withstand the high thermal stresses introduced by the steep temperature gradients within the component.

Charring ablative materials can be made of thermosetting resin, such as phenolic, epoxies, or silicones reinforced by fibers, such as glass, quartz, or carbon. Ablative effectiveness is usually proportional to the material density, whereas not while that of insulation is inversely proportional to the density. To reduce the density and thermal conductivity of most ablative materials, microballoons made of phenolic resins or glass (tiny hollow spheres about 40 μm in diameter, with a wall thickness of 1–2 μm) are utilized. Additives can also be graded so that the density can vary uniformly through the material. Some loss of char strength due to microballoons has been observed, but with a minimal effect on ablation performance.

6.2.1 Behavior of Ablative Materials

Polymeric composites have been used as ablative materials for a variety of military and aerospace applications. Thermal protection materials, such as carbon–phenolic and carbon–carbon composites, are used for spacecraft heat shields during the re-entry stage, thermal protection of missile launching system, and in SRMs as insulation and nozzle assembly materials. Thermosetting polymer matrix composites are constructed of layers of woven carbon, quartz, silica, or glass fibers impregnated with a resin matrix. Phenolic resin is the most commonly used polymer matrix for ablatives. These TPSs are exposed to a thermochemical flow and subjected to high temperatures in excess of 3000 °C (5432 °F) with very high heating rates.

Thermochemical ablation refers to the phenomenon of surface recession of an ablative due to severe thermal attack by an external heat flux. The initial heat transfer into the ablative occurs by pure conduction, and the resulting temperature rise causes material expansion (swelling), which may be attributed to pure thermal expansion as well as vaporization of any traces of moisture. When the material reaches a sufficiently high temperature, thermochemical degradation or pyrolysis of the polymer matrix begins. The pyrolysis reactions result in the production of decomposition gases and solid carbonaceous char residue. Thermal expansion and the disappearance of solid material due to decomposition result in an increase in porosity and permeability of the polymeric material. Pyrolysis gases begin to escape through the polymeric material. For the thermal protection to be effective, all these reactions must be strongly endothermic. The gases that flow through the char structure remove energy by convection, thus attenuating the conduction of heat to the reaction zone. This ablation phenomenon is depicted in Figure 6.1 [16].

The gases that flow through the virgin material will add energy by convection and accelerate the heat transfer toward the substrate, thus increasing the conduction heat transfer to the back surface. Based on experimental observations and numerical studies, the effects that result from the backflow of decomposition gases are small enough to be neglected. At a temperature region between 1000 and 1300 °C (1832–2372 °F), the carbonaceous char and the carbon (in carbon-filled polymer matrix composites) can react chemically, resulting in additional expansion or contraction and further increases the porosity and

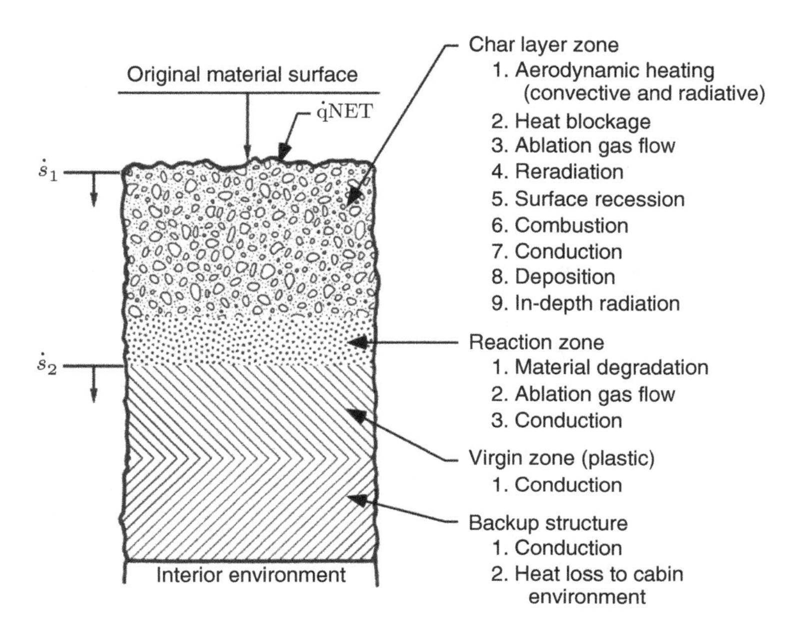

Figure 6.1 Schematic diagram illustrating the different ablation mechanisms and different zones of an ablator under thermal attack, as adapted from the NASA-JSC charring ablation model. Diagram courtesy of NASA.

permeability of the polymeric material. Due to sufficient incident energy, these reactions may be driven to completion, leaving only an inert fragile solid residue. Finally, the char layer will be removed by continued heating, and surface recession results. At this point, temperature at the surface is roughly maintained at the so-called melt or failure temperature of the ablative material.

When an organic resin is heated, it begins to decompose and release gaseous products, leaving a porous, carbonaceous char. Pyrolysis temperature is a function of the local pressure and ablation rate, and is relatively low, ranging from 227 to 527 °C (441–981 °F). As the heating continues, the pyrolysis zone proceeds into the material, and the decomposition occurs below the material surface. Gaseous products diffuse through the porous char to the surface, absorbing energy from the char while undergoing further decomposition (cracking). They finally exit into the boundary layer, where they act as a transpirant, and may undergo additional chemical reaction with the boundary layer gas. The char is primarily carbonaceous and continues to absorb heat until it reaches the temperature whereby it will oxidize or sublime, or until it is mechanically removed. For lifting and moderate ballistic entries in regard to re-entry vehicle applications, oxidation is the dominant thermochemical char-removal mechanism. At surface temperatures below 827 °C (1521 °F), oxidation is limited to reaction-rate kinetics. In this regime, surface recession can be reduced appreciably by incorporating oxidative-resistant additives, such as silica. As the surface temperature increases, the oxidation rate increases exponentially until the oxygen at the surface begins to be depleted. At even higher temperatures, surface recession is limited to the rate at which oxygen can diffuse through the boundary layer.

In this regime, the mass rate of char oxidation is virtually independent of the material properties. At temperatures of about 3327 °C (6020 °F), the char sublimes. A thick char provides an insulation barrier, radiates a large amount of heat from the surface, and is an effective ablator. However, the char forms a homogenous polymer that is usually weak and brittle; thus, the material is susceptible to rapid removal by mechanical shear or spallation (due to thermal stresses and the buildup of internal gas pressure). This reduces the insulation effectiveness of the char, exposes the cooler internal material to the hotter surface, and results in less radiative cooling. To improve the char-retention characteristics of ablative resins, reinforcing fibers (organic or inorganic) are usually added. The fibers add strength to the char until they reach their own melting or decomposition temperature. Fiber reinforcements and additives also provide flexibility in the fabrication of ablative materials for specific applications. Fibers usually possess a higher thermal conductivity (k) than the resin matrix; fibers that are normal to the surface will increase the overall k of the composite. When the fibers are placed parallel to the surface, the k approaches that of the resin, the char shear strength is reduced and the material is subject to delamination. The fiber orientation can be selected on the basis of the particular shear stress and heat conduction requirements.

6.3 Advanced Polymer Nanocomposite Ablatives

One of the most important and pioneering research studies on the use of the montmorillonite (MMT) polymer (caprolactam, a precursor to Nylon 6) to a 2 and 5 wt% (percentage by weight) loaded layered silicate/polymer (caprolactam) nanocomposite was performed by Vaia *et al.* [17] in 1999. These materials were synthesized by *in situ* polymerization of ε-caprolactam in the presence of dispersed, organically modified MMT nanoclay. Koo *et al.* [7, 8] conducted a comprehensive review on the research and development of PNCs used as TPS for reentry vehicles, and as nozzle and internal insulation materials for propulsion systems. This article summarizes the significant R&D efforts on the studies of nanostructured ablative materials by scientists from China, India, Italy, Iran, and the United States for the past decade. Thermoset, thermoplastics, elastomers, and thermoplastic elastomers were used as polymer matrices. Nanomaterials, such as montmorillonite organoclays, carbon nanofibers (CNFs), polyhedral oligomeric silsesquioxanes (POSS), nanosilicas, nanoalumina, carbon black, and carbon nanotubes (CNTs) were incorporated into the different resin matrices as nanofillers. Conventional fibers, such as asbestos, carbon, glass, and silica, were used as reinforcements. Ablation mechanisms were studied and proposed by the researchers on their nanostructured ablatives. A summary of Koo *et al.*'s review can be found in Refs [7–9], and this summary is divided into two sections: (i) PNCs designed for rocket motor nozzles and (ii) PNCs designed for motor internal insulation.

6.3.1 Polymer Nanocomposites for Motor Nozzle

6.3.1.1 Phenolic Nanocomposites Studies by The University of Texas at Austin

The University of Texas at Austin group has been involved in the traditional R&D of ablative materials since the early 1990s, as reported by Koo *et al.* [18–22], Cheung *et al.* [23], Wilson *et al.* [24], and Shih *et al.* [25], and began the development of PNC ablatives since 2002. Koo *et al.* [26–33], Ho *et al.* [34, 35], Bruns *et al.* [36], Lee *et al.* [37–41], and Allcorn *et al.* [42, 43] reported two new classes of PNCs (SRM nozzle and insulation materials) that are lighter and exhibit better ablation performance and insulation characteristics than current state-of-the-art (SOTA) ablative materials. These PNC materials exploit the ablation resistance of both resin and nanoparticles. For SRM nozzle ablatives, MMT nanoclay, CNF, and POSS® nanomaterials were incorporated into a resole (formaldehyde to phenol ratio greater than one) phenolic resin, impregnated into rayon carbon fibers, and fabricated into phenolic-based polymer matrix composites [28–32]. These PNCs demonstrated significantly improved ablation and insulative performance versus the SOTA MX-4926 ablatives as demonstrated by Koo Research Group. For SRM internal insulation, detailed results are presented in Section 6.3.2. Selective properties improvement was observed with these thermoplastic polyurethane elastomer (TPU) nanocomposites, as compared to industry standard Kevlar-filled EPDM insulation materials. In this section, the phenolic nanocomposite studies are discussed.

The NRAM (nanocomposite rocket ablative material) program is aimed at developing materials for SRM nozzles [26–30]. Both wide-angle X-ray diffraction (WAXD) and transmission electron microscopy (TEM) methods are used to determine dispersion uniformity before full ablation testing, which is expensive. SC-1008 resole phenolic resin in isopropanol alcohol (IPA) is the baseline resin system. The commercial material, MX-4926, is a rayon carbon fabric impregnated with SC-1008 containing carbon black (CB) particles. On dispersing nanoclay into the resole phenolic resin with CB, followed by cure and examination by WAXD and TEM, it was concluded that CB particles in the baseline MX-4926 caused interference during dispersion of the nanoparticles in the resole phenolic resin. As a result, CB was eliminated in subsequent blending experiments. Use of nanoclay, Cloisite® 30B from Southern Clay Products, in loadings of 5, 10, and 15 wt% indicated that the nanoparticles dispersed satisfactorily into the SC-1008. Results of the WAXD of the three nanoclay loadings in SC-1008 indicated desirable dispersibility of the nanoclay into SC-1008. On the other hand, TEM images of nanoclay with SC-1008 indicate intercalation and not exfoliation of the nanoclay in the resin system. The TEM analyses were very convenient in determining the degree of dispersion/exfoliation and were a cost-effective and efficient technique for screening different formulations before committing to a large quantity of material.

The SC-1008 with 5, 10, and 15 wt% Cloisite® 30B replaced the 15 wt% CB in the original MX-4926 formulation. Table 6.1 illustrates the different loadings of Cloisite® 30B [$(HE)_2MT$], and trisilanolphenyl-POSS® [SO-1458] in the rayon fabric reinforced SC-1008 system. When a small amount of CNF (<1.5 wt%) was added in the SC-1008 phenolic resin, the viscosity of the resin increased significantly. It was impossible to process carbon prepregs under these conditions. It was decided that a molding compound would be fabricated to include large loadings of CNF dispersed in the SC-1008 resin. As a result, three loadings of PR-24-PS CNF at 20, 24, and 28 wt% were dispersed into SC-1008 without the rayon fabric reinforcement. Compositions and densities of the NRAMs are shown in Table 6.1. These nanomodified composites were manufactured and fabricated by Cytec Engineered Materials. The nanomodified materials are lower in density than the baseline material, MX-4926 (Table 6.1).

Table 6.1 Specimen configuration for composition laminates fabrication.

Material ID	Density (g/cc)	Rayon carbon fiber reinforcement (wt%)	Resin SC-1008 phenolic (wt%)	Filler (wt%)
MX-4926 (Control)	1.44	50	35	15 Carbon black (CB)
MX-4926 ALT Clay 5%	1.42	50	47.5	2.5 Cloisite® 30B [$(HE)_2MT$]
MX-4926 ALT Clay 10%	1.43	50	45	5 Cloisite® 30B [$(HE)_2MT$]
MX-4926 ALT Clay 15%	1.43	50	42.5	7.5 Cloisite® 30B [$(HE)_2MT$]
PR-24-PS 20%/SC-1008	1.35	None	80	20 PR-24-PS CNF
PR-24-PS 24%/SC-1008	1.38	None	76	24 PR-24-PS CNF
PR-24-PS 28%/SC-1008	1.41	None	72	28 PR-24-PS CNF
MX4926 ALT SO-1458 2%	1.41	50	49	1 Trisilanolphenyl-POSS® [SO-1458]
MX4926 ALT SO-1458 6%	1.38	50	47	3 Trisilanolphenyl-POSS® [SO-1458]
MX4926 ALT SO-1458 10%	1.40	50	45	5 Trisilanolphenyl-POSS® [SO-1458]

The simulated solid rocket motor (SSRM) is a small-scale supersonic liquid-fueled rocket motor burning a combination of kerosene and oxygen. It was used to study the ablation and insulative characteristics of the ablatives. The SSRM has been demonstrated to be a very cost-effective laboratory device to evaluate ablatives under identical conditions for initial material screening and development [18–24]. The ablation rates (Figure 6.2a) of MX-4926 and all three nanoparticle-modified composites indicate that only 7.5% nanoclay composition shows a lower ablation rate than MX-4926, which is about 0.4 mm/s. All three CNF compositions are lower than MX-4926 with the 28% CNF-NRAM being the lowest. All three POSS® compositions have lowered ablation rates, with the 5% POSS®-NRAM being the lowest. Residual mass (Figure 6.2b) of MX-4926 is about 92%. The POSS®-NRAMs exhibited the highest of all the nanoparticles in residual mass at about 93%. The clay-NRAMs were comparable to MX-4926 while whereas the CNF-NRAMs

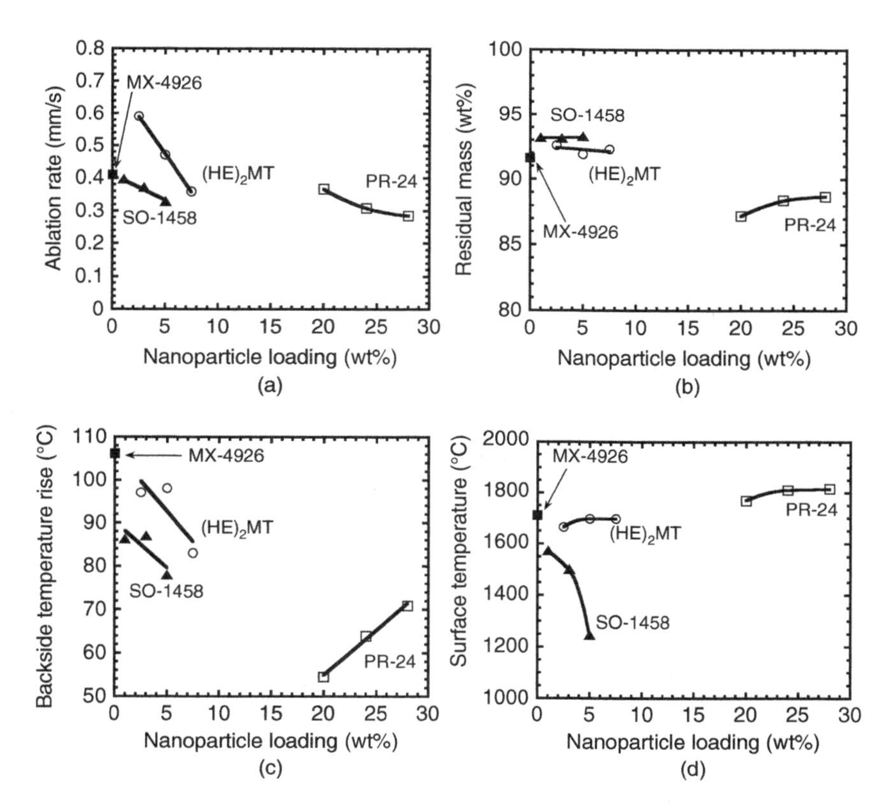

Figure 6.2 (a) Ablation rate, (b) residual mass, (c) backside temperature, and (d) surface temperature of MX-4926 and NRAMs with different types of nanoparticles [$(HE)_2$ MT-Cloisite® 30B, PR-24-PS CNF, and SO-1458 POSS®] at various loading levels.

were lower. A summary of backside temperature rise is shown in Figure 6.2c with all the nanoparticles compositions exhibiting lower maximum backside heat-soaked temperature rise than MX-4926. The value of MX-4926 is 106 °C with CNF-NRAMs being the lowest, from 54 to 72 °C. The POSS®-NRAMs were intermediate, from 75 to 86 °C and clay-NRAMs being the highest of the nanoparticles, from 82 to 98 °C, but still below the MX-4926 value of 106 °C.

An IR pyrometer was used to measure the surface temperatures of all materials during SSRM firings. Figure 6.2d shows the surface temperatures of MX-4926 (1700 °C) and the NRAMs. Surface temperatures of the CNF-NRAMs samples were higher than those of MX-4926, the clay-NRAMs, and the POSS-NRAMs. This finding suggests that better radial heat transfer is occurring rather than axial heat transfer. This is supported by the glowing heat of the material surface observed during and after material testing. This phenomenon was observed by other researchers, such as Patton *et al.* [44], and needs further study. The surface temperatures of clay-NRAMs and POSS-NRAMs were lower than that of MX-4926. The amount of nanoclay in the clay-NRAMs had essentially no effect. The amount of POSS in POSS-NRAMs had a significant effect on the surface temperature of the POSS-NRAMs.

The feasibility of using NRAMs in rocket nozzle assemblies was clearly demonstrated using SSRM subscale ablation testing. MMT nanoclay, CNF, and POSS can be implemented into the existing semiproduction line at Cytec Engineered Materials to manufacture fiber-reinforced prepregs and compression molded into laminates. Higher loadings of MMT, POSS, and CNF improve erosion resistance, and 28% CNF has the lowest erosion rate in the absence of carbon fiber reinforcements. Backside temperatures of all NRAMs were lower than that of baseline MX-4926, and CNF-NRAM as a group had lower temperatures than the MMT-NRAM and POSS-NRAM groups. Peak erosion of POSS-NRAM is 20% lower than MX-4926 at very low loading (5% POSS). Peak erosion of CNF-NRAM (28% CNF) without rayon fabric is 42% lower than MX-4926. Backside temperatures of CNF-NRAM, MMT-NRAM, and POSS-NRAM are 68%, 28%, and 26% lower than MX4926, respectively.

6.3.1.2 Phenolic-MWNT Nanocomposites Studies by Texas State University-San Marcos

This research at TSU-SM was conducted in collaboration with Cytec Engineered Materials (CEM) as an attempt to enhance properties of the SOTA ablative, MX-4926. Therefore, test panels were manufactured using processes and materials similar to those used by CEM. MX-4926 MC (Molding Compound) is rayon precursor-based carbon fabric with a MIL-R-9299 phenolic resin. Composite panels were manufactured using rayon-based carbon fabric supplied by CEM, which is an 8-harness satin weave. The fabric has a weight of

261 g/m^2, a specific gravity of 1.84, and a thickness of 0.48 mm. Break strength in the wrap direction is 0.496 and 0.599 MPa in the fill direction, with 1.96 picks/mm wrap and 2 picks/mm fill. SC-1008 phenolic resin that meets the MIL-R-9299, was supplied by Momentive Specialty Chemicals. The viscosity is roughly 180–300 cps depending on storage conditions. SC-1008 contains roughly 20–25% IPA as a solvent. Graphistrength™ C100 MWCNTs were supplied by Arkema, Inc. Typical diameters of these MWCNTs are 10–15 nm with lengths between 1 and 10 μm. Graphistrength™ C100 MWCNTs tend to be agglomerated with dimensions of roughly 50–900 μm. High shear mixing and sonication have been used in previous studies to provide good dispersion and exfoliation of nanoparticles [45–48]. Tate *et al.* [49] investigated the dispersion characteristics using sonication, high shear mixer, and combination of sonication and high shear mixer in previous studies. TGA results and SEM analysis suggested that dispersion of MWCNTs within IPA using sonication followed by high shear mixing produces optimal dispersion characteristics for use in ablative panels. This method was adapted to produce composites panels with MWCNT loading of 0.5, 1, and 2 wt%.

An oxyacetylene test bed (OTB) was developed at The University of Texas at Austin to perform a small-scale testing of newly developed ablatives. Samples of size $12.7 \times 12.7 \times 10$ mm^3 were prepared and two holes of approximately 1.59 mm (1/16 in.) diameter drilled at depth of 2.7 and 7.7 mm for thermocouple placement. Samples were exposed to a heat flux of approximately 1000 W/cm^2 (6 mm distance from the OTB) for 45 s [50]. Six samples were tested in each category. Ablative performance of the material was analyzed based on percentage mass loss, recession, peak temperatures at different depth in ablative samples and SEM to analyze dispersion of MWCNTs in phenolic resin. There was a gradual decrease in percentage mass loss as wt% of MWCNTs increased. For the control samples, average mass loss was 26% and for 2 wt% MWCNT, average mass loss was 23%. For 0.5 and 1 wt% MWCNT, average mass losses were 25.5% and 25%, respectively. There is considerable reduction in mass loss as wt% of MWCNTs increases. Typically, mass loss is because of pyrolysis of material and through erosion of char. Similarly, recession was also high for low wt% formulations and decreased gradually as wt% of MWCNTs increased. Average recession was 0.83 mm for the control samples, whereas it was 0.38 mm for 2 wt% samples. Average peak temperatures were reduced considerably with the addition of MWCNT to the composite (Figure 6.3 [50]).

6.3.2 Polymer Nanocomposites for Internal Insulation

Several elastomeric matrices are commonly selected for SRM internal insulation, namely EPDM, natural rubber (NR), and TPU. In this section, EPDM, NR, and TPUN nanocomposites used as SRM internal insulation are discussed.

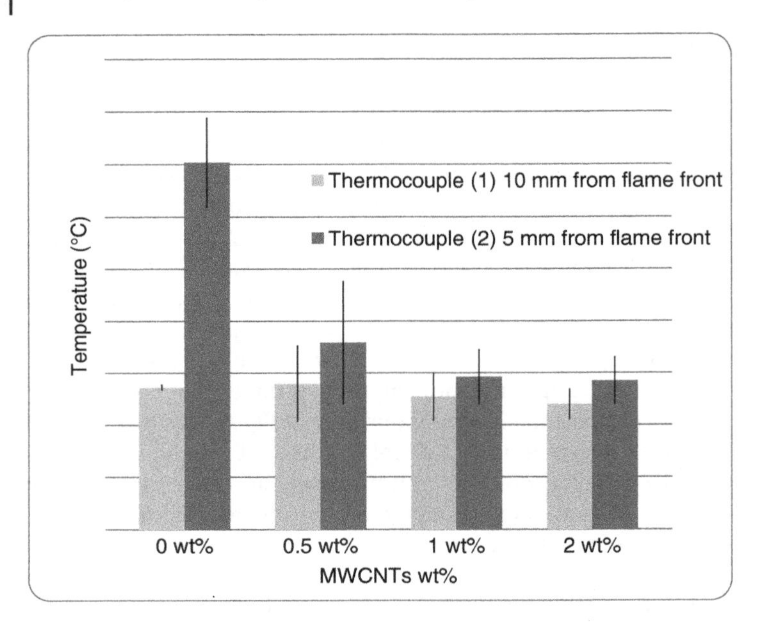

Figure 6.3 Average peak temperatures at two in-depth locations of the control sample and different phenolic-MWCNT nanocomposites.

6.3.2.1 Thermoplastic Polyurethane Nanocomposite (TPUN) Studies by The University of Texas at Austin

For SRM internal insulation, Koo and coworkers [31–43] used two TPUs as the base polymer and incorporated them with different amounts of MMT clay, CNF, POSS®, and multiwalled carbon nanotubes (MWNT). Improvement in select properties was observed with these TPU nanocomposites (TPUNs) as compared to industry standard Kevlar®-filled EPDM internal insulation material. The effects of weight loadings of layered clay, CNF, and MWNT on the thermal and flammability performance of this novel class of materials have been explored using a variety of test protocols and methods, such as thermogravimetric analysis (TGA), radiant panel experimental apparatus, UL 94, and cone calorimetry. Selected TPUNs were then tested using scaled SRMs, an oxyacetylene torch, as well as scaled hybrid rocket motors, with the intent to develop them for SRM insulation applications. Two commercially available TPUs were melt-blended with various loadings of these nanofillers using twin-screw extrusion. The morphological, physical, thermal, flammability, thermophysical properties, and kinetic parameters of these three families of TPUNs were characterized. The *processing–structure–property* relationships of this class of novel TPUNs were established. In the following section, the TPUN results are discussed.

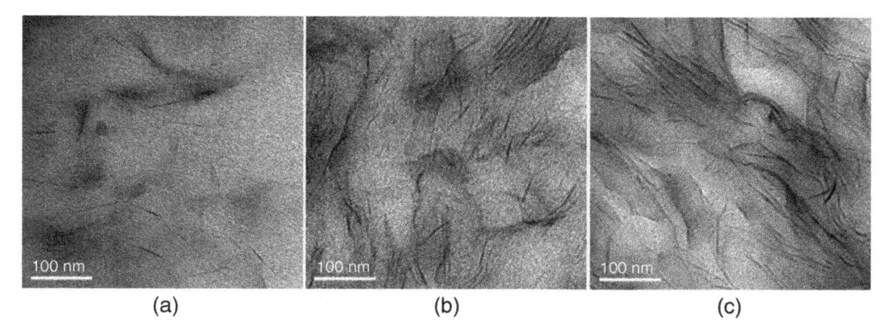

Figure 6.4 TEM images of (a) 2.5%, (b) 5%, and (c) 10% Cloisite 30B in TPUN.

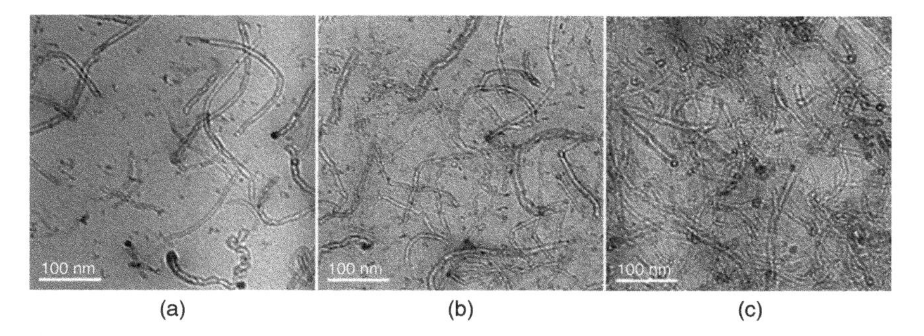

Figure 6.5 TEM images of (a) 2.5%, (b) 5%, and (c) 10% MWNT in TPUN.

Morphological results: TEM images of both the MWNT and nanoclay TPUs were performed in order to determine the visual degree of dispersion (Figures 6.4 and 6.5). In the 2.5 wt% nanoclay sample, the individual platelets are easily identifiable and good dispersion is observed. The 5 and 10 wt% images show some remaining stacks; however, the majority of the images show individual nanoclay platelets. It should also be noted that it does not appear to have any directional orientation from the injection molding process. The MWNT agglomerates are clearly debundled as shown in the two lower weight percentages. In the 10 wt% MWNT TEM image, there is a high density of MWNTs. These are not in bundle form. Bulk material dispersion is assessed by observing TEM images at different locations and view similar images. Again, there does not appear to be any preferred orientation.

OTB results: High heat flux experiments have been conducted. The material was exposed to the hottest portion of the flame, with a 7 mm flame diameter, for 25 s. Samples of 12.7 × 12.7 × 50 mm were cut out of a 12.7 mm × 10.16 cm × 10.16 cm (½″ × 4″ × 4″) compression pressed sheet. The torch was directed toward the square face (12.7 × 12.7 mm). In this orientation, the sample may

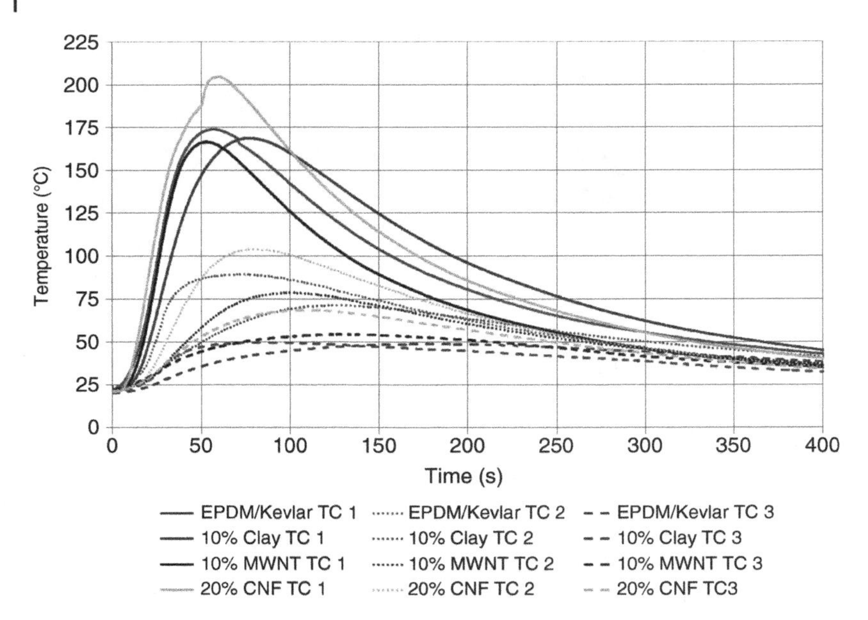

Figure 6.6 Representative temperature profiles of EPDM/Kevlar profile and PNCs tested using the OTB.

be tested twice. At 5 mm increments, three thermocouples were embedded into the center of the specimen to measure the in-depth heat soak temperature during testing. Thermocouples were drilled into the specimen at both ends for the possibility of testing a single specimen twice depending on the ablation rates.

Thermal results: A representative temperature profile of each specimen is shown in Figure 6.6. The specific materials are shown in gray scale and the thermocouples (TC) are distinguished as TC 1 (solid line), TC 2 (dotted line), and TC 3 (dashed line) in the graph. After 25 s, the oxyacetylene flame was removed and heat was transferred through the specimen. In this set of tests, the peak temperature of TC 1 in the CNF sample is 205 °C and occurs at 60 s. CNF has the highest heat soak temperature. MWNT and nanoclay TC 1 peaks at 166 and 174 °C and occur at 50 and 64 s, respectively. The EPDM/Kevlar TC 1 peak, 168 °C, shows the most delay, at 75 s. Figure 6.7 shows the peak in-depth temperatures of EPDM/Kevlar and PNCs between thermocouples. The nanoclay, MWNT, and CNF PNCs lose 13.5%, 33.7%, and 46.5% more mass, respectively, compared to the mass loss of Kevlar/EPDM.

Postfiring images: The samples after firing are shown in Figure 6.8. A 1-mm compact char layer is left at the end of the firing of Kevlar/EPDM. Although compact, the virgin and char material does not have good adhesion.

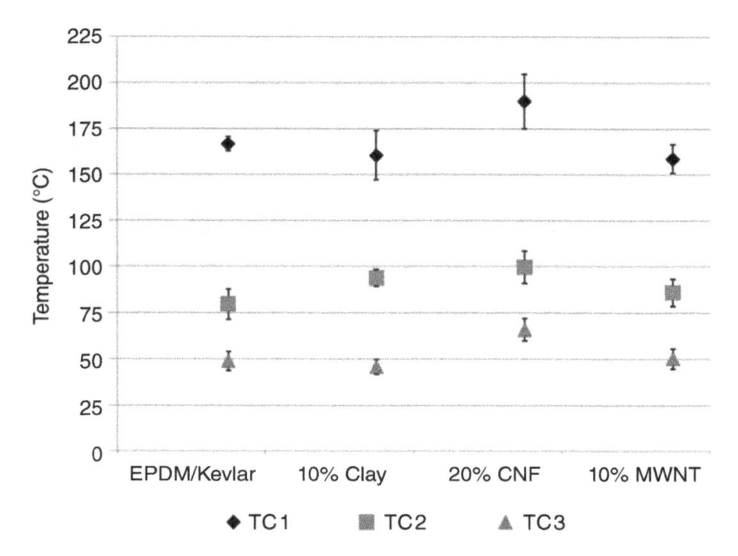

Figure 6.7 Peak in-depth temperatures of EPDM/Kevlar and PNCs.

Figure 6.8 (a) Kevlar/EPDM, (b) 10% Clay, (b) 10% MWNT, and (c) 20% CNF PNC post oxyacetylene torch burn.

The nanoclay PNC post-test material has about twice the char thickness of the Kevlar/EPDM material. Although the char has a crack, the char was maintained well within the virgin material. Similar to the results found in the vertical UL 94 test, many cracks are observed in the MWNT char surface and result in a weak 2-mm-thick char. The CNF PNC char is very thick, greater than 5 mm, as the first thermocouple was in the char layer. The CNF char surface is flat without cracks; however, it is still very weak. The results from this set of experiments are consistent with the flammability and hybrid rocket tests [39, 40].

SEM images of the post-test samples also were taken. The Kevlar/EPDM char material is compact and shows that the individual Kevlar fibers hold the material together as shown in Figure 6.9 [39, 40]. The nanoclay char is also a compact composition. A granular morphology is also observed,

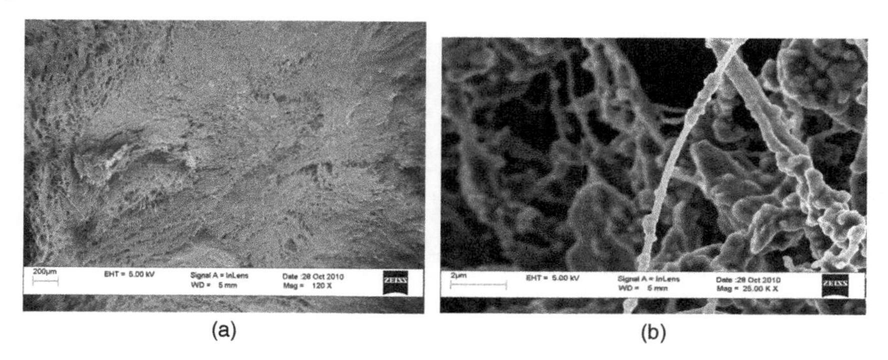

(a) (b)

Figure 6.9 SEM of Kevlar/EPDM char post oxyacetylene torch burn (unit bar at left is 200 μm and at right is 2 μm).

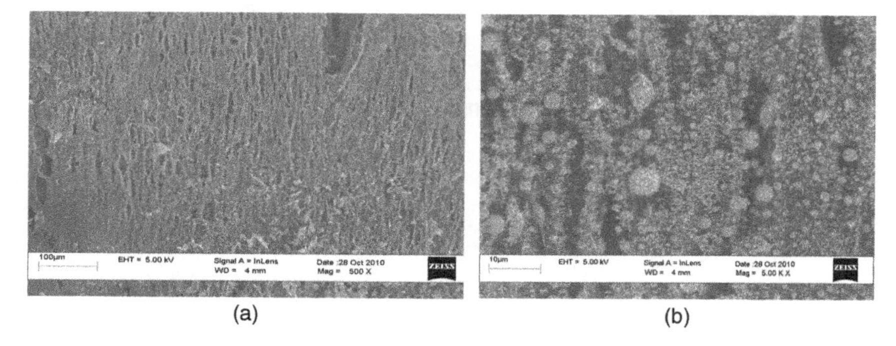

(a) (b)

Figure 6.10 SEM of 10% nanoclay PNC char post oxyacetylene torch burn in progressive magnification (unit bar at left is 100 μm and at right is 10 μm).

as per Figure 6.10. Bubble formations observed in SEM images found in the UL 94 burn tests are observed on the MWNT char surface as well as crack formations, as per Figure 6.11. An in-depth SEM image shows the bundle form of MWNT that is also observed in the UL 94 burn test. The weak CNF char is shown to have a very porous structure, as shown in Figure 6.12. Individual CNFs are also able to be observed.

In summary, the results from the oxyacetylene torch test show that both the nanoclay PNC and MWNT char layers are about twice the thickness of the Kevlar/EPDM char. However, the MWNT char layer is weak compared to the nanoclay PNC char. The CNF char is also very weak and was over five times the thickness of the Kevlar/EPDM char. However, the mass loss from the CNF sample is almost twice that lost from the Kevlar/EPDM sample. In-depth thermal measurements also show that the thermal protection of the PNCs is comparable to that provided by the Kevlar/EPDM, except for the CNF sample,

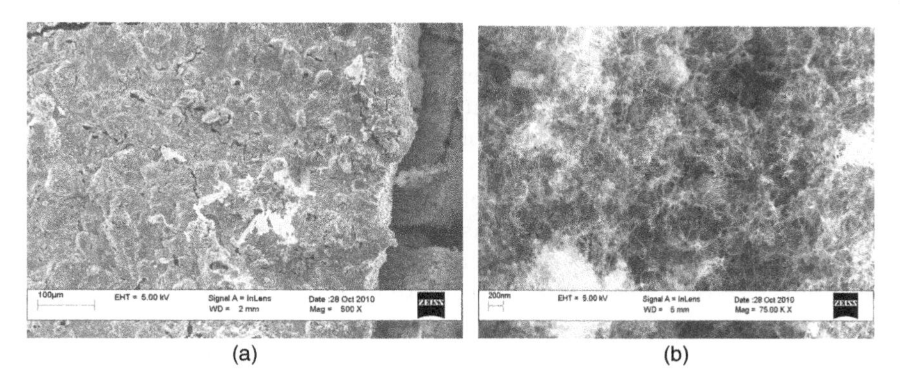

Figure 6.11 SEM of 10% MWNT PNC char post oxyacetylene torch burn in progressive magnification (unit bar at left is 100 μm and at right is 200 nm).

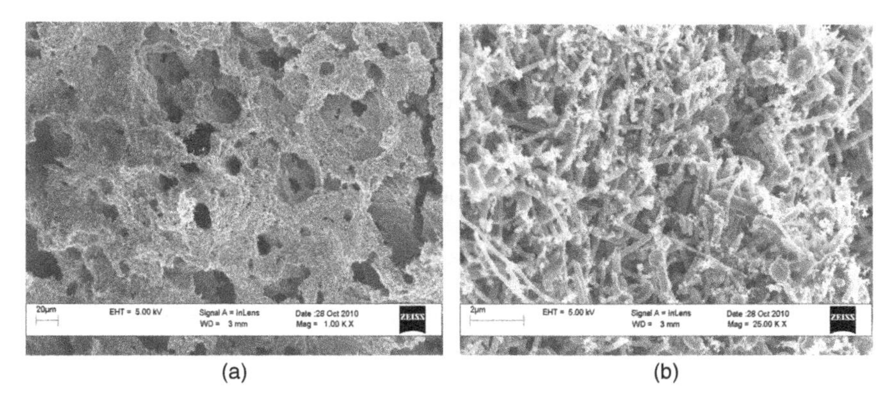

Figure 6.12 SEM of 20% CNF PNC char post oxyacetylene torch burn in progressive magnification (unit bar of left is 20 μm and right is 2 μm).

which had a much higher peak temperature due to the fact that the first embedded thermocouple ended up in the char region. A conventional advanced fiber is therefore needed to boost the PNC ablation performance. More detailed ablation test results of these TPUNs can be found in Allcorn *et al.* [42, 43].

6.4 New Sensing Technology

Two new sensing technologies developed by the Koo Research Group at The University of Texas at Austin and KAI, LLC, Austin, TX aim to understand (i) *in situ* ablation and thermal behavior of ablative materials during ablation testing and (ii) char strength of post-test ablative materials.

6.4.1 *In situ* Ablation Recession and Thermal Sensors

During the past 40 years, many different ablation recession sensors have been implemented for entry vehicle applications; most of them were primarily with high density ablating TPS materials. Some intrusive sensors have been based on the use of breakwire arrangements. In a breakwire sensor arrangement, some pairs of wires are embedded in a TPS plug with their junctions located at different prescribed depths (Figure 6.13). As the TPS surface erodes, the increasing heat melts each wire, opening an electric circuit. However, since the charred material is electrically conductive, it can promote false signals, leading to an improper reading of the TPS status.

One of the most successful efforts in the development of an ablation sensor is the previously mentioned ARAD sensor [51]. Many different version of the ARAD were developed to accommodate different TPS materials. The Galileo ARAD used a carbon/phenolic core material based on a polyimide-tape wrapped around a conductive core for electrical insulation. At high temperatures, the polyimide ribbon (or polyimide tube) charred and became electrically conductive, closing the electrical circuit between the wires. A variation in the measured resistance was related to the length of the sensor and thus to the thickness of the TPS into which the ARAD was installed. For the installation of this sensor, a hole was drilled into the TPS plug. The ARAD sensors that flew on the Galileo probe experienced multiple failures [52]: arcjet test results indicated that the resurrected ARAD occasionally lost its internal electrical connections or had a large measurement uncertainty even when the internal connection remained intact. The study of the Galileo ARAD continued at NASA ARC for the purpose of evolving its design and implementing the sensor in modern low density TPS materials, such as SLA-561V and PICA [53]. An evolved version of the ARAD sensor was recently tested on The Mars Science Laboratory Entry Descent and Landing Instrumentation (MEDLI) Project [54]. The collected data relayed by the Mars lander are under analysis. The ARAD is able to monitor up to 13 mm of depth with an accuracy of ±0.5 mm.

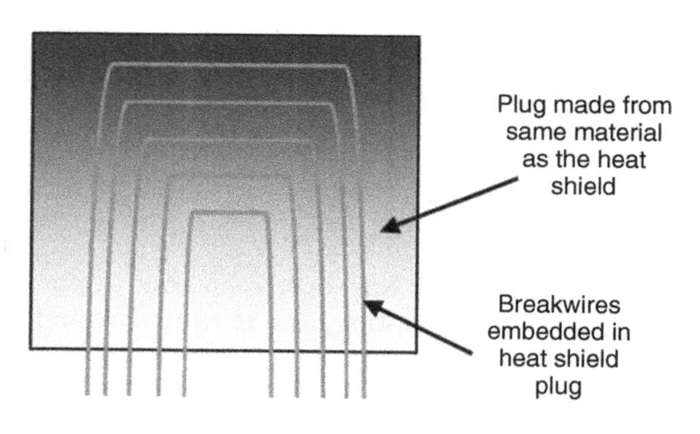

Figure 6.13 Scheme of a breakwire-like ablation recession sensor (plug).

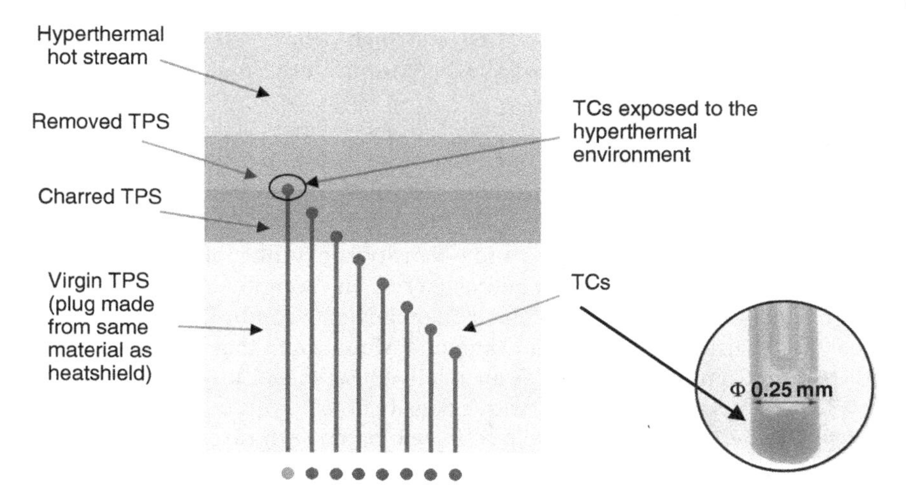

Figure 6.14 Scheme of the proposed thermocouple-based recession sensor.

Natali and Koo [55] through KAI, LLC envisioned a low intrusiveness sensor based on the breakwire-like method with the aim to be cost-effective, easy to manufacture, and scalable. In this approach, the metal wires typically used in this arrangement were replaced with ultrafine commercial thermocouples (TCs). Even if thermocouples were successfully used to instrument TPS materials, the approach introduced in this research presents some distinctiveness, which deserves to be described in detail. In fact, only commercially available processing techniques and raw materials were considered; this is a more affordable and reliable approach as compared to other equivalent solutions in which each single constituent of the measurement chain was produced ad hoc as for the ARAD sensor. Low intrusiveness K-type TCs with a stainless steel outer sheath and a diameter of only 250 μm (0.25 mm) were chosen for the initial phase of this research. The TCs were embedded in the TPS, perpendicularly at the TPS surface. Each sensing head of the TC was positioned at a well-defined depth from the surface (Figure 6.14).

During the heating of the ablator, first the TCs would work as a temperature sensor acquiring data about the state of the TPS. When the temperature of the plug would rise above the melting point of the metal sheath and of the Seebeck junction, the TC would experience a break. Due to this twofold nature of the sensing heads – as a Seebeck junction and as a position marker – it would be possible to obtain a wide range of data on the recession state of the TPS. This technology was tested on several SOTA ablative materials, such as the following:

Low Density Carbon/Carbon (LDCC) – A low-density 2D carbon/carbon composite manufactured by SMJ Carbon was used with a density of 1.34 g/cc.

High Density Carbon/Carbon (HDCC) – A high-density 2D carbon/carbon composite manufactured by DACC, Changwon, South Korea, was used with a density of 1.7 g/cc.

Phenolic Impregnated Carbon Ablator (PICA) – Fiber Materials, Incorporated (FMI) produced PICA material from a low density carbon fiber preform impregnated with a phenolic polymer. The carbon fiber preform, Fiberform®, is also produced by FMI. PICA has a density of about 0.3 g/cc.

AVCOAT 5026-39 – AVCOAT is a low-density epoxy-phenolic matrix filled with silica fibers and phenolic microspheres. The compound was placed in the cavities of a fiberglass-reinforced phenolic honeycomb. This is a material that NASA used on the Apollo Command Module and that will be used on the Orion Command Module as an ablative heat shield. It is manufactured by Textron Specialty Materials with a density of 0.53 g/cc.

Carbon/Phenolic (C/Ph) – Phenolic resin reinforced with rayon carbon fibers MX-4926 is an industry standard ablative for solid rocket nozzles with a density of 1.45 g/cc. It is manufactured by Cytec-Solvay.

Silicon Carbide Fiber-Reinforced Ceramic Matrix (PyroSic® 4686) – PyroSic 4686 is 2D laminated twill woven. The PyroSic® composite (SiRC) is a silicon-carbide-fiber-reinforced glass ceramic composite, with a 42% fiber volumetric ratio and 12% porosity. It is manufactured by Pyromeal Systems in France. It has a density of 1.83 g/cc.

Silicon Carbide Fiber-Reinforced Ceramic Matrix (PyroSic® 2704) – PyroSic 2704 is an enhanced formulation, which Pyromeal Systems suggested has better ablation and thermal properties than PyroSic 4686. It is manufactured by Pyromeal Systems in France. It has a density of 1.83 g/cc.

The authors also used 0.50-mm-diameter K-type TCs to extend study of all these materials.

6.4.1.1 Production of the C/C Sensor Plugs

This sensing technique was first tested on a set of carbon/carbon cubes (CCCs) with dimensions of $15 \times 15 \times 15\ mm^3$. This cube was machined with a waterjet cutter starting from a slab of C/C composite. Miniature K-type TCs having a stainless steel sheath and with an outer diameter of 0.25 mm were chosen [55].

First of all, it was necessary to identify a suitable technology to drill the cubic plug to produce a series of blind holes at an increasing depth from the back face of the cube. Using blind holes and knowing the depth at which the TC must be inserted, the desired arrangement could be produced. Drilled samples were produced using two techniques: a laser source and the electrical discharge machining (EDM) [55]. In the first part of the research, 4 levels of depth were considered: this means that four TCs were embedded in the CCC plugs. Then, the technology was scaled up to 8 levels. EDM proved to be more precise than laser in terms of producing holes having a defined depth. Moreover, EDM was able to produce straight holes, whereas laser produced curved holes

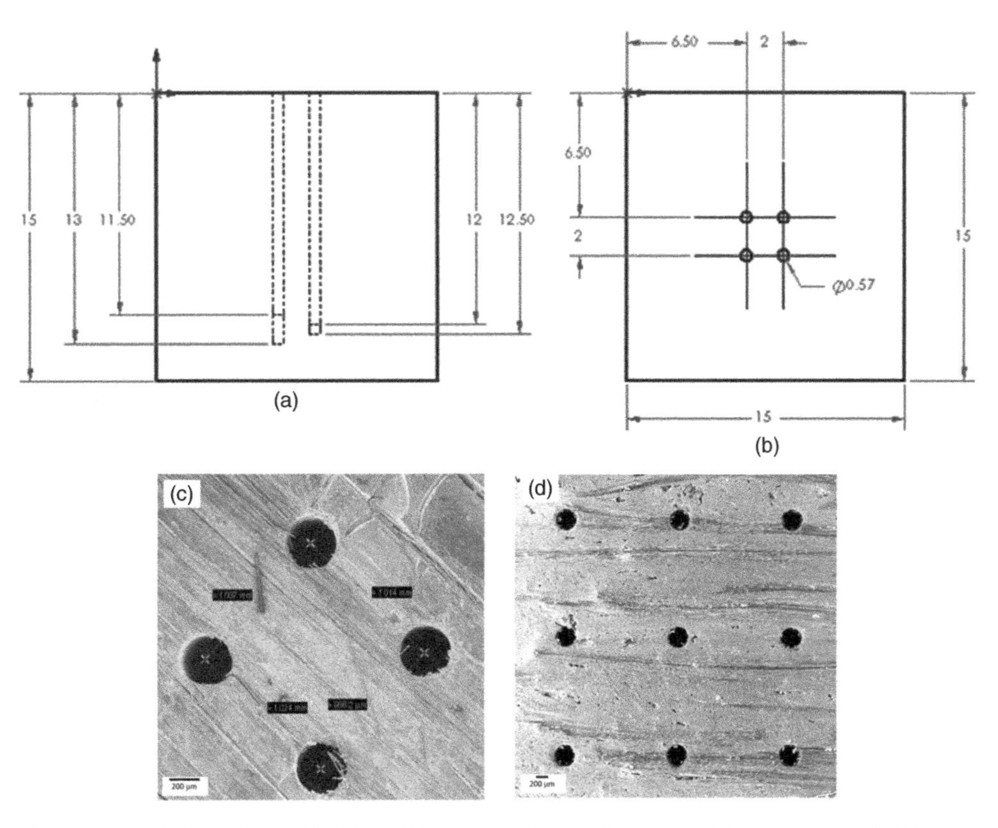

Figure 6.15 Drilling layout for the 4- (a) and 4-level (b) plugs. SEM analysis of the surface of the 4- (c) and 8-level (d) CCC plugs.

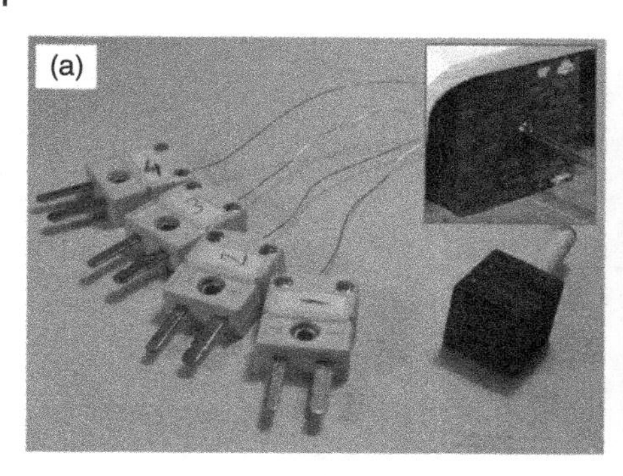

Figure 6.16 Fully assembled 4-level plug (a). CCC plug inserted in the oxyacetylene test bed (b).

(Figure 6.15a,b). Due to this twofold nature of these sensing heads – as a traditional Seebeck junction and as a position marker – it is possible to obtain a wide range of data on the recession state of the TPS. Two types of carbon/carbon materials were considered: a 2D low density material (LDCC, 1.34 g/cc) and a 2D high density material (HDCC, 1.7 g/cc).

Plugs with 4 levels of depths (Figure 6.15a) were first considered using both the LDCC and HDCC; then the scaling up the technology from 4 to 8 levels was considered. A series of 8-level CCC plugs produced using the HDCC material were prepared (Figure 6.15d). The surface of the drilled sample was studied via scanning electron microscopy (SEM). Figure 6.15c reports the surface of the 4-level plug, whereas Figure 6.15d shows the appearance of the holes for the 8-level plug. Once the holes were drilled in the cubic sample, the TCs were inserted and then glued into the CCC plug using a high temperature adhesive suitable for carbon/carbon composites [55]. Figure 6.16a shows the CCC assembled in the sample holder whereas Figure 6.16b shows the fully integrated experiment layout.

6.4.1.2 Ablation Test Results of Carbon/Carbon Sensors

The testing of the sensing technique based on the 0.25-mm diameter TCs embedded in the CCCs was first carried out by means of an oxyacetylene test bed (OTB) designed and built at the University of Perugia (Italy). Subsequent ablation testing on the other materials listed in Section 6.4.1 was conducted at The University of Texas at Austin (UT) using a similar design. The CCC was tested at a heat flux of (930 ± 30) W/cm^2. The heat flux output of the OTB depends on the oxygen/acetylene ratio and also on the distance of the torch nozzle from the sample. The calibration of the torch was carried out by a slug calorimeter in which a copper slug of known mass is exposed to the torch and the power into the slug is derived from the measure of the time versus temperature curve of the slug. Then the power is divided over the measured flame area to provide the calibrated heat flux at each distance [55]. Once all parameters are set up, this OTB is able to produce a flame with a very high repeatability; the maximum error on the measured heat flux was shown to be equal to 3%. In addition to the oxyacetylene torch, the OTB setup is composed of a mechanical setup to hold the sample and ensure consistent test conditions as well as a data acquisition system to measure the *in situ* temperature of test samples using embedded thermocouples. For the 4-level CCC plugs, an HBM Spider 8 data acquisition system was used; since it showed some limitations, the following 8-level CCC plugs were studied using a DAQ NI model 9205 combined with the software NI SignalExpress.

In order to establish the length of each test and depending on the type of material, it was necessary to measure the recession rate of both CCC materials under the selected test conditions. As a result, a series of pristine CCCs were burnt to obtain an estimation of the ablation recession rate: Table 6.2 reports the ablation rate of LDCC and HDCC. To minimize the length of the test and increase the ablation rate, a strong oxidizing flame was used (oxygen:acetylene volume ratio equal to 5:1). For each material and for each drilling layout, a series of pristine CCCs were burnt up to produce a crater deep enough to ensure the burning of the TC placed at the deeper depth. At the end of the tests, once the depth of the craters was obtained and considering the corresponding exposure times, an estimation of the mean ablation rates was also obtained (indicated as $\overline{\text{ARR}}$) (in Table 6.2). Once the depth of the crater was established for all materials and drilling configurations, the ablation rate was shown to depend both on the density of the CC, as well as on the drilling layout. As an example, the 4-level plugs produced with the HDCC exhibited a recession rate equal to 0.023 ± 0.005 mm/s, whereas the 8-level plugs produced with the same material showed a recession rate equal to 0.043 ± 0.003 mm/s. Accordingly, the recession rate also depended on the depth of the produced crater. Such a situation can be explained considering that the piercing capability of the torch changed as a function of the shape of crater, and increased with the exposure time (Table 6.2). This phenomenon is also related to the fact that, with the increase of the time, the torch produced a blind hole that trapped the combustion gas and confined them, producing a higher recession rate [55].

Four-level LDCC plugs Figure 6.16a shows the fully integrated experiment layout. A representative temperature profile acquired during the experiment on the 4-level plug (namely sample identified as LDCC4L1) is shown in Figure 6.17.

All temperature patterns displayed a behavior characterized by three simple steps. In the first part, all TCs properly worked as a temperature sensor; the smaller the distance of the TC from the surface, the higher the temperature displayed. Unfortunately, once a temperature of 1360 °C was reached by each TC, the data acquisition system cut the data and it did not allow study of the signal of the TCs. After about 39 s, the Seebeck junction of the first TC experienced a break. After this period of time, the steel sheath of the TC melted and then the

Table 6.2 Recession rate (RR) for the different materials and configurations using 0.25-mm TCs.

			First approach				Second approach						
Plug tag	# TCs	Material	RR (mm/s)[a]	$\overline{RR}$ (mm/s)	SD (mm/s)	$\overline{SD}$ (mm/s)	RR (mm/s)[b]	$\overline{RR}$ (mm/s)	SD (mm/s)	$\overline{SD}$ (mm/s)	$\overline{ARR}$ (mm/s)[c], [d]	$\overline{SD}$ (mm/s)	Notes
LDCC4L1	4	LDCC	0.0713	0.0645	*0.0055*	*0.0038*	0.0575	0.0576	*0.0012*	*0.0012*	0.056	*0.007*	d)
LDCC4L2	4	LDCC	0.0611		*0.0028*		0.0552		*0.0011*				
LDCC4L3	4	LDCC	0.0610		*0.0030*		0.0602		*0.0012*				
SUM	12	LDCC	0.0584		*0.0090*		0.0571		*0.0029*				
HDCC4L1	4	HDCC	0.0137	0.0262	*0.0008*	*0.0013*	0.0229	0.0251	*0.0005*	*0.0005*	0.023	*0.005*	d)
HDCC4L2	4	HDCC	0.0387		*0.0017*		0.0273		*0.0005*				
SUM	8	HDCC	0.0184		*0.0062*		0.0254		*0.0027*				
HDCC8L1	8	HDCC	0.0510	0.0479	*0.0008*	*0.0024*	0.0336	0.0327	*0.0004*	*0.0004*	0.043	*0.003*	e)
HDCC8L2	8	HDCC	0.0421		*0.0057*		0.0308		*0.0003*				
HDCC8L3	8	HDCC	0.0506		*0.0008*		0.0335		*0.0004*				
SUM	24	HDCC	0.0454		*0.0037*		0.0299		*0.0010*				

a) Recession rate calculated with the first approach.
b) Recession rate calculated with the second approach.
c) $\overline{ARR}$: Average recession rate measured as a function of the final depth of the crater and of the exposure time. SUM is the recession rate calculated based on linear regression with all ablation test data.
d) Measured on a 4-mm-depth crater.
e) Measured on an 8-mm-depth crater.

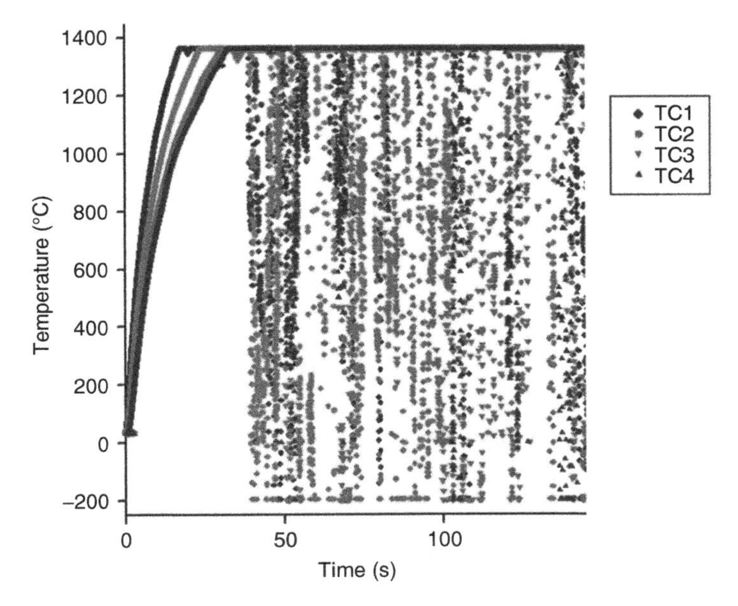

Figure 6.17 Temperature profiles for the 4-level plug LDCC4L1 acquired during the test with the OTB.

junction directly exposed to the flame lost its integrity causing a loss of electric contact between the two terminals of the TCs. Consequently, this point marked the exact time at which TC1 was broken. The strong discontinuity in the temperature profile clearly marked this event. Starting from the first break of TC1, the Seebeck junction started to work improperly, providing an intermittent signal. This mechanism occurred for all four TCs but, due to their different depths in the CCC plug, the first rupture of each TC happened at different times. The working principle of this recession sensor is exactly based on this behavior. The Seebeck junction of the first TC exhibited the loss of integrity after about 39.3 s from the beginning of the experiment. TC2 broke less than 1 s later than TC1. TC3 and TC4 broke at about 46.5 and 50 s, respectively, from the beginning of the test. The first break of each TC precisely indicated the recession, and by means of dedicated software able to find an abrupt discontinuity in the temperature profile and establishing a certain threshold in the TC signal, it is possible to neglect the signal produced after the first break of each TC.

In order to get a quantitative evaluation of the real ablation rate provided by the sensor, the actual position of the spatial markers – the TC heads – was plotted as a function of time at which it experienced the first break. This calculation was carried out by two methods. In the first approach, the four positions of each TC were considered and correlated with the time of the first break (Figure 6.18a). In the second approach, the recession rate was calculated while

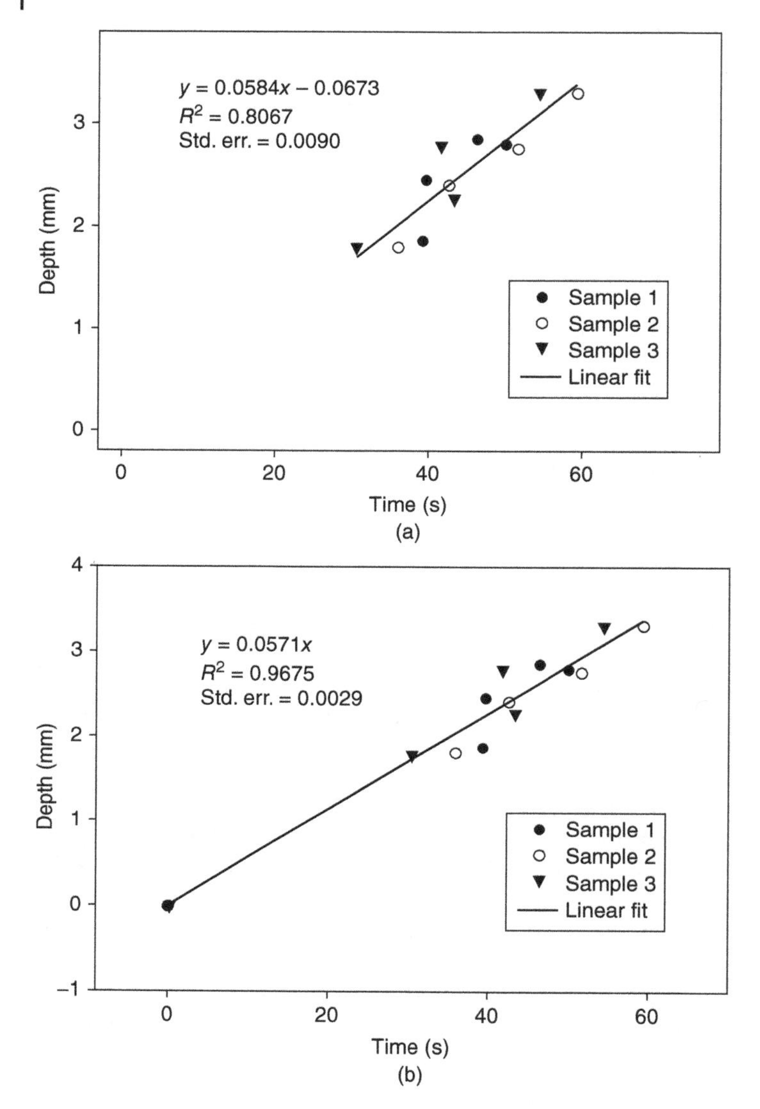

Figure 6.18 Time at which the TCs experienced the first break versus position of each TC head for the 4-level LDCC plugs: interpolation based on the first (a) and second (b) method.

considering the surface as a TC experiencing a break at time 0 s at a depth of 0 mm (Figure 6.18b).

Figure 6.18 reports these plots for all 4-level LDCC samples, whereas Table 6.2 summarizes the numerical results of the mean ablation rate ($\overline{RR}$) and its standard deviation ($\overline{SD}$) for each individual ablation test using the first and second approaches. When all three 4-level LDCC plugs ablation data were combined and indicated as SUM, and given the total erosion divided by the exposure is represented by $\overline{\text{ARR}}$ as in Table 6.2, it was possible to establish that the proposed approaches provided a very good indication of the actual recession speed of the material. These approaches, used to correlate the position of the TCs with their breaking times, were proven to be very effective.

Four-level HDCC plugs In order to confirm the robustness and flexibility of the proposed recession rate sensor approach, a different CC material having a higher density (HDCC, 1.7 g/cc) than the material used to prepare the previous LDCC-based samples (1.34 g/cc) was used to prepare new plugs. The same approach mentioned in the previous paragraphs for the LDCC 4-level plugs was used.

Figure 6.19 shows the plots of the time at which each TC experienced the first break versus the position of each TC head, for all 4-level HDCC samples, whereas Table 6.2 summarizes the numerical results of the recession rate ($\overline{\text{RR}}$) and its standard deviation ($\overline{\text{SD}}$) for individual tests as well as SUM and $\overline{\text{ARR}}$ data. These results show that the system is able to detect the recession rate of the higher density carbon/carbon material. It is also reasonable to believe that the error on the measured RR is related to the inability of the HBM Spider 8 DAQ to acquire the temperatures above 1360 °C, influencing the proper detection of the exact time in which the TCs experienced the first break. In fact, above a temperature of 1360 °C there was no possibility to study the response of the TCs and thus, in the case that a TC had experienced a break above this upper limit, it would have been impossible to detect it correctly. This problem was corrected by introducing a new DAQ (DAQ NI model 9205), which was used on the 8-level plugs. It was used as a voltage DAQ and not as a temperature reader as for the HBM Spider 8 DAQ.

Eight-level HDCC plugs A series of 8-level CCC plugs produced using the HDCC material were tested. The layout used for these experiments is shown in Figure 6.15d. The temperature profiles acquired during the experiment on the plug HDCC8L1 are shown in Figure 6.20. All temperature patterns displayed a behavior highlighted for the 4-level plugs. At the beginning, all TCs properly worked as a temperature sensor; the smaller the distance of the TC from the surface, the higher the temperature displayed. After about 100 s, the first TC reached the melting point of the Seebeck junction and then it experienced a break. Consequently, this point marked the exact time at which TC1 was broken.

This event is clearly marked by a strong discontinuity in the temperature profile. Figure 6.20b shows the details of the temperature profiles in the range of time at which all TCs experienced the first break. After the first failure of TC1, the junction provided an intermittent signal; due to the presence of

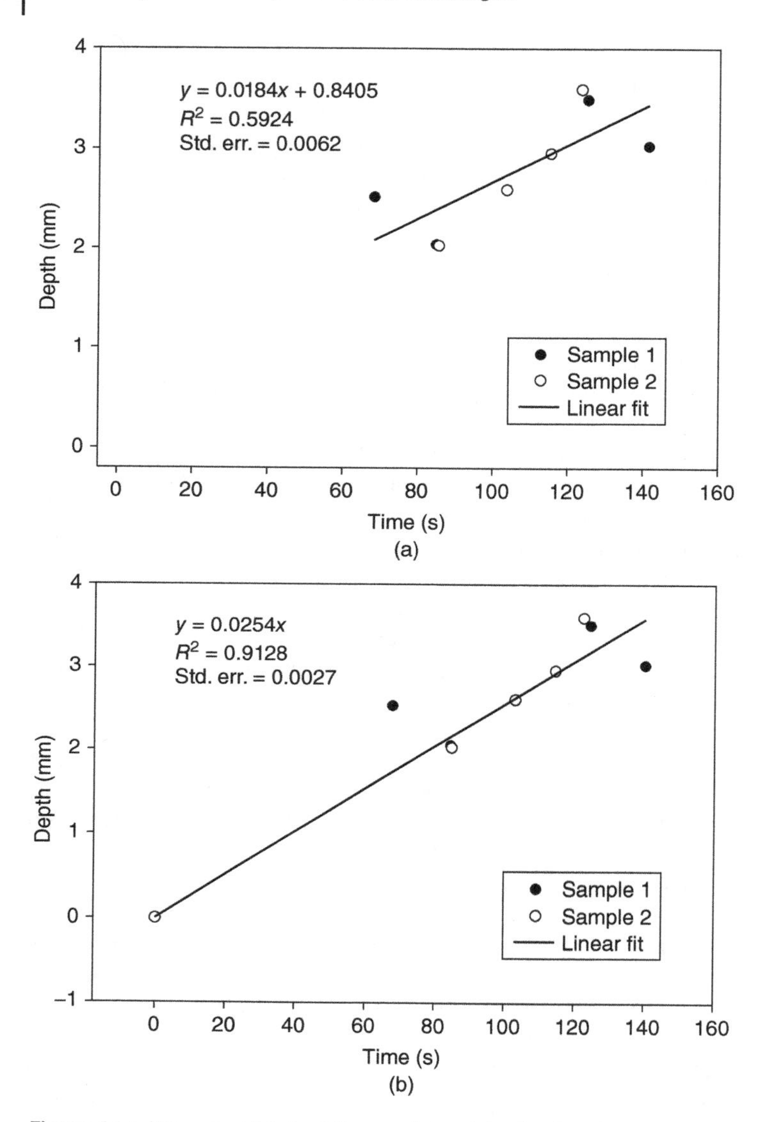

Figure 6.19 Time at which the TCs experienced the first break versus position of each TC head for the 4-level HDCC plugs: interpolation based on the first (a) and second (b) method.

the direct flame, after the TC experienced the first loss of electrical contact, the terminals of the TC continued to melt and the resulting droplets of melt metal were able to intermittently close the electrical contacts between terminals, producing a detectable signal. This process was regulated by several uncontrollable factors producing random and continuous opening or closing

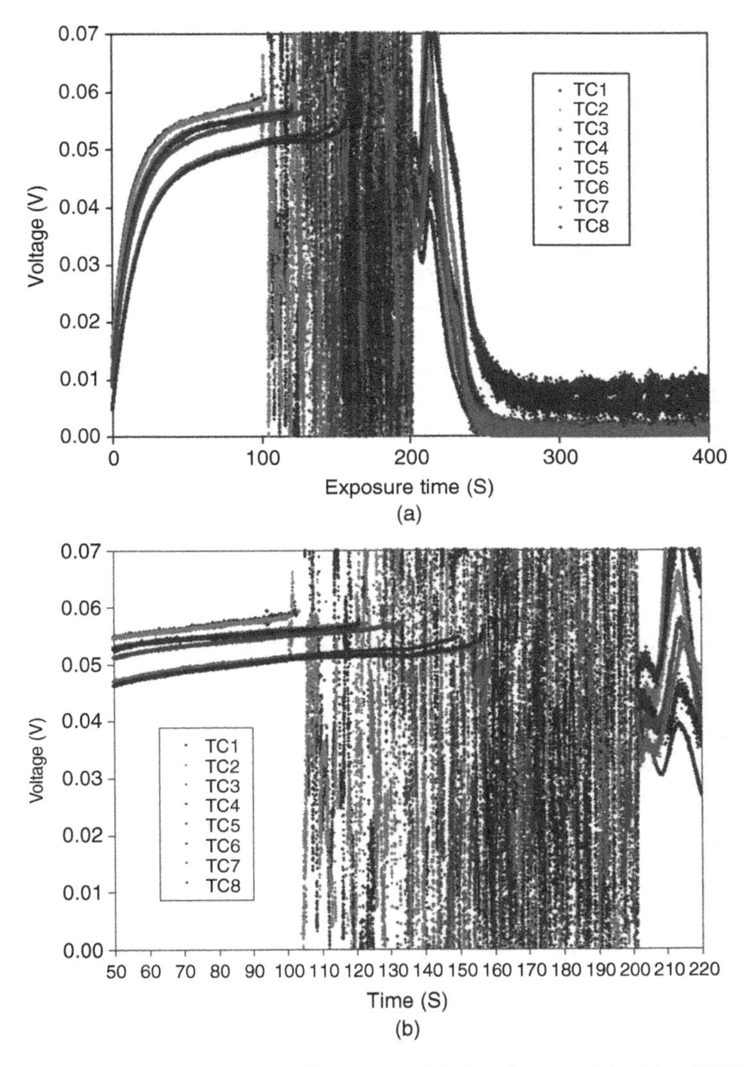

Figure 6.20 Temperature profiles acquired during the test of the 8-level CCC plug with the OTB: sample HDCC8L1 (a). Detail of the temperature profiles over the range of time at which all TCs of sample HDCC8L1 experienced the first break (b).

of the circuit. The above-described mechanism involved all eight TCs but, due to the different depths at which each junction was embedded in the CCC, the first break of each TC happened at different times. In general, the TCs broke according to their distance from the surface marking the recession layer of the material; the working principle of this recession sensor is exactly based on this behavior. In the second step, it is possible to see that once the TCs experienced the first break, all of them continued to display the previously mentioned intermittent signal.

As for the 4-level plugs, in order to get a quantitative evaluation of the real ablation rate provided by the sensor, the actual positions of the eight spatial markers (the TC heads) were plotted as a function of time at which they experienced the first break. This calculation was carried out in two ways and the numerical results are summarized in Table 6.2. In the first method, the eight positions of each TC were considered and correlated with the time of their first break (Figure 6.21a). In the second approach (Figure 6.21b), the mean $\overline{RR}$ (and its standard deviation $\overline{SD}$) was calculated considering the (0;0) point as if there were a virtual TC placed on the surface of the plug experiencing a breaking at the time 0 s. Table 6.2 shows that both approaches provided excellent agreement with the manually evaluated ablation rate of $\overline{ARR}$ data (0.043 ± 0.003 mm/s).

The extrapolated ablation recession rate provided by the approach introduced by Natali and Koo for carbon/carbon composites resulted in very good agreement with the recession rate of the tested material. These tests showed the effectiveness and scalability of the proposed technology as well as the possibility to apply the approach to different types of ablative materials. These results are quite important both from scientific as well as industrial points of view; the authors also suggested testing the proposed system to monitor the ablation recession rate of the nozzle throats of SRMs, which are typically produced with carbon/carbon materials.

In Table 6.2, it was demonstrated that four different approaches can be used to evalute the recession rate (RR) of the LDCC4L, HDCC4L, and HDCC8L sensors. For simplicity and brevity, linear regression based on the combined ablation test approach (SUM) will be used to calculate the recession rate for the remaining of the ablative sensors in the chapter.

Results based on the 0.5-mm-diameter TCs The *in situ* ablation recession sensing technology based on the 0.25-mm diameter TCs was tested using larger thermocouples (0.50 mm). The aim was to verify the possibility of using different types of TCs and drilling techniques. In fact, the use of larger TCs would allow simplification of the drilling process and reduction of the cost of each sensor; on the other hand, a larger TC would introduce a higher intrusiveness. Nonetheless, this study allowed us to determine the intrinsic limitations of these sensors. Moreover, for some applications, the use of 0.5 mm large TCs could promote a higher affordability of the technology. On the basis of these premises, Yee and Koo *et al.* [56] successfully conducted a series of tests on 4- and 9-level CCC plugs using 0.50-mm-diameter TCs. Six 4-level and three 9-level tests were conducted. Table 6.3 reports the data related to the 9-level CCC plugs using 0.50-mm TCs.

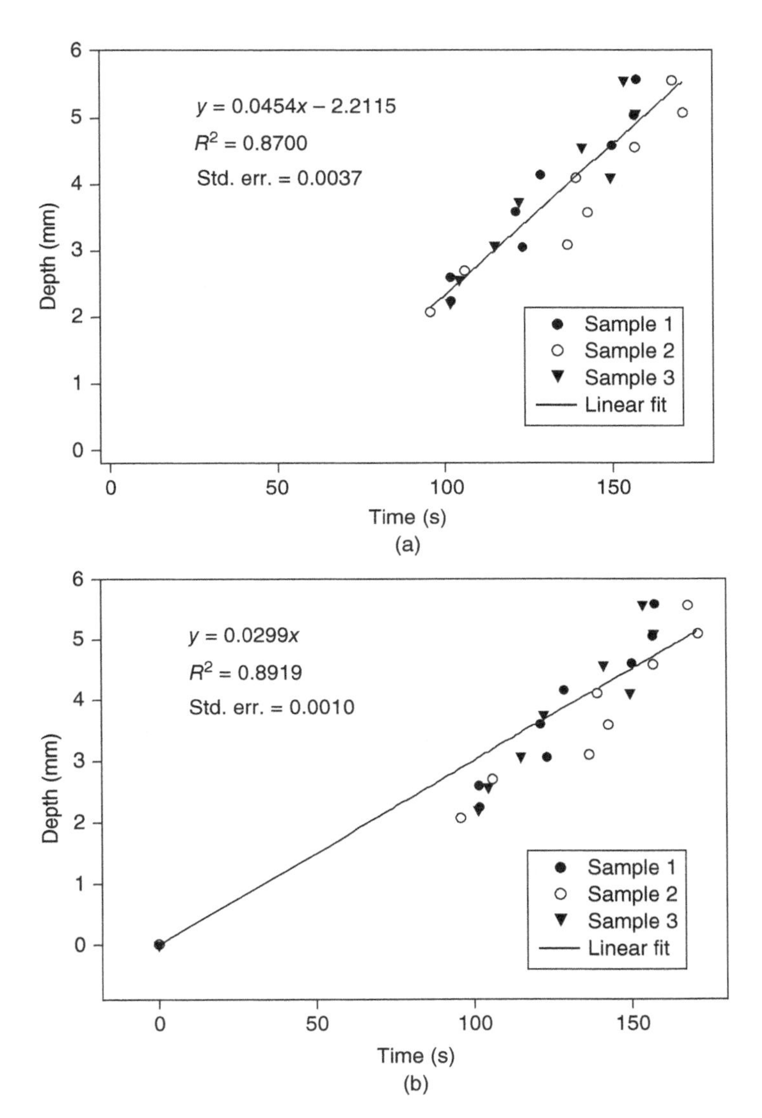

Figure 6.21 Time at which the TCs experienced the first break versus position of each TC head for the 8-level HDCC plugs: interpolation based on the first (a) and second (b) method.

6.4.1.3 Ablation Test Results of Carbon/Phenolic Carbon Sensors

A set of four plugs was constructed using the 0.50-mm type-K TC sensors with carbon/phenolic (C/Ph) as the base material. The primary purpose of these

Table 6.3 Recession rate (RR) for the CCC materials and different configurations using 0.5-mm TCs.

Plug tag	# TCs	Material	RR (mm/s)[a)]	$\overline{RR}$ (mm/s)	SD (mm/s)	$\overline{SD}$ (mm/s)	$\overline{ARR}$ (mm/s)[b),c)]	$\overline{SD}$ (mm/s)	Notes
CCC4L1	9	CCC	0.0303	0.0378	*0.0002*	*0.0003*	0.033	*0.007*	c)
CCC4L2	9	CCC	0.0455		*0.0003*				
CCC4L3	9	CCC	0.0375		*0.0003*				
SUM	27	CCC	0.0353		*0.0007*				

a) Recession rate calculated with the second approach.
b) $\overline{ARR}$: Average recession rate measured as a function of the final depth of the crater and of the exposure time. SUM is the recession rate calculated based on linear regression with all ablation test data.
c) Measured on a 6-mm-depth crater.

tests was to determine how well the ablation detection method performed with a different ablative, such as C/Ph [56]. The break temperature method worked well for predicting average ablation recession rates, although the temperature profile generated in the C/Ph plugs differed significantly from the C/C plugs. This was due to differences in the thermal conductivity and ablation mechanisms between the two materials.

When compared to the processing of the CCC plugs, the production of the sensor system with C/Ph plugs proved to be difficult, since the phenolic matrices were drilled (the EDM was not tested on C/Ph composites). This process caused excessive friction and resulted in erratic drilling. Increased friction resulted in tool embrittlement and subsequent breaking [56]. The C/Ph Tests 1 and 2 both utilized only three TCs as one of the TC holes on each was plugged by broken drill fragments during the fabrication process. Even with three TCs, the sensor plug provided a good ablation recession rate prediction. In Test 2, two of the three TCs were installed at similar depths because thermal expansion and melted chips made precise depth measurement more difficult. The C/Ph TC sensors were tested under a heat flux of about 940 W/cm^2. They performed similarly to the C/C TC sensor, although the average break temperature was not similar to the C/C plugs. This could possibly have been due to different char and combustion products from the C/Ph plug interacting with the melting TCs. Figure 6.22 shows the linear regression analysis when the four sets of C/Ph sensor data were used. The ablation recession rate is 0.0548 mm/s with an R^2 value of 0.8805. Table 6.4 reports the data related to the 4-level C/Ph plugs using the 0.50-mm TCs.

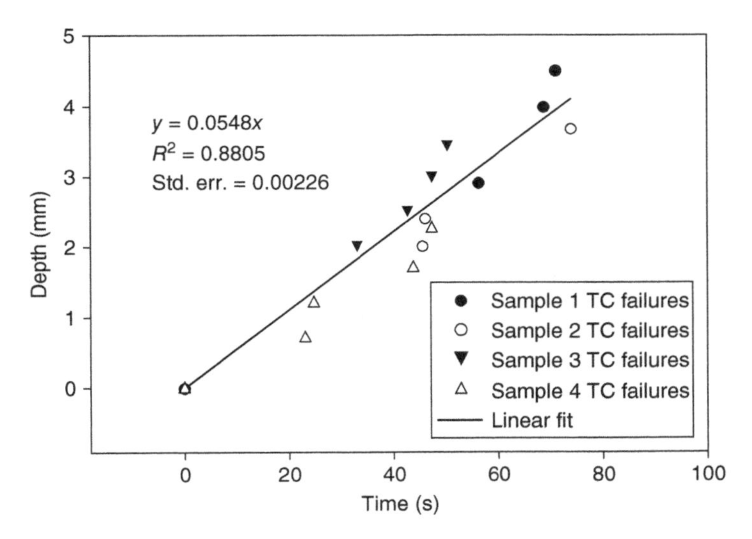

Figure 6.22 Recession data for all C/Ph sensors.

Table 6.4 Recession rate (RR) for the C/Ph and configurations using 0.5-mm TCs.

Plug tag	# TCs	Material	RR (mm/s)[a]	$\overline{\text{RR}}$ (mm/s)	SD (mm/s)	$\overline{\text{SD}}$ (mm/s)	$\overline{\text{ARR}}$ (mm/s)[b),c)]	$\overline{\text{SD}}$ (mm/s)	Notes
CPh4L1	3	C/Ph	0.0586	0.0537	*0.0026*	*0.0022*	0.060	*0.003*	c)
CPh4L2	3	C/Ph	0.0490		*0.0015*				
CPh4L3	4	C/Ph	0.0641		*0.0017*				
CPh4L4	4	C/Ph	0.0430		*0.0030*				
SUM	14	C/Ph	0.0548		*0.0007*				

a) Recession rate calculated with the second approach.
b) $\overline{\text{ARR}}$: Average recession rate measured as a function of the final depth of the crater and of the exposure time. SUM is the recession rate calculated based on linear regression with all ablation test data.
c) Measured on a 5-mm-depth crater.

6.4.1.4 Other Ablation Sensors Results

Other *in situ* ablation recession and thermal sensors data are reported elsewhere using PICA [57, 58], AVCOAT [57, 58], and two SiC CMC (PyroSic 4684 and PyroSic 2704) [58, 59] ablatives. Table 6.5 shows a summary of the ablation recession rate and the density of the different types of ablatives evaluated in this research. The developed *in situ* ablation recession and thermal sensing

Table 6.5 Summary of ablation recession rate of different ablatives for 4-level TC sensors based on combined linear regression of experimental data.

#TCs	Material	Density (g/cc)	RR (mm/s)[a]	SD (mm/s)	$\overline{\text{ARR}}$ (mm/s)[b),c)]	$\overline{\text{SD}}$ (mm/s)
12	PICA	0.3	0.2670	*0.0140*	0.283	0.047
12	AVCOAT	0.53	0.2640	*0.0110*	0.365	*0.200*
14	C/Ph	1.34	0.0548	*0.0007*	0.060	*0.003*
12	LDCC	1.45	0.0584	*0.0090*	0.056	*0.007*
8	HDCC	1.7	0.0184	*0.0062*	0.025	*0.003*
24	HDCC	1.7	0.0454	*0.0037*	0.030	*0.001*
27	HDCC	1.7	0.0353	*0.0007*	0.033	0.007
12	PyroSic® 2704	1.83	0.0100	*0.0019*	N/A	N/A
12	PyroSic® 4686	1.83	0.0220	*0.0009*	N/A	N/A

a) Recession rate calculated with the second approach.
b) $\overline{\text{ARR}}$: Average recession rate measured as a function of the final depth of the crater and of the exposure time.
c) Linear regression analysis based on 3–4 sets of ablation tests.

technology has demonstrated its adaptability to a variety of ablatives under a severe hyperthermal environment for ablative TPS application. Table 6.5 also shows how the low density materials tended to exhibit a higher error on the estimation of the recession rate. This evidence also supports the general conclusion that lower density materials tend to display spallation phenomena and, as a result, they are more prone to produce a higher uncertainly in the recession rate.

6.4.1.5 Summary and Conclusions

In summary, the 0.25-mm TC-based ablation recession sensors of 4 and 8 levels were demonstrated successfully using two types of carbon/carbon composites. The 0.50-mm TC-based ablation recession sensors of 4 and 9 levels were also demonstrated using two types of carbon/carbon composites. Other 0.50-mm TC-based ablation recession sensors of 4 levels were fabricated and tested using carbon/phenolic, PICA, AVCOAT, and two types of SiC CMC materials.

This ablation and thermal sensing technique was easily applied to a variety of ablators: (i) low density ablators (PICA has a density of 0.3 g/cc and AVCOAT has a density of 0.53 g/cc), (ii) medium density ablators (low density carbon/carbon has a density of 1.34 g/cc and carbon/phenolic, MX-4926 has a density of 1.45 g/cc), and (iii) high density ablators (high density carbon/carbon has a density of 1.7 g/cc and two SiC CMCs [PyroSic 4686 and 2704] have densities of 1.83 g/cc).

It is easy to drill holes for low and medium density ablators. The EDM technique is very useful to drill holes for C/C composites, and microdrilling of SiC CMC proved to be the most challenging task. Other techniques, such as laser drilling and microdrilling, have been investigated for the 0.25-mm TC-based sensor using PICA and AVCOAT materials.

This sensing technology is useful for our PNC ablatives development programs to understand this next generation of PNC ablatives. The surface temperature, in-depth thermal, and ablation recession data [55–61] will be used to validate our current numerical modeling efforts using our inverse heat conduction code [62], NASA's *FIAT* [63] and *CHAR* [64] ablation codes and ATK's *HERO* [65] ablation code.

6.4.2 Char Strength Sensor

This research encompasses the study of testing protocols and the design of sensors for evaluating the compressive strength of the char layer of ablative materials used in SRMs [66–68]. The testing protocol that has been developed is the continuation of previous work for determining the compressive strengths between different SRM insulation materials. A crushing test method was further developed and a sensor platform was assembled to perform the tests. The test procedure consists of measuring the amount of force required to crush a given area of the charred sample for a specified depth. The test was repeated for the industry standard Kevlar®-filled ethylene propylene diene monomer (EPDM) rubber and TPU nanocomposite (TPUN) with different weight loadings of MWNT, montmorillonite nanoclay (MMT), and CNF. The energy of destruction or energy dissipated was quantified to determine which ablative exhibited the best performance. Maximum force was also recorded as a secondary quantity to determine char strength. The proposed test method is fully automated to ensure repeatability of each measurement and to remove the potential for human-induced error. Since char layer thickness varies depending on the material, a method of differentiating neat material from char was proposed and explored. The proposed procedure also represents a novel and unique approach to solve the problem of the determination of the char strength.

The schematic of the compression char strength apparatus is shown in Figure 6.23. As the apparatus penetrates the char layer, it applies a force on the char. Since the apparatus is compressing the char as it moves, there is a distance that it travels to accomplish this. Figure 6.23 shows the testing setup and how the apparatus approaches the char sample. Since the apparatus also records the distance that it compresses the char layer, the second method of evaluating the char strength was also implemented, calculating the work done by the apparatus on the char in order to compress it.

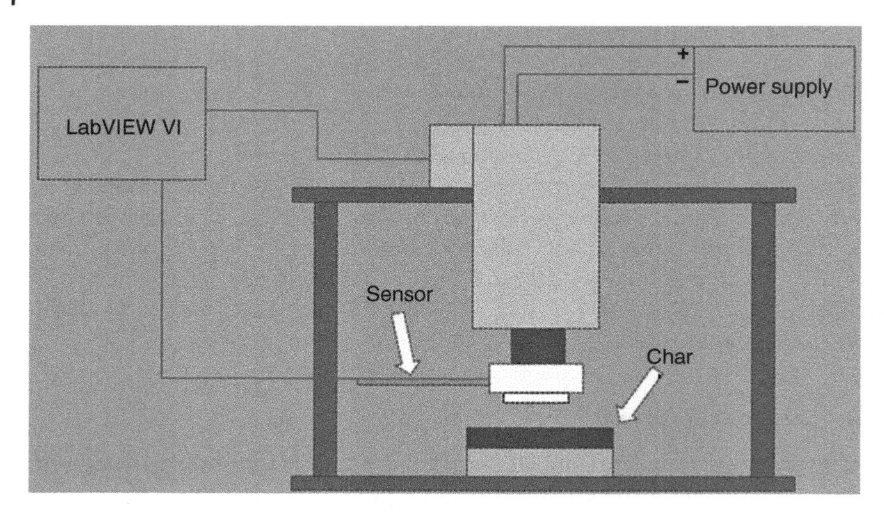

Figure 6.23 Char compressor schematic diagram.

Work (W) is defined as force (F) times the distance (D cos θ).$W = F \times D$ cos θ, and since the force is parallel to the compression, therefore $W = F \times D$. The work done can be calculated after every measurable movement of the apparatus into the char layer. Since the area under a force versus distance plot also indicates the work done, it follows that the sum of the area under the force versus distance curve can be used to calculate the work done by the apparatus on the charred material.

6.4.2.1 Setup and Calibration of Compression Sensor

The compression char strength apparatus consists of three main components: the sensor, the power supply, and the stand to hold the sensor. In order to generate a correlation for the LabVIEW-based program, a potentiometer and a force sensor are calibrated. The potentiometer is calibrated to give a correlation of the actuator's probe tip position as a function of voltage across the potentiometer. The force sensor inside the probe tip is calibrated to determine the force output as a function of voltage. The potentiometer calibration data is obtained using calipers to measure the distance traveled by the actuator and a multimeter to measure the voltage drop across its potentiometer. The force sensor is calibrated using an apparatus made up of four springs sandwiched by two metal plates. As the probe presses down on the top metal plate, the four springs are compressed, and the force sensor inside the probe tip sends a voltage signal to the control program in LabVIEW. Using the potentiometer in the linear actuator, the distance that the top plate was pressed down is known. From the distance that the springs are compressed, then the stiffness of the springs the

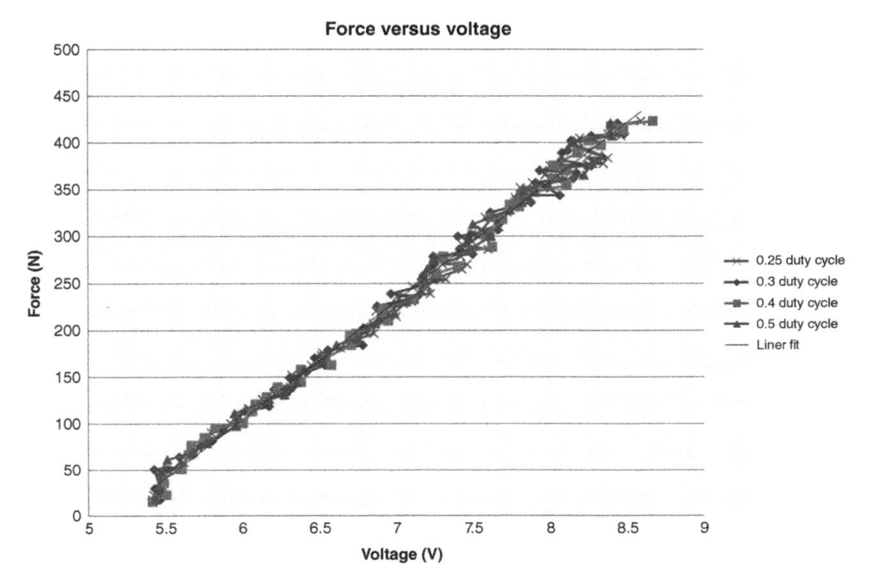

Figure 6.24 Sensor force versus voltage using several duty cycles with a linear fit for a correlation.

force can be calculated. The final step is then to generate a correlation of force as a function of voltage. Sample data of the calibration are shown in Figure 6.24.

6.4.2.2 Analysis Method

Since the criteria for the strength of the char material could consist of a variety of factors, the sensor gathers force and distance (depth) data as the probe compresses the charred material. Using these data, two methods were developed to determine the strength of the char. In the first method, the force data can be used to observe the yield strength of the char. As the piezoresistive load cell implemented in the design changes its resistance in response to pressure, the applied force is recorded. A peak in the force followed by a significant decrease in the applied force on the char will indicate the ultimate strength has been reached, revealing the maximum allowable stress for the material. Using this method, one can rank the strength of the different samples, the strongest being the sample that has the largest ultimate stress. The second method to determine the strength of the char material is to generate a plot of force versus distance (depth). If one integrates the force versus distance plot to find the area under the curve ($W = F \times D$), one can determine the energy required to destroy the char. The strongest char will produce the highest energy of destruction or energy dissipated.

Post-test samples of TPUN-5 wt% CNF, TPUN-5 wt% MWNT, and TPUN-5 wt% MMT specimens were tested using the char strength apparatus to determine their respective strength. More details on the tested materials can be found in Allcorn *et al.* [42, 43]. These TPUN specimens were exposed to the OTB [59, 61] with a heat flux of 600 W/cm^2 for duration of 5 s. Five repeated tests were conducted for each TPUN material.

6.4.2.3 Char Compressive Strength Results

Post-test TPUN samples were used to evaluate the proposed technology. Figures 6.25–6.27 show the force history of the TPUN-5 wt% CNF material, the TPUN-5 wt% MWNT material, and the TPUN-5 wt% MMT material, respectively. Each figure shows five tests for each type of material. After examining the char force data, the 5 wt% MWNT and 5 wt% MMT proved to be stronger than the 5 wt% CNF samples, but close in strength to each other. However, the 5 wt% MWNT force was not resulted to be higher than 5 wt% MMT force for most of the performed tests and, therefore, may be concluded to be slightly stronger.

The first way to establish the stronger char among tested candidate materials was based on the determination of the maximum force needed to penetrate the charred region. Since five tests were performed for each type of char, the average maximum force was calculated. Figure 6.28 shows the results of the average maximum force (measured in N) for all three charred TPUN materials. In order of stronger to weaker char, 5 wt% MWNT (183 ± 31 N) has the strongest char, followed by 5 wt% MMT (164 ± 55 N), then the 5 wt% CNF (66 ± 25 N).

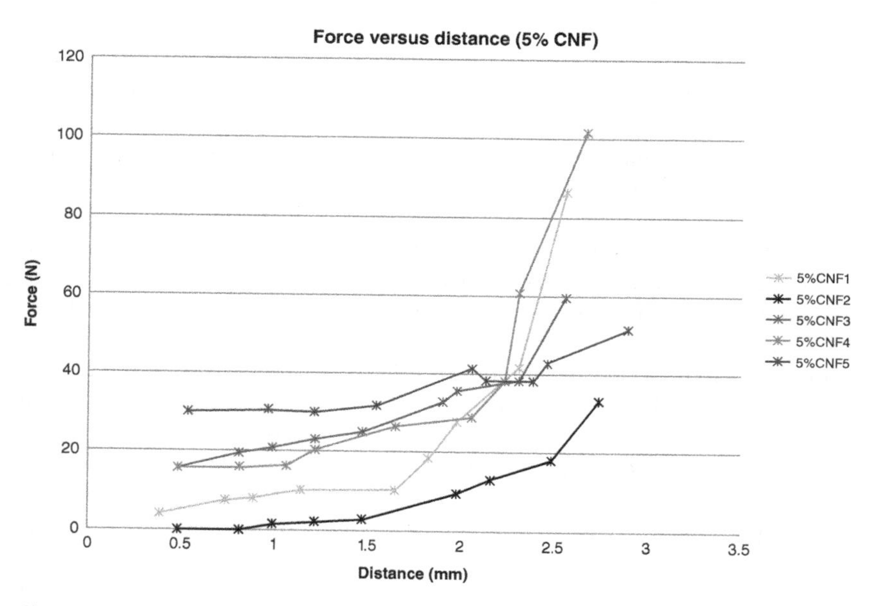

Figure 6.25 Char force data of 5 wt% CNF thermoplastic elastomer nanocomposite.

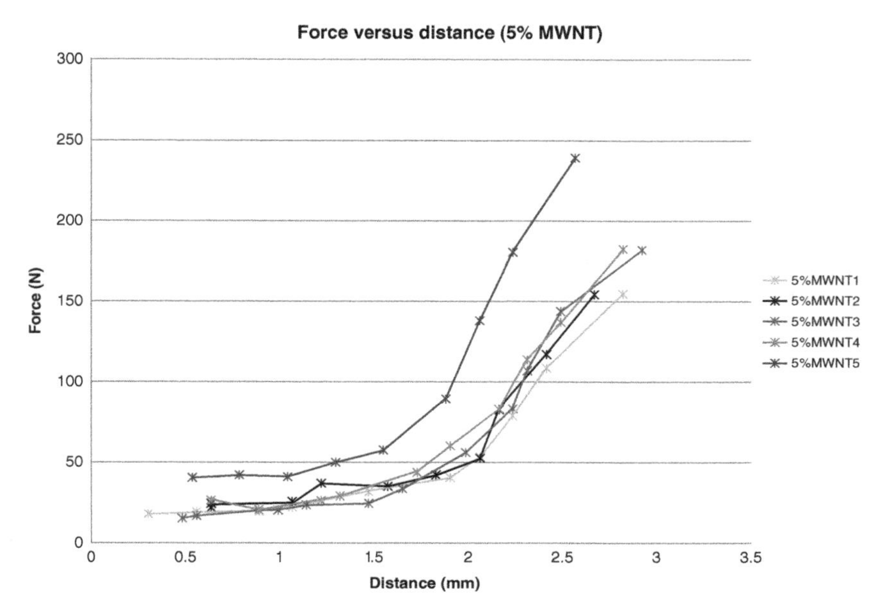

Figure 6.26 Char force data of 5 wt% MWNT thermoplastic elastomer nanocomposite.

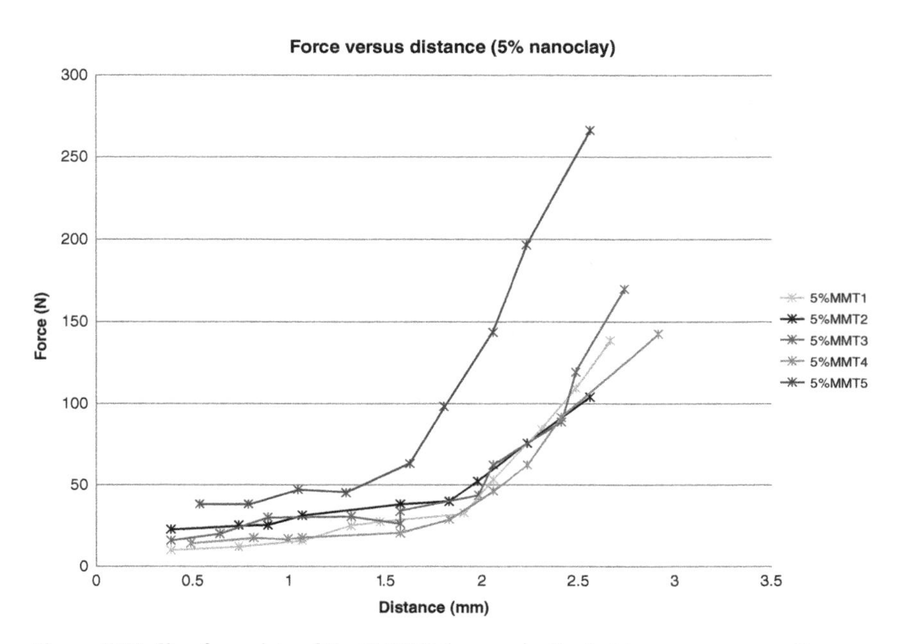

Figure 6.27 Char force data of 5 wt% MMT thermoplastic elastomer nanocomposite.

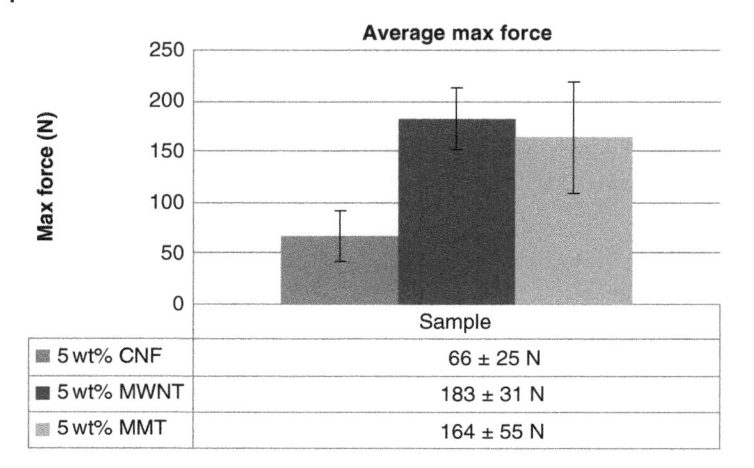

Figure 6.28 Average maximum force of each type of thermoplastic elastomer nanocomposites.

However, in addition to the history of the force experienced by the sensor, the average energy dissipated (measured in N-mm) was defined and calculated. The average energy dissipated was defined as the energy needed to penetrate or destroy the charred layer. Using the force and distance plot, the area under the curve was calculated and averaged among all the tests conducted. Figure 6.29 shows the average energy dissipated for the three types of TPUN samples tested. Within the experimental data, it can be concluded that 5 wt% MWNT is the stronger of the three types of TPUNs with about 155 ± 26 N-mm, followed by 5 wt% MMT (131 ± 42 N-mm), and then 5 wt% CNF (59 ± 26 N-mm). The conclusion from the average maximum force was reaffirmed as the average energy dissipated shows a similar trend.

Thus, the average maximum force for the three charred TPUNs decreases according to the following order: 5 wt% MWNT>5 wt% MMT>5 wt% CNF. In light of our results on TPUNs – here presented and those generated by other researchers – and considering the literature on nanostructured ablatives, this order can be explained by the following considerations. First of all, in view of Figures 6.28 and 6.29, we immediately understand that, in terms of average maximum force, the worst filler is CNF and that, at the same time, the difference between the CNTs and MMTs – considering the standard deviation – is marginal. In other words, CNFs, which are structured on a larger scale than CNTs and MMTs, seem unable to reinforce the charred matrix when compared to the other two fillers. The most plausible explanation is that once the polymeric matrix (the precursor of the char) is completely degraded, the marginal residual char produced by the TPU – which is the same for all different formulations at 5% – is unable to work effectively as a binder for CNF-TPUNs.

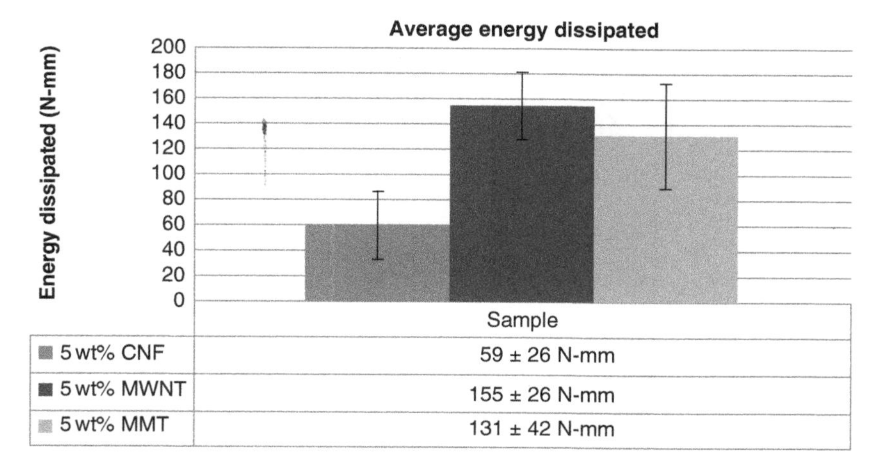

Figure 6.29 Average energy dissipated of each type of thermoplastic elastomer nanocomposites.

In-depth SEM analysis on the burnt compositions also shows the nature of the different chars. Figure 6.30a shows the representative appearance of the in-depth surface of a burnt MMT-based composition. For the MMT-based composition, the char is a hybrid organic/inorganic material in which the platelets of the nanoclay underwent ablative reassembly. Increasing the magnification (Figure 6.30b) shows the nanoclay platelets. The compactness of the wall of the porous char appears to be high, thus leading to relatively good mechanical properties of the medium. The major factor leading to improved ablation performance and char toughness relative to the neat resin or traditional micron-scale filled polymers is the nanoscopic distribution of the silicate layers. This result was shown in the pioneering work of Vaia and coworkers for ablative PNCs [17].

This mesoscopic mixing of inorganic and organic drastically alters the ablative properties. The spatially uniform distribution of aluminosilicate layers on the nanoscale results in the formation of a uniform inorganic char layer at a relatively low fraction of inorganic additive. This nanoscopic morphology is comparable to the length scale of the decomposition and char-forming reactions determined by the temperature profile and the diffusivities of the reactants: a uniform supply of inorganic precursor to the char is available during decomposition. In contrast, the localization of inorganic on the micron-scale associated with traditional filled systems requires a higher inorganic additive loading for the formation of a uniform inorganic char at the surface.

Figure 6.30c shows a typical appearance of the in-depth char of the MWNT-based composition and at high magnification (Figure 6.30d), analogous to the MMT-based formulation, the compactness of the char is clearly visible. On the

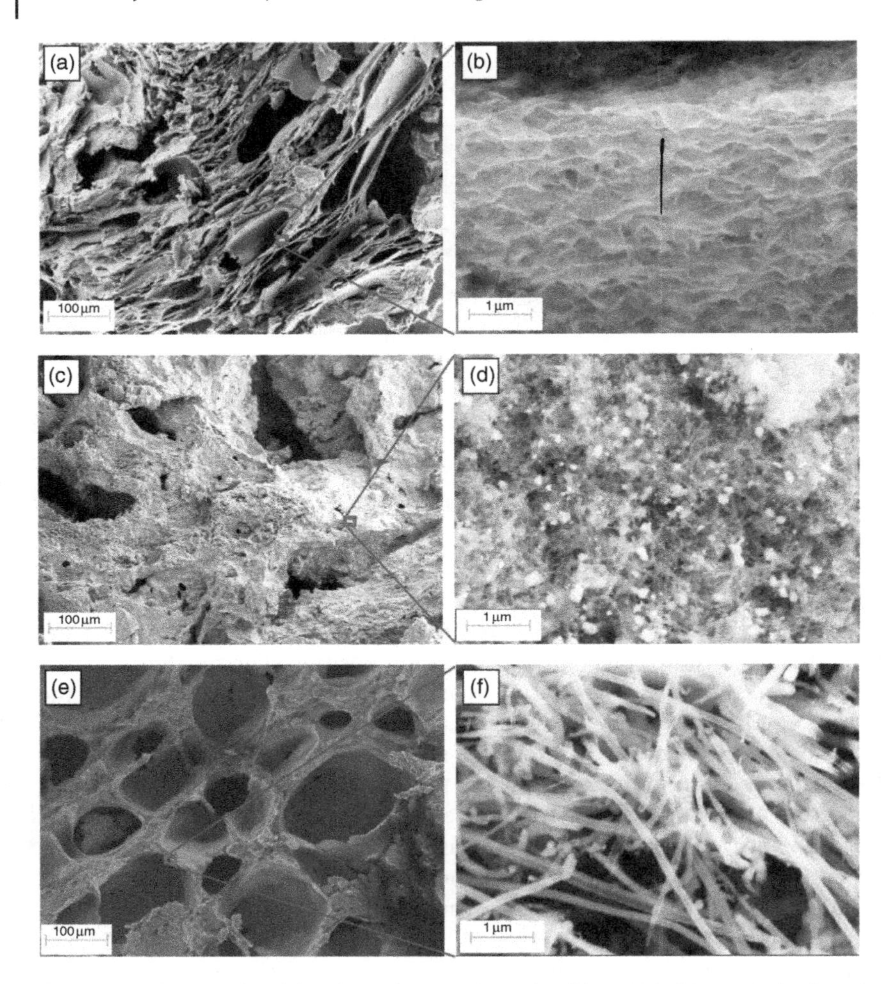

Figure 6.30 SEM results of the charred materials (in-depth). MMT (a, b), MWNT (c, d), and CNF (e, f)-based TPUNs.

nanoscale, no voids can be identified in the charred structure; instead, it is visible as a dense network of burnt MWNTs embedded in the residue of the matrix. However, it is not possible to distinguish the char residue from the fluffed mass of MWNT filaments.

Figure 6.30e reports the appearance of the CNF-based composition; on the nanoscale (Figure 6.30f), it is visible as a networked structure constituted by CNFs and in which large voids are present in the medium. In light of Figure 6.30f, it could be inferred that the carbon nanofilaments tend to stay entangled together only by virtue of the ablative reassembly. In fact, on the external surface of the CNFs, it is not possible to clearly identify the presence of a carbonaceous residue. In light of this analysis, since the carbonaceous interface plays a vital role on the mechanical properties of the resulting char, it is possible. It is possible to conclude that the nanoscopic nature of the charred medium is responsible for the macro response of the medium in terms of char strength and force necessary to crush the burnt material.

In order to further support the above considerations and highlight the different response between MWNTs and CNFs when embedded in a charred matrix, studies on the use of these carbon nanofilaments as flame retardant additives can also be very helpful. Particularly, Kashiwagi *et al.* [69–72] found that CNTs effectively act as flame retardant fillers. They attributed the improved flame resistance to the formation of a continuous, dense protective nanotube network structure able to work as a heat shield. Such a protective barrier slows down the mass loss rate and the material flammability. Consistent with this mechanism, the resultant flame retardancy improved at higher loading of CNTs, and with nanotubes of higher interfacial area (aspect ratio). The use of buckypapers also was shown to constitute a very effective way to exploit CNTs. The dense CNTs network and the small pore size within the buckypaper provide low gas- and mass-permeability. Buckypaper may act as an inherent flame retardant shield when applied onto the polymeric material surface. CNF-based papers were also studied as fire retardant sheets [73]. Wu *et al.* [74] compared the flame retardancy properties of buckypapers produced with MWNTs and CNFs placed on the surface of an epoxy/carbon fiber-reinforced composite. Cone calorimetry testing was conducted at a heat flux of 5 W/cm^2. MWNT-based buckypaper acted as an effective fire shield to reduce heat, smoke, and toxic gases generated during fire combustion. In the case of CNF-paper-based composites, the big pore size of the network resulted in high gas permeability, leading to a poor flame retardant efficiency. Thus, the presence of a CNF paper placed on the surface of a glass fiber-reinforced composite laminates was not effective in improving the fire performance of the composites.

The parallel between the CNT/CNF buckypapers and the charred TPUNs can now be drawn. In presence of a porous medium (when CNFs are used), the degradation and the ablation of the charred matrix can penetrate into the inner layers of the material – as a volume phenomenon – thus, promoting a depth removal of the carbonaceous medium and compromising the mechanical integrity of the charred fluffed medium (Figure 6.31a). Moreover, in this situation, the thermo-oxidation of the char is also favored. Instead, in the presence of a dense and compact network of MWNTs, the char is protected by a continuous, protective layer in which the nanotubes are intimately integrated with the carbonaceous residue (Figure 6.31b).

To understand the response of the porous char of the CNF-TPUNs and, in particular, the effect of the removal of the carbonaceous binder on the fluffed

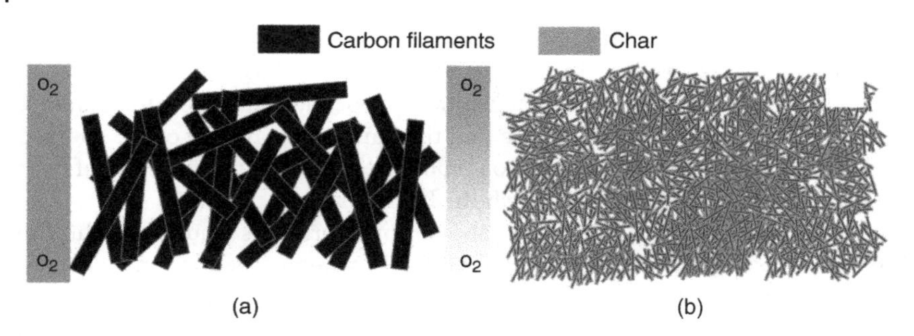

Figure 6.31 Schematization of the different types of porous char: CNF (a) and CNT (b) based char. The carbonaceous interface plays a vital role on the mechanical properties of the resulting char. Moreover, increasing the char compactness, the capability of the oxidizing species to diffuse into the inner layers of the char tends to decrease.

mass of carbon filaments and its relation with the compression strength of the medium, it is worthwhile to consider the ablation mechanism of Phenolic Impregnated Carbon Ablators (PICAs) [75–77] once exposed to an oxidizing hyperthermal environment. PICA is generally based on the use of carbon felt and a phenolic resin as a binder: the felt is produced by chopping carbon fibers with a length of a few millimeters in a water slurry. A water-soluble phenolic resin is used as infiltrant. Standard density PICA showed that at a heat flux of about 500 W/cm^2, the ablation mechanism was oxidation rate controlled; accordingly, the higher the stagnation pressure, the higher the concentration of oxygen diffusing into the boundary layer thus, increasing the oxidation rate and consequently the recession. For PICA-like materials, chemical reactions, phase changes, and even spallation may occur in volume, that is, in the voids present in the medium.

From a microscopic point of view, when oxygen can penetrate into the porous material, inhomogeneous in-depth oxidation and recession of the matrix and fibers can occur; the recession due to oxidation of the charred resin is faster for the matrix than for the fibers. Lachaud *et al.* [78] considered the oxidation of PICAs with the assumption that oxygen may diffuse through the pores of the char and ablate the material in depth. The local surface recession of the carbon fibers and of the carbonized matrix were individually modeled and coupled with the oxygen transport in the porous char. Numerical simulations showed that ablation was controlled by the competition between reaction and diffusion and that it could be either a volume or surface process. The macroscopic model showed that when the oxidizer concentration becomes homogeneous inside the porous char, ablation occurs in volume, that is, the ablation zone involves many fiber layers. The microscopic scale simulations showed that volume ablation weakens the structure of the material because the carbonized matrix was deeply ablated thus, leaving the fibers without protection and binder. It was also concluded that the upper layer of the ablation zone can undergo spallation, under the effect of the shear stress exerted at the surface by the surrounding flow.

All these considerations help us to understand the response of the CNF-TPUN porous char once exposed to a compression stress and, particularly, in the presence of a fluffed fiber mass weakly bound by a limited amount of char.

6.4.2.4 Additional Considerations on the Interpretation of the Data

Although the distance (depth) for all the tests conducted used to determine the char strength of the TPUNs was kept the same for strength comparison purposes, the thickness of the char layer varied from material to material. As an example, some char could be observed to be as thick as 10 mm. If one is truly going to use the energy dissipated to crush the char as the measure to determine strength, then one is compelled to test the entire thickness of char observed. However, one immediate obstacle is evident: how does one determine without any bias where the char layer ends and the neat material begins? It was suggested by Natali that the neat material should be tested in order to observe the rate of change, that is, the slope of the graph. Once the rate of change of the neat material is determined, the charred material can be tested even past the charred layer.

According to this hypothesis, to determine where the neat material starts on the data chart, all one needs to do is to compare it to the neat material chart and determine if the rate of change is the same. Tests were conducted to check if this theory was plausible, that is, that the char data would be significantly different from the neat data. The outcome can be seen in Figures 6.32–6.34. The industry standard Kevlar®-filled EPDM, 7.5 wt% MWNT, and 10 wt% MWNT were used to test this hypothesis. These new formulations were used because, according to our previous results, the MWNTs-based TPUNs – which preliminary tests seemed to produce the best char – also exhibited their best performance in terms of mass loss at a filler weight somewhere between 5% and 10%. In light of the experimental outcomes based on this new approach, in which the response of the neat material was isolated from the real, charred region, it resulted that data of these two zones were significantly different. Thus, this evidence confirmed the hypothesis and allowed us to obtain force data from the entire char layer on all specimens [66–68].

The average maximum force results obtained with the new approach are shown in Figure 6.35; the 10 wt% MWNT ranks higher with a value of 170 ± 53 N, followed by Kevlar®-filled EPDM with a value of 136 ± 62 N, and lastly 7.5 wt% MWNT again ranks lowest with a value of 112 ± 66 N. According to Figure 6.35, considering the standard deviation, the difference between Kevlar/EPDM and TPUN-7.5 wt% MWNT is quite limited; in light of the experimental errors, only the TPUN-10 wt% MWNT showed a higher average maximum force than the other materials. This result can be

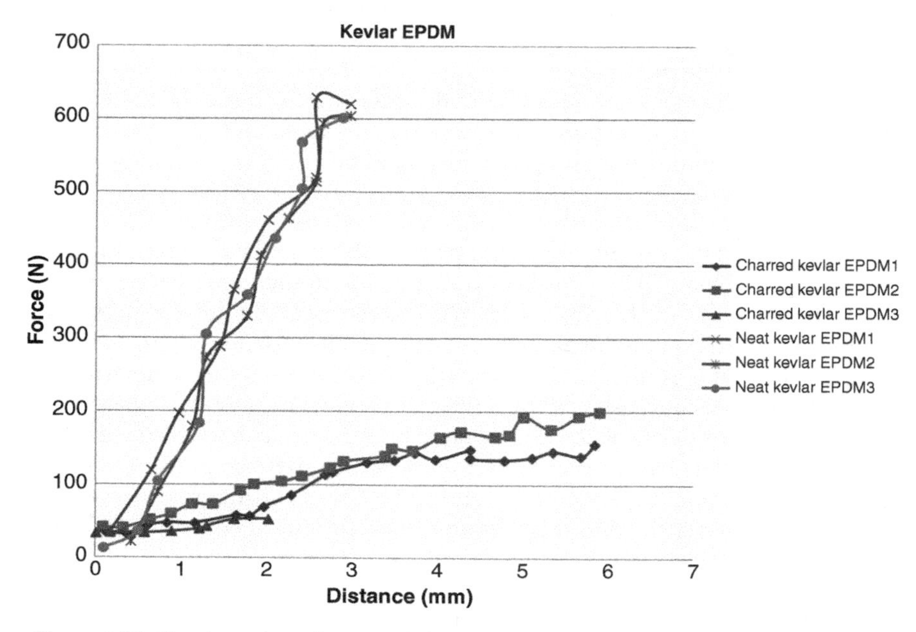

Figure 6.32 Char force data of neat and charred Kevlar-filled EPDM.

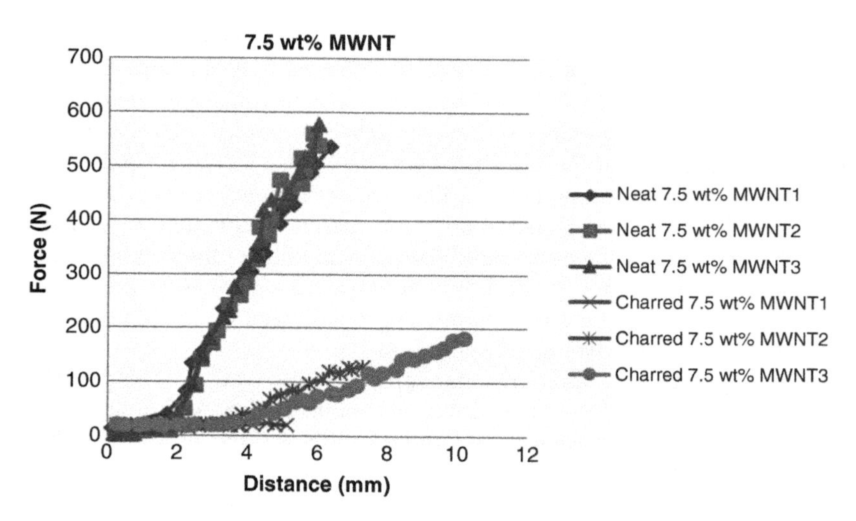

Figure 6.33 Char force data of neat and charred 7.5 wt% MWNT thermoplastic elastomer nanocomposite.

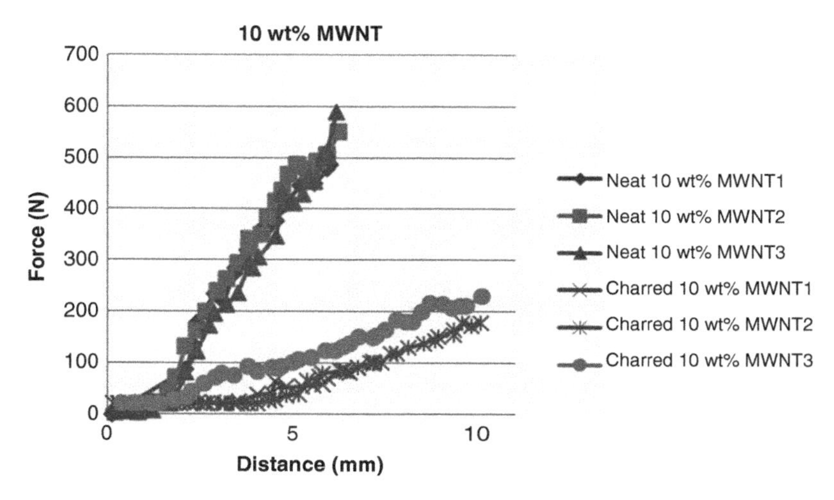

Figure 6.34 Char force data of neat and charred 10 wt% MWNT thermoplastic elastomer nanocomposite.

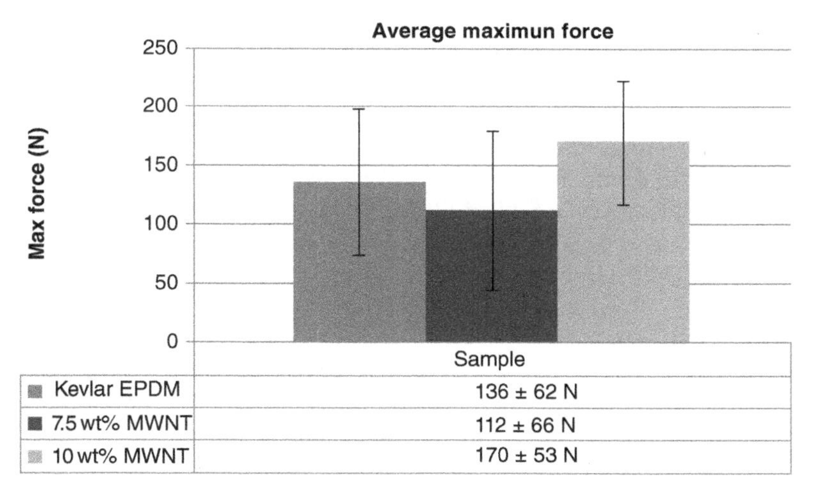

Figure 6.35 Average maximum force of Kevlar EPDM, 7.5 wt% MWNT, and 10 wt% MWNT.

explained as follows: increasing the MWNTs percentage, the resulting char could exhibit a decreased porosity and, in the case where oxidizing reactants were present, slowing down the thermo-oxidation of the burnt material. Thus, an improved tendency to retain a higher amount of char was exhibited. An increased amount of char could better bind the carbon fiber fluffed mass and

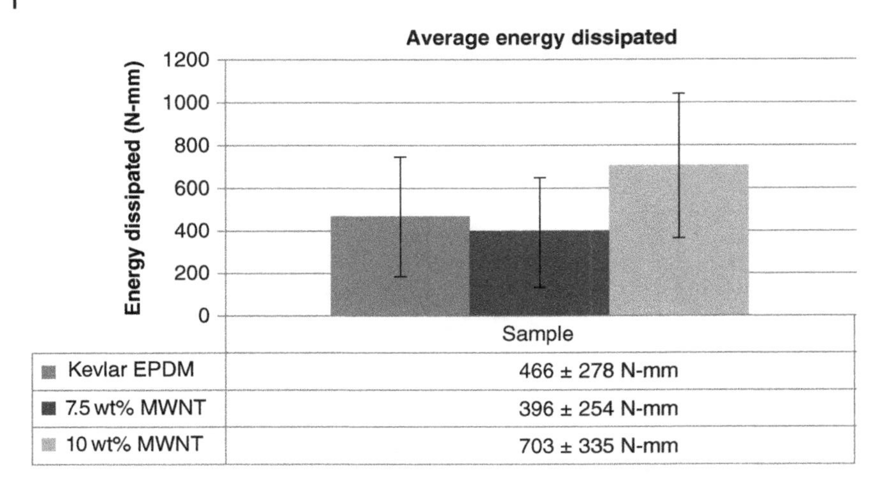

Figure 6.36 Average energy dissipated of Kevlar EPDM, 7.5 wt% MWNT, and 10 wt% MWNT.

consequently, from a macroscopic point of view, a higher average maximum force of the carbonaceous residue could be measured.

The average energy dissipated results are also displayed in Figure 6.36; the 10 wt% MWNT ranks higher with a value of 703 ± 335 N-mm, followed by Kevlar®-filled EPDM with a value of 466 ± 278 N-mm, and lastly 7.5 wt% MWNT ranks lowest with a value of 396 ± 254 N-mm. Thus, even considering that the high standard deviation on the data reduced the differences among the formulations, in terms of char strength, the TPUN-10 wt% MWNT has the potential to replace Kevlar®-filled EPDM. In fact, TPUN-10 wt% MWNT is shown to rank higher in average energy dissipated and average maximum force than Kevlar®-filled EPDM.

6.4.2.5 Concluding Remarks

In this study, a novel system designed to quantitatively study the charred regions of elastomeric heat shielding materials has been envisioned and introduced; the designed char apparatus can be considered a remarkable and unique attempt to systematically study charring elastomeric ablatives. The obtained data showed the reliability and the potential of the proposed approach. Although the compression apparatus is fully functional and capable of performing repeatable tests with relatively few user inputs, char strength knowledge can be further researched. Currently, the shear strength of char from TPUNs is being further tested on different materials and different load cells. The results of the shear tests will be presented in a different study in the near future. Moreover, the possibility to tune and optimize the apparatus is currently under evaluation.

As an example, the development of a geometrically dynamic clamping mechanism for charred samples of any shape and size (within the limits of the compression apparatus) is recommended. The clamping mechanism should not only adapt to the sample shape but also create little or no bending stress on the sample as the material is very fragile. The clamping mechanism concept has been implemented on the shear strength version of the developed apparatus [68, 79, 80]. New protocols to elaborate the data could be considered. As an example, since a material that has thicker char layers will inherently require more energy to crush the char, then in order to compare the char strength of different specimens more accurately, the energy can be normalized by the thickness of the char. Normalizing the energy dissipated by the char thickness would allow a comparison of the char strength of different materials per millimeter, similar to any other property of the material. Thus, the efforts spent on this apparatus could certainly allow the optimization of the formulation of TPUNs and in general of elastomeric heat shielding materials (EHSMs), as well as to improve the comprehension of ablation processes.

6.5 Technologies Needed to Advance Polymer Nanocomposite Ablative Research

The two research areas that are urgently needed in order to advance PNC ablative research are (i) thermophysical properties characterization and (ii) ablation modeling. These two research areas are entirely coupled and are briefly discussed in this section.

6.5.1 Thermophysical Properties Characterization

Ablative material thermal response modeling of polymeric composites requires a multitude of material properties to be characterized with respect to temperature in order to provide accurate and reliable results. This section summarizes and recommends the best methods for collecting data on various thermophysical properties. A total of 11 thermophysical properties are outlined in this section. These properties include (i) thermal conductivity, (ii) thermal expansion, (iii) density, (iv) microstructure, (v) elemental composition, (vi) char yield, (vii) specific heat capacity, (viii) heat of combustion, (ix) optical properties, (x) porosity, and (xi) permeability of *ablative material characterized in the virgin and charred states.* These properties are used as input parameters for numerical models to simulate ablation material thermal response. The NASA PICA Material Property Report provided an excellent source of information for the authors to prepare this section [81].

6.5.1.1 Thermal Conductivity

There are a variety of methods for measuring thermal conductivity that are currently being used. The preferred method is largely dependent on the specific needs of the researcher performing the measurements. Although it is possible to modify some of these methods to perform thermal conductivity measurements with respect to temperature, the only one found in this literature review was the 3ω method performed in a temperature-controlled cryostat chamber by Liang *et al.* [82]. The ASTM C 177 Steady-State Heat Flux Measurements and Thermal Transmission Properties by Means of Guarded Hot Plate (GHP) Apparatus seems to be the most widely used method, but its limited temperature range and low testing rate make it impractical for certain applications. The modified transient plane source method is clearly preferable to the older transient plane source method due to its standardized timing and power parameters. It is quick and easy to use and can perform thermal conductivity measurements up to 500 °C. Laser flash diffusivity has the advantage of being able to test at much higher temperatures than the other methods, up to 2800 °C, but it is limited by a high price tag and its small test sample size. All things considered, the modified transient plane source method will be the preferred method for the widest range of applications, but other methods such as the ASTM E 1225 Standard Test Method for Thermal conductivity of Solids by Means of the Guarded-Comparative-Longitudinal Heat Flow Technique may also be useful method [81].

6.5.1.2 Thermal Expansion

A dilatometer is a proven, reliable way to measure thermal expansion per ASTM E 228 Standard Test Method for Linear Thermal Expansion of Solid Materials with a Vitreous Silica Dilatometer. It is a relatively simple instrument and can be acquired at a lower price than thermomechanical analyzers. Thermomechanical analyzers, however, have the advantage of being much more versatile instruments capable of measuring creep, stress relaxation, strain, and other mechanical properties with respect to temperature in addition to thermal expansion. Both instruments use essentially the same principles to determine coefficient of thermal expansion (CTE).

6.5.1.3 Density and Composition

The two most popular methods for determining density seem to be theoretical determination based on manufacturing process and machining measurements to certain dimensions, measuring them, and dividing mass by volume. Theoretical determinations should not be relied upon when density must be known at different temperatures. If measuring the volume by straightforward methods, such as calipers or a ruler is not feasible, a variation of Archimedes' principle must be used. No descriptions of methods used to measure density with respect to temperature were found. Using ASTM B962-15 Standard Test Methods for Density of Compacted or Sintered Metallurgy (PM) Products Using Archimedes' Principle for determining the density of irregularly shaped samples with surface-connected porosity and altering the procedure as needed is recommended. If it is possible to use calipers to measure length, width, and height, simply weighing the sample and dividing mass by volume is the preferred method.

6.5.1.4 Microstructure

The microstructure of the material can be observed using SEM and TEM techniques. SEM and TEM images before ablation testing will allow the researcher to determine the degree of dispersion of the nanoparticles within the polymer matrix. This will allow the researcher to determine if the material processing techniques were successful. SEM and TEM images can be taken of the neat polymer and of the post-test ablative. This will allow the researcher to compare the images and to determine if the nanoparticles provided ablation-resistant performance. Observing the microstructure of the material is just one piece of the puzzle to determine if the candidate material can be used as an effective ablative.

6.5.1.5 Elemental Composition

Elemental composition is determined using Energy Dispersive X-ray Spectroscopy (EDS). Only a small area of a surface can be analyzed at a time. Because of this, it is important that the researcher takes proper care in cutting a sample with the cross-sectional area of the material to be tested.

6.5.1.6 Char Yield

Pour *et al.* conducted TGA experiments of PC/ABS and PC/ABS/GNP in their study [83]. The results showed that the residual mass, activation energies, and peak temperatures of PC/ABS/GNP composites were higher than those of PC/ABS. The higher temperatures and activation energies are an indicator that GNP reduced the amount of decomposition at higher temperatures. The increased residual mass indicated an increase of char. The char residue acts as an insulating layer and a barrier that reduces the escape of volatile decomposition by-products.

TGA analysis can be conducted in accordance with ASTM E1131 Standard Test Method for Compositional Analysis by Thermogravimetry [84]. The current model available for use at The University of Texas at Austin is the Shimadzu TGA-50 series. This model is capable of reaching temperatures up to 1000 °C with heating rates at 5, 10, 20, and 40 °C/min in an oxygen or nitrogen atmosphere. These TGA data allow us to calculate the activation energy of the polymer using an isoconversion technique.

6.5.1.7 Specific Heat

Per ASTM E1269 Standard Test Method for Determining Specific Heat Capacity by Differential Scanning Calorimeter (DSC) [85], specific heat capacity of virgin ablative material can be found as a function of temperature. For precise data, a heating rate of the test specimen at 20 °C/min in an inert atmosphere (nitrogen) is recommended. The current model available for use at The University of Texas at Austin is the Shimadzu DSC-60 series.

6.5.1.8 Heat of Combustion

The heat of combustion of 49 polymers were measured and calculated by Walters *et al.* [86] using an oxygen bomb calorimeter. The experimental values for the gross and net heat of combustion were compared to the theoretical heat of combustion. Theoretical quantities were found using thermochemical calculations from heat of formation and oxygen consumption. All values agreed within 4.2–4.4%, confirming the oxygen bomb calorimeter produces a reasonably accurate value. Overall, lower heat of combustions indicate a more thermally stable material. The ASTM D 4809 Standard Test Method for Heat Combustion of Liquid Hydrocarbon Fuels Calorimeter (Precision Method) is recommended.

6.5.1.9 Optical Properties

Two widely used methods for determining optical properties are Fourier transform infrared spectroscopy and spectrophotometry. The former is capable of collecting high-resolution data over a large spectral range, whereas the latter is better suited for reflection or transmission properties as a function of wavelength. Ideally, both methods would be employed when performing a complete optical characterization of a material.

6.5.1.10 Porosity

Although several methods for measuring porosity are mentioned in the literature, ultrasonic attenuation is the most relevant method. It has the advantage of being nondestructive, and should not be difficult to use for obtaining porosity with respect to temperature. If surface porosity or surface void density is of interest, deflectometry will likely be the best method. For some materials and applications, using a microvolumeter to measure porosity may work quite well. However, the experimental setup for this method is quite complex and it has been primarily used in geological applications to date. Other instruments, such as the PoreMaster 60 automated mercury porosimeter and the helium pycnometry can be used to characterize this property [81].

6.5.1.11 Permeability

Permeability measurements for porous rocks are typically measured by geologists using the pore pressure oscillation method. This is most likely the best method to use for other materials as well, but some modification may be required depending on the material being studied. A computational method for characterizing the permeability of fabric preforms based on the Lattice Boltzmann method has been developed and shown to be reliable at predicting effective preform permeability based on the number of fabric layers in the preform, the shear angle, and the compaction force. This or a similar program may be useful for predicting permeability in a specific set of circumstances. SRI International (SRI) and Southern Research Institute (SR) are two independent contractors in the United States who can conduct this measurement [81].

6.5.2 Ablation Modeling

Most ablatives are composed of multiple components that erode by a combination of thermal decomposition, chemical reaction, and mechanical removal. At high temperature, the resin components will pyrolyze and flow out into the external stream where they will react chemically as additional ingredients (this section was prepared based on two texts: M. Salita's *Basic Analytical and Numerical Methods for Propulsion and Aerodynamic Analysis of Solid Propellant Rockets* [87] and G. Duffa's *Ablative Thermal Protection Systems Modeling* [88]). Other components ("reinforcing fiber") will not pyrolyze but will react directly with the external stream once the surrounding resin is gone. Much of reinforcing fibers will not react completely, and the remaining material is known as "char." Thus, the material will decompose from an initial ("virgin") density ρ_v to a final ("char") density ρ_c. The char can be solid (e.g., carboneous char) or liquid (melt layer), and may be removed either mechanically by shearing of the boundary layer, or by chemical reaction with the oxidizing species in the adjacent boundary layer. The "pyrolysis depth" is often defined as the depth where the material density has decreased by 2% from the ρ_v, whereas the "char depth" S_{ch} is often defined as the depth where the material density has decreased to within 2% of ρ_v. The chemical reaction of the boundary layer gas with the char and pyrolysis products is known as "surface thermochemistry." Since the total heat flux depends on the surface thermochemistry, the decomposition rate is coupled to the surface thermochemistry.

The ablation modeling depends on whether the material is decomposing or non-decomposing. All PNC ablatives are decomposing materials. This section discusses model for decomposing materials with or without charring.

For most materials, density decreases continuously as a function of temperature only: $d\rho/dt = f(T)$. Figure 6.37 shows different models of

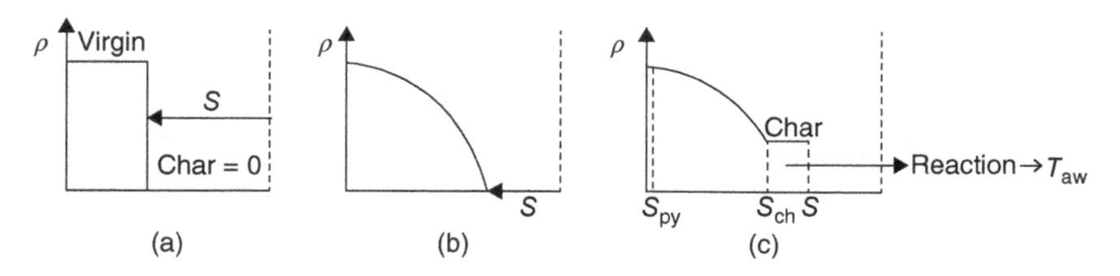

Figure 6.37 Models of material erosion: (a) non-decomposing, (b) decomposition of noncharring material, and (c) decomposition plus chemical reaction of charring material [87]. T_{aw} is the adiabatic wall temperature.

material erosion where (a) is for non-decomposing material, (b) is for decomposition of noncharring material, and (c) is for decomposition plus chemical reaction of charring material [87]. If the char density of the material was zero, then the profile of density would eventually resemble Figure 6.37b. The surface recession depth S and pyrolysis depth S_{py} would be the same, and equal to the distance from the initial surface to the location where the material density is zero. This process involves a reduction of material density at constant volume, $dm = Vd\rho$.

If the density of the char is nonzero, a layer of constant-density char will exist that is fully pyrolyzed but continues to react chemically with the boundary layer gas. The outer location of this layer defines the recession depth, whereas the inner location defines the char depth S_{ch} (Figure 6.37c). In summary, severe heating of walls of case/nozzle/insulation can cause two types of material erosion:

a) Chemical ablation (reactive) – Part of the material will pyrolyze and percolate into the boundary layer where it reacts. Part of material will react directly with boundary layer gas known as "charring" resulting in surface recession.
b) Mechanical removal (inert) – Surface melt failure (inert) and shear removal of weakened char.

The analysis of this process requires two-part coupling:

Surface thermochemistry
gas temperature T_{sat} adjacent to the wall (in boundary layer)
rate of reaction (char rate) of the adjacent gas with surface material
heat flux into the wall
Material response
propagation of thermal wave into the material
decomposition and pyrolysis of part of the material
Coupling
T_{sat} depends on the pyrolysis rate and char rate, but pyrolysis and char rates depend on T_{sat}.

Two industry-standard codes that can provide this coupled analysis are as follows:

- ACE [89, 90]: Calculate surface thermochemistry that provides table of T_{sat} versus pyrolysis rate B'_g and char rate B'_c, where B'_g is the normalized mass flow rate of pyrolysis gas from surface and B'_c is the normalized mass flow rate of char from surface
- CMA [91, 92]: Run material response code and interpolate in T_{sat} table.

Surface Thermochemistry – The lowest temperature at which no condensed products appear is defined as the "saturation temperature" T_{sat}. The calculation of T_{sat} is implemented differently in thermochemical codes ACE [89, 90] and CET93 [93] as follows:

- ACE: T_{sat} is determined internally.
- CET93: User must iterate manually for T_{sat}.

The Aerotherm Chemical Equilibrium (ACE) code was originally developed by Kendall in the late 1960s [89]. Salita developed a user-friendly version of ACE in 1999 [90], which is currently used by Koo Research Group.

1D Material Ablation: CMA stands for Charring Material Thermal Response and Ablation Program, originally developed by Moyer and Rindal of Aerotherm in 1970 [91]. The code was improved over the following four decades, with new versions released in 1987, 1990, 1992, and 2004. Many organizations have written their own codes patterned after CMA that still solve the original energy equation, such as Thiokol (ABLATE), TRW (CARE), NASA Ames (FIAT), and Northrop Grumman (CMA92, and CMA04). Salita developed a user-friendly version of CMA92 in 1992 [92], which is currently used by Koo Research Group.

2D Material Ablation: A 2D thermal and ablation solver TITAN has recently been developed at NASA Ames Research Center [94] and is coupled with a Navier–Stokes solver GIANTS for external flow and MEIT for the boundary layer flow. The code solves the energy balance on a moving body-fitted grid to calculate the shape due to surface recession. The equations are discretized with a finite-volume approximation. A time-accurate solution is achieved using an implicit time-marking technique with Gauss–Seidel line relaxation with alternating sweeps. Benchmark solutions have been calculated and compared with available solutions to check code consistency and accuracy.

1D/2D/2D-axi/3D Material Ablation: The CHAR (*CH*arring *A*blator *R*esponse) code developed by NASA-JSC [64] is a 1D/2D/2D-axi/3D thermal and ablation analysis tool. This work concentrates on the derivation and verification of many terms and boundary conditions in the equations that govern three-dimensional heat and mass transfer for charring ablating TPSs including pyrolysis gas flow through the porous char layer. The governing equations are discretized according to the Galerkin finite element method with first- and second-order implicit time integrators. The governing equations are fully coupled and are solved in parallel via Newton's method, and the fully implicit linear system is solved with the Generalized Minimal Residual method. Verification results from problems with analytical solutions and the Method of Manufactured Solutions are presented to show spatial and temporal orders of accuracy as well as nonlinear convergence rates. Modeling details and verification of CHAR's thermoelastic and inverse heat conduction and ablation capabilities are also presented [64]. This code is currently used by Koo Research Group.

The Porous-material Analysis Toolbox based on Open-FOAM (PATO) is a fully portable OpenFoam library developed at NASA-Ames [95]. It is implemented to test innovative multiscale physics-based models for reacting program materials that undergo recession. Current developments are focused on ablative materials. The ablative material response module implemented in the Porous-material Analysis Toolbox relies on an original high-fidelity ablation model. The governing equations are volume-averaged forms of the conservation equations for gas mass, gas species, solid mass, gas momentum, and total energy. It may be used as a SOTA ablation model when the right model options are chosen. Three physical applications were initially analyzed: (i) volume-averaged study of the oxidation of a carbon-fiber preform under dry air, (ii) 3D analysis of the pyrolysis gas flow in a porous ablative material sample facing arcjet, and (iii) comparison of the SOTA and a high-fidelity model for the thermal and chemical response of a carbon/phenolic ablative material.

The PATO code coupled with thermodynamics and chemistry library Mutation++ is used as a third-party library to compute equilibrium compositions, gas properties, and solve the SOTA boundary layer approximation to provide the ablation rate and the element mass fractions at the surface of the material [96]. The model is applied to the detailed analysis of boundary layer and pyrolysis gas flows within a porous carbon/phenolic ablator characterized in a SOTA arcjet test. The selected configuration consists of an iso-flux ellipsoid-cylinder sample subjected to a 2.5 MW/m^2 heat flux with a decreasing pressure gradient from the stagnation point to the cylinder's side. During the first tenths of a second of the test, boundary layer gases percolate through the sample. As the sample heats up, the internal pressure increases inside the sample due to pyrolysis-gas production. The resulting pressure gradient blocks the boundary layer gases and leads to a pyrolysis gas flow that separates into two streams: one going toward the upper surface and another going toward the lower pressure side under the shoulder of the sample. The sample's subshoulder zone is significantly cooled down, while a temperature increase is observed in depth. Implementing this model of this study in space agency codes will allow improved ground-test analyses and provide more accurate material properties for design. This code is currently used by Koo Research Group.

An excellent review of the governing equations and boundary conditions used to model the response of ablative materials exposed to a high-enthalpy flow was recently presented by Lachaud *et al.* [97]. At least 25 codes are currently in use or in development, with an active community both maintaining SOTA capability and seeking to increase the fidelity of the SOTA model. Table 6.6 shows the currently available simulation tools for the ablation model according to Lachaud *et al.* [97]. Design-rated material response codes currently in use implement a heritage model (from the 1960s) in which the equation parameters may be modified to model different materials or conditions. Research and development codes developed for analysis – at least in a first stage – are generally more advanced but often not as robust. Current research efforts undertaken in the community are various and complementary;

Table 6.6 The currently available simulation tools for ablation modeling [97].

Name	Contact	Owner	Users	Applications	References
Amaryllis	T. van Eekelen	Samtech, Belgium	EADS Astrium, ESA	Design	[21]
CAMAC	W.-S. Lin	CSIST, Taiwan	Taiwan Ins. of Sci. Tech.	Unknown	[22]
CAT	N. N. Mansour	NASA ARC, USA	NASA ARC	Analysis	[23]
CHALEUR	B. Blackwell	SNL, USA	SNL	Design	[24]
CHAP	P. Keller	Boeing, USA	Boeing	Design	[25]
CMA	R. Beck	Aerotherm, USA	NASA, SNL	Design	[26]
CMA/SCMA	C. Park	Tokyo Univ., Japan	JAXA	Design	[27]
CMA/KCMA	P. Reygnier	ISA, France	ISA/ESA	Analysis	[28]
CODE-JSC	A. Amar	NASA JSC, USA	NASA	Analysis	[29]
CODE-LaRC	J. Dec	NASA LaRC, USA	NASA LaRC	Analysis	[30]
FABL	J. Merrifield	Fluid Grav. Eng. Ltd., UK	ISA/ESA/FGE	Analysis	[31]
FIAT	Y.-K. Chen	NASA ARC, USA	NASA, SpaceX	Design	[19]
3DFIAT	Y.-K. Chen	NASA ARC, USA	NASA ARC	Analysis	[32]
HERO	M. E. Ewing	ATK, USA	ATK	Analysis	[33]
ITARC	M. E. Ewing	ATK, USA	ATK	Design	[33]
libAblation	R. R. Upadhyay	Univ. of Tex. Aust., USA	UTA	Analysis	[34]
MIG	S. Roy	Univ. of Flo., USA	Univ. of Florida	Analysis	[35]
MOPAR	A. Martin	Univ. of Mich., USA	UKY/Univ. of Michigan	Analysis	[36]
NEQAP	J. B. Scoggins	N. Carol. St. Univ., USA	NCSU	Analysis	[37]
NIDA	G. C. Cheng	Univ. Alab. Birm., USA	UAB	Analysis	[38]
PATO	J. Lachaud	NASA ARC, USA	Univ. Calif. Santa Cruz	Analysis	[39]
STAB	B. Remark	NASA JSC, USA	NASA, FGE	Design	[40]
TITAN	F. S. Milos	NASA ARC, USA	NASA	Analysis	[41]
TMU	A. R. Bahramian	T. Modares Univ., Iran	TMU	Analysis	[42]
US3D	G. Candler	Univ. of Minn., USA	UM	Analysis	[43]

they include detailed pyrolysis modeling, finite-rate chemistry mechanism development, mass transport in porous media in the rarefied regime, in-depth ablation and coking, radiative heat transfer analysis, spallation modeling, and boundary layer-material coupling. The capabilities of these 25 codes along with research and development efforts currently in progress are summarized in a color-code table by Lachaud *et al.* [97].

6.6 Summary and Conclusion

Fundamentals of SRM nozzles and internal insulation materials for thermal protection have been reviewed in this chapter. New thermal protection material development is needed to capitalize on our advances in PNCs research. Fundamental research will continue to provide a basic understanding of PNCs to eventually enable full exploitation of their *multifunctionality.* Effects of different chemistry, size, shape, and combined nanoparticles for these novel material systems need to be considered. Understanding of *processing–structure–property–performance* relationships for these PNC systems must be expanded. Characterizations of physical, mechanical, and thermal properties of PNCs are important. There are five interdependent areas to note: constituent selection, processing, fabrication, property, and performance of PNCs. PNCs can provide lightweight alternatives to conventional polymeric materials, with additional functionality associated with nanoscale-specific value-added properties. Opportunities to extend PNC concepts to other nanoparticles and polymer matrix resin approaches are vast, resulting in tailor-made materials that circumvent current limitations, and expand our knowledge of materials science. Development of new *in situ* ablation recession and thermal sensing techniques and of the char strength measurement technique will advance understanding of the thermal behavior of the next generation of PNC ablatives. The two much needed areas of research, thermophysical properties characterization and adapting ablation modeling to PNCs, are identified and briefly discussed.

Nomenclature

D distance
F force
T temperature
W work
θ angle of applied force

Acronyms

AVCOAT	low density ablator name
C/C	carbon/carbon composite
C/Ph	carbon/phenolic composite
CEM	Cytec Engineered Materials
CNF	carbon nanofibers
CNT	carbon nanotubes
EHSM	elastomeric heat shielding material
EPDM	ethylene propylene diene monomer
FMI	Fiber Materials Inc.
HDCC	high density carbon/carbon composite
IR	infrared
LDCC	low density carbon/carbon composite
MWNT	multiwalled carbon nanotubes
NASA	National Aeronautics and Space Administration
NRAM	nanocomposite rocket ablative material
MMT	montmorillonite
OTB	oxyacetylene test bed
PICA	phenolic impregnated carbon ablator
PNC	polymer nanocomposite
POSS	polyhedral oligomeric silsesquioxane
SEM	scanning electron microscopy
SOTA	state-of-the-art
SSRM	solid rocket motor simulator
TC	thermocouple
TPUN	thermoplastic polyurethane nanocomposites

Acknowledgments

J. H. Koo would like to acknowledge the continuous funding from different government agencies, such as AFOSR, AFRL, AMRDEC, DTRA, MDA, NASA, NAVAIR, NSF, ONR, and other private companies to support the "*Ablation, Flammability & Additive Manufacturing Research at UT Austin & KAI.*" Special thanks are to Stan A. Bouslog, Steven Del Papa, Dr Randy Lillard, and other NASA-JSC researchers for their continuous support and guidance to our research group in numerous research areas. Thanks are also due to Dr Salita and Dr Stefani and his students in the material response modeling research area.

References

1 D'Alelio, G.F. and Parker, J.A. (eds) (1971) *Ablative Plastics*, Marcel Dekker, New York.
2 Duffa, G. (2013) *Ablative Thermal Protection Systems Modeling*, AIAA Publications, Reston, VA.
3 Natali, M., Kenny, J.M., and Torre, L. (2016) Science and technology of polymer ablative materials for thermal protection systems and propulsion devices: a review. *Progress in Materials*, **84**, 192–275.
4 Koo, J.H., Ho, W.K., and Ezekoye, O.A. (2006) *A Review of Numerical and Experimental Characterization of Thermal Protection Materials - Part I. Numerical Modeling*, 42nd AIAA/ASME/SAE/ASEE Joint Propulsion Conference, AIAA-2006-4936, Sacramento, CA.
5 Koo, J.H., Ho, W.K., Bruns, M., and Ezekoye, O.A. (2007) *A Review of Numerical and Experimental Characterization of Thermal Protection Materials - Part II. Material Properties Characterization*, 48th AIAA/ASME/ASCE/AHS Structures, Structural Dynamics, and Materials Conference, AIAA-2007-2131, Honolulu, HI.
6 Koo, J.H., Ho, W.K., Bruns, M., and Ezekoye, O.A. (2007) *A Review of Numerical and Experimental Characterization of Thermal Protection Materials - Part III. Experimental Testing*, 43rd AIAA/ASME/SAE/ASEE Joint Propulsion Conference, AIAA-2007-5773, Cincinnati, OH.
7 Koo, J.H., Natali, M., Tate, J., and Allcorn, E. (2013) Polymer nanocomposites as advanced ablatives – a comprehensive review. *International Journal of Energetic Materials and Chemical*, **12** (**2**), 119–162.
8 Koo, J.H., Lisco, B., Schellhase, K. *et al.* (2017) *Ablative Polymer Nanocomposites - Further Review*, 2017 AIAA SciTech Forum, AIAA-2017-0347, Grapevine, TX.
9 Koo, J.H. (2016) Ablative properties of polymer nanocomposites, in *Fundamentals, Properties, and Applications of Polymer Nanocomposites*, Chapter 10, , Cambridge University Press, Cambridge, pp. 425–520.
10 Sutton, G.P. and Biblarz, O. (1992) *Rocket Propulsion Elements*, 7th edn, Wiley, New York, pp. 550–563.
11 Natali, M. and Torre, L. (2012) Composite materials: ablative, in *Wiley Encyclopedia of Composites*, 2nd edn (eds L. Nicolais and A. Borzacchiello), John Wiley & Sons, New York.
12 Anonymous (1976) *Solid Rocket Motor Internal Insulation*, NASA SP-8093.
13 Truchot, A. (1998) Design and analysis of solid rocket motor internal insulation, in *Design Methods in Solid Rocket Motors*, Lecture Series LS 150 (ed. D. Reydellet), AGARD/NATO, Loughton, Essex, Chap. 10.
14 Anonymous (1975) *Solid Rocket Motor Nozzle*, NASA SP-8115.
15 Anonymous (1968) *Entry Thermal Protection*, NASA SP-8014.

16 Curry, D.M. and Tillian, D.J. (2006) *Apollo Thermal Protection System Revisited*, Proc. 2006 National Space & Missile Materials Symposium, Orlando, FL.
17 Vaia, R.A., Price, G., Ruth, P.N. *et al.* (1999) Polymer/layer silicate nanocomposites as high performance ablative materials. *Applied Clay Sciences*, **15**, 67–92.
18 Koo, J.H., Kneer, M. *et al.* (1992) A cost-effective approach to evaluate high-temperature ablatives for military applications. *Naval Engineers Journal*, **104** (**3**), 166–177.
19 Koo, J.H., Lin, S. *et al.* (1992) Performance of high-temperature polymercomposite ablatives under a hostileenvironment. *Science of Advanced Materials and Process Engineering Series*, **37**, SAMPE, Covina, CA, 506–520.
20 Koo, J.H., Miller, M. *et al.* (1993) Evaluation of fiber-reinforced composites ablatives for thermal protection. *Science of Advanced Materials and Process Engineering Series*, **38**, SAMPE, Covina, CA, 1085–1098.
21 Koo, J.H., Miller, M.J., Weispfenning, J., and Blackmon, C. (2011) Silicone polymer composites for thermal protection system: fiber reinforcements and microstructures. *Journal of Composite Materials*, **45** (**13**), 1363–1380.
22 Koo, J.H., Miller, M.J., Weispfenning, J., and Blackmon, C. (2011) Silicone polymer composite for thermal protection of naval launching system. *Journal of Spacecraft and Rockets*, **48** (**6**), 904–919.
23 Cheung, F.B., Koo, J.H. *et al.* (1993) Modeling of one-dimensional thermo-mechanical erosion of high-temperature ablatives. *Journal of Applied Mechanics*, **60**, 1027–1032.
24 Wilson, D., Beckley, D., and Koo, J.H. (1994) Development of silicone matrix-based advanced composites for thermal protection. *High Performance Polymer*, **6** (**2**), 165–181.
25 Shih, Y.C., Cheung, F.B., and Koo, J.H. (2003) Numerical study of transient thermal ablation of high-temperature insulation materials. *Journal of Thermophysics and Heat Transfer*, **17** (**1**), 53–61.
26 Koo, J.H. *et al.* (2003) Nanocomposites rocket ablative materials: processing, characterization, and performance. *International SAMPE Symposium and Exhibition (Proceedings)*, **48**, SAMPE, Covina, CA, 1156–1170.
27 Koo, J.H., Pilato, L., and Wissler, G.E. (2005) Polymer nanostructured materials for high-temperature applications. *SAMPE Journal*, **41** (**2**), 7.
28 Koo, J.H. (2006) *Polymer Nanocomposites: Processing, Characterization, and Applications*, McGraw-Hill, New York.
29 Koo, J.H. and Pilato, L.A. (2006) Thermal properties and microstructures of polymer nanostructured materials, in *Nanoengineering of Structural, Functional, and Smart Materials* (eds M.J. Schulz, A. Kelkar, and M.J. Sundaresan), CRC Press, Boca Raton, FL, pp. 409–441.

30 Koo, J.H., Pilato, L., and Wissler, G. (2007) Polymer nanostructured materials for propulsion systems. *Journal of Spacecraft and Rockets*, **44** (**6**), 1250–1262.
31 Koo, J.H., Ezekoye, O.A., et al. (2009) Characterization of polymer nanocomposites for solid rocket motor – recent progress. International SAMPE Symposium and Exhibition (Proceedings), Proceedings published in 2009 by SAMPE, Covina, CA.
32 Koo, J.H. *et al.* (2010) Flammability studies of a novel class of thermoplastic elastomer nanocomposites. *Journal of Fire Sciences*, **28** (**1**), 49–85.
33 Koo, J.H., Ezekoye, O.A., Lee, J.C. *et al.* (2011) Rubber-clay nanocomposites based on thermoplastic elastomers, in *Rubber-Clay Nanocomposites* (ed. M. Galimberti), Wiley and Sons, Hoboken, NJ, pp. 489–521.
34 Ho, W.K., Koo, J.H., and Ezekoye, O.A. (2009) Kinetics and thermophysical properties of polymer nanocomposites for solid rocket motor insulation. *Journal of Spacecraft and Rockets*, **46** (**3**), 526–544.
35 Ho, W.K., Koo, J.H., and Ezekoye, O.A. (2010) Thermoplastic polyurethane elastomer nanocomposites: morphology, thermophysical, and flammability properties. *Journal of Nanomaterials*, **Article D**, 583224.
36 Bruns, M.C., Koo, J.H., and Ezekoye, O.A. (2009) Population-based models of thermoplastic degradation: using optimization to determine model parameters. *Polymer Degradation and Stability*, **94**, 1013–1022.
37 Lee, J.C., Koo, J.H., and Ezekoye, O.A. (2009) *Flammability Studies of Thermoplastic Polyurethane Elastomer Nanocomposites*, 50th AIAA/ASME/ASCE/AHS/ASC Structures, Structural Dynamics, and Materials Conference, Palm Spring, CA, AIAA-2009-2544.
38 Lee, J.C., Koo, J.H., Ezekoye, O.A. *et al.* (2009) *Heating Rate and Nanoparticle Loading Effects on Thermoplastic Polyurethane Elastomer Nanocomposite Kinetics*, AIAA Thermophysics Conference, San Antonio, TX, AIAA-2009-4096, AIAA.
39 Lee, J.C., Koo, J.H., and Ezekoye, O.A. (2009) *Thermoplastic Polyurethane Elastomer Nanocomposites: Density, Hardness, and Flammability Properties Correlations*, AIAA Joint Propulsion Conference, Denver, CO, AIAA-2009-5273, AIAA.
40 Lee, J.C. (2010) Characterization of ablative properties of thermoplastic polyurethane elastomer nanocomposites. PhD Dissertation, Mechanical Engineering Department, The University of Texas at Austin, Austin, TX.
41 Lee, J.C., Koo, J.H., and Ezekoye, O.A. (2011) *Thermoplastic Polyurethane Elastomer Nanocomposite Ablatives: Characterization and Performance*, 47th AIAA/ASME/SAE Joint Propulsion Conference, San Diego, CA, AIAA-2011-6051, AIAA.
42 Allcorn, E., Natali, M. and Koo, J.H. (2011) Ablation performance and characterization of thermoplastic elastomer nanocomposites. International

SAMPE Symposium and Exhibition (Proceedings), 2011 ISTC, Proceedings published in 2013 by SAMPE, Fort Worth, TX.

43 Allcorn, E., Natali, M., and Koo, J.H. (2013) Ablation performance and characterization of thermoplastic elastomer nanocomposites. *Composites: Part A*, **45**, 109–118.

44 Patton, R.D., Pittman, C.U. Jr., Wang, L., and Hill, J.R. (1999) Ablation, mechanical and thermal conductivity properties of vapor grown carbon fiber/phenolic matrix composites. *Composites Part A*, **30** (**9**), 1081–91.

45 Rahatekara, S.S., Zammarano, M., Matko, S. *et al.* (2010) Effect of carbon nanotubes and montmorillonite on the flammability of epoxy nanocomposites. *Polymer Degradation and Stability*, **95**, 870–9.

46 Thostenson, E.T., Li, C., and Chou, T.W. (2005) Nanocomposites in context. *Composites Science and Technology*, **65**, 491–516.

47 Cheng, J. (2006) Polycyanate ester/small diameter carbon nanotubes nanocomposite. Master thesis, Mechanical Engineering Department, The University of Texas at Austin, Austin, TX.

48 Safadi, R.A. (2002) Multiwalled carbon nanotube polymer composites: synethis and characterization of thin films. *Journal of Applied Polymer Science*, **84**, 2660–2669.

49 Tate, J.S., Jacobs, C.J., and Koo, J.H. (2011) Dispersion of MWCNT in Phenolic Resin Using Different Dispersion Techniques and Evaluation of Thermal Properties, in *Proceedings of 2011 SAMPE ISSE*, SAMPE, Long Beach, CA.

50 Tate, J.S., Gaikwad, S., Theodoropoulou, N. *et al.* (2013, Article ID 403656) Carbon/phenolic nanocomposites as advanced thermal protection material in aerospace applications. *Journal of Composites*, **2013**, 9.

51 Legendre, P.J. (1975) *Reentry Vehicle Nosetip Instrumentation*, Instrument Society of America, ASI 5208, pp. 1–17.

52 Laub, B. and Venkatapathy, E. (2004) Thermal protection system technology and facility needs for demanding future planetary missions. *European Space Agency, ESA SP*, **544**, 239–247.

53 Oishi, T. and Martinez, E.R. (2008) *Development and Application of a TPS Ablation Sensor for Flight*, AIAA 2008-1219, AIAA, Reston, VA.

54 Gazarik, M.J., Little, A., Cheatwood, F.N. *et al.* (2008) *Overview of the MEDLI Project*, Aerospace Conference.

55 Natali, M., Koo, J.H., Allcorn, E., and Ezekoye, O.A. (2014) In situ ablation recession sensor for carbon/carbon ablatives based on commercial ultra-miniature thermocouples. *Sensors and Actuators B: Chemical*, **196**, 46–56.

56 Yee, C., Ray, M., Tang, F. *et al.* (2014) In situ ablation recession sensor for ablative materials based on ultraminiature thermocouples. *Journal of Spacecraft and Rockets*, **51** (**6**), 1789–1796.

57 Lisco, B., Yao, E., Pinero, D. *et al.* (2014) *In-Situ Ablation and Thermal Sensing for Two Low Density Ablators – Revisited*, Proc. CAMX, SAMPE, Orlando, FL.
58 Koo, J.H., Natali, M., Lisco, B. *et al.* (2015) *A Versatile In-situ Ablation Recession and Thermal Sensor Adaptable to Different Types of Ablatives*, AIAA SciTech 2015, Kissimmee, FL, AIAA-2015-1122.
59 Grantham, T., Duong, N.-M., Molina, R. *et al.* (2015) *Ablation, Thermal, and Morphological Properties of SiC Fibers Reinforced Glass Ceramic Matrix Composites*, AIAA SciTech 2015, Kissimmee, FL, AIAA-2015-1581.
60 Allcorn, E., Robinson, S., Tschoepe, D. *et al.* (2011) *"Development of an Experimental Apparatus for Ablative Nanocomposites Testing", 47th* AIAA/ASME/SAE Joint Propulsion Conference, *AIAA-2011-6050*, San Diego, CA.
61 Gutierrez, L., Koo, J.H. *et al.* (2015) *Design of Small-scale Ablative Testing Apparatus with Sample Position and Velocity Control*, AIAA-2015-1584, AIAA SciTech 2015, Kissimmee, FL.
62 Kurzawski, A., Ezekoye, O.A., Koo, H. *et al.* (2013) *Recession Experiments and Modeling for Carbon Surface Oxidation Process*, Proc. of ASME 2013 Summer Heat Transfer Conference, ASME, Minneapolis, MN.
63 Milos, F. *et al.* (2012) Nonequilibrium ablation of phenolic impregnated carbon ablator. *Journal of Spacecraft and Rockets*, **49 (5)**, 894–904.
64 Amar, A.J., Kirk, B.S., and Oliver, A.B. (2013) *Development and Verification of the CHarring Ablation Response (CHAR) Code*, NASA-JSC, Report, Houston, TX.
65 Ewing, M. *et al.* (2013) *Heat Transfer and Erosion Analysis Program (Hero 4)*, ATK Aerospace Systems, Theory Manual, , Brigham City, UT.
66 Forinash, D.M., Alter, R.J., Clatanoff, S.B. *et al.* (2012) *Development of an Apparatus for Measuring the Shear Strength of Charred Ablatives*, Proc. SAMPE TECH 2012 [CD-ROM], Covina, CA, SAMPE.
67 Jaramillo, M., Forinash, D., Wong, D. *et al.* (2013) *An Investigation of Compressive and Shear Strength of Char from Polymer Nanocomposites for Propulsion Applications*, AIAA-2013-3864, 49th AIAA/ASEM/SAE/ASEE Joint Propulsion Conference, San Jose, CA.
68 Jaramillo, M., Koo, J.H., and Natali, M. (2014) Compressive char strength of thermoplastic polyurethane elastomer nanocomposites. *Polymer for Advanced Technology*, **25** (77), 742–751.
69 Kashiwagi, T., Du, F., Winey, K.I. *et al.* (2005) Flammability properties of polymer nanocomposites with single-walled carbon nanotubes: effects of nanotube dispersion and concentration. *Polymer*, **46**, 471–81.
70 Kashiwagi, T., Du, F., Douglas, J.F. *et al.* (2005) Nanoparticle networks reduced the flammability of polymer nanocomposites. *Nature Materials*, **4**, 928–33.

71 Cipiriano, B.H., Kashiwagi, T., Raghavan, S.R. *et al.* (2007) Effects of aspect ratio of MWNT on the flammability properties of polymer nanocomposites. *Polymer*, **48**, 6086–6096.
72 Kashiwagi, T., Mu, M., Winey, K. *et al.* (2008) Relation between the viscoelastic and flammability properties of polymer nanocomposites. *Polymer*, **49**, 4358–4368.
73 Zhao, Z.F., Gou, J.H., Bietto, S. *et al.* (2009) Fire retardancy of clay/carbon nanofiber hybrid sheet in fiber reinforced polymer composites. *Composites Science and Technology*, **69**, 2081–7.
74 Wu, Q., Zhu, W., Zhang, C. *et al.* (2010) Study of fire retardant behavior of carbon nanotube membranes and carbon nanofiber paper in carbon fiber reinforced epoxy composites. *Carbon*, **48**, 1799–1806.
75 Tran, H.K. (1994) *Development of Lightweight Ceramic Ablators and Arc-Jet Test Results*, NASA Ames Research Center, NASA Technical Memorandum 108798, Moffett Field, CA.
76 Tran, H.K., Esfahani, L., and Rasky, D.J. (1994) Thermal response and ablation characteristics of lightweight ceramic ablators. *Journal of Spacecraft and Rockets*, **31** (**6**), 993–998.
77 Tran, H.K., Johnson, C.E., Rasky, D.J. *et al.* (1997) *Phenolic Impregnated Carbon Ablators (PICA) as Thermal Protection Systems for Discovery Missions*, NASA Ames Research Center, NASA Technical Memorandum 110440, Moffett Field, CA.
78 Lachaud, J., Cozmuta, I., and Mansour, N.N. (2010) Multiscale approach to ablation modeling of phenolic impregnated carbon ablators. *Journal of Spacecraft and Rockets*, **47** (**6**), 13.
79 Wong, D., Pinero, D., Jaramillo, M. *et al.* (2013) *Ablation and Combustion Characteristics of Thermoplastic Polyurethane Nanocomposites*, 49th AIAA/ASEM/SAE/ASEE Joint Propulsion Conference, San Jose, CA, AIAA-2013-3862.
80 Lewis, J., Koo, J. H. *et al.* (2015) Sensor to measure the shear strength of ablative polymer nanocomposites. Proceedings of CAMX 2015, Dallas, TX.
81 NASA PICA Material Property Report (2009) CEV TPS ADP, C-TPSA-A-DOC-158. Rev. 1.
82 Liang, J., Saha, M., and Altan, M. (2013) Effect of carbon nanofibers on thermal conductivity of carbon fiber reinforced composites. *Procedia Engineering*, **56**, 814–820. doi: 10.1016/j.proeng.2013.03.201
83 Pour, R., Soheilmoghaddam, M., Hassan, A., and Bourbigot, S. (2015) Flammability and thermal properties of polycarbonate /acrylonitrile-butadiene-styrene nanocomposites reinforced with multilayer graphene. *Polymer Degradation and Stability*, **120**, 88–97.
84 ASTM E1131-08(2014) (2014) *Standard Test Method for Compositional Analysis by Thermogravimetry*, ASTM International, West Conshohocken, PA, www.astm.org.

85 ASTM E1269-11 (2011) *Standard Test Method for Determining Specific Heat Capacity by Differential Scanning Calorimetry*, ASTM International, West Conshohocken, PA, www.astm.org.

86 Walters, R., Hackett, S.M., and Lyon, R.E. (2000) Heats of combustion of high temperature polymers. *Fire Mater*, **24**, 245–252.

87 Salita, M. (2011) Chapter VII: heat conduction and material ablation, in *Basic Analytical and Numerical Methods for Propulsion and Aerodynamic Analysis of Solid Propellant Rockets*, Self-Published, Sun Lakes, AZ.

88 Duffa, G. (2013) Chapter 13: an example: Apollo, in *Ablative Thermal Protection Systems Modeling*, American Institute of Aeronautics and Astronautics, Reston, VA, pp. 353–366.

89 Kendall, R. (1968) *An Analysis of the Coupled Chemically Reacting Boundary Layer and Charring Ablator, Part V, A General Approach to the Thermochemical Solution of Mixed Equilibrium-Non-Equilibrium Homogeneous or Heterogeneous Systems*, NASA CR-1064.

90 Salita, M. (1999) *User-Friendly PC Version of Aerotherm Chemical Equilibrium Code ACE*, TRW IOC 2ETM, MS.99-009.

91 Moyer, C.B. and Rindal, R.A. (1968) *An Analysis of the Coupled Chemically Reacting Boundary Layer and Charring Ablator, Part II, Finite Difference Solution for the In-Depth Response of Charring Materials Considering Surface Chemical and Energy Balances*, NASA CR-1061.

92 Salita, M. (2006) *A User-Friendly PC Version of the Heating and Ablation Code CMA92*, Northrop Grumman Report PCFT08.MS.06-012.

93 Salita, M. (1991) *Understanding ACE Charring Thermochemistry using NASA Lewis Thermochemical Code or the 14-Species Simplified Model SIMPACE*, Thiokol Report TWR-40314.

94 Chen, Y.-K. and Milos, F.S. (2001) Two-dimensional implicit thermal response and ablation program for charring materials. *Journal of Spacecraft and Rockets*, **38** (**4**), 473–481.

95 Lachaud, J. and Mansour, N.N. (2014) Porous-material analysis toolbox based on OpenFOAM and application. *Journal of Thermophysics and Heat Transfer*, **28** (**2**), 191–202.

96 Lachaud, J., van Eekelen, T., Scoggins, J.B. *et al.* (2015) Detailed chemical equilibrium model for porous ablative materials. *International Journal of Heat and Mass Transfer*, **90**, 1034–1045.

97 Lachuad, J., Magin, T.E., Cozmuta, I., and Mansour, N.N. (2011) *A Short Review of Ablative-Material Response Models and Simulation Tools*, 7th Aerothermodynamics Symposium, Brugge, Belgium, European Space Agency, Noorwijk, The Netherlands, pp. 1–8.

7

Manufacture of Multiscale Composites

David O. Olawale[1], Micah C. McCrary-Dennis[2], and Okenwa O. Okoli[3]

[1] R.B. Annis School of Engineering, Shaheen College of Arts and Sciences, University of Indianapolis, Indianapolis, IN, USA
[2] High-Performance Materials Institute, FAMU-FSU College of Engineering, Tallahassee, FL, USA
[3] Intel Corporation, Ronler Acres, Hillsboro, OR, USA

7.1 Introduction

The need for new materials and structures that simultaneously perform (i) multiple structural functions, (ii) combined nonstructural and structural functions, or (iii) both resulted in the development of multiscale composites [1]. Multiscale composites are produced by combining traditional fiber reinforcements, such as glass and carbon fibers, along with nanoscale fillers, such as carbon nanotubes (CNTs). The concept is to use traditional reinforcements for in-plane load carrying, and nanoscale reinforcements for improving through-thickness performance, as well as other resin-dominated properties of the resultant composites [2]. This is achieved by (i) replacing the neat resin polymer matrix with a nanocomposite matrix (Figure 7.1) and/or by (ii) growing nanoreinforcements such as CNTs on the surface of the fibers (Figure 7.2). Figure 7.3 illustrates the concept of a multiscale composite employing single-walled carbon nanotubes (SWNTs) and traditional fibers [2].

7.1.1 Multifunctionality of Multiscale Composites

The desirable multifunctionality possible with multiscale composites includes development of composite structures with high strength, high stiffness, high fracture toughness, and high damping. Another example would be a load-bearing structure with noise and vibration control, self-repair, *in situ* damage sensing, thermal insulation, and energy harvesting/storage capabilities [1]. The addition of very small amounts of CNTs to nonconducting polymers and polymer composites transforms them into conducting materials, thereby

Nanotechnology Commercialization: Manufacturing Processes and Products, First Edition.
Edited by Thomas O. Mensah, Ben Wang, Geoffrey Bothun, Jessica Winter, and Virginia Davis.

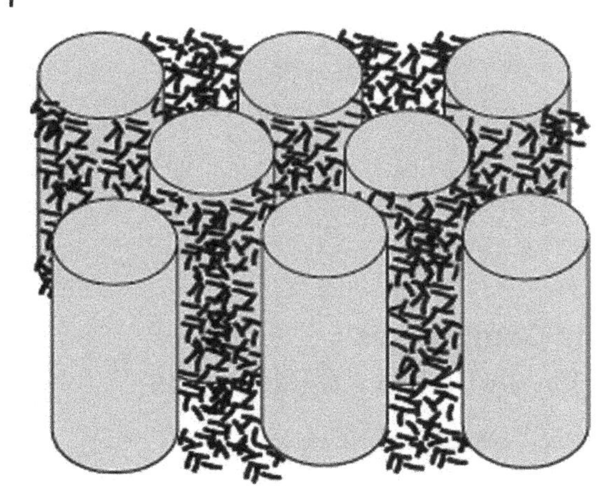

Figure 7.1 Nanoparticle reinforcement of the matrix in a unidirectional fiber composite. Vlasveld *et al.* 2005 [3]. Reproduced with permission of Elsevier.

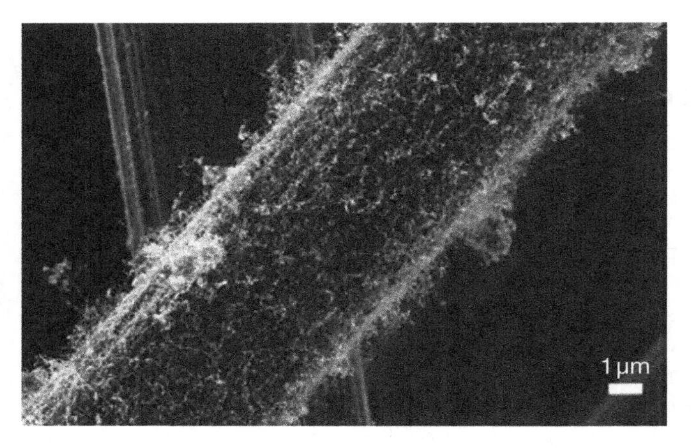

Figure 7.2 Multiwalled carbon nanotubes grown on the surface of carbon fibers. Zhao *et al.* 2005 [4]. Reproduced with permission of Elsevier.

enhancing their multifunctionality. Similarly, the use of nanoreinforcements in polymer composites has produced unprecedented improvements in mechanical properties of the composites [1]. Koratkar *et al.* [5] reported over 1000% increases in the loss modulus of polycarbonate (PC) without significant reductions in the storage modulus when the PC was enhanced by 2 wt% of SWNTs. The use of a silica nanoparticle-enhanced epoxy as the matrix material in a unidirectional E-glass/epoxy composite resulted in significant improvement in the longitudinal compressive strength and modulus [6].

In other applications, aligned CNT forests improved interlaminar strength and toughness, thereby addressing the concerns about conventional composite laminates because of the weak matrix resin-rich regions that exist

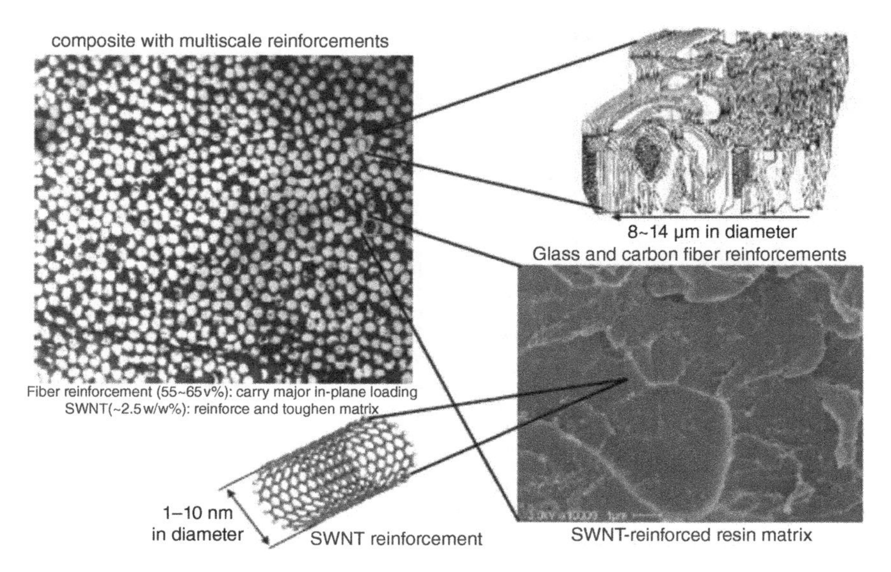

Figure 7.3 Concept of multiscale reinforcement composites. Ware *et al.* 2007 [2]. Reproduced with permission of Society for the Advancement of Material and Process Engineering.

between the composite laminae. The vertically aligned CNT forests can bridge and strengthen this interlaminar region as shown in Figure 7.4 [7]. The CNT-modified interfaces increased the Mode I interlaminar fracture toughness of aerospace grade carbon/epoxy laminates by a factor of 1.5–2.5 and the corresponding Mode II value by a factor of 3 [7].

7.1.2 Nanomaterials

Nanomaterials can be broadly classified into three types based on their geometries [8, 9] as highlighted in Figure 7.5: *particulate materials* such as carbon black, silica nanoparticle, and polyhedral oligomeric silsesquioxanes (POSS); *fibrous materials* such as nanofibers and CNTs; and *layered nanomaterials* such as an organosilicate that has a nanometer thickness and a high aspect ratio (30–1000) plate-like structure.

The inclusion of nanoscale materials in a fiber-reinforced polymer system yields dramatic changes in various properties of the multiscale composite. Nanoscale materials have a large surface area for a given volume [11]. A nanostructured material can therefore have substantially different properties from a larger dimensional material of the same composition because many important chemical and physical interactions are governed by surfaces and surface properties [12]. Figure 7.5 shows the common particle geometries and their respective surface area-to-volume ratios. For the fiber and layered

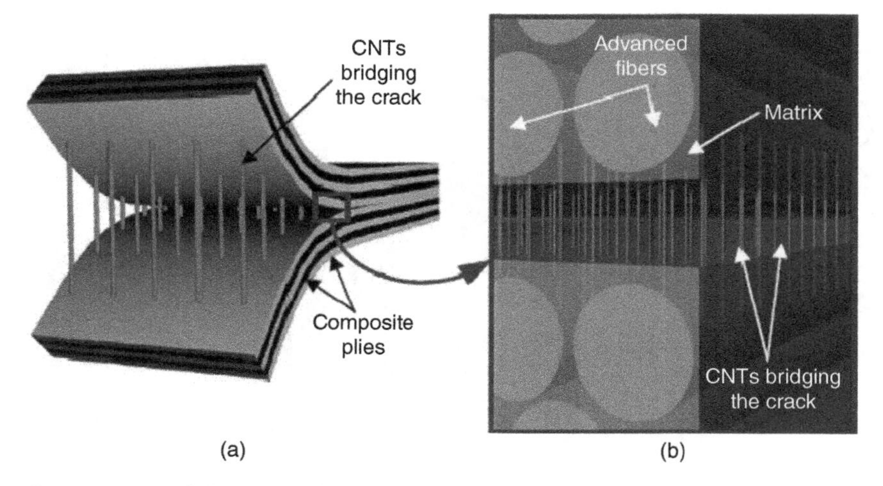

Figure 7.4 Use of aligned CNT forests to strengthen interlaminar region in composite laminates. Garcia *et al.* 2008 [7]. Reproduced with permission from Elsevier.

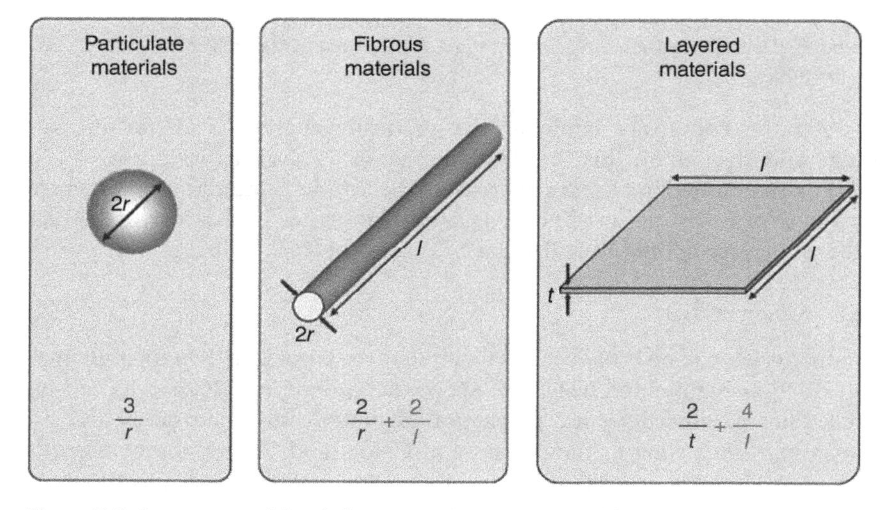

Figure 7.5 Common particle reinforcements/geometries and their respective surface area-to-volume ratios. Thostenson *et al.* 2005 [10]. Reproduced with permission of Elsevier.

material, the surface area/volume is dominated, especially for nanomaterials, by the first term in the equation. Consequently, a change in particle diameter, layer thickness, or fibrous material diameter from the micrometer to nanometer range will affect the surface area-to-volume ratio by three orders of magnitude [10]. This chapter focuses on multiscale composites with CNTs as the nanomaterial.

7.2 Nanoconstituents Preparation Processes

The properties of multiscale composites depend not only on the properties of their individual components but also on their morphology and interfacial characteristics [13]. Their properties are greatly influenced by the size scale of the component phases and the degree of mixing between the phases. Depending on the nature of the components used and the method of preparation, significant differences in composite properties may be obtained [14]. Three key processes required to get the nanomaterial (CNT) component ready for enhanced multiscale composite manufacture are discussed.

7.2.1 Functionalization of CNTs

The performance of a CNT/polymer nanocomposite depends on the dispersion of CNTs in the matrix and interfacial interactions between the CNT and the polymer [15]. CNTs are inert because the carbon atoms on CNT walls are chemically stable due to the aromatic nature of the bond. Consequently, the reinforcing CNTs can only interact with the surrounding matrix mainly through van der Waals interactions and are unable to provide an efficient load transfer across the CNT/matrix interface. To enhance the interaction between the CNT and the matrix, a number of methods have been developed to modify the surface properties of CNTs. The methods can be classified into chemical and physical methods based on the interactions between the active molecules and carbon atoms on the CNTs.

7.2.1.1 Chemical Functionalization

Chemical functionalization involves the covalent linkage of functional entities onto carbon scaffolds of CNTs at the termini of the tubes or at their sidewalls [15]. Chemically functionalized CNTs can produce strong interfacial bonds with many polymers, allowing CNT-based nanocomposites to possess high mechanical and functional properties [15]. Chemical functionalization can be broadly classified into direct covalent sidewall functionalization and defect functionalization.

Direct covalent sidewall functionalization involves a change of hybridization from sp^2 to sp^3 and a simultaneous loss of p-conjugation system on graphene layer (Figure 7.6a). This is achieved through reaction with high chemical reactivity molecules such as fluorine. The fluorination of purified SWCNTs occurred at temperatures up to 325 °C and the process was reversible with anhydrous hydrazine [16]. Fluorinated CNTs have C—F bonds that are weaker than those in alkyl fluorides [17], thereby providing substitution sites for additional functionalization [16] through replacements of the fluorine atoms by amino, alkyl, and hydroxyl groups [18].

Defect functionalization takes advantage of chemical transformation of defect sites on CNTs (Figure 7.6b). Defect sites include the open ends and/or

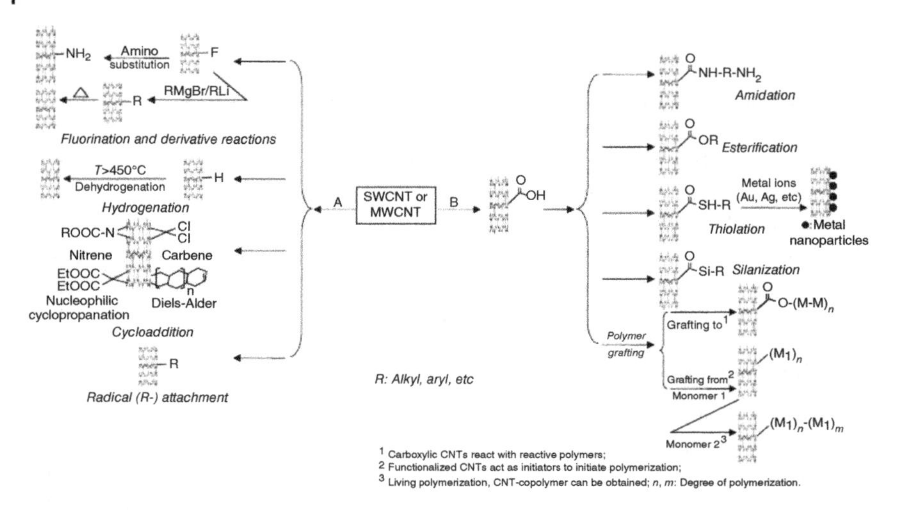

Figure 7.6 Strategies for covalent functionalization of CNTs (a: direct sidewall functionalization; b: defect functionalization). Ma *et al.* 2010 [15]. Reproduced with the permission of Elsevier.

holes in the sidewalls, pentagon or heptagon irregularities in the hexagon graphene framework, and oxygenated sites. Defects can be created on the sidewalls and at the open ends of CNTs by an oxidative process with strong acids such as HNO_3, H_2SO_4, or a mixture of both [19], or with strong oxidants such as $KMnO_4$ [20], ozone [21, 22], and reactive plasma [23, 24]. Defects created on CNTs by oxidants are stabilized by bonding with functional groups with rich chemistry such as carboxylic acid (—COOH) or hydroxyl (—OH). The CNTs can be used as precursors for further chemical reactions, such as silanation, polymer grafting [25], esterification, thiolation, alkylation, and arylation [26], and even some biomolecules. These functionalized CNTs are soluble in many organic solvents because the hydrophobic nature of CNTs is changed to hydrophilic one due to the attachment of the polar groups [15].

7.2.1.2 Physical (Noncovalent) Functionalization

Although the functionalization of CNTs using covalent methods can provide useful functional groups on the CNT surface, there are two main drawbacks associated with these methods. First, the functionalization reaction coupled with the ultrasonication process introduces a large number of defects on the CNT sidewalls. In some extreme cases, the CNTs are fragmented into smaller pieces. These damaging effects result in severe degradation in mechanical properties of CNTs as well as disruption of π electron system in nanotubes [15]. Disruption of π electrons is detrimental to transport properties of CNTs because defect sites scatter electrons and phonons that are responsible for the electrical and thermal conductions of CNTs, respectively. Second, the concentrated acids or strong oxidants usually used for the CNT functionalization are not environmentally friendly. The drawbacks associated with covalent methods of CNT functionalization led to the development of noncovalent functionalization methods that are convenient to use, of low cost, and less damaging to the CNT structure. Some of the noncovalent functionalization methods are discussed next.

Polymer wrapping is achieved through the van der Waals interactions and $\pi-\pi$ stacking between CNTs and polymer chains containing aromatic rings. Suspension of CNTs in the presence of polymers, such as poly(phenylene vinylene) [27] or polystyrene [28], lead to the wrapping of polymer around the CNTs to form supermolecular complexes of CNTs (Figure 7.7a).

Surfactant functionalization of CNTs (Figure 7.7b) can be with (i) nonionic surfactants, such as polyoxyethylene 8 lauryl ($CH_3(CH_2)_{11}(OCH_2-CH_2)_7OCH_2CH_3$) [31], nonylphenol ethoxylate (Tergitol NP-7) [32], and polyoxyethylene octylphenylether (Triton X-100) [29, 33]; (ii) anionic surfactants, such as sodium dodecylsulfate (SDS), sodium dodecylbenzenesulfonate (SDBS), and poly(styrene sulfate) (PSS) [34]; (iii) cationic surfactants, such as dodecyl tri-methyl ammoniumbromide (DTAB) [35] and cetyltrimethylammounium 4-vinylbenzoate (CTVB) [36]. The physical adsorption of surfactant on the CNT surface lowered the surface tension of CNT, thereby preventing the formation of aggregates [37]. Surfactant-treated CNTs are also able to overcome the van der Waals attraction by electrostatic/steric repulsive forces. The efficiency of this method depends strongly on the properties of surfactants, medium chemistry, and polymer matrix [15].

In endohedral functionalization of CNTs (Figure 7.7c), guest atoms or molecules (nanoparticles) such as C_{60}, Ag, Au, and Pt are stored in the inner cavity of CNTs through the capillary effect [15]. The insertion often takes place at defect sites localized at the ends or on the sidewalls. Small biomolecules, such as proteins and DNA, can also be entrapped in the inner hollow channel of nanotubes by simple adsorption, forming natural nano-test tubes [15]. The combination of the CNTs and guest molecules is particularly useful to integrate the properties of the two components in hybrid materials for use in

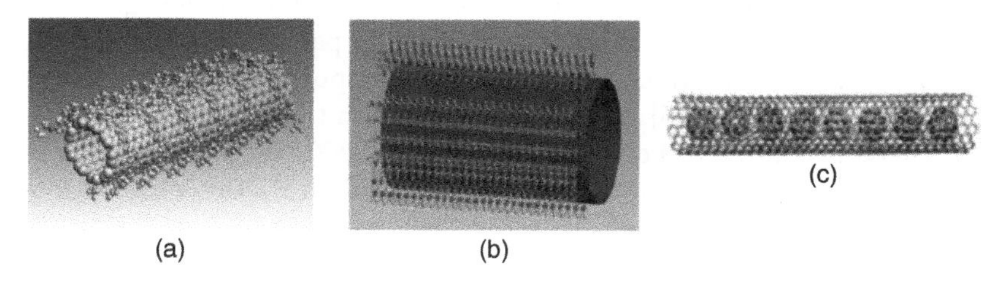

Figure 7.7 Schematics of CNT functionalization using noncovalent methods (a: polymer wrapping; b: surfactant adsorption; c: endohedral method) [29, 30].

Table 7.1 Advantages and disadvantages of various CNT functionalization methods.

Method		Principle	Possible damage to CNTs	Easy to use	Interaction with polymer matrix	Reagglomeration of CNTs in matrix
Chemical method	Side wall	Hybridization of C atoms from sp^2 to sp^3	V	X	S	V
	Defect	Defect transformation	V	V	S	V
Physical method	Polymer wrapping	Van der Waals force, π–π stacking	X	V	V	X
	Surfactant adsorption	Physical adsorption	X	V	W	X
	Endohedral method	Capillary effect	X	X	W	V

S, strong; W, weak; V, variable according to the miscibility between matrix and polymer on CNT; X, not applicable.
Source: Vaisman *et al.* 2006 [33]. Reproduced with permission of John Wiley & Sons.

catalysis, energy storage, nanotechnology, and molecular scale devices [38]. Table 7.1 provides a summary of the advantages and disadvantages of the various functionalization methods discussed.

7.2.2 Dispersion of Carbon Nanotubes

The full potential of employing CNTs as reinforcements has been severely limited because of the difficulties associated with dispersion of entangled CNTs during processing and poor interfacial interaction between CNTs and polymer matrix [15]. The problem of CNT dispersion is compounded compared to other conventional fillers, such as spherical particles and carbon fibers, because CNTs possess small diameter in nanometer scale with high aspect ratio (>1000) and extremely large surface area [15]. The dimensions of commonly used fillers, namely Al_2O_3 particles, carbon fibers, graphite nanoplatelets (GNPs), and CNTs, and the number of particles corresponding to a uniform filler volume fraction of 0.1% in a composite of 1.0 mm^3 cube are as shown in Table 7.2 [15]. In the composite with the same filler volume fraction, there are only 2 pieces of Al_2O_3 particles, 200 pieces when carbon fiber is added, and some 442 million pieces of CNTs [15]. This observation highlights why the dispersion of CNTs in a polymer matrix is more difficult than with the other fillers. The large quantity of particles and their size effect will lead to an exceptionally large surface area of nanoscale fillers in the composite with corresponding large interface or interphase area between

Table 7.2 Dimension and corresponding number of particles in composites for different fillers.

Filler	Description			
	Average dimension of filler	Density (g/cm^3)	N	S
Al_2O_3 particle	100 μm in diameter (d)	4.0	1.9	$S = \pi d^2$
Carbon fiber	5 μm in diameter (d) × 200 μm in length (l)	2.25	255	$S = \pi dl + \pi d^2/2$
Graphite nanoplatelet	45 μm in length (square, l), 7.5 nm in thickness (t)	2.2	6.58×10^4	$S = 4l^2 + 2lt$
CNT	12 nm in diameter (d) × 20 μm in length (l)	1.8	4.42×10^8	$S = \pi dl + \pi d^2/2$

N, number of particles in 1.0 mm^3 with 0.1 vol% filler content; S, surface area of individual particles.
Source: Ma *et al.* 2010 [15]. Reproduced with the permission of Elsevier.

the filler and matrix [15]. The "interface" in composites is a surface formed by a common boundary of reinforcing fillers and matrix that is in contact and maintains the bond in between for load transfer [39]. The "interphase" is defined as the region with altered chemistry, altered polymer chain mobility, altered degree of cure, and altered crystallinity that are unique from those of the filler or the matrix [15].

In addition to the size effect, as-produced CNTs are held together by van der Waals force in bundles or entanglements of fifty to a few hundred individual CNTs as shown in Figure 7.8 [40]. These bundles and agglomerates are inherently difficult to disperse and they result in diminished mechanical

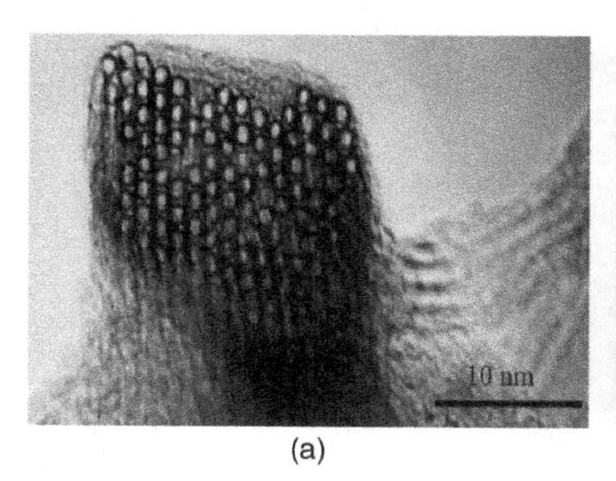

(a)

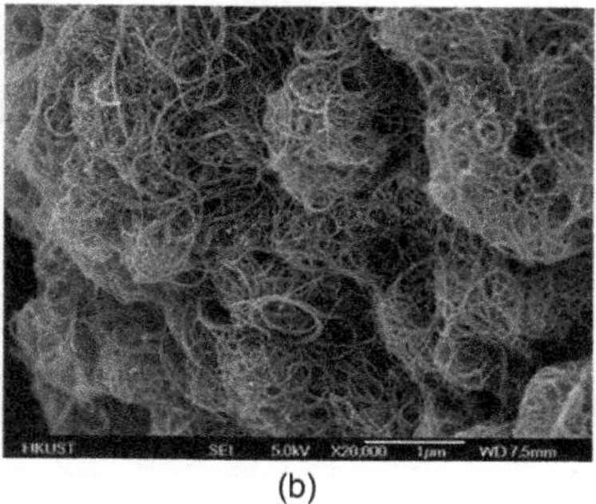

(b)

Figure 7.8 Electronic microscope images of different CNTs: (a) TEM image of SWCNT bundle. Thess *et al.* 1996 [40]. Reproduced with permission of The American Association for the Advancement of Science. (b) SEM image of entangled MWCNT agglomerates. Ma *et al.* 2010 [15]. Reproduced with the permission of Elsevier.

and electrical properties of composites compared with theoretical predictions for individual CNTs [39, 41, 42]. Dispersion of CNTs is not only a geometrical problem involving the length and size of the CNTs, but it also relates to a method on how to separate individual CNTs from CNT agglomerates and stabilize them in polymer matrix to avoid secondary agglomeration [15]. Methods for CNT dispersion are hereby discussed.

7.2.2.1 Ultrasonication

Ultrasonication is the most frequently used method for nanoparticle dispersion [15]. The method uses ultrasound energy to agitate particles in a solution for various purposes. When ultrasound propagates via a series of compression, attenuated waves are induced in the molecules of the medium through which it passes. The shock waves produced promote the "peeling off" of individual nanoparticles located at the outer part of the nanoparticle bundles, or agglomerates, resulting in the separation of individualized nanoparticles from the bundles [15]. Ultrasonication is done with an ultrasonic bath or an ultrasonic probe (Figure 7.9a,b). Standard laboratory sonicators (in a water bath) run at 20–23 kHz with a power less than 100 W, while commercial probe sonicators have an adjustable amplitude ranging from 20% to 70% and a power of 100–1500 W. The method of ultrasonication is effective for dispersing CNTs in low viscosity liquids. Most polymers are, however, in either a solid or viscous liquid state, which requires the polymer to be dissolved or diluted using a solvent such as water, acetone, and ethanol to reduce the viscosity before dispersion of CNTs.

There are critical issues to consider during sonification. The high energy intensity at the sonicator's tip during sonification can generate substantial

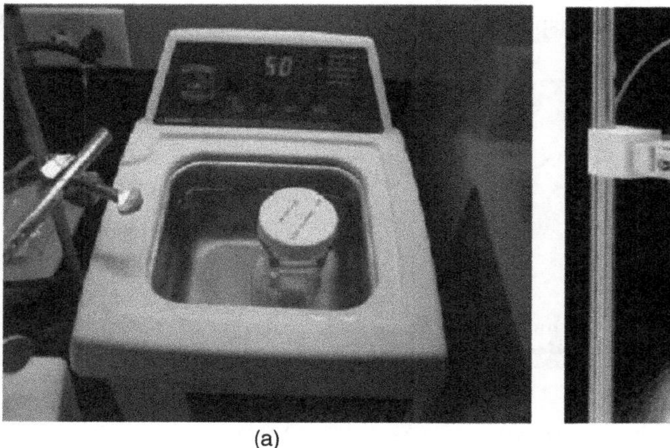

(a)

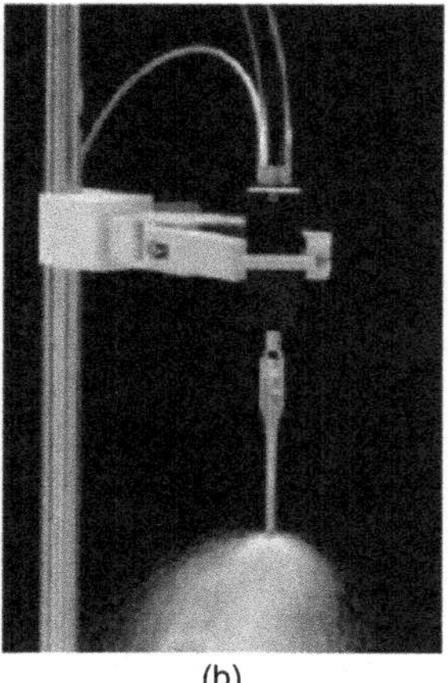

(b)

Figure 7.9 Sonicators with different modes for CNT dispersion (a) water bath sonicator; (b) probe/horn sonicator), and the effect of sonication on the structure of CNTs. Ma *et al.* 2010 [15]. Reproduced with the permission of Elsevier.

heat rapidly. In order to minimize the evaporation of the volatile solvents in which the CNTs are dispersed during sonication, the samples should be kept cold (e.g., using an ice bath) and the sonication must be done in short intervals. In addition, the sonification time must be kept short because if the sonication treatment is too aggressive and/or too long, the CNTs can be easily and seriously damaged, especially when using a probe sonicator. Localized damage to CNTs degrades both the electrical and mechanical properties of the CNT/polymer composites [15].

7.2.2.2 Calendering Process

The calender (Figure 7.10a), also known as three roll mills, is a machine tool that employs the shear force created by rollers to mix, disperse, or homogenize viscous materials [15]. The first and third rollers are called the feeding and apron rollers (n1 and n3 in Figure 7.10b), respectively, and they rotate in the same direction, while the center roller rotates in the opposite direction. The three rollers move at different velocities. The principle of operation is as described by Ma *et al.* [15]: The material to be mixed is fed into the hopper, where it is drawn between the feed and center rollers. When predispersed, the material sticks to the bottom of the center roller from where it is transported into the second gap. The material is dispersed to the desired degree of fineness in the second gap. Upon exiting, the material that remains on the center roller moves through the second gap between the center roller and apron roller, which subjects it to even higher shear force due to the higher speed of the apron roller. A knife blade scrapes the processed material off the apron roller and transfers it to the apron. This milling cycle is repeated several times to maximize dispersion. The adjustable narrow gaps (500 to about 5 μm) between the rollers and the mismatch in angular velocity of the adjacent rollers result in locally high shear forces with a short residence time.

There are, however, a number of concerns with the use of the calendering process [15]. The dimensional disparities between the roller gap (1–5 μm)

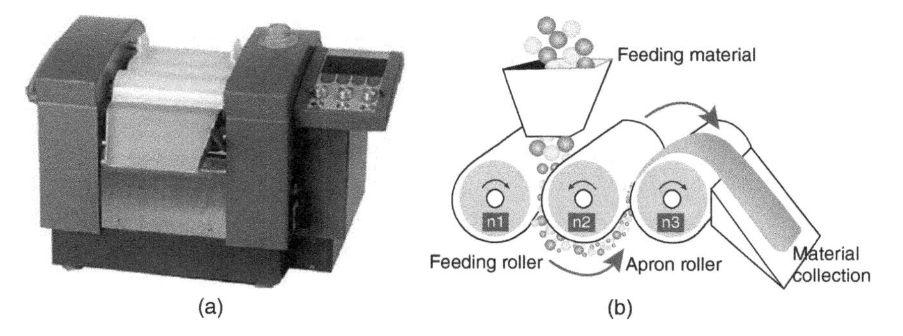

Figure 7.10 (a) Calendering (or three roll mills) machine used for particle dispersion into a polymer matrix and (b) corresponding schematic showing the general configuration and its working mechanism. Ma *et al.* 2010 [15]. Reproduced with the permission of Elsevier.

and the CNT dimensions suggest that calendering can better disperse the large agglomerated CNTs into small ones at submicron level, although some individual CNTs may be disentangled out from the agglomerates. In addition, calendering requires that the feeding materials should be in the viscous state when mixing with nanoparticles. This limits its use for dispersing CNTs into thermoplastic matrices, such as polyethylene, polypropylene, and polystyrene.

7.2.2.3 Ball Milling

Ball milling is a grinding method in which a high pressure is generated locally due to the collision of the tiny, rigid balls in a concealed container as schematically shown in Figure 7.11. An internal cascading effect of balls reduces the material to fine powder as small as 100 nm with high-quality ball mills. Ball milling has been successfully used to transform CNTs into nanoparticles [43], generate highly curved or closed-shell carbon nanostructures from graphite [44], enhance the saturation of lithium composition in SWCNTs [45], modify the morphologies of cup-stacked CNTs [46], and generate different carbon nanoparticles from graphitic carbon for hydrogen storage applications [47]. In addition, ball milling of CNTs in the presence of chemicals enhances their dispersibility and also introduces some functional groups onto the CNT surface. Such a chemomechanical method has been used to achieve in situ amino functionalization of CNTs using ball milling [48, 49]. The ball milling process may, however, damage the CNTs.

7.2.2.4 Stir and Extrusion

Stir is a technique commonly used to disperse particles in liquid systems and can also be used to disperse CNTs in a polymer matrix. The key factors that determine the dispersion result are the size and shape of the propeller as well as the mixing speed. The technique can more easily disperse MWCNTs than SWCNTs although the MWCNTs tend to reagglomerate due to frictional

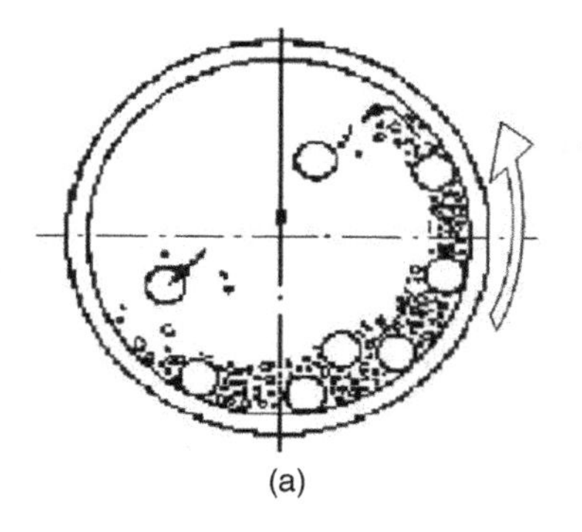

(a)

(b)

Figure 7.11 (a) Schematics of ball milling technique and (b) container. Ma *et al.* 2010 [15]. Reproduced with the permission of Elsevier.

contacts and elastic interlocking mechanisms [50]. For some thermosetting polymers, such as epoxy, obvious CNT reagglomerations were observed after several hours of curing reaction [51]. Higher shear forces are needed to achieve a fine dispersion of severely agglomerated CNTs in the polymer matrix. A high-speed shear mixer (Figure 7.12a) at a speed of up to 10,000 rpm can be used to achieve this.

Extrusion is a popular technique to disperse CNTs into solid polymers similar to most thermoplastics, where thermoplastic pellets mixed with CNTs are fed into the extruder hopper. Twin screws rotating at a high speed create high shear flow (Figure 7.12b), which results in the dispersion of the CNT agglomerates and mixing with the polymer melt. This technique is particularly useful to produce CNT/polymer nanocomposites with a high filler content [30].

A comparison of the characteristics of the various CNT dispersion techniques is provided in Table 7.3. This can serve as a general guideline for selecting appropriate dispersion technique to prepare CNT/polymer nanocomposites. The techniques for CNT dispersion are, however, not limited to those described. Many recent studies are based on the use of a combination of aforementioned techniques, such as ultrasonication plus ball milling, and ultrasonication plus extrusion. Many factors, such as physical (solid or liquid) and chemical (thermoplastic or thermoset) states of polymer matrix, dimensions and content of CNTs to be added, availability of techniques and fabrication processes, should be taken into account when selecting a proper technique for CNT dispersion [15].

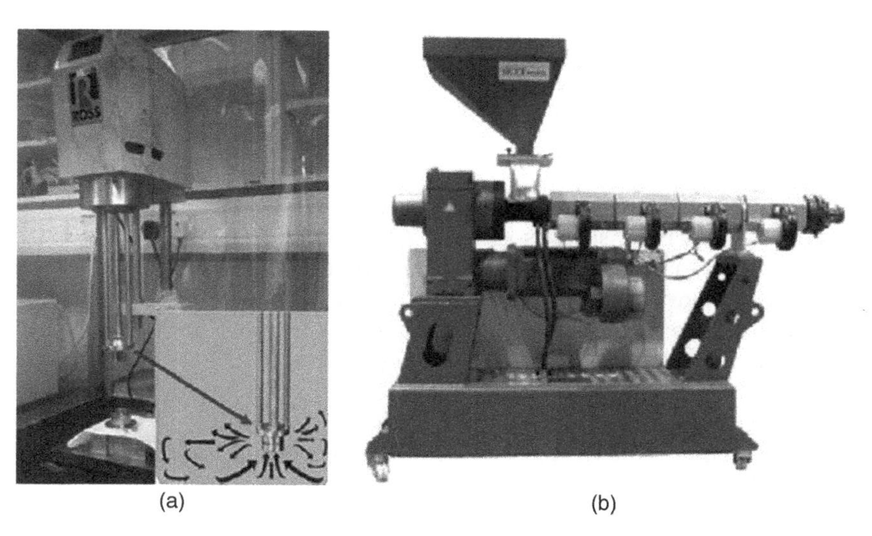

Figure 7.12 (a) Shear mixer and (b) extruder used for CNT dispersion. Ma *et al.* 2010 [15]. Reproduced with the permission of Elsevier.

Table 7.3 Comparison of various techniques for CNT dispersion in polymer composites.

Technique	Factor			
	Damage to CNTs	**Suitable polymer matrix**	**Governing factors**	**Availability**
Ultrasonication	Yes	Soluble polymer, low viscous polymer or oligomer, monomer	Power and mode of sonicator, sonication time	Commonly used in lab, easy operation and cleaning after use
Calendering	No. CNTs may be aligned in matrix	Liquid polymer or oligomer, monomer	Rotation speed, distance between adjacent rolls	Operation training is necessary, hard to clean after use
Ball milling	Yes	Powder (polymer or monomer)	Milling time, rotation speed, size of balls, balls/CNT ratio	Easy operation, need to clean after use
Shear mixing	No	Soluble polymer, low viscous polymer or oligomer, monomer	Size and shape of the propeller, mixing speed and time	Commonly used in lab, easy operation and cleaning after use
Extrusion	No	Thermoplastics	Temperature, configuration, and rotation speed of the screw	Large-scale production, operation training is necessary, hard to clean after use

Source: Ma *et al.* 2010 [15]. Reproduced with the permission of Elsevier.

7.2.3 Alignment of CNTS

Mechanical properties, such as stiffness and strength, as well as functional properties, such as electrical, magnetic and optical properties, of polymer/CNT nanocomposites are linked directly to the alignment of CNTs in the matrix [52]. Some of the effective techniques for aligning the CNTs in the polymer matrix are discussed in the following sections.

7.2.3.1 *Ex situ* Alignment

With these methods, the CNTs are aligned in advance, then compounded with the polymer matrix by in situ polymerization of some monomers. Feng *et al.* [53] prepared well-aligned polyaniline (PANI)/MWCNT composite films by in situ polymerization of aniline in the presence of aligned MWCNTs, as shown

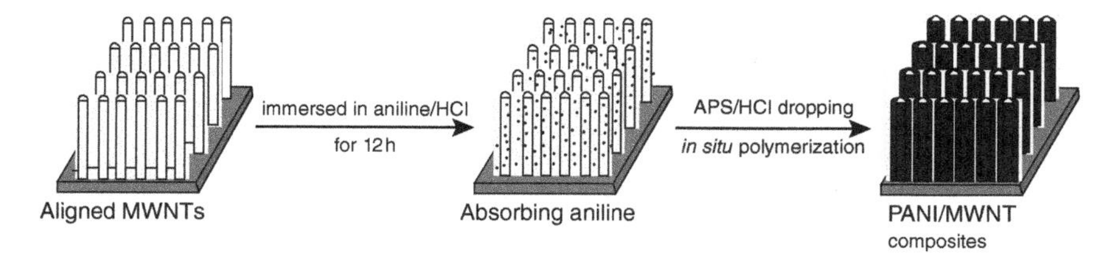

Figure 7.13 Schematic of preparing well-aligned PANI/MWCNT composites. Qin *et al.* 2004 [54]. Reproduced with permission of American Chemical Society.

in Figure 7.13. The CNTs, however, need to be first aligned by using methods such as filtration, plasma-enhanced chemical vapor deposition (PECVD), and use of templates.

7.2.3.2 Force Field-Induced Alignment of CNTs

There are many variants of the force field-induced alignment of CNTs approach. In one instance, SWCNTs were dispersed in a surfactant solution (sodium dodecyl sulfate, SDS) and slowly injected through a syringe needle into a PVA solution [55]. Because the latter is more viscous than the SWCNT dispersion, and there is a shear contribution in the flow at the tip of the syringe needle, the flow-induced alignment is maintained by the PVA solution, and SWCNTs are rapidly stuck together as they are injected out from the syringe. By pumping the polymer solution from the bottom, meter-long ribbons are easily drawn, and well-oriented PVA/CNT composite fibers and ribbons are formed by a simple process [52]. In another instance, CNTs in composites were aligned by uniaxially stretching polymer/CNT composite films at 100 °C [56]. CNTs in PMMA/CNT composites have also been aligned by melt extrusion [57]. This approach provides an easy and effective method to develop high-performance composites for the materials industry.

7.2.3.3 Magnetic Field-Induced Alignment of CNTs

This approach was first used by Kimura *et al.* [58] whereby they dispersed MWCNTs in monomer solution of unsaturated polyester and then applied a constant magnetic field of 10 T to align the nanotubes. Polymerization of the MWCNT-monomer dispersion under the applied magnetic field freezes the alignment of MWCNTs in the polyester matrix. The method can be used to prepare special composites with anisotropic electric and mechanical properties based on the anisotropic nature of MWCNTs. Aligned MWepoxy/CNT nanocomposites have also been prepared under a 25 T magnetic field [56]. The thermal and electrical properties along the magnetic field alignment direction are increased by 10% and 35%, compared with those epoxy/MWCNT nanocomposites without the application of a magnetic field.

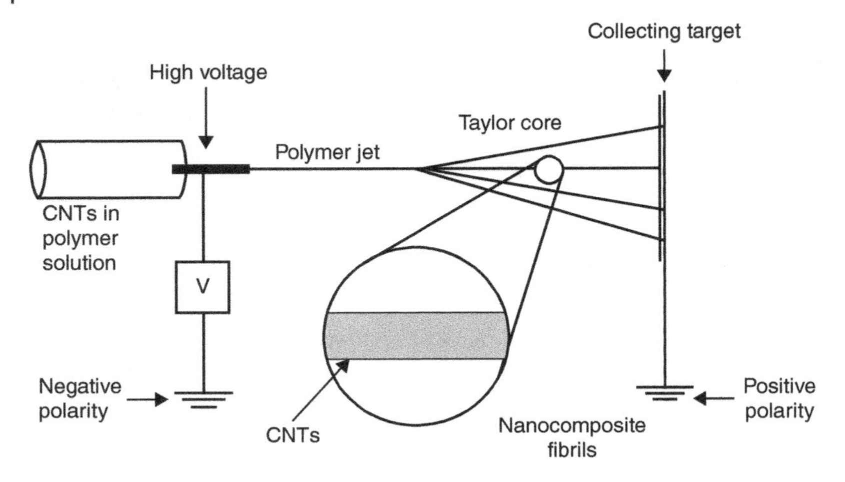

Figure 7.14 Schematic of the electrospinning process. Ko *et al.* 2003 [59]. Reproduced with permission of John Wiley & Sons.

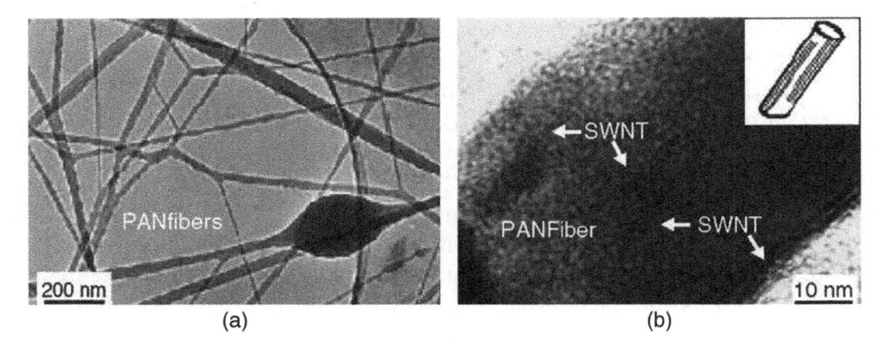

Figure 7.15 TEM images of PAN/SWCNT nanocomposite fibrils (the average diameter of SWCNTs is 1.3 nm) Ko *et al.* 2003 [59]. Reproduced with permission of JohnWiley & Sons.

7.2.3.4 Electrospinning-Induced Alignment of CNTs

Electrospinning is an electrostatic method for the fabrication of long organic fibers [52]. The process of electrospinning-induced alignment of CNTs (Figure 7.14) involves generating a high direct current (DC) voltage (e.g., 25 kV) between a negatively charged polymer fluid and a metallic fiber collector for random orientation or nanoscale fibril alignment so that a continuous yarn is manufactured along with the fiber mats [59]. The polyacrylonitrile (PAN)/SWCNT nanofibers are smooth and uniform, and SWCNTs are aligned along the direction of the nanofibers as shown in Figure 7.15.

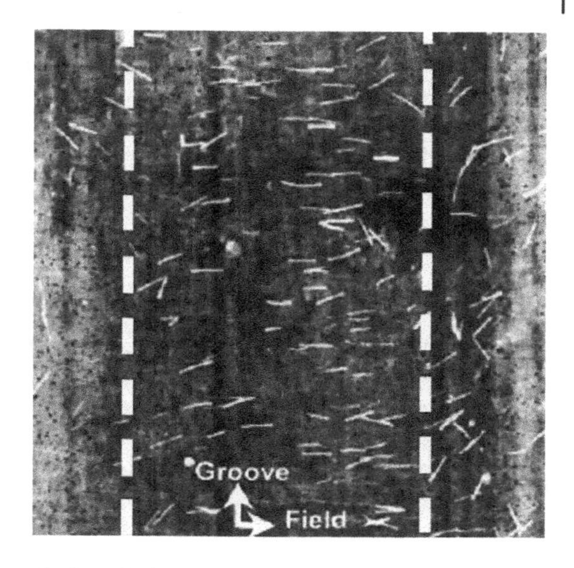

Figure 7.16 AFM image of LC/MWCNTs under 1.8 V/μm electric field (5 μm × 5 μm). Lee *et al.* 2005 [60]. Reproduced with permission of Springer.

7.2.3.5 Liquid Crystalline Phase-induced Alignment of CNTs

Due to the unique molecular structure of liquid crystals (LCs), the liquid crystalline phase is easy to orient along the applied force, electric or magnetic field [52]. This principle was used by Lynch and Patrick [60] to orient the nematic low molar mass LCs in an electric field, and the matrices were used to align the suspended MWCNTs (Figure 7.16). A 1.8 V/μm electric field was strong enough to overcome the orientational effect of the grooves, which were perpendicular to the electric field.

7.3 Liquid Composites Molding (LCM) Processes for Multiscale Composites Manufacturing

After the CNTs have been prepared through any or a combination of the processes described in the earlier section, the multiscale composites can be manufactured. Figure 7.17 provides an overview of a typical process for the manufacture of multiscale composites with a thermosetting polymer such as epoxy. The solvent containing already dispersed CNTs is added to the polymer (epoxy monomer), and the mixture is mixed and the CNTs redispersed. The mixture is distilled to remove the solvent and also placed in the vacuum oven to remove any remaining solvent or trapped air. After this, the curing agent (hardener) is added and the mixture infused into prepared fiber layup and allowed to cure.

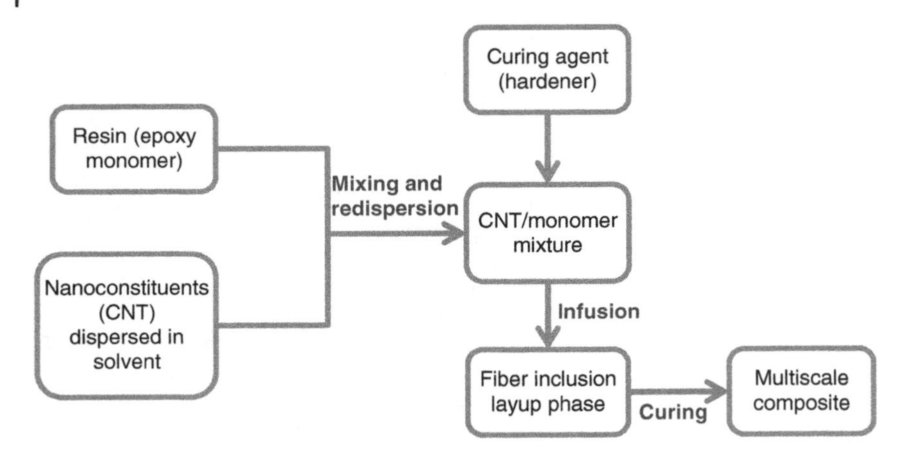

Figure 7.17 Overview of multiscale composite manufacturing process.

Many of the techniques or methods for manufacturing composites can be used for manufacturing the multiscale composite at the final stage. Liquid composite molding (LCM) techniques have gained wider interest among the various processes for manufacturing advanced polymer composites because of their relatively short cycle time, low labor requirement, and low equipment cost. LCM methods have been widely used in aerospace, automobile, marine, infrastructure, and other structural material fields [61–65]. They are expected to provide a viable manufacturing approach for multiscale CNTs reinforced polymer/fiber composites.

In LCM multiscale composite manufacturing processes such as vacuum-assisted resin transfer molding (VARTM) and resin transfer molding (RTM), CNT filtration by the fiber fabric structures has been observed. CNTs cluster around the resin inlet, while the rest of the flow paths lack CNTs, as shown in Figure 7.18. Filtration leads to an inhomogeneous microstructure of multiscale composites and finally results in inhomogeneous physical properties.

7.3.1 Resin Transfer Molding (RTM)

Figure 7.19 is a schematic representation of the RTM process. RTM is a closed mold operation [68] that uses doped resin containing dispersed CNTs for multiscale composite manufacture. The dry fiber reinforcements are set up in the mold in the form of woven mat, random-continuous strand mat, or binder-bound chopped mat. After the fibers are installed into the mold, a premixed catalyst and CNT-doped resin is injected into the closed mold cavity to encompass or impregnate the fiber within. In this process, resin flow and fiber wet-out are critical issues. The resin flows in both the plane and the transverse directions of the preform. The degree or ease of fiber wet-out depends on the fiber architecture and permeability of the preform.

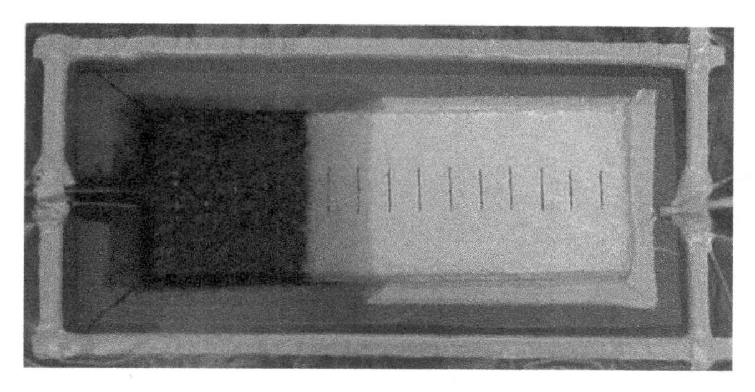

Figure 7.18 Blocking phenomenon in a VARTM experiment. Reproduced with kind permission of Dr. Qui [66].

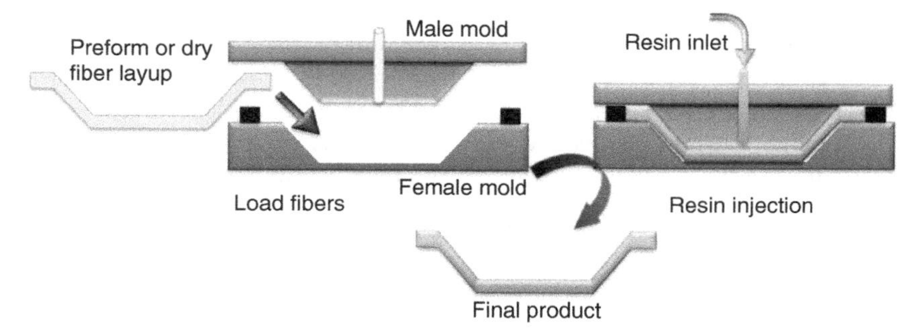

Figure 7.19 Resin transfer molding (RTM). McCrary-Dennis *et al.* 2012 [67]. Reproduced with permission of Sage publications.

Recent developments in textile and resin technology have allowed designers and manufacturers to use RTM for fabricating parts for both primary (skeletal or frame) and secondary structures (enclosures) [69]. Advanced textile technology has helped to increase the wettability of the preforms [70]. Higher toughness can be achieved by using three-dimensional weaving and stitching technology. Resin flow is, however, more difficult at higher viscosities prevalent in CNT-doped resin systems in both RTM and VARTM processes. The nanoconstituents in the matrix (resin) alter the resin viscosity and cure kinetics [71] that may result in dry spots or uneven distribution of resin over the entire volume of the reinforcement [72]. Higher CNT volume fractions tend to result in lower permeability and flow rate of the resin thus making a CNT/epoxy composite by the RTM process practically impossible.

7.3.2 Vacuum-Assisted Resin Transfer Molding (VARTM)

VARTM is one of the most commonly used low pressure closed mold composites manufacturing processes. The success of the VARTM process is due to

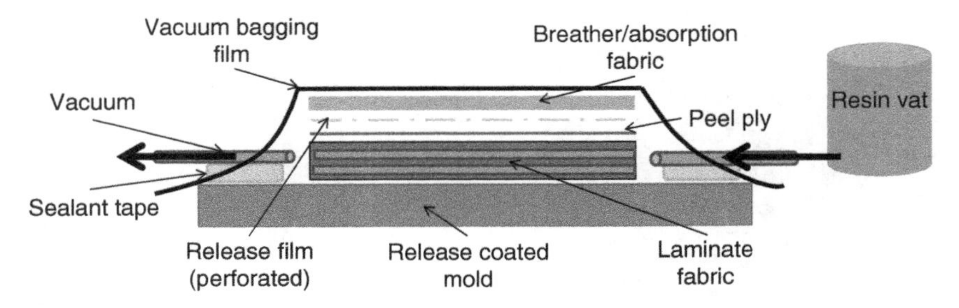

Figure 7.20 Vacuum-assisted resin transfer molding (VARTM) layup illustration. McCrary-Dennis *et al.* 2012 [67]. Reproduced with permission of Sage publications.

the controlled emission of volatile organic compounds (VOCs) [73]. VARTM is a variation of the RTM process and is a commonly used single-sided tooling process utilized to make parts employing vacuum pressure [68]. VARTM uses a sheet of flexible transparent material such as nylon or Mylar plastic that is placed over the preform on the mold and sealed (Figure 7.20).

Vacuum is applied to remove the entrapped air between the bagging film and the preform layers. Catalyst is added to the resin vat and vacuum applied to draw in the catalyzed resin in order to impregnate (infuse/wet) the fibers. The resin flow during an infusion process is given by Darcy's Law:

$$Q = \frac{kA}{\eta}\frac{dP}{dx}$$

where Q is the flow rate across a section A, η is the resin viscosity, k is the preform permeability, and dP/dx is the pressure gradient [74].

VARTM techniques utilize longitudinal and transverse flows of the resin (Figure 7.21) with porous distribution media/material on the surface. In the first case, the resin flows longitudinally between the preform layers, where the permeability is higher than in the preform itself (Figure 7.21a). This process is used for the fabrication of small parts where the injection length is short. In the second case, resin is distributed transversely through a highly permeable distribution media, called a flow media. The flow media is placed on top of the preform layers and because of its high permeability, the resin flows longitudinally through the flow media first and then saturates the preforms' inter-tow and intra-tow [22] spacing by means of a transverse flow through the thickness of the preform layers (Figure 7.21b) [75]. Very large parts may be fabricated using VARTM because of the transverse flow dynamics. This variation of the VARTM process is called the Seemann Composites Resin Infusion Molding Process (SCRIMP) [68, 76].

The SCRIMP is used in diverse applications ranging from turbine blades, boats, and bridge decks fabrication.

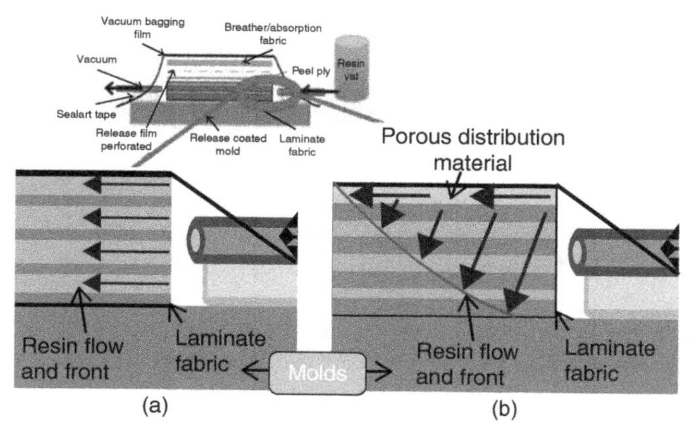

Figure 7.21 VARTM: (a) longitudinal flow and (b) transverse flow illustrating flow front through the thickness of preform laminates. McCrary-Dennis *et al.* 2012 [67]. Reproduced with permission of Sage publications.

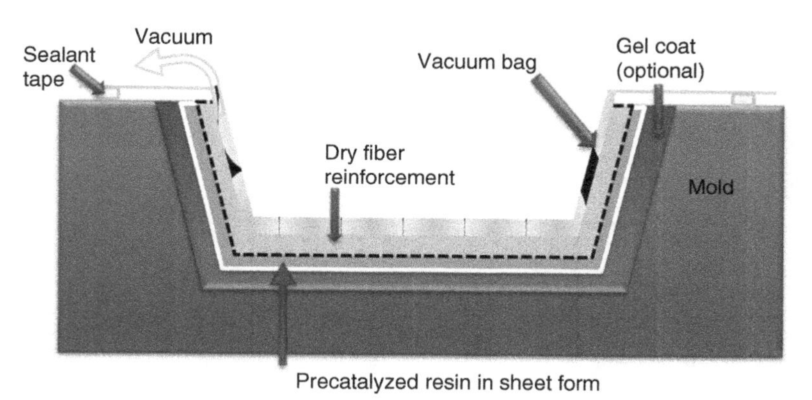

Figure 7.22 Resin film infusion diagram. McCrary-Dennis *et al.* 2012 [67]. Reproduced with permission of Sage publications.

Upon complete wet-out, the system can be clamped for curing and the part can be placed in an oven to post cure.

7.3.3 Resin Film Infusion (RFI)

Resin film infusion (RFI) utilizes a thin film or sheet of solid resin that is laid into the mold. Preform fibers are laid on top of the resin film under applied heat and pressure (Figure 7.22). Resin is impregnated into unidirectional or woven fabric. Although the method is labor intensive, resin distribution in its preforms is usually uniform [70].

7.3.4 The Resin Infusion under Flexible Tooling (RIFT) and Resin Infusion between Double Flexible Tooling (RIDFT)

The resin infusion under flexible tooling (RIFT), also referred to as "The Marco method," was considered a clean alternative to hand layup. The RIFT was designed for the manufacture of boat hulls with reduced voids and tooling costs compared to the RTM. The Marco tooling design utilized dry reinforcement that was laid up onto the solid male tool, and a semiflexible/splash female tool was used for consolidation and to provide a seal for the application of vacuum (Figure 7.23) [77].

The resin infusion between double flexible tooling (RIDFT) is a variation of the RIFT. The RIDFT (Figure 7.24) was developed to address some limitations of other LCM processes such as tool wear, high cycle time, high tooling, and equipment costs. The RIDFT is a process that facilitates an even and effortless

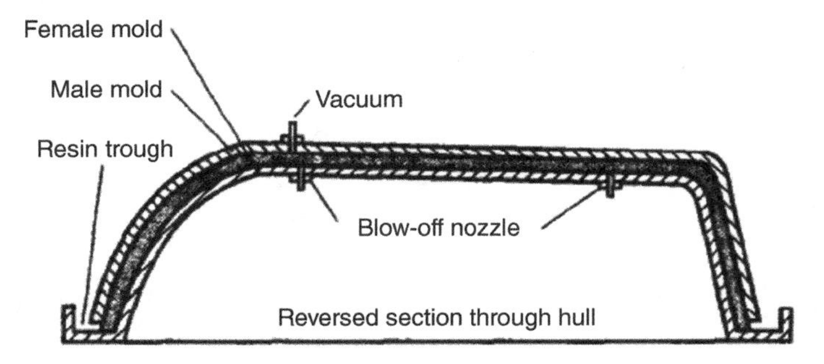

Figure 7.23 The marco method of RIFT (ca. 1950) [77].

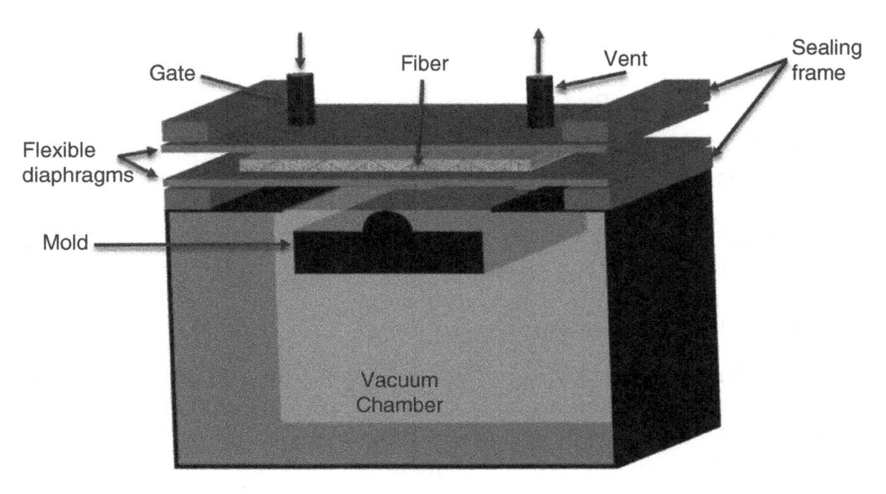

Figure 7.24 Resin infusion between double flexible tooling. McCrary-Dennis *et al.* 2012 [67]. Reproduced with permission of Sage publications.

resin flow making it a way of producing composites economically and quickly while still taking into consideration environmental concerns and safety [78]. One main advantage of the RIDFT with respect to the RTM is that the flow of resin in the RIDFT is two-dimensional, thereby eliminating the complexity of a three-dimensional flow front. The RIDFT eliminates the contact of the resin with the mold surface, which leads to longer mold life (multiple uses).

The RIDFT method is a novel two-stage process, which incorporates resin infusion and wetting with vacuum forming [67]. After placing the mold in the vacuum chamber, the bottom sealing frame and attached flexible diaphragm (silicone sheet) are set in place. The dry fiber preforms are then laid flat onto the silicone sheet. The preforms are covered with the top sealing frame and attached flexible diaphragm (silicone sheet).

The system is then infused with resin two dimensionally until complete wet-out is achieved. Subsequently, the vacuum pulls the flexible diaphragms and inter-lain fibers down to conform over the mold. After the part is cured, the vacuum is released and the part removed. The two-dimensional resin flow in the RIDFT avoids the flow complexities prevalent in the three-dimensional flow in other LCM techniques. Parker *et al.* [79] were able to fabricate sizable "C"-shaped carbon fiber parts with up to 2 wt% CNT-doped vinyl-ester resin. They were able to infuse viscous resins of about 10,000 cPs using specialized flow distribution channels.

7.3.5 Autoclave Manufacturing

Autoclave is the leading process for the manufacture of most aerospace composites [80]. Autoclave process achieves a high fiber volume fraction and low porosity levels. The autoclave enables controlled temperature and pressure for the curing of composites during the manufacture.

In preparation for autoclave manufacturing, the prepreg is removed from the storage freezer, cut into piles, and stacked together. After stacking the fibers, the bagging process is completed as depicted in Figure 7.25.

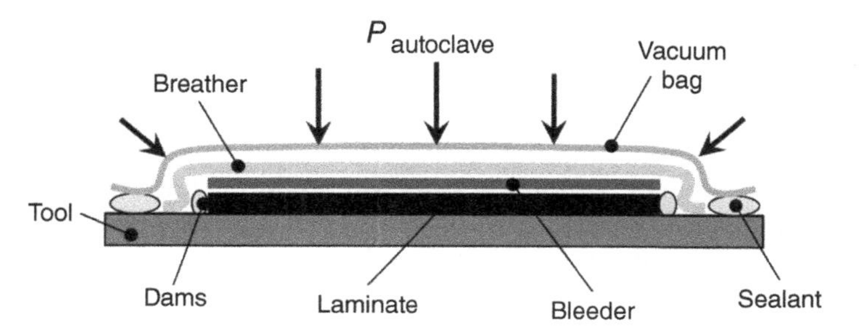

Figure 7.25 Bagging setup for autoclave. Mallow and Campbell 2000 [81]. Reproduced with permission of Hanser Publishers.

After the bagging is completed, the setup is placed in the autoclave for curing. The autoclave offers additional pressure for consolidation and as well as heat. Ramp phases of pressure and heat could be achieved to obtain the optimal quality properties for the composite. During the first ramp, the fibers are consolidated together and resin flow begins with the reduction in the viscosity of the resin. During the second ramp, cross-linking is observed as the resin reaches the gelation point. After this, the prepregs could also be taken through a postcure process that will help to increase the degree of cure.

7.3.6 Out-of-Autoclave Manufacturing: Quickset

Out-of-autoclave (OOA) processes seek solutions to some of the key challenges associated with the autoclave process. The challenges include high energy consumption and high cost [82]. In addition, the autoclave process requires constant supervision and high tooling investment [82]. There are some key factors needed to be taken into consideration when finding a replacement for autoclave production of multiscale composites. These include the need for the OOA process to produce multiscale composites with desirable fiber content (resin to fiber ratio), minimum void content, high surface finish, high vacuum force, and processing temperature range 120–230 °C. It also needs to be a closed mold system.

7.3.6.1 Quickstep

The Quickstep process is used as an OOA solution for manufacturing polymer composites. Quickstep reduces labor cost, consumes less energy, and requires less processing time [83]. Quickstep is a fluid-filled mold developed by an Australian company known as Quickstep. The process involves a suspended laminate, which is cured in a two-part mold composed of elastomer bladders that include heat transfer fluid (HTF), such as polyalkylene glycon, to allow for rapid and controlled thermal curing [84].

The main component of the Quickstep is the two-pressure chamber that includes a silicone membrane that forms on the part and the fluid pressurizing the part and transferring the heat. Figure 7.26 is a schematic of the quickstep process. An older version of the quickstep process consisted of three tanks (high, medium, and low temperature) that contain the HTF, which were preheated for curing [86].

Quickstep achieves a faster heating and cooling rate, which helps in the elimination of voids in the laminate. A faster heating will produce a lower viscosity resin faster, thereby facilitating the exit of air or voids from the laminate. Quickstep provides a faster curing cycle than autoclave process because of the more effective heat transfer by the HTF compared to heat transfer by gas in the autoclave [87].

Some of the advantages of the quickstep process include faster heating ramps, low pressure, low viscosity processing, green process, exotherm control, low

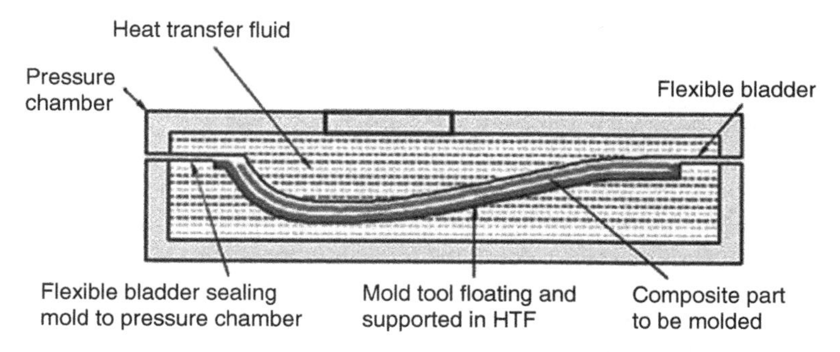

Figure 7.26 Quickstep schematic. Davies *et al.* 2007 [85]. Reproduced with permission of Elsevier.

cost, low energy consumption, low void content, and capability for processing resin films and prepregs [88]. The rapid heating produces a low viscosity resin that facilitates fiber wetting. The low viscosity results in the removal of air in the laminate, resulting in a void level of less than 1% that is equal or better than that required by the aerospace quality standard [89].

7.4 Continuous Manufacturing Processes for Multiscale Composites

7.4.1 Pultrusion

Pultrusion (Figure 7.27) is a continuous process for manufacturing composite materials with constant cross sections. This process begins as individual reinforced fiber strands from fiber creels are brought together and passed through a guide plate [67]. This conglomeration of fiber strands (fiber tow) is then guided by a series of rollers through a resin/doped resin bath and preformer before

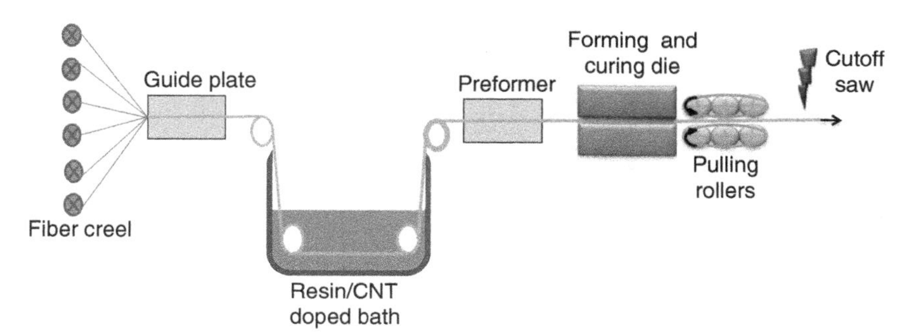

Figure 7.27 Pultrusion process. McCrary-Dennis and Okoli 2012 [67]. Reproduced with permission of Sage publications.

passing through the forming and curing die to culminate at a cutoff saw. Many resin types may be used in pultrusion including polyester, polyurethane, vinyl ester, and epoxy. Pultrusion is a low-cost continuous process with a high production rate for manufacturing linear composite parts.

The fibers in pultruded materials are generally well aligned [90], which helps to reduce fiber misalignment in the composite by optimizing the manufacturing process variables, such as pull-speed, preformer temperature, nanoparticle alignment, and/or dispersion. Along with clogging or the accumulation of materials that could occur near the die causing jams, voids may also be created if the dies run with too large an opening to the fiber volume input. Moreover, a constant cross section is a disadvantage limiting this process.

7.4.2 Filament Winding

In filament winding (Figure 7.28), resin-impregnated fibers are wrapped over a mandrel at the same or different winding angles to form a part. Complicated cylindrical parts can be manufactured using this method. The viscosity-related problems of the resin systems can be eliminated in this technique due to the absence of batch flow wetting because the process guides continuous individual fiber strands through a resin/CNT doped bath prior to cylindrical part wrapping. The critical task is programming the wet winding in this technique [70].

A method for dispersing SWNTs in epoxy was investigated by Spindler-Ranta *et al.* [92]. Arc-produced SWNTs were dispersed in bisphenol A epoxy resin and triamine hardener with the aid of a surfactant. They used a high power ultrasound to produce CNT-reinforced epoxy, which could then be used

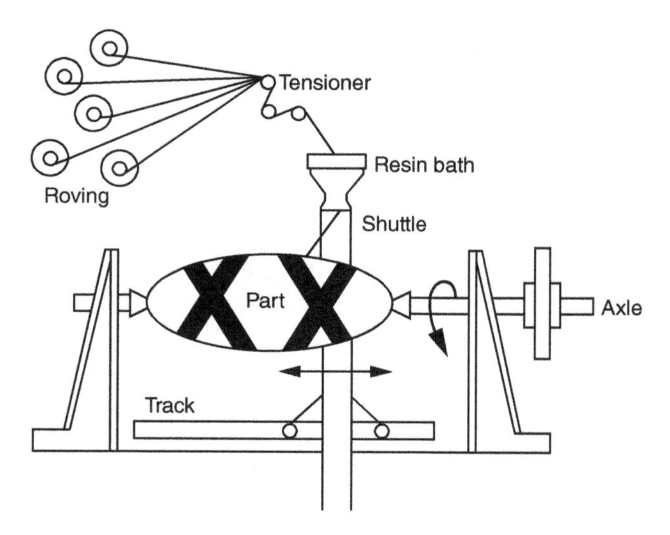

Figure 7.28 Filament winding diagram [91].

in filament winding. Filament winding is a cost-effective alternative for fabricating spherical and cylindrical parts, even those with varying diameters and surface contours. However, the drawback is that it cannot lay tow on a concave surface [93].

7.5 Challenges and Advances in Multiscale Composites Manufacturing – Environmental, Health, and Safety (E, H, & S)

Among the challenges and advances in multiscale composite manufacturing, environmental, health, and safety (E, H, & S) concerns are critical issues for production and scalability. There are many types of nanomaterials and one of the challenges is to categorize and prioritize these materials for the purposes of ecotoxicological risk assessments, for product life cycle analysis, and the potential points in the product life cycle where these materials may enter the environment [94]. A workshop cosponsored by the National Science Foundation and the US Environmental Protection Agency has identified a number of critical risk assessment issues regarding manufactured nanoparticles, such as (i) exposure assessment of manufactured nanoparticles; (ii) toxicology of manufactured nanoparticles; (iii) ability to extrapolate manufactured nanoparticle toxicity using existing particle and fiber toxicological databases; (iv) environmental and biological fate, transport, persistence, and transformation of manufactured nanoparticles; and (v) recyclability and overall sustainability of manufactured nanomaterials [95]. Occupational hazards, workplace safety issues, and hazards associated with using nano-enabled products are also points that are rigorously under investigation [96].

7.5.1 Nanoconstituents Processing Hazards

Manufactured nanomaterials are designed to achieve particular physicochemical properties that relate to the product application. The materials can be carbon based such as carbon spheres, Buckminster fullerenes, C60, and CNTs, or metal-based nanoparticles, composite nanomaterials, or multilayer nanomaterials [94]. The increase in the production and use of engineered nanoparticles makes exposure to the natural environment of these compounds increasingly likely [95, 97, 98]. Manufactured nanoparticles may contain chemically toxic components in concentrations or structural forms that do not occur naturally. If engineered-nanomaterial applications develop as projected, the increasing concentrations of nanomaterials in groundwater and soil may present the most significant exposure avenues for assessing environmental risk [98].

Information about nanoparticle exposure conditions is only useful when paired with characterization of nanomaterial biological effects. More specific information concerning nanoparticle clearance and bioavailability can be gleaned from studies aiming to develop nanomaterials for biotechnology. Because of the great interest in using engineered nanomaterials for medical applications, there is some relevant information for assessing health effects incidentally reported in this area of literature. In particular, the facile transport and association of engineered nanostructures with cells has received much attention from environmental groups who mistakenly interpret this feature as an indicator for toxicity [98].

Processing nanoconstituents to form multiscale composites also creates secondary E, H, & S issues. Sonication is one of the primary methods of dispersing nanoparticles; however, its use creates by-products that are of concern with regard to negative environmental, health, and safety emissions. An example of sonic activation of particular relevance to environmental and toxicological studies is that of water sonolysis. Sonicated water partly dissociates into hydrogen and hydroxide radicals, with concentration and lifetime dependency on the sonication parameters that subsequently recombine to form hydrogen peroxide. Any materials or chemical species present in sonicated water will thus be exposed to low (and variable) concentrations of highly reactive species, and as a result may undergo oxidative and other chemical transformations [99]. As with all chemical manufacturing processes, production of nanoparticles may give rise to exposure by inhalation, through the skin and by ingestion. Exposure may also occur for workers in downstream processes that use these materials and to consumers as these products enter the marketplace. In gas-phase production processes, exposure by inhalation may be caused by direct leakage of the reactants or products into the workplace environment. For all production methods, product recovery, subsequent processing and cleaning may result in the generation of airborne nanoparticles. It is, however, probable that these downstream activities will not generate discrete nanoparticles, due to the relatively high energies, which would be necessary to break the forces that keep particles agglomerated [100].

7.5.2 Composite Production and Processing

Composite production techniques primarily fall under open or closed molding processes. Open mold processes (i.e., spray and hand layup) are most contributory to E, H, & S issues. The nature of these methods can directly release nanoparticles, harmful toxins/by-products, and volatile organics into the environment, which can subsequently impact the health and safety of those workers who utilize them in addition to environmental concerns. Thus, continuous efforts, standardizations, and strict protocols must be adhered to. Closed mold processes (i.e., VARTM and RIDFT) are limited by their scalability but address several of the harmful concerns. These methods utilize bagging, suppression, or covering of the constituents during processing for forming and infusion containment. The aforementioned LCM processes for multiscale composite manufacturing have limited the concerns for harmful exposure.

7.5.3 Life Cycle Assessment – Use and Disposal

All stages of production processes could potentially result in dermal exposure, particularly at the powder handling, packaging, and bagging stages [100]. Life cycle assessment (LCA) is an important technique in the successful implementation of a process or product development in the context of environmental sustainability. Attempts have been made to incorporate LCA in public and corporate processes and product-related decision-making. Systems that facilitate LCA applications in processes, product evaluation, and decision-making have been investigated and reported. The findings conclude that any evaluation must be comprised of four important subindices or attributes – environmental, health and safety (E, H, & S), cost, technical feasibility, and sociopolitical factors [101]. For their wider acceptance, nanotechnology-based consumer products should be developed to offer significantly improved functional performance and reduced life cycle costs. Complete life cycle planning and control issues should be investigated as nano-enabled products pose end-of-life disposal concerns.

7.6 Modeling and Simulation Tools for Multiscale Composites Manufacture

Any production is based on materials eventually becoming components of a final product. Material properties are determined by the microstructure of the material. Predictability of materials' properties is highly important. Such predictions require tracking of microstructure and property evolution along the entire component life cycle starting from a homogeneous, isotropic, and stress-free melt and eventually ending in failure under operational load [102].

Simulation-based engineering science (SBES) is an evolving interdisciplinary research area rooted in the methods for modeling multiscale, multiphysics events. The objective in SBES is to develop methodologies that are foundational to designing multiscale systems by accounting for phenomena at multiple scales of lengths and time. Some of the key challenges faced in SBES include lack of methods for bridging various time and length scales, management of models and uncertainty associated with them, management of huge amount and variety of information, and methods for efficient decision-making based on the available models. Although efforts have been made to address some of these challenges for individual application domains, a domain-independent framework for addressing these challenges associated with multiscale manufacturing problems is not currently available in the literature. Considerations for multiscale evaluation fall into two categories of modeling and design. Multiscale modeling deals with efficient integration of information from multiscale models to gain a holistic understanding of the system. Multiscale design deals with efficient utilization of information to satisfy design objectives [103]. Both modeling and design of multiscale systems face challenges that are yet to be addressed due to the vast intricacies involved with such variable and inclusive entities.

7.6.1 Nanoparticle Modeling

Modeling and simulation tools are paramount in the fabrication, manufacture, and analysis of multiscale composites. Nanoparticles, polymer/resin matrix, and fiber reinforcement make up the three phases of multiscale composites. The initial modeling considerations begin with the incorporation of the nanoparticles into the polymer/resin matrix. Unlike traditional manufacturing, nanomanufacturing involves product characteristics and process variables at multiple length scales, ranging from atomistic-scale molecular structures to the macro product scale. This multiscale product/process variation presents the multiscale and multiphenomenon challenge in nanomanufacturing. However, control of nanomanufacturing process and product performance is hampered by the scarcity of measurement data, confounding effects during processing, and limited physical knowledge [96].

7.6.2 Molecular Modeling

Molecular modeling and simulation combines methods that cover a range of size scales in order to study material systems. These range from the subatomic scales of quantum mechanics (QM), to the atomistic level of molecular mechanics (MM), molecular dynamics (MD), and Monte Carlo (MC) methods, to the micrometer focus of mesoscale modeling. Multiscale simulation can be defined as the enabling technology of science and engineering that links phenomena, models, and information between various scales of complex systems. The idea of multiscale modeling is straightforward: one computes information at a smaller (finer) scale and passes it to a model at a larger (coarser) scale. The ultimate goal of multiscale modeling is then to predict the macroscopic behavior of an engineering process from first principles by starting from the quantum scale and passing information into molecular scales and eventually to process scales [104, 105].

7.6.3 Simulation

Among different simulation methods, continuum methods of modeling the behavior of materials have dominated the materials research and applications for a long time. Since continuum mechanics is grounded in the statistical mechanics of atomic scale process over a long time and a large length scale (larger than 10 μm), they may fail as the length scales are less than approximately 10 atoms. Molecular dynamics (MD) is an alternative approach for atomistic behavior of such a system. MD is suitable for simulating very small volumes of liquid flow with linear dimensions of 100 nm or less and for time intervals of several tens of nanoseconds. The gap between traditional atomistic methods (such as MD) and continuum methods (such as FEM) presents significant challenges. When the length scales are too small for continuum methods and too large for the atomistic methods, these two approaches become inadequate. Therefore, mesoscale modeling has proposed to bridge MD and continuum methods.

Brownian dynamics (BD) is among the most attractive mesoscale techniques because of its enormous versatility in constructing simple models for complex fluids, such as dissipative particle dynamics (DPD) and finitely extensible nonlinear elastic (FENE). Some multiscale modeling approaches of nanocomposite systems have also been reported, such as the coarse-grained molecular dynamics (CGMD) approach, MD combined FEM, MD combined boundary element method (BEM), MD combined representative volume elements (RVE), DPD, DPD combined FEM, and MD combined DPD and FENE.

These modeling methods provided valuable insight in analyzing the composite properties. Most of these mesoscale models focused on the dynamical mechanical properties (such as the stress–strain behavior, strength, modulus, and interfacial bonding), some focused on conductivity of composites, and the others focused on rheological behavior. Although great improvements have been made in multiscale modeling and simulation, to date, these theories cannot be implemented for multiscale complex flow through porous media. However, some have investigated a framework of the micro/nanoflow in multiscale composite manufacturing [105]. There are too few simulation techniques on CNT/polymer composites that can be realized in the time and length scales of the experiments. However, many efforts are being made for continuous improvement and understanding of multiscale composite systems [105].

7.7 Conclusion

CNTs-based multiscale composites possess great multifunctional capabilities. These include load-bearing structures with noise and vibration control, self-repair, *in situ* damage sensing, thermal insulation, and energy harvesting/storage capabilities. There are, however, some processing challenges that have to be overcome in order to successfully manufacture these multifunctional composites. The critical processes for preparing the CNTs for multiscale composite manufacture have been identified. These include functionalization, dispersion, and alignment of the CNTs. Some of the various techniques available for achieving these have been described and their strengths and limitations discussed. In addition, available composite manufacturing processes that can be used for manufacturing multiscale composites have also been discussed. Environmental, safety and health issues related to the handling and processing of multiscale composites and the constituent materials have also been discussed. Finally, available simulation tools to help in understanding the behavior and processing of multiscale composites have also been discussed.

The RTM, VARTM, RFI, Quickset, RIDFT, and filament winding techniques can be used to manufacture multiscale composite parts for various applications such as commercial aircraft structures for Boeing and Airbus, as well as many products in the industrial markets. Routine use of multiscale composites in automotive and aerospace industries is, however, a long-term prospect as these are risk-averse sectors. In addition, the extensive testing and characterization required before introducing these new parts take significant time and cost [70, 106].

References

1 Gibson, R.F. (2010) A review of recent research on mechanics of multifunctional composite materials and structures. *Composite Structures*, **92**, 2793–2810.

2 Ware, G. *et al.* (2007) *Processing and Characterization of Epoxy/Carbon Fiber/Carbon Nanotube Multiscale Composites Fabricated Using VARTM*. in *SAMPE 2007 Fall Technical Conference*, SAMPE, Cincinnati, OH.

3 Vlasveld, D.P.N., Bersee, H.E.N., and Picken, S.J. (2005) Nanocomposite matrix for increased fibre composite strength. *Polymer*, **46** (**23**), 10269–10278.

4 Zhao, Z.-G. *et al.* (2005) The growth of multiwalled carbon nanotubes with different morphologies on carbon fibers. *Carbon*, **43**, 663–665.

5 Koratkar, N.A. *et al.* (2005) Characterizing energy dissipation in single-walled carbon nanotube polycarbonate composites. *Applied Physics Letters*, **87**, 063102.

6 Uddin, M.F. and Sun, C.T. (2008) Strength of unidirectional glass/epoxy composite with silica nanoparticle-enhanced matrix. *Composites Science and Technology*, **68** (**7-8**), 1637–1643.

7 Garcia, E.J., Wardle, B.L., and John Hart, A. (2008) Joining prepreg composite interfaces with aligned carbon nanotubes. *Composites Part A: Applied Science and Manufacturing*, **39** (**6**), 1065–1070.

8 Schmidt, D., Shah, D., and Giannelis, E.P. (2002) New advances in polymer/layered silicate nanocomposites. *Current Opinion in Solid State and Materials Science*, **6** (**3**), 205–212.

9 Alexandre, M. and Dubois, P. (2000) Polymer-layered silicate nanocomposites: preparation, properties and uses of a new class of materials. *Materials Science and Engineering: R: Reports*, **28** (**1–2**), 1–63.
10 Thostenson, E., Li, C., and Chou, T. (2005) Review nanocomposites in context. *Journal of Composites Science & Technology*, **65**, 491–516.
11 Luo, J.-J. and Daniel, I.M. (2003) Characterization and modeling of mechanical behavior of polymer/clay nanocomposites. *Composites Science and Technology*, **63** (**11**), 1607–1616.
12 Baur, J. and Silverman, E. (2007) Challenges and opportunities in multifunctional nanocomposite structures for aerospace applications. *MRS Bulletin*, **32** (**04**), 328–334.
13 Oriakhi, C.O. (1998) Nano sandwiches. *Chemistry in Britain*, Chemical Society, London, ROYAUME-UNI, **34**, 59–62.
14 Park, C.I. *et al.* (2001) The fabrication of syndiotactic polystyrene/organophilic clay nanocomposites and their properties. *Polymer*, **42** (**17**), 7465–7475.
15 Ma, P.-C. *et al.* (2010) Dispersion and functionalization of carbon nanotubes for polymer-based nanocomposites: a review. *Composites: Part A*, **41**, 1345–1367.
16 Wei, C. (2006) Adhesion and reinforcement in carbon nanotube polymer composite. *Applied Physics Letters*, **88** (**9**), 093108.
17 Mickelson, E.T. *et al.* (1998) Fluorination of single-wall carbon nanotubes. *Chemical Physics Letters*, **296** (**1–2**), 188–194.
18 Liu, Y.J. and Chen, X.L. (2003) Evaluations of the effective material properties of carbon nanotube-based composites using a nanoscale representative volume element. *Mechanics of Materials*, **35** (**1–2**), 69–81.
19 Seyhan, A.T. *et al.* (2007) Rheological and dynamic-mechanical behavior of carbon nanotube/vinyl ester–polyester suspensions and their nanocomposites. *European Polymer Journal*, **43** (**7**), 2836–2847.
20 Fan, Z. and Advani, S.G. (2007) Rheology of multiwall carbon nanotube suspensions. *Journal of Rheology*, **51** (**4**), 585–604.
21 Hong, J.S. and Kim, C. (2007) Extension-induced dispersion of multi-walled carbon nanotube in non-Newtonian fluid. *Journal of Rheology*, **51** (**5**), 833–850.
22 Qiu, J. *et al.* (2007) Carbon nanotube integrated multifunctional multiscale composites. *Nanotechnology*, **18** (**27**), 275708.
23 Elliott, J.A. and Windle, A.H. (2000) A dissipative particle dynamics method for modeling the geometrical packing of filler particles in polymer composites. *The Journal of Chemical Physics*, **113** (**22**), 10367–10376.
24 De Fabritiis, G. (2002) Multiscale dissipative particle dynamics. *Philosophical Transactions of the Royal Society of London A: Mathematical, Physical and Engineering Sciences*, **360** (**1792**), 317–331.

25 ten Bosch, B.I.M. (1999) On an extension of dissipative particle dynamics for viscoelastic flow modelling. *Journal of Non-Newtonian Fluid Mechanics*, **83** (**3**), 231–248.

26 NIE, X.B. *et al.* (2004) A continuum and molecular dynamics hybrid method for micro- and nano-fluid flow. *Journal of Fluid Mechanics*, **500**, 55–64.

27 McCarthy, B. *et al.* (2001) Microscopy studies of nanotube-conjugated polymer interactions. *Synthetic Metals*, **121**, 1225–1226.

28 Hill, D. *et al.* (2002) Functionalization of carbon nanotubes with polystyrene. *Macromolecules*, **35**, 9466–9471.

29 Geng, Y. *et al.* (2008) Effects of surfactant treatment on mechanical and electrical properties of CNT/epoxy nanocomposites. *Composites Part A Applied Science and Manufacturing*, **39**, 1876–1883.

30 Hirsch, A. (2002) Functionalization of single-walled carbon nanotubes. *Angewandte Chemie International Edition*, **41**, 1853–1859.

31 Rausch, J., Zhuang, R.-C., and Mäder, E. (2010) Surfactant assisted processing of carbon nanotube/polypropylene composites: impact of surfactants on the matrix polymer. *Journal of Applied Polymer Science*, **117** (**5**), 2583–2590.

32 Cui, S. *et al.* (2003) Characterization of multiwall carbon nanotubes and influence of surfactant in the nanocomposite processing. *Carbon*, **41** (**4**), 797–809.

33 Vaisman, L., Marom, G., and Wagner, H.D. (2006) Dispersions of surface-modified carbon nanotubes in water-soluble and water-insoluble polymers. *Advanced Functional Materials*, **16** (**3**), 357–363.

34 Yu, J. *et al.* (2007) Controlling the dispersion of multi-wall carbon nanotubes in aqueous surfactant solution. *Carbon*, **45** (**3**), 618–623.

35 Whitsitt, E.A. and Barron, A.R. (2003) Silica coated single walled carbon nanotubes. *Nano Letters*, **3** (**6**), 775–778.

36 Kim, T.H. *et al.* (2007) Water-redispersible isolated single-walled carbon nanotubes fabricated by in situ polymerization of micelles. *Advanced Materials*, **19** (**7**), 929–933.

37 Vaisman, L., Wagner, H.D., and Marom, G. (2006) The role of surfactants in dispersion of carbon nanotubes. *Advances in Colloid and Interface Science*, **128–130**, 37–46.

38 Georgakilas, V. *et al.* (2007) Decorating carbon nanotubes with metal or semiconductor nanoparticles. *Journal of Materials Chemistry*, **17** (**26**), 2679–2694.

39 Kim, J.-K. and Mai, Y.W. (1998) *Engineered Interfaces in Fiber Reinforced Composites*, Elsevier Sciences, Amsterdam; New York.

40 Thess, A. *et al.* (1996) Crystalline ropes of metallic carbon nanotubes. *Science*, **273** (**5274**), 483–487.

41 Thostenson, E.T., Ren, Z., and Chou, T.-W. (2001) Advances in the science and technology of carbon nanotubes and their composites: a review. *Composites Science and Technology*, **61** (**13**), 1899–1912.
42 Coleman, J.N., Khan, U., and Gunko, Y.K. (2006) Mechanical reinforcement of polymers using carbon nanotubes. *Advanced Materials*, **18**, 689–706.
43 Lu, K.L. *et al.* (1996) Mechanical damage of carbon nanotubes by ultrasound. *Carbon*, **34**, 814–816.
44 Gao, B. *et al.* (2000) Enhanced saturation lithium composition in ball-milled single-walled carbon nanotubes. *Chemical Physics Letters*, **327** (**1–2**), 69–75.
45 Huang, J.Y., Yasuda, H., and Mori, H. (1999) Highly curved carbon nanostructures produced by ball-milling. *Chemical Physics Letters*, **303** (**1–2**), 130–134.
46 Kim, Y.A. *et al.* (2002) Effect of ball milling on morphology of cup-stacked carbon nanotubes. *Chemical Physics Letters*, **355** (**3–4**), 279–284.
47 Awasthi, K. *et al.* (2002) Ball-milled carbon and hydrogen storage. *International Journal of Hydrogen Energy*, **27** (**4**), 425–432.
48 Ma, P.C., Tang, B.Z., and Kim, J.K. (2008) Converting semiconducting behavior of carbon nanotubes using ball milling. *Chemical Physics Letters*, **458**, 166–169.
49 Ma, P.C. *et al.* (2009) In-situ amino functionalization of carbon nanotubes using ball milling. *Journal of Nanoscience and Nanotechnology*, **9** (**2**), 749–753.
50 Schmid, C.F. and Klingenberg, D.J. (2000) Mechanical flocculation in flowing fiber suspensions. *Physical Review Letters*, **84**, 290–293.
51 Li, J. *et al.* (2007) Correlations between percolation threshold, dispersion state and aspect ratio of carbon nanotube. *Advanced Functional Materials*, **17**, 3207–3215.
52 Xie, X.-L., Mai, Y.-W., and Zhou, X.-P. (2005) Dispersion and alignment of carbon nanotubes in polymer matrix: a review. *Materials Science and Engineering: R: Reports*, **49** (**4**), 89–112.
53 Feng, W. *et al.* (2003) Well-aligned polyaniline/carbon-nanotube composite films grown by in-situ aniline polymerization. *Carbon*, **41** (**8**), 1551–1557.
54 Qin, S.H. *et al.* (2004) Polymer brushes on single-walled carbon nanotubes by atom transfer radical polymerization of n-butyl methacrylate. *Journal of the American Chemical Society*, **126**, 170–176.
55 Vigolo, B. *et al.* (2000) Macroscopic fibers and ribbons of oriented carbon nanotubes. *Science*, **290** (**5495**), 1331–1334.
56 Jin, L., Bower, C., and Zhou, O. (1998) Alignment of carbon nanotubes in a polymer matrix by mechanical stretching. *Applied Physics Letters*, **73** (**9**), 1197–1199.

57 Bannov, A.G. *et al.* (2012) Effect of the preparation methods on electrical properties of epoxy resin/carbon nanofiber composites. *Nanotechnologies in Russia*, **7** (**3-4**), 169–177.
58 Kimura, T. *et al.* (2002) Polymer composites of carbon nanotubes aligned by a magnetic field. *Advanced Materials*, **14** (**19**), 1380–1383.
59 Ko, F. *et al.* (2003) Electrospinning of continuous carbon nanotube-filled nanofiber yarns. *Advanced Materials*, **15** (**14**), 1161–1165.
60 Lee, W., Gau, J.S., and Chen, H.Y. (2005) Electro-optical properties of planar nematic cells impregnated with carbon nanosolids. *Applied Physics B*, **81** (**2-3**), 171–175.
61 Shih, C.-H. and Lee, L.J. (1998) Effect of fiber architecture on permeability in liquid composite molding. *Polymer Composites*, **19** (**5**), 626–639.
62 Koefoed, M.S. (2003) Modeling and simulation of the VARTM process for wind turbine blades. PhD Dissertation, Aalborg University.
63 Lee, Y.J. *et al.* (2006) A prediction method on in-plane permeability of mat/roving fibers laminates in vacuum assisted resin transfer molding. *Polymer Composites*, **27** (**6**), 665–670.
64 Sanchez, F. *et al.* (2006) A process performance index based on gate-distance and incubation time for the optimization of gate locations in liquid composite molding processes. *Composite Part A: Applied Science and Manufacturing*, **37** (**6**), 903–912.
65 Ding, L. *et al.* (2003) In situ measurement and monitoring of whole-field permeability profile of fiber perform for liquid composite molding processes. *Composites Part A: Applied Science and Manufacturing*, **34** (**8**), 779–783.
66 Qiu, J. (2008) *Multifunctional Multiscale Composites: Processing, Modeling and Characterization*, in *Industrial and Manufacturing Engineering*, Tallahassee, Florida State University.
67 McCrary-Dennis, M.C. and Okoli, O.I. (2012) A review of multiscale composite manufacturing and challenges. *Journal of Reinforced Plastics and Composites*, CRC Press LLC Taylor & Francis Group, Boca Raton, FL, USA, **31** (**24**), 1687–1711.
68 Mazumder, S.K. (2002) *Composites Manufacturing Materials Product and Process Engineering*, CRC Press LLC Taylor & Francis Group, Boca Raton, FL.
69 Dexter, H.B. (1998) Development of textile reinforced composites for aircraft structures. Presented at the 4th International Symposium for Textile Composites, Kyoto Institute of Technology Kyoto, Japan, October 12-14, 1998, NASA, https://ntrs.nasa.gov/archive/nasa/casi.ntrs.nasa.gov/19990032212.pdf (accessed 26 May 2017).
70 Hussain, F. *et al.* (2006) Review article: polymer-matrix nanocomposites, processing, manufacturing, and application: an overview. *Journal of Composite Materials*, **40** (**17**), 1511–1575.

71 Fielding, J.C., Chen, C., and Borhes, J. (2004) *Vacuum Infusion Process for Nanomodified Aerospace Epoxy Resins*, in *SAMPE Symposium & Exhibition*, SAMPE (Society for the Advancement of Material and Process Engineering), Long Beach, CA.
72 Hussain, F. *et al.* (2005) S2 glass/vinyl ester polymer nanocomposites: manufacturing, structures, thermal and mechanical properties. *Journal of Composite Materials*, **37** (**20**), 1821–1837.
73 Xu, L. (2004) *Integrated analysis of liquid composite molding (LCM) processes*, The Ohio State University.
74 Whitaker, S. (1986) Flow in porous media I: a theoretical derivation of Darcy's law. *Transport in Porous Media*, **1** (**1**), 3–25.
75 Rachmadini, Y., Tan, V.B.C., and Tay, T.E. (2010) Enhancement of mechanical properties of composites through incorporation of CNT in VARTM - a review. *Journal of Reinforced Plastics and Composites*, **29** (**18**), 2782–2807.
76 TPI Technology (2005) *An Overview of the SCRIMP™ Technology*, TPI Technology, Inc.
77 Williams, C., Summerscales, J., and Grove, S. (1996) Resin Infusion under Flexible Tooling (RIFT): a review. *Composites Part A: Applied Science and Manufacturing*, **27** (7), 517–524.
78 Thagard, J.R. (2003) *Investigation and Development of the Resin Infusion between Double Flexible Tooling (Ridft) Process for Composite Fabrication*, Florida State University.
79 Parker, L. *et al.* (2011) *Enhancing Flow of CNT-Doped Resins in the Manufacture of Multiscale Composites Using the RIDFT Process*, in *Composites 2011-The Composites Exhibition and Convention*, American Composites Manufacturers Association, Fort Lauderdale, Fl USA.
80 Brillant, M. (2010) *Out-of-Autoclave Manufacturing of Complex Shape Composite Laminates*, in *Mechanical Engineering*, McGill University.
81 Mallow, A. and Campbell, F. (2000) *Autoclave processing*, in *Processing of Composites* (eds R.S. Dave and A.C. Loos), Hanser Publishers.
82 Brosius, D. (2014) *Out-of-autoclave manufacturing: the green solution*, in *High-Performance Composites*, CompositesWorld (ed. J. Sloan), Richard G. Kline Jr..
83 Zhang, J. and Fox, B.L. (2007) Manufacturing Influence on the delamination fracture behavior of the T800H/3900-2 carbon fiber reinforced polymer composites. *Materials and Manufacturing Processes*, **22**, 768–772.
84 Campbell, J.A., P. Compston, and Z.H. Stachurski, (2004) *Improved Manufacture and Productivity of a High Performance Composite OAR Blade*, Ginninderra Press, Canberra.
85 Davies, L.W. *et al.* (2007) Effect of cure cycle heat transfer rates on the physical and mechanical properties of an epoxy matrix composite. *Composites Science and Technology*, **67** (**9**), 1892–1899.

86 Herring, M.L., Mardel, J.I., and Fox, B.L. (2010) The effect of material selection and manufacturing process on the surface finish of carbon fibre composites. *Journal of Materials Processing Technology*, **210**, 926–940.

87 Zhang, J., Guo, Q., and Fox, B.L. (2009) Study on thermoplastic-modified multifunctional epoxies: influence of heating rate on cure behaviour and phase separation. *Composites Science and Technology*, **69**, 1172–1179.

88 Quickstep Technology, Quickstep Curing Process, http://www.quickstep.com.au/Capabilities/Quickstep-Curing-Process (accessed 12 May 2017).

89 Quickstep. The out-of-autoclave process for high performance autoclave grade materials.

90 Roy, S. *et al.* (2005) *Characterization and Modeling of Strength Enhancement Mechanism in Polmer Clay Nanocomposites, AIAA Conference Proceedings*, Texas.

91 OSHA (1999) *Polymer Matrix Materials: Advanced Composites*, OSHA, Washington, D.C.

92 Spindler-Ranta, S. and Bakis, C.E. (2002) *Carbon nanotube reinforcement of a filament winding resin, 47th International SAMPE Symposium and Exhibition*, Long Beach, CA, USA, pp. 1775–1787.

93 Mahfuz, H., Baseer, M.A., and Zeelani, S. (2005) Fabrication, characterization and mechanical properties of nano phased carbon prepreg laminates. *SAMPE Journal*, **41** (**2**), 40–48.

94 Handy, R., Owen, R., and Valsami-Jones, E. (2008) The ecotoxicology of nanoparticles and nanomaterials: current status, knowledge gaps, challenges, and future needs. *Ecotoxicology*, **17** (**5**), 315–325.

95 Dreher, K.L. (2004) Health and environmental impact of nanotechnology: toxicological assessment of manufactured nanoparticles. *Toxicological Sciences*, **77** (**1**), 3–5.

96 Bukkapatnam, S. *et al.* (2012) Nanomanufacturing systems: opportunities for industrial engineers. *IIE Transactions*, **44** (7), 492–495.

97 Mueller, N.C. and Nowack, B. (2008) Exposure modeling of engineered nanoparticles in the environment. *Environmental Science & Technology*, **42** (**12**), 4447–4453.

98 Colvin, V.L. (2003) The potential environmental impact of engineered nanomaterials. *Nature biotechnology*, **21** (**10**), 1166–1170.

99 Taurozzi, J.S., Hackley, V.A., and Wiesner, M.R. (2011) Ultrasonic dispersion of nanoparticles for environmental, health and safety assessment - issues and recommendations. *Nanotoxicology*, **5** (**4**), 711–729.

100 Aitken, R.J. *et al.* (2006) Manufacture and use of nanomaterials: current status in the UK and global trends. *Occupational Medicine*, **56** (**5**), 300–306.

101 Khan, F.I., Sadiq, R., and Veitch, B. (2004) Life cycle iNdeX (LInX): a new indexing procedure for process and product design and decision-making. *Journal of Cleaner Production*, **12** (**1**), 59–76.

102 Schmitz, G.J. and Prahl, U. (2009) Toward a virtual platform for materials processing. *The Journal of The Minerals, Metals & Materials Society*, **61** (**5**), 19–23.

103 Panchal, J.H. *et al.* (2005) *A strategy for simulation-based multiscale, multi-functional products and associated design processes*, in *ASME 2005 International Design Engineering Technical Conferences and Computers and Information in Engineering Conference*, American Society of Mechanical Engineers.

104 Fermeglia, M. and Pricl, S. (2007) Multiscale modeling for polymer systems of industrial interest. *Progress in Organic Coatings*, **58** (**2–3**), 187–199.

105 Qiu, J. (2008) *Multifunctional Multiscale Composites: Processing, Modeling and Characterization*, ProQuest.

106 Askland, D. and Phule, P. (2003) in *The Science and Engineering of Materials*, 4th edn (ed. P. Phule), Brooks/Cole-Thompson Learning, Pacific Grove, CA.

8

Bioinspired Systems

Oluwamayowa Adigun[1], Alexander S. Freer[1], Laurie Mueller[2], Christopher Gilpin[2], Bryan W. Boudouris[1], and Michael T. Harris[1]

[1] *School of Chemical Engineering, Purdue University, West Lafayette, IN, USA*
[2] *Life Sciences Microscopy Facility, Purdue University, West Lafayette, IN, USA*

8.1 Introduction and Literature Overview

Bioinspired systems have increasingly been studied for the fabrication of nanomaterials due to their advantages over conventional chemical or electrochemical methods [1–5]. For instance, biomolecules have numerous surface functionalities that can promote inorganic mineralization at very mild reaction conditions and in nontoxic aqueous solutions [6–11]. In addition, biomolecules, especially proteins in viruses, assemble very stringently, to form precise nanometric arrangements. Thus, they are very attractive for highly controlled nanoparticle syntheses. To this end, multiple biological molecules have been used as templates in the synthesis of metal nanoparticles [4, 12–15]. The biotemplate approach utilizes the naturally occurring properties of molecules to promote the formation of nanostructures of desired sizes and shapes through an electroless deposition process [16]. The highly uniform nanorod dimensions (i.e., a material of cylindrical shape that has a vertical axis that is much longer in length than the diameter of the cylinder) of certain viruses afford this biotemplate synthetic strategy many advantages over traditional methods of nanowire synthesis. One of the advantages of greatest import is the simple solution chemistry associated with the nanowire fabrication [17, 18]. Furthermore, the self-assembly of the nano-sized building blocks guarantees the uniformity of the nanorod template and allows for the controllable placement of desired functional groups on the surface via genetic engineering [19–21].

The tobacco mosaic virus (TMV) is a primary example of a nanorod-shaped biotemplate. TMV is formed through the naturally driven assembly of identical coat proteins around a single RNA core [19]. Once assembled, the TMV is

Nanotechnology Commercialization: Manufacturing Processes and Products, First Edition.
Edited by Thomas O. Mensah, Ben Wang, Geoffrey Bothun, Jessica Winter, and Virginia Davis.

stable, maintaining a 300 nm in length by 18 nm diameter nanorod structure across a wide basicity ($3 < \mathrm{pH} < 9$) and temperature ($25\,^{\circ}\mathrm{C} < T < 70\,^{\circ}\mathrm{C}$) range [2]. These properties allow for the use of TMV as a biotemplate in a large number of electroless deposition conditions [22]. In fact, Shenton *et al.* [6] reported the first case of the formation of an inorganic–organic nanotube containing TMV. In this seminal work, four different substances, Fe, SiO_2, CdS, and PbS, were coordinated to a wild-type TMV surface under varying conditions. Since that time, TMV has been investigated as a template, through coordination with different molecules. Most of these are sol compounds [23, 24], organic molecules [25–27], or inorganic materials, particularly metals [6–9, 14–16, 28–35]. In particular, the studies involving metal mineralization, which are the subject of this paper, have yielded the most prolific results.

Following the astute work by Shelton *et al.*, Dujardin *et al.* [15] mineralized Pt and Au nanoparticles on the outer surface of TMV and Ag nanoparticles in the internal channel of TMV. Along with the novel nanoparticle decoration, the authors also attempted to gain insight into the mineralization process by reducing the surface charge on TMV via genetic engineering. Their results led to the postulate that, below the isoelectric point (pI) of TMV, the positive charge held by some amino acids (arginine and lysine) allows them to serve as a template for adsorption of inorganic anions, which are subsequently reduced by addition of an external reducing agent. Following these hypotheses, Knez *et al.* [12] adsorbed Pd ions for the first time unto TMV by controlling the pH of the incubation solution. However, while the pH can be controlled to promote metal uptake onto the surface of naturally harvested TMV, the native surface functionalities of the coat proteins can also be manipulated in order to promote the uptake of metal onto the virus surface [36]. Lee *et al.* [8], with the innovation of TMV1Cys (created by insertion of Cys on TMV surface and first used by Yi *et al.* [37]), became the first of a series of works that have sought to elucidate, improve, diversify, and apply inorganic mineralization to TMV1Cys.

In initial applications, it was conceived that the added cysteine, which provides a thiol group, should complex readily with the metals and certain salts in solution. The uptake capacity of the virus will increase this way. Accordingly, in the first work applying TMV1Cys to inorganic mineralization, Lee *et al.* [8] showed that the mutation increased the coverage of platinum particles on the virus. Closely following this work, the same authors demonstrated the utility of a double cysteine mutant TMV [16] (TMV2Cys) to form nanowires decorated with an increasing amounts of gold, silver, and palladium. By relating increased uptake with more decoration, these works established the expediency of extra thiol groups as a means of producing higher metal content nanowires. However, the capability to acquire both high metal density and controlled uniform morphology on the TMV template still proved to be elusive, as demonstrated in the quantitative study by Lim *et al.* [7]

It is important to note that all metal-virus syntheses were performed hitherto with a conventional synthetic scheme. This involved the incubation of the metal precursor with virus in an aqueous salt solution, followed by addition of an external reducer. The quantitative study showed that this synthetic scheme is not as efficient at higher metal precursor incubation concentrations, particularly in the cases of Pd and Au. Under such conditions, the presence of additional metal precursor did not significantly enhance metal coating but instead exacerbated both the dispersity of sizes and uniform spatial distribution of nanoparticles on the template surface (Figure 8.1). It was inferred that at lower concentrations (Figure 8.1a) the system mostly comprised biosorbed precursor molecules. However, as the concentration is significantly increased in an effort to increase the intensity of biosorption, the unreduced system comprised both biosorbed molecules and unadsorbed precursor molecules in equilibrium. Recent small angle X-ray scattering (SAXS) results by Yi and coworkers [30] showed that the average nanoparticle size was significantly different when reduction occurs from unadsorbed precursor molecules instead

Figure 8.1 TEM images of palladium-deposited TMV1Cys at 25 °C with different molar ratio of palladium-ion to TMV1Cys. This ratio is (a) 3.4×10^{-6} mol/mg of palladium-ion to TMV1Cys and (b) 10.8×10^{-6} mol/mg of palladium-ion to TMV1Cys. The scale bars represent 100 nm. Lim *et al.* 2010 [7]. Reproduced with permission of Elsevier.

of from those biosorbed on the virus surface. At high concentration, both these sites serve as potential nucleation sites for reduction, thus leading to (i) a large distribution of sizes after reduction and (ii) a lack of spatial uniformity on the template surface, specifically solution-based nanoparticles (from the unadsorbed phase) attach to the virus–metal nanorod structure. Both of these are evident in Figure 8.1b. It is therefore impractical to achieve both efficiently controllable uniform coating and high coating density even when assuming that only the external reducer drives metal reduction. Moreover, the nonideality of the conventional synthesis was further revealed when Lim *et al.* [29] discovered the hydrothermal synthesis, which indicates the presence of a dual reduction mechanism (external reducer driven and template driven) when a reducer is used.

In the aforementioned work, Lim *et al.* [29] circumvented the inherent encumbrance in the conventional synthesis by synthesizing palladium nanowires on wild-type TMV and TMV2cys without the addition of an external reducing agent using the hydrothermal synthesis. The nanowires synthesized from this work were more uniform than any other biotemplated metal nanowires reported in the literature. In comparison with the conventional reducing agent synthesis, the nanowires produced were completely covered with smaller seemingly uniform-sized palladium nanoparticles throughout the entire virus surface (Figure 8.2). The method also allowed for a unique ability to control the size and degree of mineralization of the nanowires. This is performed by periodic washing and droplet addition of precursor to control the mineralization reaction. The mechanism for the hydrothermal reduction has been further elucidated elsewhere [38]. Five additional coatings of the virus are achieved in

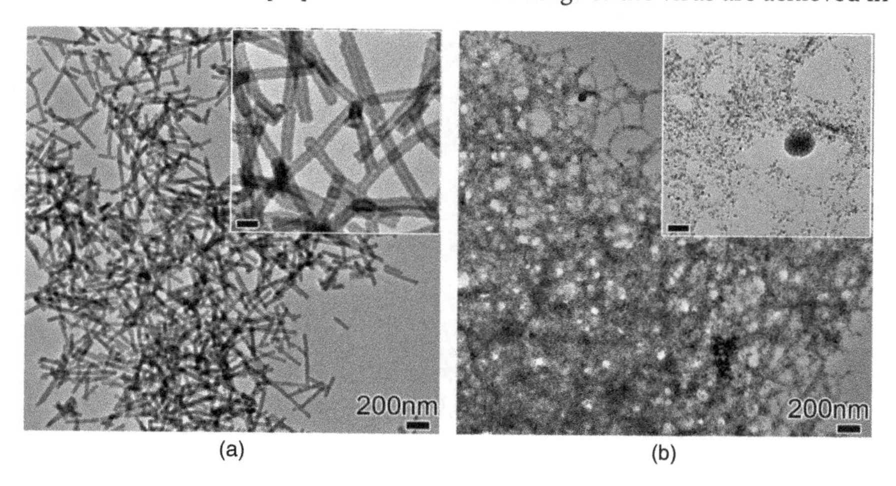

Figure 8.2 Palladium biomineralization on the TMV by (a) self-mineralization process and by (b) adding an external reducing agent (DMAB). Lim *et al.* 2010 [29]. Reproduced with permission of American Chemical Society.

this manner. Freer *et al.* [31] thereafter continued the studies by using SAXS analysis to study the statistically significant changes in the diameter of the TMV-Pd nanowires as more coating cycles were performed. By utilizing a model-dependent analysis, it was determined that the nanowires retain low dispersity values along their diameters throughout the solution even as more coats were added to the TMV-Pd.

This uniform palladium-coated tobacco mosaic virus (TMV-Pd) could prove to be a much more useful tool in nanoelectronic applications when compared to other similarly synthesized biotemplated materials. Particularly, they represent the first TMV-metal nanowires with a unique combination of improved nanomaterial quality and synthetic simplicity. They are (i) certainly continuous and not merely decoration, (ii) facilely synthesized because they do not require a long incubation time or a finely tuned synthetic window for reagents, (iii) controllable over their nanorod diameters from (i.e., $30\,\text{nm} \leq d \leq 60\,\text{nm}$), (iv) uniformly coated, (v) uniform in size, and (vi) easily reproduced, which offers the promise of convenient scalability. In addition, these materials have the ability to align vertically on Au substrates under mild conditions purportedly via an exposed thiol on a Cys at the end of the virus [8, 39–41].

However, the electrical properties of these types of coated nanostructures (or of any TMV-Pd nanomaterials) have not been determined in a reliable manner. As such, it is critical to establish the electronic properties of these biotemplated materials on the nanoscale. The following section is a successful attempt at such an endeavor. We show that the biotemplated nanostructures produced from this simple, yet robust, synthetic method lead to uniform materials with electrical conductivity values of about $2.5 \pm 0.5 \times 10^2$ S/cm, which is reasonable for organic–inorganic-type nanowires. As such, these bio-based materials present themselves as sustainable, high-performance materials, which can now be applied to organic-based electronic applications.

8.2 Electrical Properties of a Single Palladium-Coated Biotemplate

Focused ion beam (FIB) platinum-deposited electrodes are used in nanowire conductivity measurements [42–45]. Generally, a gallium FIB is incident on the desired area of the substrate and an organometallic gas molecule is introduced into the sample chamber. The ion bombardment decomposes the organometallic precursor leaving a metal deposit on the sample [46]. Typically, the precursor gas is trimethylcyclopentadienyl-platinum ($(CH_3)_3CH_3C_5H_4Pt$). Studies have shown that the ion beam-deposited platinum has a much lower resistance than the electron beam-deposited platinum due to decreased carbon contamination in the final platinum deposit [47]. In most cases, the nanowires being studied with this type of two or four probe platinum contact setup are over 2 μm in length, allowing space to deposit the contacts. Due to the fact that the TMV-Pd nanowires are only 300 nm in length, a combination of techniques has been used to determine the resistivity of these particles. The particles are deposited across a gold nanogap electrode, and then FIB-deposited platinum contacts are used to ensure the minimization of contact resistance between the gold and the TMV-Pd particle.

8.3 Materials and Methods

The facile TMV biotemplating synthesis technique was performed in an aqueous solution in the absence of external reducing agents to obtain uniform TMV-Pd nanowires. The procedure was as follows. The TMV1Cys (0.22 g/l) and palladium precursor solution (0.75 mM) were placed in deionized water at 50 °C for 20 min, and the brown precipitates were collected and reintroduced to a second solution of water and palladium precursor (0.75 mM). These twice-coated TMV-Pd particles are shown in Figure 8.3. The transmission electron microscopy (TEM) image shows the uniformity of the Pd coating on the TMV using this synthesis technique over a number of TMV-Pd nanowires. Following the nanowire synthesis, they were deposited onto the electrical characterization platforms as shown graphically in the procedure in Figure 8.4.

In brief, a ~100 nm gold film was initially sputtered onto a glass slide to form the substrates used in these experiments. Then the nanofabrication abilities of the FIB were used to mill a pattern of nanogaps through the gold film. The Ga^+ ions were accelerated to 30 kV at 10 pA, and the widths of the nanogaps were varied between 50 and 250 nm. All gaps are 100 μm in length and ~300 nm

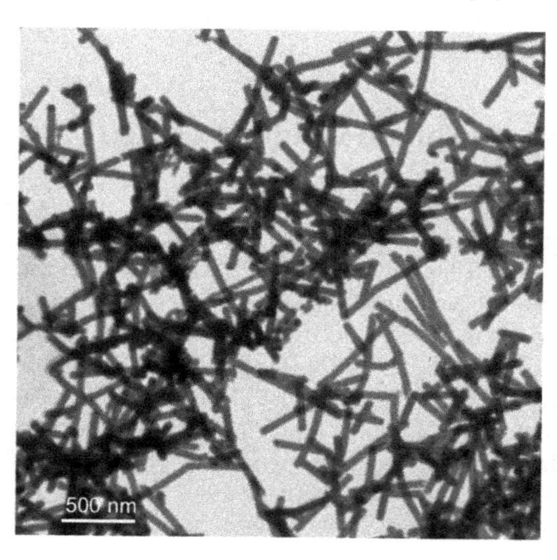

Figure 8.3 TEM image of a large number of tobacco mosaic virus-templated nanostructures formed through electroless deposition of palladium onto the viral surface in the absence of external reducing agents. The Pd-coated TMV nanowires are 350 nm in length and 50 nm in diameter. In some instances the nanowires are on top of one another, which causes a cross-like pattern to be seen, and in other instances the nanowires are aligned end to end. These instances can cause two well-aligned rods to appear as a single rod with twice the rod length.

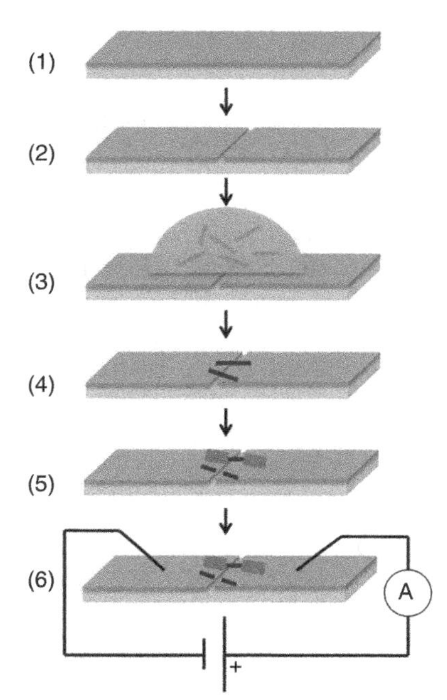

Figure 8.4 Summary of electrical device fabrication characterization procedure. (1) The process starts with a thin film of gold on a glass substrate. (2) The FIB is used to mill a 100 μm × 100 nm gap in the gold. (3) A solution containing TMV-Pd particles is deposited onto the substrate. (4) TMV-Pd is deposited across the nanogap through evaporation of the solvent. (5) The FIB assists Pt deposition on the edges of the TMV-Pd and ensures no other particles are crossing the gap. (6) Microprobes are used to apply a voltage and measure the current between the two electrodes.

deep to ensure all of the gold is removed from them. Finally, a 200 μm × 100 μm rectangular outline was milled around the outside to electrically isolate the two gold electrodes from each other (Figure 8.5). The FIB current was increased when milling the larger rectangular outline in the interest of milling time.

A dilute solution of the TMV-Pd nanowires was then drop cast onto the substrate. Following this deposition, platinum contacts needed to be added to reduce any contact resistance between the particles and substrate. However,

Figure 8.5 SEM image of the two nanogap electrodes electrically separated from each other and the rest of the gold substrate by gold milling with a FIB. (1) indicates the nanogap and (2) shows one of the ~100 μm^2 gold electrodes.

the prolonged exposure of the sample to the ion beam (in order to find the correct locations to deposit the pads) destroyed the nanowires. To circumvent this problem, dual beam scanning electron microscope and FIB systems were utilized. The two beams were aligned precisely at the desired magnification to give the same field of view. In this way, the stigmation of the FIB was properly tuned and focused away from the nanowires at a desired magnification, and the SEM was then used to image the nanogaps to locate sections where a TMV-Pd particle was bridging the gap between two electrodes.

When an acceptable particle was located, the FIB was used to deposit Pt first onto the substrate close to the particle while keeping the particle out of the path of the ion beam. Then the sample was reimaged with the SEM to ensure the platinum was deposited in the desired location. Next the Pt was deposited from the edge of the biotemplated nanoparticle to the edge of the nanogap while limiting the FIB exposure of the sample to only the short times that the platinum is being deposited. An SEM image of the TMV-Pd tethered with the Pt pads is shown in Figure 8.6. Eliminating the need to search and align with the FIB at high magnification helped in reducing damage to the imaged nanoparticles.

After the platinum was deposited successfully on both ends of the TMV-Pd, the FIB was again used for its milling capabilities. That is, any sites across the nanoscale channel where TMV-Pd nanowires were connecting the electrodes, as in Figure 8.6, were removed with the FIB to ensure that the only path through which charge could be transferred between the electrodes was the platinum-contacted TMV-Pd. The nanowires in the upper left of Figure 8.6 with a more disperse Pd coating are examples of particle degradation due to the FIB exposure, as they were used to quickly focus the FIB for the platinum

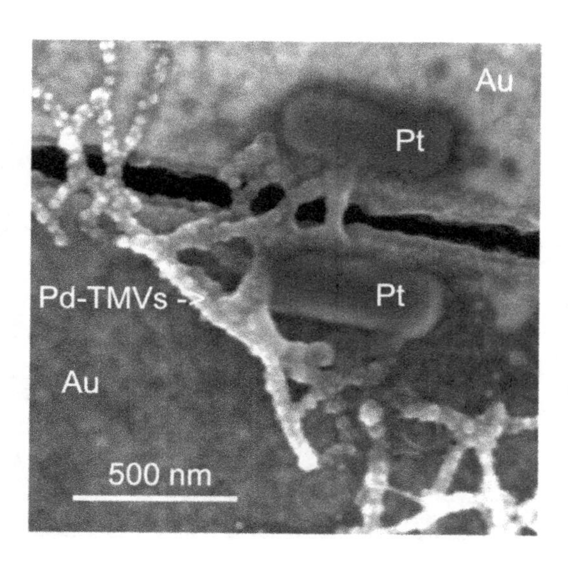

Figure 8.6 SEM image of TMV-Pd particles spanning the nanogap between the gold electrodes. Platinum is deposited onto the edges of one TMV-Pd nanorod. The other TMV-Pd particles spanning the gap are later removed with the FIB. The TMV-Pd in the upper left corner of the image show the degradation from FIB exposure as they are no longer uniform nanowires and instead appear as discrete particles.

deposition on the desired particle. The particles spanning the gap directly next to the nanorod of interest were milled accurately with the FIB through careful alignment, stigmation, and focusing of the ion beam. Micromanipulators and a light microscope were used as part of a Lake Shore Model CPX Probe Station to contact the 100 μm^2 gold electrodes. Once in contact, a Keithley voltage source was used to apply a −0.1 to 0.1 V sweep across the sample while monitoring the current passing between the electrodes.

In the TMV-Pd system, X-ray absorption spectroscopy (XAS) is a convenient method to monitor both the degree of mineralization and the size of palladium nanoparticles attached to TMV1Cys. A 2.0×10^{-1} mg/ml solution of TMV1Cys was placed in a 2 ml reaction vessel at 60 °C with a 5 mM sodium tetrachloropalladate aqueous solution. The concentrations here were increased significantly from the original synthesis in order to reach the detectable limit for the XAS analysis. The reaction was monitored via XAS measurements taken every 4 min for the next 18 h at the palladium K-edge (24.4 KeV) until there was no noticeable change in spectra. The final spectrum was analyzed for relevant extended X-ray absorption fine structure (EXAFS) parameters using WINXAS 3.1 fitting software. A palladium foil and palladium(II) chloride were used as references to obtain experimental phase shift and backscattering amplitudes for Pd–Pd (12 at 2.75 Å) and Pd–Cl (4 at 2.28 Å). EXAFS parameters were obtained by performing a least square fit of Fourier transformed k^2-weighted normalized absorption data in R-space.

8.4 Results and Discussion

After testing multiple TMV-Pd nanowires with varying cross-sectional areas and gold electrode gap lengths, a conductivity of $2.5 \pm 0.5 \times 10^2$ S/cm^1 was determined for these biotemplated palladium nanowires. A representative current–voltage (I–V) curve for a single TMV-Pd nanorod is shown in Figure 8.7a. To demonstrate that the contact resistance of the gold/platinum/TMV-Pd junction was not the highest contributing factor in the measurement, the resistance multiplied by the cross-sectional area was plotted versus the length of the nanowire (Figure 8.7b). The resulting curve should be linear in nature with the slope corresponding to the resistivity and an intercept of zero. Any intercept greater than zero would correspond to a contact resistance within the system. The intercept of the linear regression, shown in Figure 8.7b, is $1.3 \times 10^{-12}\ \Omega\ m^2$. Taking an average contact area of the TMV-Pd with the gold substrate as $10^{-13}\ m^2$, the contact resistance can be calculated as ~15 Ω, a value much less than the overall resistance of the nanowires. In order to further investigate the reasoning for the electrical properties of the synthesized TMV-Pd particles, they were examined in more detail using TEM and XAS.

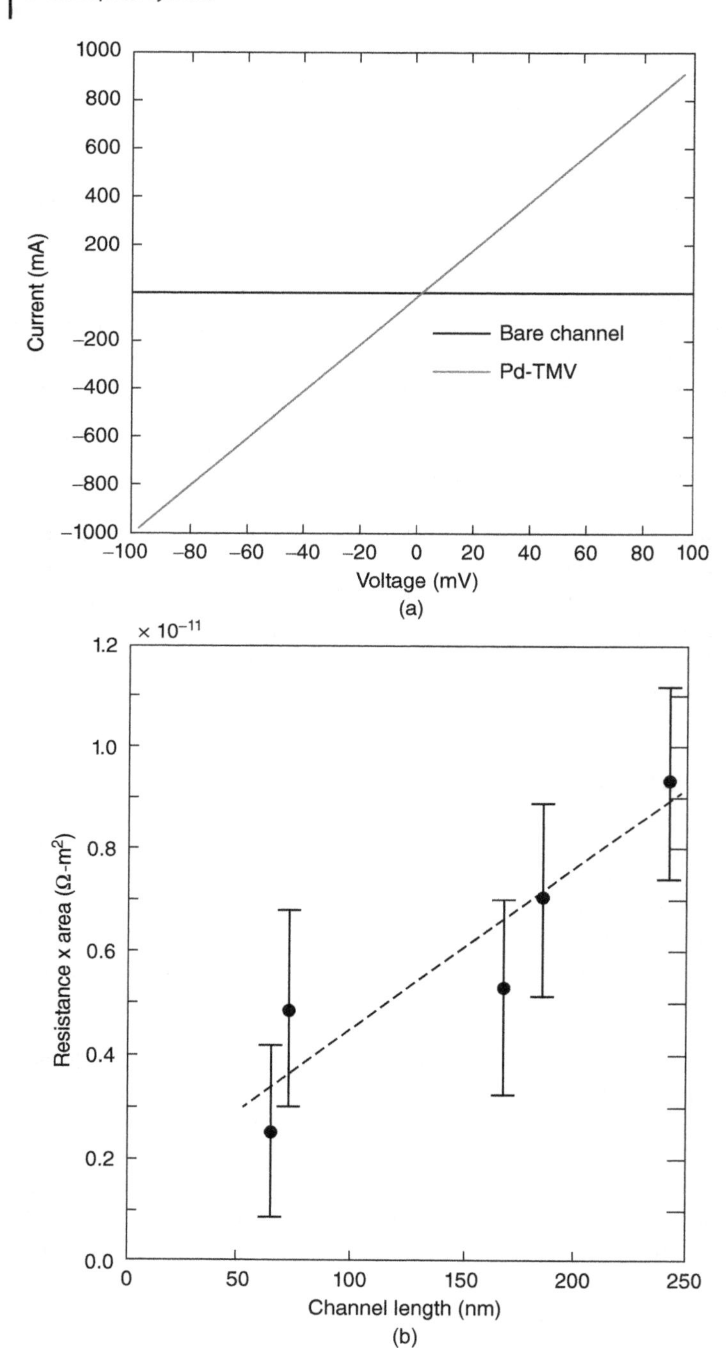
1000
800
600
400
200
–200
–400
–600
–800
–1000
Current (mA)
–100 –80 –60 –40 –20 0 20 40 60 80 100
Voltage (mV)
Bare channel
Pd-TMV
(a)
× 10⁻¹¹
1.2
1.0
0.8
0.6
0.4
0.2
0.0
Resistance x area (Ω-m²)
0 50 100 150 200 250
Channel length (nm)
(b)

Figure 8.7 (a) *I–V* profiles for the TMV-Pd nanowire across the gold nanoscale channel (light gray) and across the bare (i.e., in the absence of the TMV-Pd nanorod) nanoscale channel (dark gray) with an applied bias of $-0.1\,V \leq V \leq 0.1\,V$. (b) A linear plot of the experimentally determined resistance values of TMV-Pd particles spanning nanoscale gaps between gold electrodes of varying channel lengths. The resistivity equation, $RA = \rho L + R_{contact}$, describes the linear behavior where ρ is the resistivity, and a vertical axis intercept ($R_{contact}$) greater than zero relates to the contact resistance of the system.

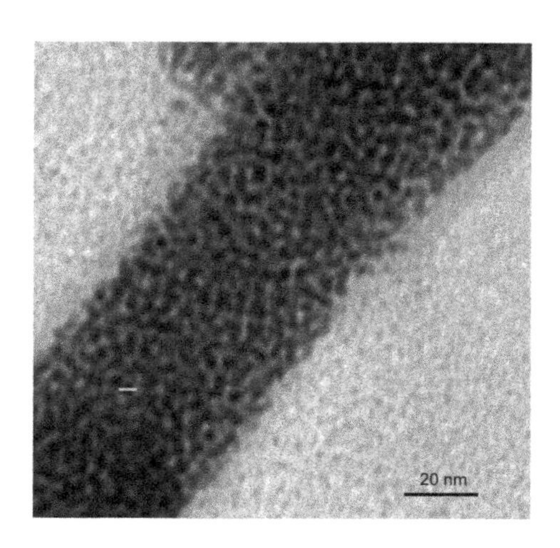

Figure 8.8 TEM image of a TMV-Pd nanowire displaying the individual grain size of the Pd particles that form on the surface of the TMV during the aqueous coating and reduction process. The gray (~5 nm) line indicates the width of a single Pd nanoparticle on the outer surface of the uniform palladium coating.

The grain size (the diameter of the individual particles that form the final nanowire) of the palladium on the biotemplated nanowires was determined to be ~5 nm through TEM imaging (Figure 8.8), indicating a large number of grain boundaries on a single TMV-Pd nanowire. As the grain size of a nanowire decreases, the resistance increases due to the increased number of grain boundaries [48, 49]. In a nonbiotemplated example, a palladium nanowire electrochemically grown between two gold electrodes has a palladium grain size that appears to be ~25 nm [50]. This larger grain size equates to a lower number of grain boundaries and a higher conductivity. Therefore, the large number of grain boundaries of the TMV-Pd nanowire agrees with the lower conductivity result of the TMV-Pd system when compared to the nonbiotemplated system.

This ~5 nm domain size was confirmed by XAS as well, and the spectroscopic measurements indicated that physical grain boundaries, as opposed to chemical charge trap sites, were responsible for the reduced conductivity relative to pristine Pd nanowires. In fact, the palladium on the TMV template was determined to be completely reduced with an average nanoparticle size of 7–8 nm. The final palladium K-edge spectrum (Figure 8.9a) along with X-ray absorption near edge spectroscopy (XANES) spectra in the final stages of analysis (not shown) indicated that the Pd(II) has been reduced to Pd(0) completely. EXAFS fitting analysis to produce Fourier transformed

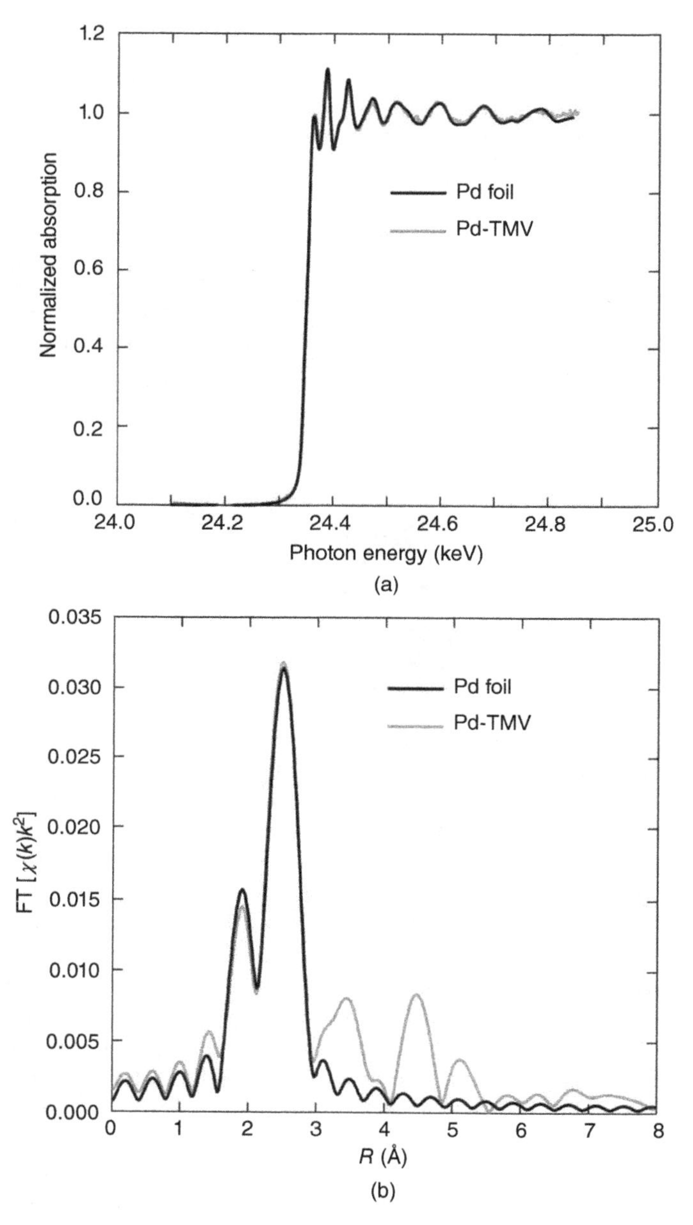

Figure 8.9 (a) The normalized absorption spectra of the XAS experimental data (light gray line) compared to palladium foil (black line). The overlap indicates presence of Pd–Pd bonds. (b) The Fourier transform of k^2-weighted normalized absorption data in R-space for TMV-Pd compared to Pd foil indicates a Pd coordination number of 11, which corresponds to a particle diameter of ~7 nm, which is in good agreement with the TEM image of Figure 8.8.

k^2-weighted normalized absorption data (Figure 8.9b) confirmed that the palladium atoms on TMV1Cys had an average coordination number of 11, and that the Pd–Pd bond distance was ~2.75 Å with a Debye–Waller factor fixed at a reasonable value of 0.001 $Å^2$. The coordination number correlated to an average particle diameter of 7–8 nm, according to the model developed by Miller *et al.*, which uses hydrogen chemisorption to relate particle size (via dispersion) to coordination number of platinum [51]. Therefore, the organic core of the TMV-Pd and its significantly smaller nanoparticle size, when compared to bulk, accounts for the increase in resistivity of the synthesized nanowires. The higher interfacial surface area, which is due to smaller grain sizes, can be beneficial in sensor applications due to probable higher sensitivity of nanowire conductivity to interfacial processes (e.g., hydrogen adsorption).

8.5 Conclusion and Outlook

In summary, genetically engineered TMV was used as a biotemplate in the synthesis of palladium nanowires. The palladium was mineralized on the surface of the TMV1Cys in the absence of an external reducing agent to form a uniform coating of Pd on the surface of the virus. XAS studies showed that complete mineralization of the palladium was obtained, indicating the potential use of these particles in nanoelectronic devices. The resulting nanowires were deposited across a 100 nm gap between two gold electrodes and FIB-assisted platinum deposition was used to minimize the contact resistance between the particles and the electrodes. The resulting conductivity ($2.50 \pm 0.5 \times 10^2$ S/cm) of the TMV-Pd particles demonstrated the tangible ability of these biotemplated materials to be used in nanoelectronic devices. Using XAS and TEM characterization, it has been shown that these electrical properties are reasonable for a bio-based organic–inorganic system when compared to pure palladium-based nanowire systems that contain a significantly greater amount of the precious palladium metal with less polycrystallinity.

It is useful to assess the significance of this characterization by comparing it with other systems. Even with the current perpetual miniaturization of nanocircuits, this paper and our work in Freer *et al.* [52] are, to our knowledge, the first works to demonstrate the utility of a single Pd-coated TMV for nanocircuit applications. A few additional studies have attempted to establish the electrical properties in other metalized TMV systems. These are summarized in Table 8.1. Gorzny *et al.* successfully reported properties of Pt-coated TMV nanowires (conventionally synthesized) obtained using a four point nanoprobe measurement [53]. Results are on the same order of magnitude as the conductivity obtained in the above experiment. Nevertheless, a high variance in electrical measurements was observed even along a single wire. Such results are not unexpected given the lack of topological uniformity across

Table 8.1 Summary of comparable biotemplate-metal systems suitable for applications in nanocircuits.

Biotemplate	Metal	Approximate conductivity (relative to bulk)	Additional synthetic requirement	References
TMV	Pt	10–100 times bulk	Traditional conventional synthesis	Gorzny *et al.* [53]
TMV	Pt	30 times bulk	Alcohol reducer	Gorzny *et al.* [33]
TMV	Au	20 times bulk	Traditional conventional synthesis followed by annealing at 290C	Wnek *et al.* [32]
DNA	Pd	10 times bulk	Traditional Conventional synthesis followed by annealing at 200 °C	Richter *et al.* [54–56]
DNA	Pd	30 times bulk	Very slow precipitation of PdO	Nguyen et al. [57]
Microtubules	Pd	Not visually continuous	Traditional conventional synthesis	Habicht *et al.* [58] Fu *et al.* [59] Behrens *et al.* [60, 61]
Microtubules	Pd	Not visually continuous	MT catalyzed autoreduction	Kirsch *et al.* [62]

a wire synthesized via the conventional synthesis. Furthermore, the authors reported that using the conventional technique with Pt, different morphologies in metal coating were obtained. Thus, only those with seemingly continuous coatings could be electrically characterized. In a more recent study by the same researchers, the synthetic method was adjusted in an effort to attain more continuous nanowires. Alcohol reduction was employed by utilizing methanol as the cosolvent/reducer [33]. More control over the populating nanoparticle size was obtained this way and the uniformity was reported to be far greater than those obtained with the previous method. The electrical continuity was therefore demonstrated via a two nanoprobe measurement. Unfortunately, the authors did not perform such measurements on the available range of nanowires with obviously different granular sizes to confirm the hypothesis that grain size is a significant property affecting electrical conductivity of the wires. Also, demonstrations of further coating using this technique will potentially increase the critical current density obtainable in the TMV-Pt system.

In a TMV-Au metallized system, one study [32] has successfully been able to confirm synthesis of electrically continuous TMV-Au. However, nanowires synthesized by Bromley *et al.*, [9] which were not electrically characterized, appear continuous enough for nanocircuit applications. In this single study, [32] TMV nanotubes with polyHis tags were first decorated with commercial gold nanoparticles (GNPs) before being used in the conventional synthesis. The nanoparticles then needed to be annealed above 290 °C to produce an electrically continuous coating. The resistivity obtained was impressively just ~20 times higher than that of bulk gold. However, this property is achieved at considerable cost. Along with the energy required for annealing, possible high temperature-induced oxidation of Au and discontinuity along the wire are cited as other possible deficiencies of the produced nanomaterials. Moreover, the adaptability of such a scheme to create a variety of nanowires with predefined properties (e.g., length, thickness, and conductivity) may also prove to be challenging. Therefore, although the finally synthesized nanowires are of relatively high quality for nanocircuitry, the synthetic route is more complex and energetically consuming.

In systems with other biotemplates, DNA [54–57, 63, 64] and microtubules (MTs) [59, 60, 62, 65] are prominent rod-shaped biotemplates, which have been applied to creating conductive Pd nanowires in many metal systems. DNA-metal nanoparticle coatings comprise the 2 nm nucleic acid strand surrounded by nanoparticle clusters with thicknesses up to 60 nm. Richter *et al.* [55, 56] was the first to successfully synthesize DNA templated Pd nanowires >1 μm in length, which have been shown to be continuous via electrical characterization. The nanowires are impressively only 10 times less conductive than bulk Pd, albeit likely mechanically frail. Notable continued progress on these materials has been established. Particularly, Nguyen *et al.* [57] through gradual precipitation of PdO have circumvented the initial flakiness of previous synthesized DNA-Pd to create more controllable and mechanically robust DNA-Pd nanowires with similar electrical conductivity values. Nevertheless, the synthesis is prolonged in favor of nanowire quality; the most ideal/optimized syntheses require very long incubation times (1–2 days) for the very slow precipitation of PdO onto the DNA.

Attempts to produce metallized MTs have shown promise, as is seen in apparent mineralization by functionalities on both tube- and sheet-like structures of tubulin [58–61]. Despite this, none of the studies have attempted to demonstrate electrical continuity of materials, possibly due to the lag in quality of materials relative to viruses and DNA. A study on Ag [65] has produced plausibly (as inspected through electron microscopy) continuous materials that were not electrically characterized. On the other hand, studies involving Pd, which have employed conventional electroless deposition [58–60] and even incubation in the absence of a reducer [62], have resulted in low-quality nanowires. They consist of mechanically weak nanotubes decorated with singularly isolated nanoparticles and no apparent continuity across their surface. While these may find utility in explaining fundamental chemistry (affinity relationships on the nanoparticle/precursor-biotemplate interface), and catalyzing reactions, they currently lack the electrical properties and mechanical robustness for incorporation in modern electronics.

In the context of other promising biotemplates, TMV-metal systems offer the ability to produce comparable electrical properties while retaining advantages in terms of mineralization capacity, mechanical robustness, and simplicity of synthesis. Accordingly, the materials are currently being incorporated into nanodevices [39–41]. The hydrothermal synthesis of Pd on TMV in particular shows the attainability of more controllable and conductive nanowires. Studies in understanding the important functionalities and adsorption/reaction mechanisms conducive to such an ideal synthesis are necessary, especially in order to apply the technique to other metals where expedient scalable synthesis are still elusive (e.g., TMV-Au and MT-metal systems). The chemical malleability of the TMV surface, via genetic engineering, makes the aforementioned goal feasible. Furthermore, from the standpoint of applications, the incorporation of the TMV-metal nanomaterials into nanocircuits requires that their assembly and alignment be precisely controlled. Along with the fortunate advantageous ability to align vertically on Au, TMV has also been known to shorten via breakage, lengthen due to shifts in intermolecular bonds, become less stiff after reactions and align end to end. The elucidation of these processes will become pertinent as the materials are applied more widely to a variety of nanocircuits. The ability to understand, harness, control, and prevent them is the next step in the application of TMV-metal systems. Notwithstanding, with refinement and large-scale application, these nanomaterials offer the promise of decreased capital cost with high performance in current and next-generation electronic devices.

Acknowledgments

B.W.B. gratefully acknowledges the National Science Foundation (NSF) Energy for Sustainability program (Grant Number: 1336731, Program Manager: Dr Gregory Rorrer). The authors thank Dr James Culver at the University of Maryland Biotechnology Institute, Center for Biosystems Research for providing the TMV1Cys. This research used resources of the Advanced Photon Source, a U.S. Department of Energy (DOE) Office of Science User Facility operated for the DOE Office of Science by Argonne National Laboratory under Contract No. DE-AC02-06CH11357. XAS data was collected the X-ray Operations and Research beamline 10-ID-B at the Advanced Photon Source, Argonne National Laboratory.

References

1 Zhou, H., Fan, T., and Zhang, D. (2011) Biotemplated materials for sustainable energy and environment: current status and challenges. *ChemSusChem*, **4**, 1344–1387.

2 Douglas, T. and Young, M. (2006) Viruses: making friends with old foes. *Science*, **312**, 873–875.

3 Namba, K., Pattanayek, R., and Stubbs, G. (1989) Visualization of protein-nucleic acid interactions in a virus. Refined structure of intact tobacco mosaic virus at 2.9. A resolution by X-ray fiber diffraction. *Journal of Molecular Biology*, **208**, 307–325.
4 Flynn, C.E., Lee, S.-W., Peelle, B.R., and Belcher, A.M. (2003) Viruses as vehicles for growth, organization and assembly of materials. The golden jubilee issue—selected topics in materials science and engineering: past, present and future, Edited by S. Suresh. *Acta Materialia*, **51**, 5867–5880.
5 Lee, S.-Y., Lim, J.-S., and Harris, M.T. (2012) Synthesis and application of virus-based hybrid nanomaterials. *Biotechnology and Bioengineering*, **109**, 16–30.
6 Shenton, W., Douglas, T., Young, M. *et al.* (1999) Inorganic-organic nanotube composites from template mineralization of tobacco mosaic virus. *Advanced Materials*, **11**, 253–256.
7 Lim, J.S., Kim, S.M., Lee, S.Y. *et al.* (2010) Quantitative study of Au(III) and Pd(II) ion biosorption on genetically engineered tobacco mosaic virus. *Journal of Colloid and Interface Science*, **342**, 455–461.
8 Lee, S.Y., Choi, J., Royston, E. *et al.* (2006) Deposition of platinum clusters on surface-modified tobacco mosaic virus. *Journal of Nanoscience and Nanotechnology*, **6**, 974–981.
9 Bromley, K.M., Patil, A.J., Perriman, A.W. *et al.* (2008) Preparation of high quality nanowires by tobacco mosaic virus templating of gold nanoparticles. *Journal of Materials Chemistry*, **18**, 4796.
10 Nam, Y.S., Magyar, A.P., Lee, D. *et al.* (2010) Biologically templated photocatalytic nanostructures for sustained light-driven water oxidation. *Nature Nanotechnology*, **5**, 340–344.
11 Yang, C., Jung, S., and Yi, H. (2014) A biofabrication approach for controlled synthesis of silver nanoparticles with high catalytic and antibacterial activities. *Biochemical Engineering Journal*, **89**, 10–20.
12 Knez, M., Sumser, M., Bittner, A.M. *et al.* (2004) Spatially selective nucleation of metal clusters on the tobacco mosaic virus. *Advanced Functional Materials*, **14**, 116–124.
13 Yang, C., Manocchi, A.K., Lee, B., and Yi, H. (2011) Viral-templated palladium nanocatalysts for suzuki coupling reaction. *Journal of Materials Chemistry*, **21**, 187.
14 Manocchi, A.K., Horelik, N.E., Lee, B., and Yi, H. (2010) Simple, readily controllable palladium nanoparticle formation on surface-assembled viral nanotemplates. *Langmuir*, **26**, 3670–3677.
15 Dujardin, E., Peet, C., Stubbs, G. *et al.* (2003) Organization of metallic nanoparticles using tobacco mosaic virus templates. *Nano Letters*, **3**, 413–417.
16 Lee, S.Y., Royston, E., Culver, J.N., and Harris, M.T. (2005) Improved metal cluster deposition on a genetically engineered tobacco mosaic virus template. *Nanotechnology*, **16**, S435–S441.

17 Liu, W.L., Alim, K., Balandin, A.A. *et al.* (2005) Assembly and characterization of hybrid virus-inorganic nanotubes. *Applied Physics Letters*, **86**, 1–3.
18 Singh, P., Gonzalez, M.J., and Manchester, M. (2006) Viruses and their uses in nanotechnology. *Drug Development Research*, **67**, 23–41.
19 Durham, A.C.H., Finch, J.T., and Klug, A. (1971) States of aggregation of tobacco mosaic virus protein. *Nature: New Biology*, **229**, 37–42.
20 Dawson, W., Beck, D.L., Knorr, D.A., and Grantham, G.L. (1986) cDNA cloning of the complete tobacco mosaic virus and production of infectious transcripts. *Proceedings of the National Academy of Sciences*, **83**, 1832–1836.
21 Lu, B., Stubbs, G., and Culver, J.N. (1996) Carboxylate interactions involved in the disassembly of tobacco mosaic tobamovirus. *Virology*, **225**, 11–20.
22 Knez, M., Bittner, A.M., Boes, F. *et al.* (2003) Biotemplate synthesis of 3-Nm nickel and cobalt nanowires. *Nano Letters*, **3**, 1079–1082.
23 Royston, E.S., Brown, A.D., Harris, M.T., and Culver, J.N. (2009) Preparation of silica stabilized tobacco mosaic virus templates for the production of metal and layered nanoparticles. *Journal of Colloid and Interface Science*, **332**, 402–407.
24 Fowler, C.E., Shenton, W., Stubbs, G., and Mann, S. (2001) Tobacco mosaic virus liquid crystals as templates for the interior design of silica mesophases and nanoparticles. *Advanced Materials*, **13** (**16**), 1266–1269.
25 Schlick, T.L., Ding, Z., Kovacs, E.W., and Francis, M.B. (2005) Dual-surface modification of the tobacco mosaic virus. *Journal of the American Chemical Society*, **127**, 3718–3723.
26 Bruckman, M.A., Liu, J., Koley, G. *et al.* (2010) Tobacco mosaic virus based thin film sensor for detection of volatile organic compounds. *Journal of Materials Chemistry*, **20**, 5715.
27 Yi, L., Shi, J., Gao, S. *et al.* (2009) Sulfonium alkylation followed by "click" chemistry for facile surface modification of proteins and tobacco mosaic virus. *Tetrahedron Letters*, **50**, 759–762.
28 Lim, J.-S., Kim, S.-M., Lee, S.-Y. *et al.* (2010) Formation of Au/Pd alloy nanoparticles on TMV. *Journal of Nanomaterials*, **2010**, 1–6.
29 Lim, J.S., Kim, S.M., Lee, S.Y. *et al.* (2010) Biotemplated aqueous-phase palladium crystallization in the absence of external reducing agents. *Nano Letters*, **10**, 3863–3867.
30 Manocchi, A.K., Seifert, S., Lee, B., and Yi, H. (2011) In situ small-angle X-ray scattering analysis of palladium nanoparticle growth on tobacco mosaic virus nanotemplates. *Langmuir*, **27**, 7052–7058.
31 Freer, A.S., Guarnaccio, L., Wafford, K. *et al.* (2013) SAXS characterization of genetically engineered tobacco mosaic virus nanorods coated with palladium in the absence of external reducing agents. *Journal of Colloid and Interface Science*, **392**, 213–218.

32 Wnek, M., Górzny, M.L., Ward, M.B. *et al.* (2013) Fabrication and characterization of gold nano-wires templated on virus-like arrays of tobacco mosaic virus coat proteins. *Nanotechnology*, **24**, 025605.
33 Górzny, M.Ł., Walton, A.S., and Evans, S.D. (2010) Synthesis of high-surface-area platinum nanotubes using a viral template. *Advanced Functional Materials*, **20**, 1295–1300.
34 Yang, C., Choi, C.H., Lee, C.S., and Yi, H. (2013) A facile synthesis-fabrication strategy for integration of catalytically active viral-palladium nanostructures into polymeric hydrogel microparticles via replica molding. *ACS Nano*, **7**, 5032–5044.
35 Yang, C., Meldon, J.H., Lee, B., and Yi, H. (2014) Investigation on the catalytic reduction kinetics of hexavalent chromium by viral-templated palladium nanocatalysts. *Catalysis Today*, **233**, 108–116.
36 Lim, J.S., Kim, S.M., Lee, S.Y. *et al.* (2011) Surface functionalized silica as a toolkit for studying aqueous phase palladium adsorption and mineralization on thiol moiety in the absence of external reducing agents. *Journal of Colloid and Interface Science*, **356**, 31–36.
37 Yi, H., Nisar, S., Lee, S.-Y. *et al.* (2005) Patterned assembly of genetically modified viral nanotemplates via nucleic acid hybridization. *Nano Letters*, **5**, 1931–1936.
38 Adigun, O.O., Freer, A.S., Miller, J.T. *et al.* (2015) Mechanistic study of the hydrothermal reduction of palladium on the tobacco mosaic virus. *Journal of Colloid and Interface Science*, **450**, 1–6.
39 Royston, E., Ghosh, A., Kofinas, P. *et al.* (2008) Self-assembly of virus-structured high surface area nanomaterials and their application as battery electrodes. *Langmuir*, **24**, 906–912.
40 Ghosh, A., Guo, J., Brown, A.D. *et al.* (2012) Virus-assembled flexible electrode-electrolyte interfaces for enhanced polymer-based battery applications. *Journal of Nanomaterials*, **2012**, 1–6.
41 Gerasopoulos, K., McCarthy, M., Royston, E. *et al.* (2008) Nanostructured nickel electrodes using the tobacco mosaic virus for microbattery applications. *Journal of Micromechanics and Microengineering*, **18**, 104003.
42 Bernal, R.A., Filleter, T., Connell, J.G. *et al.* (2013) In situ electron microscopy four-point electromechanical characterization of freestanding metallic and semiconducting nanowires. *Small*, **10**, 725–733.
43 Nam, C.Y., Tham, D., and Fischer, J.E. (2005) Disorder effects in focused-ion-beam-deposited Pt contacts on GaN nanowires. *Nano Letters*, **5**, 2029–2033.
44 Motayed, A., Davydov, A.V., Vaudin, M.D. *et al.* (2006) Fabrication of GaN-based nanoscale device structures utilizing focused ion beam induced Pt deposition. *Journal of Applied Physics*, **100**, 024306.
45 Cronin, S.B., Lin, Y., Rabin, O. *et al.* (2002) Making electrical contacts to nanowires with a thick oxide coating. *Nanotechnology*, **13**, 653–658.

46 Reyntjens, S. and Puers, R. (2001) A review of focused ion beam applications in microsystem technology. *Journal of Micromechanics Microengineering*, **11**, 287–300.

47 Vilà, A., Hernández-Ramirez, F., Rodríguez, J. *et al.* (2006) Fabrication of metallic contacts to nanometre-sized materials using a focused ion beam (FIB). *Materials Science and Engineering: C*, **26**, 1063–1066.

48 Yun, M., Myung, N.V., Vasquez, R.P. *et al.* (2004) Electrochemically grown wires for individually addressable sensor arrays. *Nano Letters*, **4**, 419–422.

49 Mayadas, A.F. (1969) Electrical resistivity model for polycrystalline films: the case of specular reflection at external surfaces. *Applied Physics Letters*, **14**, 345.

50 Bangar, M.A., Ramanathan, K., Yun, M. *et al.* (2004) Controlled growth of a single palladium nanowire between microfabricated electrodes. *Chemistry of Materials*, **16**, 4955–4959.

51 Miller, J.T., Kropf, A.J., Zha, Y. *et al.* (2006) The effect of gold particle size on AuAu bond length and reactivity toward oxygen in supported catalysts. *Journal of Catalysis*, **240**, 222–234.

52 Freer, A.S., Mueller, L., Gilpin, C. *et al.* (2014) A novel method to determine the resistance of biotemplated nanowires. *Chemical Engineering Communications*, **202**, 1216–1220.

53 Górzny, M.L., Walton, A.S., Wnęk, M. *et al.* (2008) Four-probe electrical characterization of Pt-coated TMV-based nanostructures. *Nanotechnology*, **19**, 165704.

54 Richter, J., Mertig, M., Pompe, W., and Vinzelberg, H. (2002) Low-temperature resistance of DNA-templated nanowires. *Applied Physics A*, **74**, 725–728.

55 Richter, J., Mertig, M., Pompe, W. *et al.* (2001) Construction of highly conductive nanowires on a DNA template. *Applied Physics Letters*, **78**, 536–538.

56 Richter, J., Seidel, R., Kirsch, R. *et al.* (2000) Nanoscale palladium metallization of DNA. *Advanced Materials*, **12**, 507–510.

57 Nguyen, K., Monteverde, M., Filoramo, A. *et al.* (2008) Synthesis of thin and highly conductive DNA-based palladium nanowires. *Advanced Materials*, **20**, 1099–1104.

58 Habicht, W., Behrens, S., Wu, J. *et al.* (2004) Characterization of metal decorated protein templates by scanning electron/scanning force microscopy and microanalysis. *Surface and Interface Analysis*, **36**, 720–723.

59 Bin, F.Y., De, Z.L., Zheng, J.Y., and Fy, S.G. (2004) Site-specific deposition of colloidal Pd nanoparticles on self-assembled microtubles from biolipid. *Chinese Journal of Chemistry.*, **22**, 1142–1147.

60 Behrens, S., Habicht, W., Wenzel, W., and Böhm, K.J. (2009) Deposition of palladium nanoparticles on self-assembled, zinc-induced tubulin macrotubes and sheets. *Journal of Nanoscience and Nanotechnology*, **9**, 6858–6865.

61 Behrens, S., Habicht, W., Wu, J., and Unger, E. (2006) Tubulin assemblies as biomolecular templates for nanostructure synthesis: from nanoparticle arrays to nanowires. *Surface and Interface Analysis*, **38**, 1014–1018.
62 Kirsch, R., Mertig, M., Pompe, W. *et al.* (1997) Three-dimensional metallization of microtubules. *Thin Solid Films*, **305**, 248–253.
63 Lund, J., Dong, J., Deng, Z. *et al.* (2006) Electrical conduction in 7 Nm wires constructed on Λ-DNA. *Nanotechnology*, **17**, 2752–2757.
64 Deng, Z. and Mao, C. (2003) DNA-templated fabrication of 1D parallel and 2D crossed metallic nanowire arrays. *Nano Letters*, **3**, 1545–1548.
65 Behrens, S., Wu, J., Habicht, W., and Unger, E. (2004) Silver nanoparticle and nanowire formation by microtubule templates. *Chemistry of Materials*, **16**, 3085–3090.

9

Prediction of Carbon Nanotube Buckypaper Mechanical Properties with Integrated Physics-Based and Statistical Models

Kan Wang[1], Arda Vanli[2], Chuck Zhang[1], and Ben Wang[1]

[1] *School of Industrial and Systems Engineering, Georgia Tech Manufacturing Institute, Georgia Institute of Technology, Atlanta, GA, USA*
[2] *Department of Industrial and Manufacturing Engineering, Florida State University, Tallahassee, FL, USA*

9.1 Introduction

Since discovering them in 1991, researchers have found that carbon nanotubes (CNTs) have exceptional mechanical, chemical, and electrical properties [1]. Those properties arise from the quasi-one-dimensional nature and extraordinary nanostructure of CNTs. Because of their covalent sp^2 bonds between the individual carbon atoms, single-walled carbon nanotubes (SWCNTs) are the strongest and stiffest materials yet discovered in terms of tensile strength and elastic modulus [2]. The experimental records of tensile strength and Young's modulus for multiwalled carbon nanotubes (MWCNTs) are as high as 63 GPa [3] and 900 GPa [4], respectively. Although MWCNTs are 1–2 magnitudes stronger than steel or Kevlar fibers and several times stiffer, the superior mechanical properties of individual MWCNTs cannot be fully transferred to MWCNT bundles due to the weak, shear interactions between adjacent tubes. Considering their density is only 1.3–1.4 g/cm^3, CNTs are the perfect engineering materials from a mechanical engineering perspective. Desirable features include excellent electrical and thermal properties and an extremely high surface area to volume ratio. Theoretically, CNTs are the ideal materials for numerous engineering applications. In a laboratory environment, CNTs are used as a raw material to fabricate amazing materials and devices. Numerous researchers consider CNTs an essential building block of future nanoelectromechanical systems (NEMS) that will revolutionize the way humans shape the physical world and provide a possible solution for the energy crisis. With such high expectations for CNTs, researchers have been drawn to working with them during the past decade. One example is

Nanotechnology Commercialization: Manufacturing Processes and Products, First Edition.
Edited by Thomas O. Mensah, Ben Wang, Geoffrey Bothun, Jessica Winter, and Virginia Davis.

the development of high-performance nanocomposites that use CNTs as reinforcement materials.

In reality, the exceptional properties of CNTs have not been achieved in the macroscale because of the small size of individual CNTs. The nanoscale dimensions and exceptionally large surface areas (e.g., 1000–1350 m^2/g for SWNTs) of CNTs lead to strong interactions between tubes due to the van der Waals forces causing the CNTs to form into ropes, bundles, or aggregates. This results in poor dispersion and is unlike conventional fibrous reinforcements. It takes a tremendous effort to disperse CNTs into matrix materials. Even if direct dispersion is achieved, the volume fraction of CNTs in the resin cannot be very high because the viscosity of the resin will rapidly increase with the amount of CNTs [5]. Thus, the direct mixture of CNTs with matrix materials lacks the ability to effectively align and interconnect CNTs to achieve the desired directional and conductance properties of the resulting materials [6]. For more than a decade, researches have attempted to develop methods that effectively transfer the exceptional properties of CNT from the nanoscale to the macroscale.

Buckypaper (BP), a membrane composed of entangled CNTs (Figure 9.1), is one of the most promising discoveries for overcoming the problem of handling CNTs. Since its discovery in 2003 [7], BP has shown promising capabilities in practical and potential applications such as high-current carrying [8], microwave shielding [9], electromagnetic interference shielding [10], fire retardancy [11], and mechanical reinforcement materials in composites [5]. The most commonly used technique to fabricate BP is filtration of well-dispersed CNT suspension [12, 13], which usually requires three basic steps: (i) dispersing raw CNTs into solvents to form a suspension, (ii) filtering the CNT suspension, and (iii) drying the suspension to remove solvents. The challenge of this technique is that CNTs tend to agglomerate in solvents, which results in cracking during the drying process [14]. Forming a well-dispersed suspension

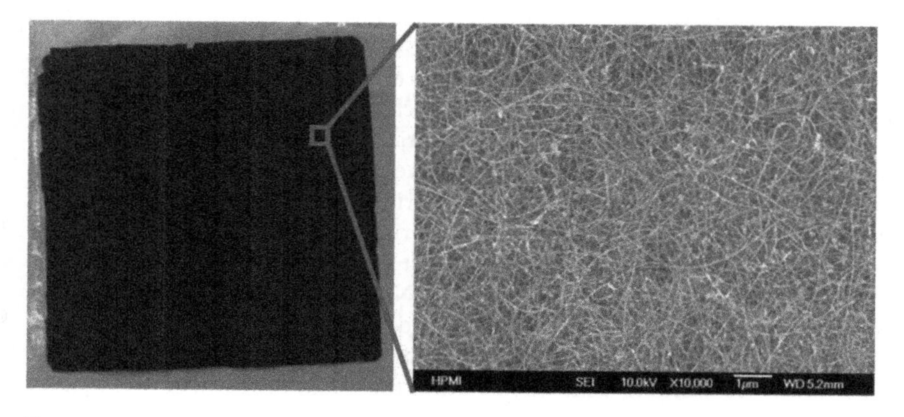

Figure 9.1 A sample of BP and its nanostructure with a scale factor of ×10,000.

with the assistance of a surfactant [15, 16] can alleviate this problem. However, this requires anywhere from 90 to 120 min of sonication and a small concentration of CNT suspension. The time-consuming nature of the conventional filtration-based synthesis process to create BP hinders scaling up the industrial production of BP. Furthermore, there is a high amount of spoilage during transferring and handling BP due to the random nature of the CNT network that can create defects in the nanostructure. To improve the ability to handle BP, a common practice is to improve its mechanical properties by inserting layers of polymeric adhesives [17]. The resulting polymer-treated BP is significantly stiffer and stronger and is an ideal material to be used as reinforcement in nanocomposites.

The idea of using BP as reinforcement material in nanocomposite manufacturing has been explored and studied for years. In a typical BP/polymer nanocomposite, BP is primarily used as load bearer and distributor. Resin fully surrounds the CNTs in BP. The weight percentage of CNT content in this kind of nanocomposite ranges from 20% to 39%, and the Young modulus of the composite can be as high as 3–5 times the Young modulus of the matrix polymer, or around 10 times of that of the BPs used in those systems [5, 18]. Using BP is somewhat limited in this application even with the good mechanical performance of the nanocomposites.

Recently, the multifunctionality of BP has become the focus of research in various fields. The high surface area of BP is the desired feature in many applications such as with sensors or fire retardant materials. However, the porous structure of BP that is filled with polymer resin prevents it from being much help in these kinds of applications. Thus, the real challenge is to improve the maneuverability of BP and keep most of the porous structure of BP. Some studies have attained a 32% and 66% void volume, respectively, in polymer-impregnated BP using vacuum-aided filtration and hot compression [19]. This is still far lower than the theoretic porosity one can achieve. The Young modulus of CNTs is 2–3 orders of magnitude higher than that of BP, which proves that the interconnections between tubes are the weak spots in this CNT network.

Another technical challenge is to model the mechanical properties of BP. Although BP is usually treated as homogeneous material in bulk scale, the nanostructure of BP is random in nature. The mechanical properties of BP are closely related to the characteristics of the CNT network, that is, the lengths and diameters of the CNTs, the volume fraction of CNTs, the alignment of CNTs, and how those CNTs connect to each other. Due to the large number of CNTs in BP, it is far beyond the computational capacity to model the mechanical properties of BP from nanoscale directly. On the other hand, a pure data-driven statistical model requires that a large set of experimental data be accurate, especially when there are many input parameters. In the BP manufacturing process, as with any process in nanomanufacturing, the costs are high to build an accurate data-driven model that will produce experimental data. Those models are not accurate, and explanations and treatments for discrepancy are missing.

This chapter describes a finite element-based physical model to predict the mechanical properties of BP that uses advanced statistical methods to adjust the model even when there is limited experimental data. With this multiresolution modeling strategy, the statistically enhanced model is more accurate and less costly than both the physical model and the pure statistical model.

The performance of BP is determined by its nanostructure, which is related to the manufacturing process. The sonication stage, in which the CNTs are dispersed into deionized water, is the most decisive stage for the final nanostructure of BP. Due to the lack of an on-line characterization technique, how the nanostructure changes during sonication is difficult to describe. In addition, CNTs' tubular geometry makes it difficult to quantify the extent of dispersion by classic definition for spherical nanoparticle suspensions [20]. The fragmentation theory can be used to determine the length distribution of CNTs after sonication and to describe the nanostructure using CNT length distributions, CNT clusters size distribution, and CNT clusters number distribution. This information provides the necessary inputs for the physical model.

9.2 Manufacturing Process of Buckypaper

Most researchers regard CNTs as nanoscale building blocks with great promise for creating high-performance multifunctional materials. The major drawback of using CNT is that is they are difficult to handle or maneuver. Methods of assembling CNTs into a desired geometry and structure need to be developed in order to use their outstanding properties.

Researchers have had great interest in refining and improving the manufacturing process for BP ever since the Smalley group first fabricated BP. As an attractive candidate for transferring the outstanding properties of CNT's at a bulk scale, there is an urgent need to find an industry-level manufacturing process for BP. The filtration-based deposition process shows great promise in scaling up in comparison with all existing manufacturing processes of BP, including filtration-based deposition, direct deposition, and controlled growth. The filtration-based manufacturing process is energy efficient, has a fast production cycle, and its open process line is easy to modify. The governing mechanism analogizes the manufacturing process of paper and is a successful example of industrialized production system.

However, even in the laboratory circumstances, the filtration-based manufacturing process yields inconsistent final products. Various failure modes are observed in the final products. Some BPs cannot be separated from filtration membranes. Some are fragile under shear force. Some have wrinkles and warpage over the surface. In the most serious situation, microcracks emerge

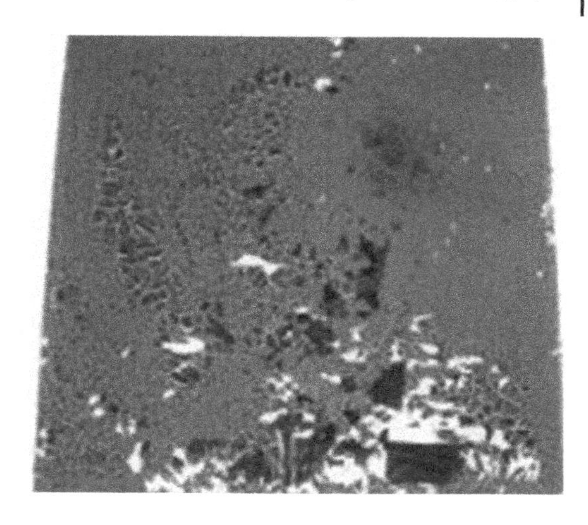

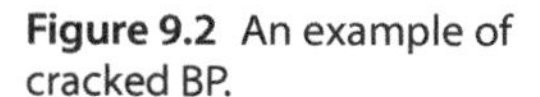

Figure 9.2 An example of cracked BP.

in the BP upon drying (Figure 9.2). The fundamental source of these issues is the relatively weak interactions between CNTs. The van der Waals attraction between tubes can only provide a small amount of friction to prevent sliding. Compared to natural cellulosic fibers in regular paper that normally has rough surfaces, CNTs are atomic, smooth, and rigid.

In order to increase the tube-to-tube interactions in BP, one can either break the CNT surface smoothness by functionalization or add binding materials. Since intentionally introducing defects to CNTs would reduce their mechanical performance and require complex procedures, the easier way to enhance BP's mechanical properties is to add binding materials into the CNT network. A typical manufacturing process of the PVA-treated BP is described in this section. PVA is a widely used industrial adhesive. This process and its final product are taken as a case study of the mechanical modeling and statistical analysis methodologies.

The manufacturing process for creating BP is modified to create PVA-treated BP. Immediately after the filtration of the CNT suspension, another filtration step for the PVA solution is added. The PVA solution flows through the CNT network. PVA molecules tend to wrap around the CNTs and strengthen the intersectional joints between CNTs. Figure 9.3 shows a molecular dynamic simulation result of PVA wrapping around a CNT [21].

The manufacturing process for PVA-treated BP involves CNT suspension preparation, film forming, PVA treatment, and film release. The CNT suspension preparation step is achieved in the sonication stage and film forming. PVA treatment steps are achieved in the filtration stage. The film release step is not believed to affect the nanostructure of PVA-treated BP, so it is not included in the modeling process. For the modeling purpose, we used x_k to represent the state variable, that is, nanostructure after each step ($k = 0,1,2,3,$

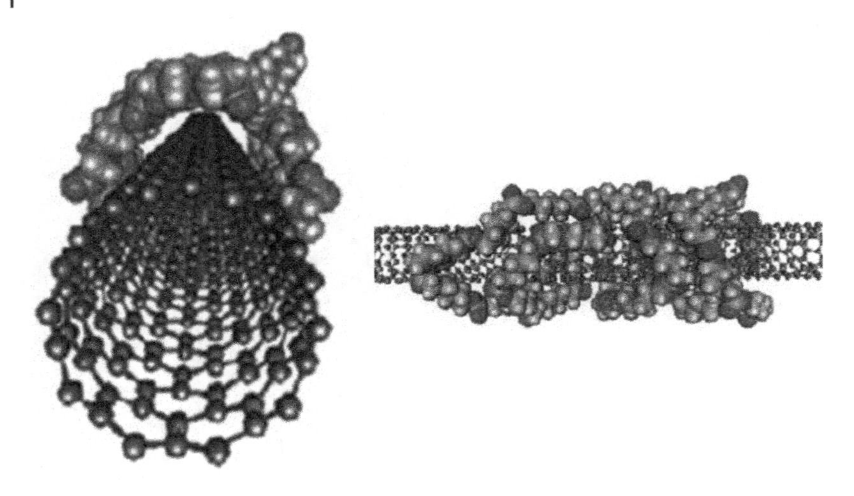

Figure 9.3 MD simulation result of PVA wrapping around a CNT. Rahmat and Hubert 2011 [21]. Reproduced with permission of Elsevier.

x_0 is the nanostructure information of the raw material), y_k to represent the output variable, that is, Young's modulus, u_k to represent the controllable process variables, w_k to represent the noise variables, and v_k to represent the measurement error. The modified manufacturing process is shown in Figure 9.4.

The process model can be expressed as

$$x_k = A_{k-1}x_{k-1} + B_k u_k + w_k$$
$$y_k = C_k x_k + v_k$$

where A_k, B_k, and C_k are coefficient matrices and can be estimated from design of experiment and regression techniques.

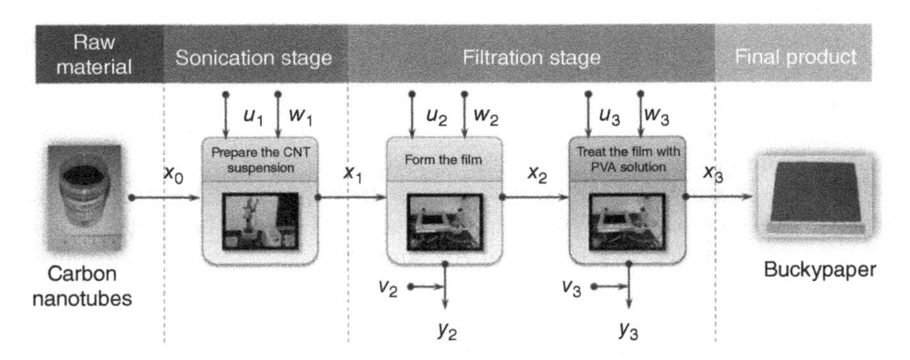

Figure 9.4 The multistage manufacturing process of PVA-treated BP.

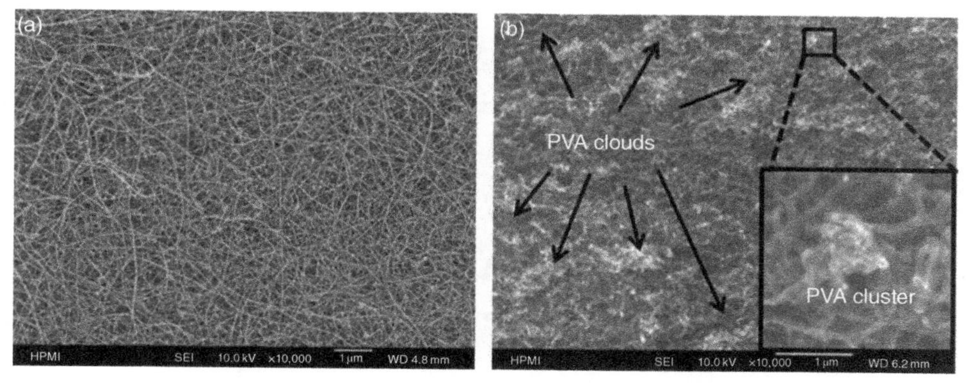

Figure 9.5 The comparison of SEM images of pristine BP (a) and PVA-treated BP (b).

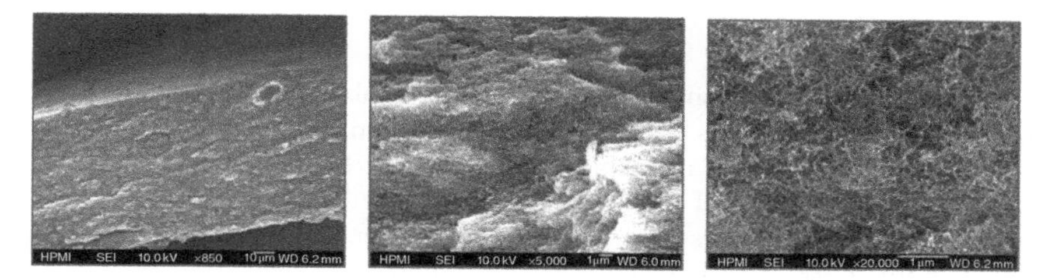

Figure 9.6 Cross-sectional images of PVA-treated BP.

The result of PVA treatment is shown in SEM images (Figure 9.5). The pristine BP shows clean CNTs in SEM, while the PVA-treated BP brighter PVA clouds over the surface of CNT network.

SEM image of cross sections of PVA-treated BP shows that PVA molecules not only treat the CNTs on the surface but also permeate through to treat CNTs in lower layers (Figure 9.6).

9.3 Finite Element-Based Computational Models for Buckypaper Mechanical Property Prediction

A critical task is image analysis for nanostructure characterization because most inputs for the FEA model are directly measured from SEM images of BP. The properties of BP are determined by nanostructure characteristics such as length, diameter, orientation of CNTs, and how they interconnect with each other. However, direct measurement from SEM images of a CNT sheet is difficult [22–24] due to its multilayer nature (Figure 9.7). In order to extract accurate characteristics of CNTs, the SEM images need to be processed and analyzed using image enhancement tools.

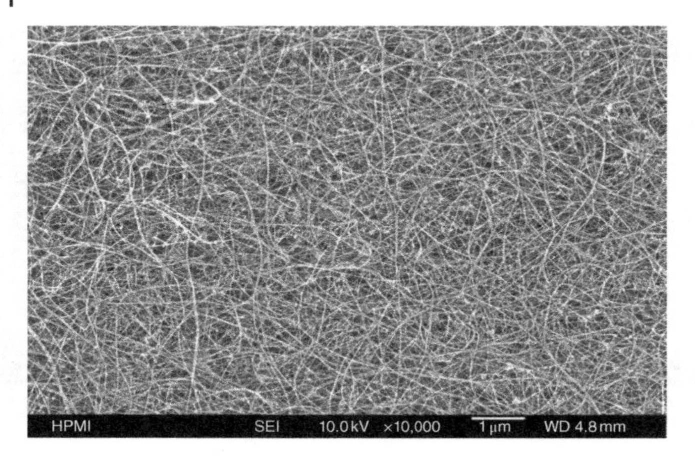

Figure 9.7 Typical SEM image of CNT sheet.

Various image processing algorithms are available to handle this need [24–26]. Wang (2013) has developed an easy-to-implement algorithm that is designed particularly for analyzing SEM images of BPs [24]. This algorithm captures the boundaries of CNTs in the image. Then it removes the noise from the lower layer and repairs the tubes on the top layer. The resulting image has much more contraction between top layer tubes and lower layer noises, as shown in Figure 9.8.

SIMAGIS, an advanced image analysis software, is used to extract the nanostructure information from the SEM images. A comparison of the analysis

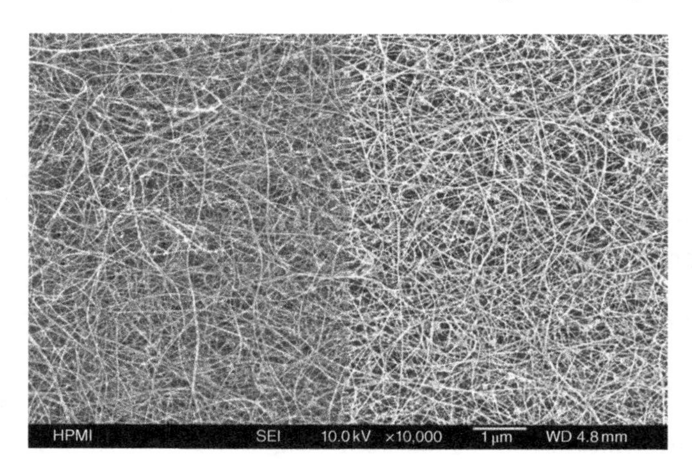

Figure 9.8 A comparison between original image (a) and processed image (b). The brightness and sharpness of top layer CNTs are enhanced, providing a more distinguishable input for the analysis software.

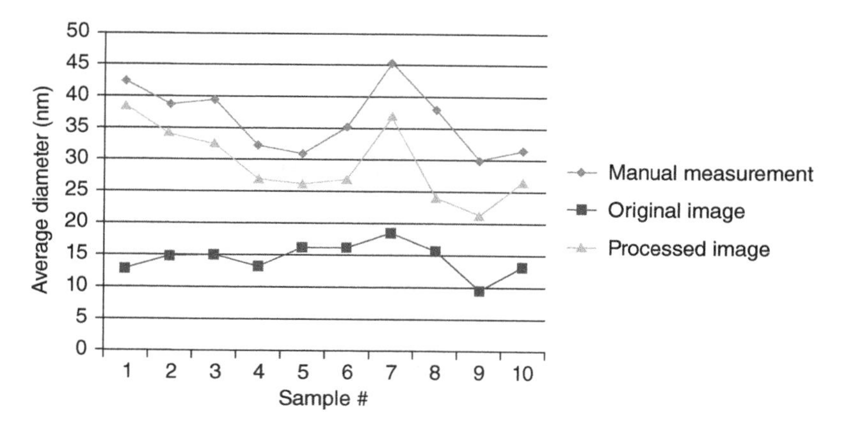

Figure 9.9 The comparison of diameter measurement results. The results from the processed image not only reduced the discrepancy between manual measurement and software analysis but also better captured the trend.

results from both the original image and the processed image to the manual measurement of 100 CNTs in each SEM image is shown in Figure 9.9. The comparison indicates that the processed image gives automatic analyzed results much closer to the manual measurements.

Similarly, the length distribution and orientation distribution can be measured from the SEM images. An example of such characterizations of a random BP is shown in Figure 9.10.

The critical mechanism underlying its mechanical properties is the effective load transferring at tube interconnections in a BP [27]. In order to provide a strong interaction between tubes, the joints will be treated by a certain method, such as e-beam sintering or polymeric binding. Assuming all joints are fixed after the CNTs achieve the network structure, the mechanical properties can be predicted based on the nanostructure of the CNT sheet. A finite element-based mechanical model for CNT sheets is used for this purpose.

The network of CNT in the BP can be viewed as a truss structure [28]. The representative CNT network is generated by adding CNTs into a control cell. Assuming that the BP is macroscopically homogeneous, the Young modulus of the network in this control cell should be the same everywhere on the BP and is equal to the Young modulus of the entire BP. First, we randomly generate a point in the control area as the center point of the additional CNTs. Then, we generate the length (L), diameter (D), and orientation (θ) of this CNT and add it into the control cell, which is represented by a line segment (Figure 9.11). The length and diameter are generated from their distributions given by image analysis or another physics-based model. For random BP, the orientation is generated from a uniform distribution, uniform $(0, \pi)$. We repeat this procedure of adding CNTs until the volume fraction (V_f) of CNTs reaches the preset value.

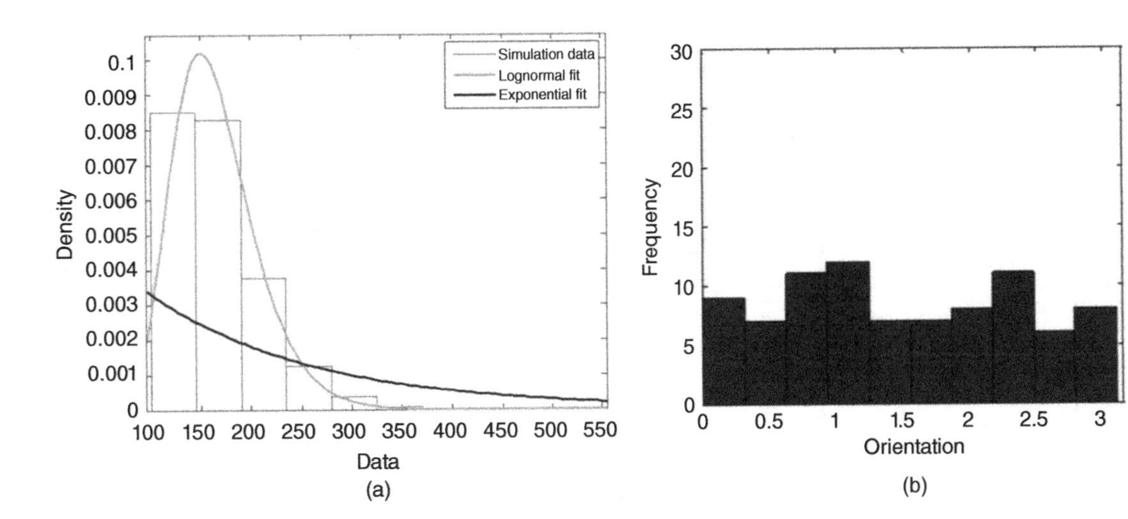

Figure 9.10 (a) Length measurements and (b) orientation measurements of a random BP.

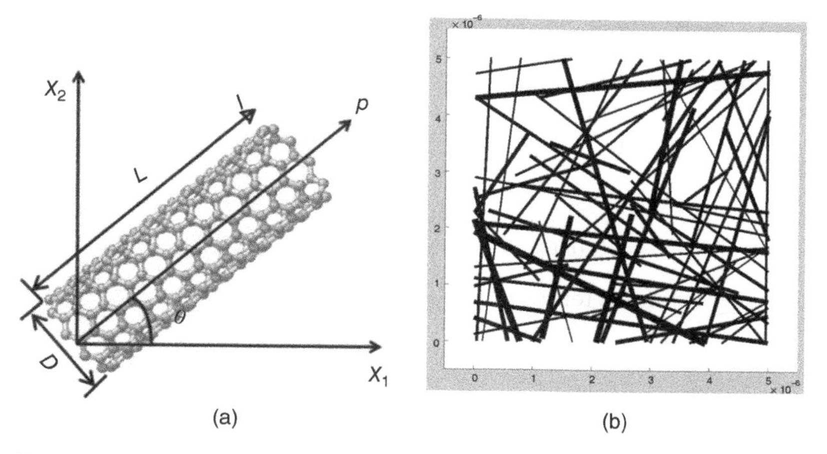

Figure 9.11 The parameters for generating individual CNTs (a) and a generated CNT network sample with a control cell size of 5 μm × 5 μm and a CNT volume fraction of 0.3 (b).

The volume fraction of CNT is calculated by

$$V_\mathrm{f} = \frac{\sum_{i=1}^{N_\mathrm{tube}} L_i D_i}{A}$$

where N_tube is the current number of tubes in the control cell and A is the area of control cell.

The joints in this truss structure are identified by finding the intersections of elements, as shown in Figure 9.12. Each section between any two nodes is a

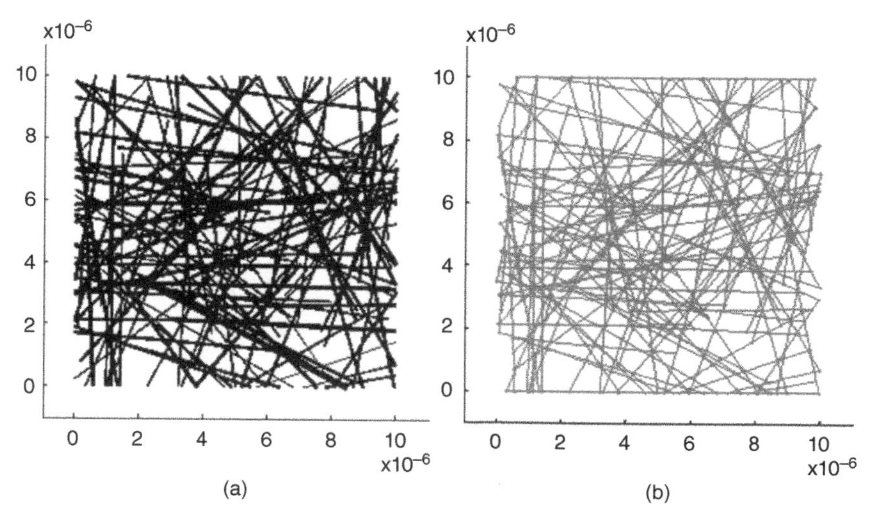

Figure 9.12 The randomly generated CNT network (a) and the corresponding truss structure (b).

rod element. Each element has a length (l) that equals to the distance between its two nodes, a diameter (d) and an orientation (θ) that equals to those of the CNT it belongs to, and a modulus (E) that can be calculated by the empirical equation

$$E = 8509.2\ d^{-1.139}$$

Let

$$A = \frac{\pi d^2}{4}, \quad S = \sin(\theta), \quad C = \cos(\theta)$$

The elemental stiffness tensor can be represented by the 4×4 matrix

$$K^e = \frac{EA}{l}\begin{bmatrix} C^2 & CS & -C^2 & -CS \\ CS & S^2 & -CS & -S^2 \\ -C^2 & -CS & C^2 & CS \\ -CS & -S^2 & CS & S^2 \end{bmatrix}$$

In the simulation, one edge of the control cell is set to be fixed and a 1% strain is applied to the opposing edge. The accumulated force on the opposing edge of the control cell can be calculated by the FEA, which then can be used to derive the stress with the control cell dimensions. The CNT networks before and after the strain is applied are shown in Figure 9.13.

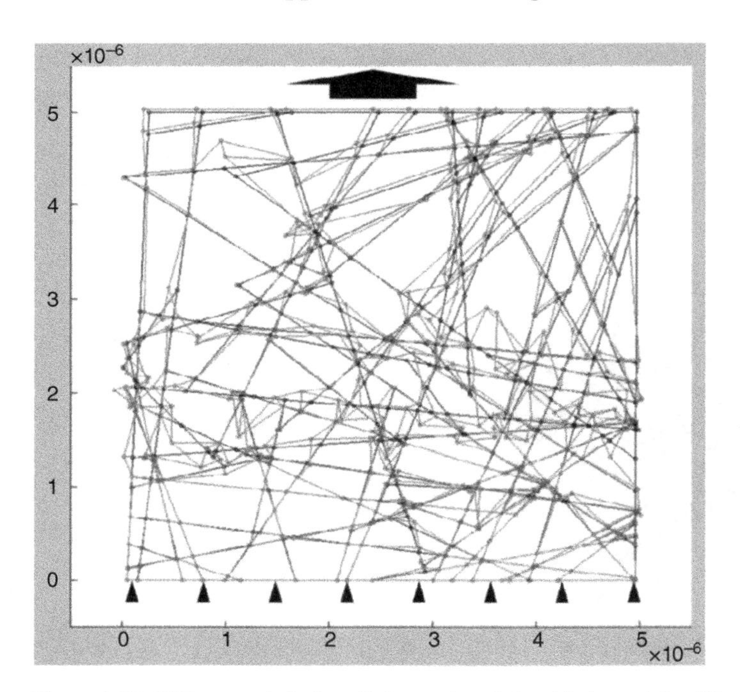

Figure 9.13 CNT networks before (light gray) and after (dark gray) the 1% strain is applied.

Suppose the number of nodes in this truss system is N_{node}. The global stiffness tensor can be represented by a $2N_{\text{node}} \times 2N_{\text{node}}$ matrix $K_{\text{Truss}} = [K_{ij}]$. The rule of assembly for all of the elemental stiffness tensors is explained as follows.

If the lth element to be assembled has two ends with the m_1th node and the m_2th node, and the corresponding stiffness matrix is

$$\begin{bmatrix} k_{11} & k_{12} & k_{13} & k_{14} \\ k_{21} & k_{22} & k_{23} & k_{24} \\ k_{31} & k_{32} & k_{33} & k_{34} \\ k_{41} & k_{42} & k_{43} & k_{44} \end{bmatrix}$$

Then update the $l-1$th global stiffness matrix as

$$\mathrm{K}^{(l)} = \mathrm{K}^{(l-1)} + \begin{bmatrix} 0 & \cdots & 0 & 0 & \cdots & 0 & 0 & \cdots & 0 \\ \vdots & \vdots & \vdots & \vdots & \vdots & \vdots & \vdots & \vdots & \vdots \\ 0 & \cdots & k_{11} & k_{12} & \cdots & k_{13} & k_{14} & \cdots & 0 \\ 0 & \cdots & k_{21} & k_{22} & \cdots & k_{23} & k_{24} & \cdots & 0 \\ \vdots & \vdots & \vdots & \vdots & \ddots & \vdots & \vdots & \vdots & \vdots \\ 0 & \cdots & k_{31} & k_{32} & \cdots & k_{33} & k_{34} & \cdots & 0 \\ 0 & \cdots & k_{41} & k_{42} & \cdots & k_{43} & k_{44} & \cdots & 0 \\ \vdots & \vdots & \vdots & \vdots & \vdots & \vdots & \vdots & \vdots & \vdots \\ 0 & \cdots & 0 & 0 & \cdots & 0 & 0 & \cdots & 0 \end{bmatrix} \begin{matrix} \\ \\ \text{The } (2m_1-1)\text{th row} \\ \text{The } (2m_1)\text{th row} \\ \\ \text{The } (2m_2-1)\text{th row} \\ \text{The } (2m_2)\text{th row} \\ \\ \\ \end{matrix}$$

The $(2m_1-1)$th column; The $(2m_1)$th column; The $(2m_2-1)$th column; The $(2m_2)$th column

Let U be the displacement vector of all the nodes and F be the loading vector on the nodes. The finite element problem can be summarized to the simple equation

$$K_{\text{Truss}} U = F$$

To solve this equation, we need to add some boundary conditions. The goal is to get the modulus, so one can either assume the loadings (stress) and solve for the displacements (strain), or assume the displacements and then solve for the loadings. The following discussion is based on the latter.

Without loss of generality, assume the stress is along the y-axis. Set the displacements of all the nodes on the bottom boundary to zero. With the assumed strain and the known side length of the control cell, the displacement along the y-axis of all the nodes on the top boundary can be easily calculated. Now for any node in the truss system and for both x and y directions, we know either its displacement along a certain direction or the loading on this node along the same direction. The known displacements are called prescribed displacements and the known loadings are called prescribed loadings. With some basic linear transformations, the equation can be written as the following form:

Prescribed loadings

$$\begin{bmatrix} k_{11} & \cdots & k_{1,m} & k_{1,m+1} & \cdots & k_{1n} \\ \vdots & \ddots & \vdots & \vdots & \ddots & \vdots \\ k_{m,1} & \cdots & k_{mm} & k_{m,m+1} & \cdots & k_{mn} \\ k_{m+1,1} & \cdots & k_{m+1,m} & k_{m+1,m+1} & \cdots & k_{m+1,n} \\ \vdots & \ddots & \vdots & \vdots & \ddots & \vdots \\ k_{n1} & \cdots & k_{nm} & k_{n,m+1} & \cdots & k_{nn} \end{bmatrix} \begin{bmatrix} u_1 \\ \vdots \\ u_m \\ u_{m+1} \\ \vdots \\ u_n \end{bmatrix} \begin{bmatrix} f_1 \\ \vdots \\ f_m \\ f_{m+1} \\ \vdots \\ f_n \end{bmatrix}$$

Prescribed displacements

Then the unknown displacements and loadings can be calculated as

$$\begin{bmatrix} u_1 \\ \vdots \\ u_m \end{bmatrix} = \begin{bmatrix} k_{11} & \cdots & k_{1,m} \\ \vdots & \ddots & \vdots \\ k_{m,1} & \cdots & k_{mm} \end{bmatrix}^{-1} \left(\begin{bmatrix} f_1 \\ \vdots \\ f_m \end{bmatrix} - \begin{bmatrix} k_{1,m+1} & \cdots & k_{1,n} \\ \vdots & \ddots & \vdots \\ k_{m,m+1} & \cdots & k_{mn} \end{bmatrix} \begin{bmatrix} u_{m=1} \\ \vdots \\ u_n \end{bmatrix} \right)$$

$$\begin{bmatrix} f_{m+1} \\ \vdots \\ f_n \end{bmatrix} = \begin{bmatrix} k_{m+1,1} & \cdots & k_{m+1,m} \\ \vdots & \ddots & \vdots \\ k_{n1} & \cdots & k_{nm} \end{bmatrix} \begin{bmatrix} u_1 \\ \vdots \\ u_m \end{bmatrix} + \begin{bmatrix} k_{m+1,m+1} & \cdots & k_{m+1,n} \\ \vdots & \ddots & \vdots \\ k_{n,m+1} & \cdots & k_{nn} \end{bmatrix} \begin{bmatrix} u_{m=1} \\ \vdots \\ u_n \end{bmatrix}$$

After all the loadings on the top boundary nodes are calculated, the stress over the control cell is simply calculated by dividing the combined force by the side length of control cell.

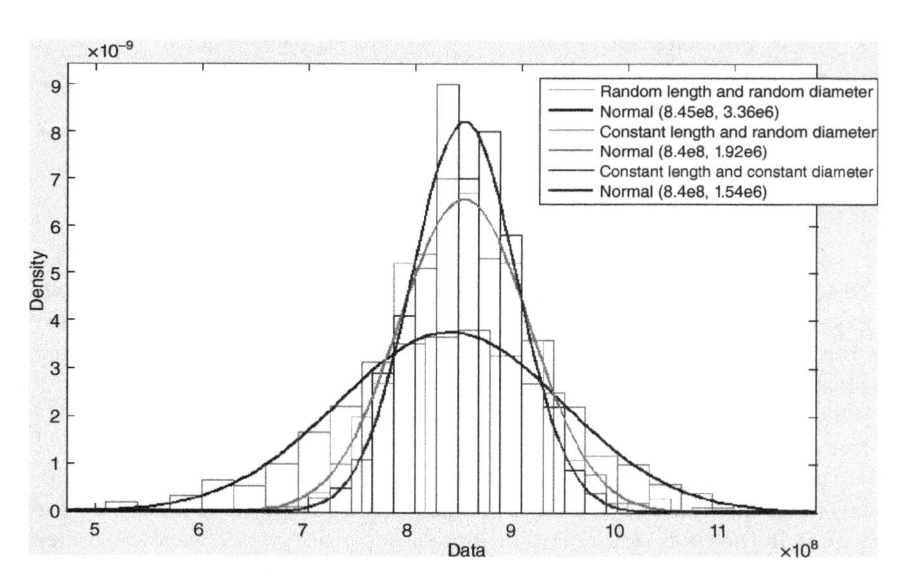

Figure 9.14 The comparison of three sets of 500 computer simulations under different settings.

Table 9.1 The source of variations in computer simulation results.

Setting	Variance	Percentage	Source	Contribution (%)
Random length and random diameter	1.12E+16	100.00	CNT length	67.05
Constant length and random diameter	3.69E+15	32.95	CNT diameter	11.88
Constant length and constant diameter	2.36E+15	21.07	Other sources	21.07

Since the CNT network is randomly generated in the FE model, and the nanostructure inputs follow their own distributions, the FE model is stochastic rather than deterministic. Monte Carlo simulations can be employed to find the variation in Young's modulus of BP [18]. By decreasing the degree of freedom in the CNT network, the source of variations can be determined. The results of the variation source analysis, based on 500 runs of computer simulations, are shown in Figure 9.14 and Table 9.1.

The fundamental trend of Young's modulus of BP is provided by the FE computer model. To capture this trend, a Kriging surrogate model has been developed with the uncertain input variable (e.g., average CNT length) and calibration variable (e.g., polymeric binder used in BP manufacturing). The PVA is used to enhance intertube interactions, which resemble the network nanostructure of the BP. Because the degree of cross-linking in the BP network is unknown, different amounts of PVA are used for different possible levels of cross-linking. The simulation results are listed in Table 9.2 and the response surface of this Kriging surrogate model is shown in Figure 9.15.

Table 9.2 Simulation results of Young's modulus under different combinations of average CNT length and PVA amount.

Length (μm)	PVA (100 wt%)					
	0.5	0.6	0.7	0.8	0.9	1.0
6	6.11E+07	8.72E+07	1.23E+08	1.71E+08	2.91E+08	6.28E+08
7	1.10E+08	1.39E+08	1.97E+08	3.19E+08	4.99E+08	8.91E+08
8	1.64E+08	2.38E+08	3.13E+08	4.68E+08	7.67E+08	1.37E+09
9	2.12E+08	2.87E+08	4.46E+08	7.05E+08	1.08E+09	1.82E+09
10	2.85E+08	3.90E+08	5.87E+08	8.45E+08	1.41E+09	2.35E+09
11	3.17E+08	4.19E+08	6.10E+08	9.43E+08	1.51E+09	2.60E+09
12	3.15E+08	4.20E+08	6.85E+08	9.97E+08	1.53E+09	2.65E+09

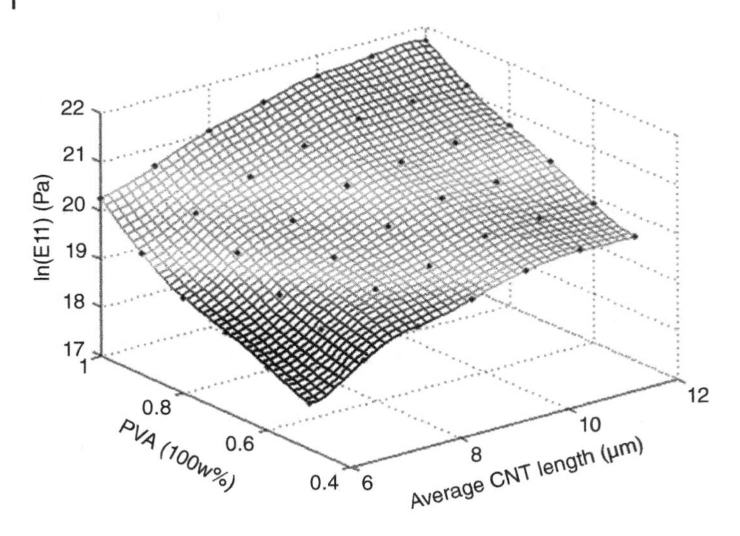

Figure 9.15 The response surface of the Kriging model.

9.4 Calibration and Adjustment of FE Models with Statistical Methods

The FE model of Young's modulus prediction for the CNT random network is usually not accurate due to simplifications employed during model development and uncertainties in model inputs. It is necessary to use a few experimental data to calibrate and adjust the model. The resulting statistics-enhanced physical model will have higher accuracy than both a pure statistical model and an FE-based physical model.

DMA measurements are used as experimental data. The data set consists of measurements from 20 PVA-treated BP samples at four different PVA levels, 0.6, 0.7, 0.8, and 0.9, with five replications at each level. The SEM images suggest that the CNTs in those samples have an average length of around 10 μm. The experimental measurements of those 20 samples are provided in Table 9.3 and the discrepancy between the experimental data and the surrogate model is shown in Figure 9.16.

The physical model underestimates Young's modulus of PVA-treated BP. The major reasons for this underestimation include the following: (i) the model only considers the PVA-induced tube-to-tube interaction, while other types of interactions exist in the real PVA-treated BP; (ii) the model only considers the in-plane tube-to-tube interactions, while interaction along through-thickness direction can also contribute the Young modulus of PVA-treated BP; and (iii) the PVA is only treated as a binder at the intersectional joints of CNTs, while the modulus of PVA itself and potential enhancement by PVA crystallization on CNT sidewalls is not considered.

Table 9.3 Experimental measurements of Young's modulus of random BP.

Replicate	PVA (100 wt%)							
	0.6		0.7		0.8		0.9	
	E11 (Pa)	ln(E11)	E11 (Pa)	ln(E11)	E11 (Pa)	ln(E11)	E11 (Pa)	ln(E11)
1	9.15E+08	20.63	1.32E+09	21.00	1.09E+09	20.81	1.80E+09	21.31
2	1.00E+09	20.73	1.04E+09	20.76	1.41E+09	21.07	1.90E+09	21.36
3	1.22E+09	20.92	8.97E+08	20.62	1.52E+09	21.14	1.74E+09	21.28
4	1.25E+09	20.94	1.22E+09	20.92	1.54E+09	21.15	1.55E+09	21.16
5	1.20E+09	20.90	1.19E+09	20.90	1.36E+09	21.03	1.21E+09	20.91

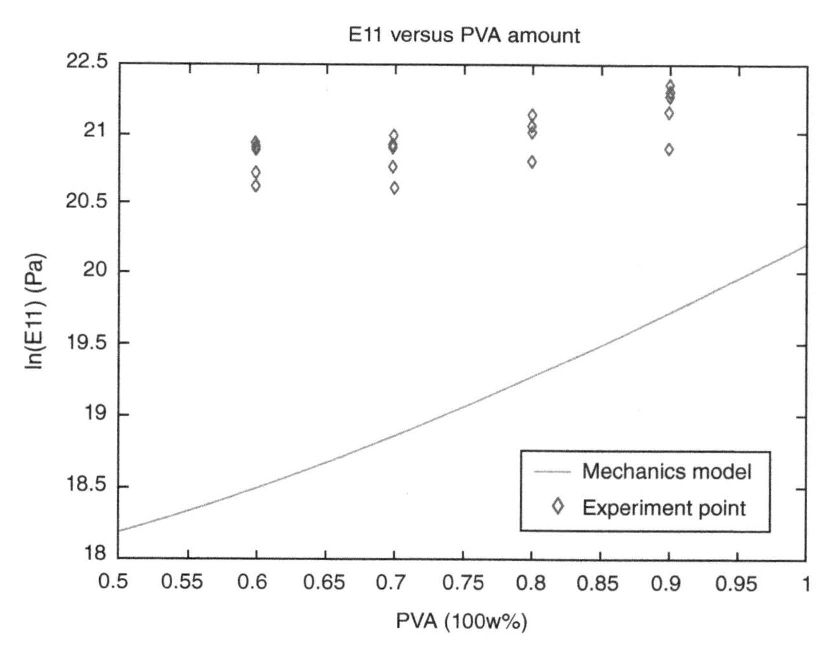

Figure 9.16 The comparison between experimental data and physical model.

One major cause of this discrepancy is the inaccurate CNT length input for the mechanical model. The average length of CNTs is calculated from a 1D segmentation model, which is based on the assumption that the initial lengths of CNTs are identical. In fact, the lengths of as-purchased CNTs have a range and follow an unknown distribution. Although the CNT lengths in BP do follow a lognormal distribution, the average length of the CNTs is not accurate. This is an important input for the mechanical model. To increase the model adequacy by calibrating and adjusting the surrogate model with experimental data, a Bayesian-based calibration and adjustment procedure can be employed to back-calculate the average CNT length.

The average CNT length is used as the calibration parameter (θ) and the PVA amount as input parameter (x) for adjustment. Let $y^e(x)$ be the prediction of adjusted model, $y^m(x, \theta)$ be the prediction of physical model, $\delta(x)$ be the model bias, and ε be a random error that follows a Gaussian distribution. The model adjustment can be formulated as

$$y^e(x) = y^m(x, \theta) + \delta(x) + \varepsilon$$

The physical model is assumed to have the same trend as the real output, so we want to keep the same trend in the adjusted model. The bias term is assumed to be linear, as

$$\delta(x) = \delta_0 + \delta_1 x$$

The calibration parameter, average CNT length, is assumed to follow a normal distribution

$$\theta \sim N(\mu_\theta, \sigma_\theta^2)$$

According to the assumptions by equations (2) and (3), the corrected and calibrated model prediction $y^e(x, \theta)$ follows the normal distribution

$$y^e(x, \theta) \sim N(\mu_y, \sigma_y^2)$$

in which the mean and the variance are functions of the bias-correction model and the calibration model parameters. The mean of $y^e(x, \theta)$ is obtained by

$$\mu_y = y^m(x, \mu_\theta) + \delta_0 + \delta_1 x$$

and the variance is found by applying the Delta method:

$$\sigma_y^2 = \left(\frac{\partial y^e}{\partial \theta}\right)^2 \sigma_\theta^2 = \left(\frac{\partial y^m}{\partial \theta}\right)^2 \sigma_\theta^2$$

The output of the 2D truss model is obtained numerically; therefore, the derivative will also be obtained numerically. The partial derivative term can be evaluated numerically using the finite difference formula

$$\left(\frac{\partial y^m}{\partial \theta}\right) = \frac{y^m(x, \theta + h) - y^m(x, \theta - h)}{2h}$$

in which a step size $h = 10^{-6}$ was chosen.

Suppose that the experimental measurements, $\mathbf{y} = [y(x_1), y(x_2), \dots, y(x_n)]'$, are observed at the input sampling site $x_1, x_2, \dots, x_n$. The probability density function of the adjusted model $y^e(x_i, \theta)$ is

$$f(x_i|\theta) = \frac{1}{\sqrt{2\pi}\sigma_y} \exp\left(-\frac{(y(x_i) - \mu_y)^2}{2\sigma_y^2}\right) \quad \text{for } i = 1,2, \dots, n$$

The likelihood function $L(\theta|x_1, x_2, \dots, x_n)$ is the joint density function of all n observations:

$$L(\theta|x_1, x_2, \dots, x_n) = f(x_1, x_2, \dots, x_n|\theta)$$
$$= \prod_{i=1}^{n} f(x_i|\theta) = \prod_{i=1}^{n} \frac{1}{\sqrt{2\pi}\sigma_y} \exp\left(-\frac{(y(x_i) - \mu_y)^2}{2\sigma_y^2}\right)$$

The log-likelihood function is

$$l(\theta|x_1, x_2, \dots, x_n) = \log L(\theta|x_1, x_2, \dots, x_n)$$
$$= -\frac{n}{2}\log(2\pi) - \frac{n}{2}\log(\sigma_y^2) - \frac{1}{2\sigma_y^2}\sum_{i=1}^{n}(y(x_i) - \mu_y)^2$$

Letting constant $c = n\log(2\pi)$, then

$$-2l(\theta|x_1, x_2, \dots, x_n) = c + n\log(\sigma_y^2) + \frac{\sum_{i=1}^{n}(y(x_i) - \mu_y)^2}{\sigma_y^2}$$

The right-hand side of this equation is a function of δ_0, δ_1, μ_θ, and σ_θ^2. To estimate those calibration and adjustment parameters, the log-likelihood function $L(\theta|x_1, x_2, \dots, x_n)$ is maximized. Therefore, the problem becomes a conditional optimization problem

$$\hat{\delta}_0, \hat{\delta}_1, \hat{\mu}_\theta, \hat{\sigma}_\theta = \arg\min\left(n\log(\sigma_y^2) + \frac{\sum_{i=1}^{n}(y(x_i) - \mu_y)^2}{\sigma_y^2}\right)$$

subject to

$$\hat{\sigma}_\theta > 0$$

This conditional optimization problem is solved in Matlab. After calibration, the average lengths of CNTs are found to be following a normal distribution

$$\overline{L} \sim N(\mu_{\overline{L}}, \sigma_{\overline{L}}^2)$$

where

$$\mu_{\overline{L}} = 9.53\ \mu\text{m}, \quad \sigma_{\overline{L}} = 0.59\ \mu\text{m}$$

The estimated bias function is

$$\delta(\text{PVA}) = 2.75 - 2.74 \times \text{PVA}$$

The performance of the calibration and adjustment approach is tested using leave-one-out (LOO) cross-validation. The standardized root-mean-squared errors (SRMSEs) of the model predictions are compared with a pure statistical model, which uses least squares regression on the training data.

The experimental data includes five replications at four different PVA levels, as previously discussed. To perform the LOO cross-validation, data from three out of four PVA levels is used as training data, and the five replications from the unused PVA level are used as prediction points. The SRMSE is defined as

$$\text{SRMSE}_i = \sqrt{\frac{1}{M}\sum_{j=1}^{M}\left(\frac{\hat{y}_i - y_{ij}}{y_{ij}}\right)}$$

where M is the number of replicates in the validation site, that is, 5, i is the index of validation site, and j is the index of replicate. The result is shown in Figure 9.17 and Table 9.4.

Another major cause of the discrepancy between the physical model and experimental data is the inherent model uncertainty. Assume there is a constant bias, B, between the physical model and experimental data, which accounts for the effects of simplifications in the physical model. Then after a constant adjustment, the discrepancy between the physical model and experimental data is caused by an unknown factor [19]. This factor may have a clear physical meaning, but it is not directly measureable in real experiments. In this project, PVA level is defined as the weight percentage of PVA over CNTs used to treat BP. The effectiveness of PVA is assumed to be the same at different PVA levels, which may not be the real case in practical experiments. According to the engineering knowledge of this process, the higher amount of PVA used in the treatment, the more PVA exists in the filtrate. The effectiveness of PVA should be a monotone decreasing function of the total amount of PVA used in the treatment. The effectiveness of PVA is not measureable in real experiments, but it has a direct impact on the final output, Young's modulus of PVA-treated BP.

In this example problem, two sources of information are provided. The computer simulations, S, which are based on the physical model, are relatively less expensive and also less accurate. The physical experiments, P, although expensive, are very accurate. In order to use the information from those two levels of resolution, a latent variable, α, is introduced to reflect the effectiveness of PVA. Both the physical model and experiments can be viewed as functions of PVA level and α. The value of α cannot be observed in the experiments. However, it can be employed to consolidate the separate pieces of information in the two sources of data.

The goal is to predict Young's modulus of PVA-treated BP after measuring the amount of PVA used in future manufacturing process. This prediction can be made by solving the nonlinear problem

$$\min_{\alpha_i, B} \sum_i \sum_j \left(y_{ij} - S\left(\text{PVA}_i, \alpha_i\right) - B\right)^2$$

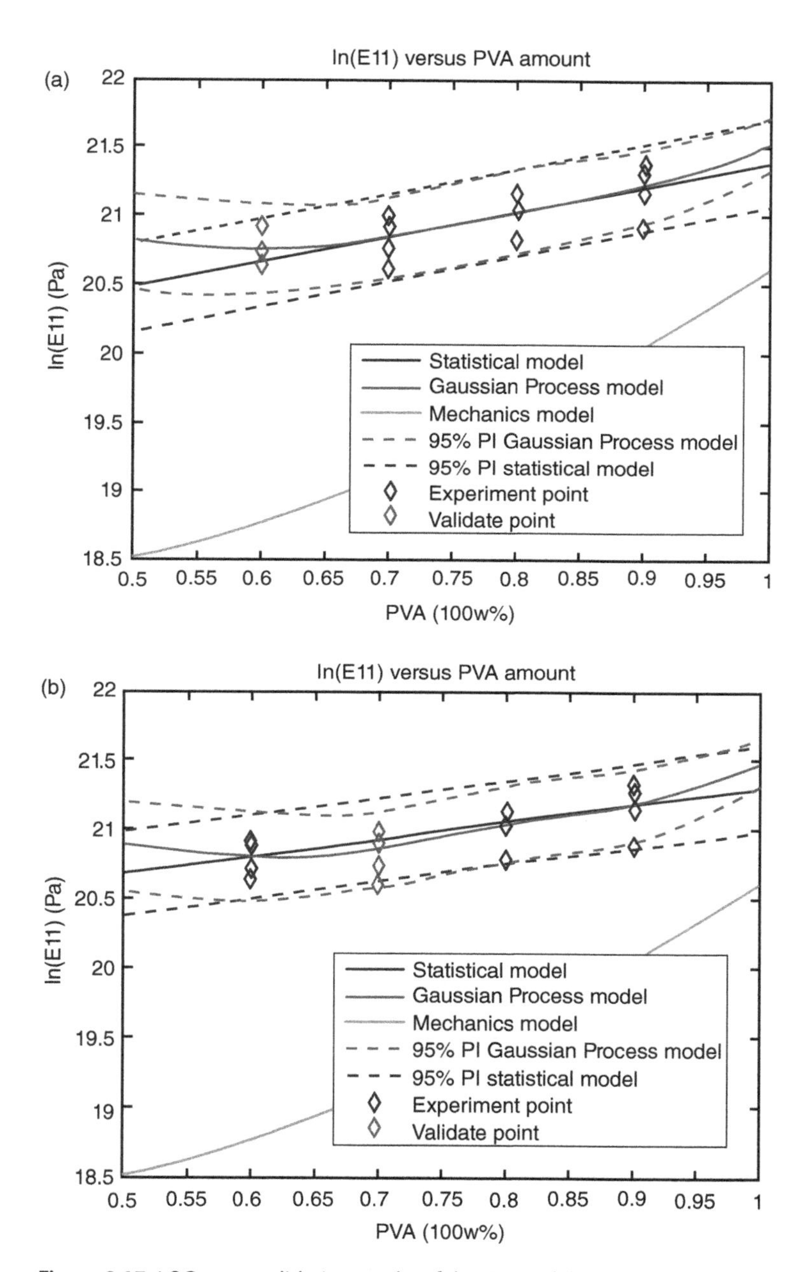

Figure 9.17 LOO cross-validation results of the GP model at PVA level of (a) 0.6, (b) 0.7, (c) 0.8, and (d) 0.9.

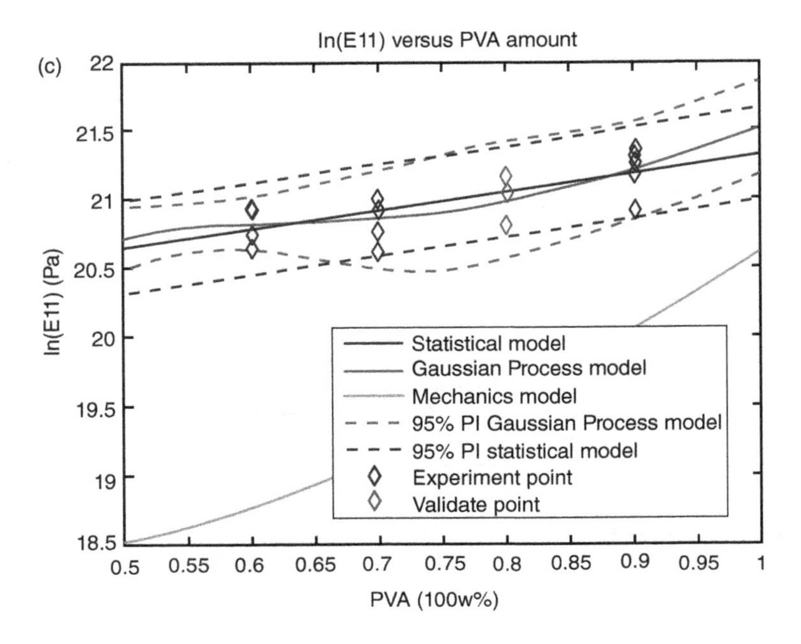

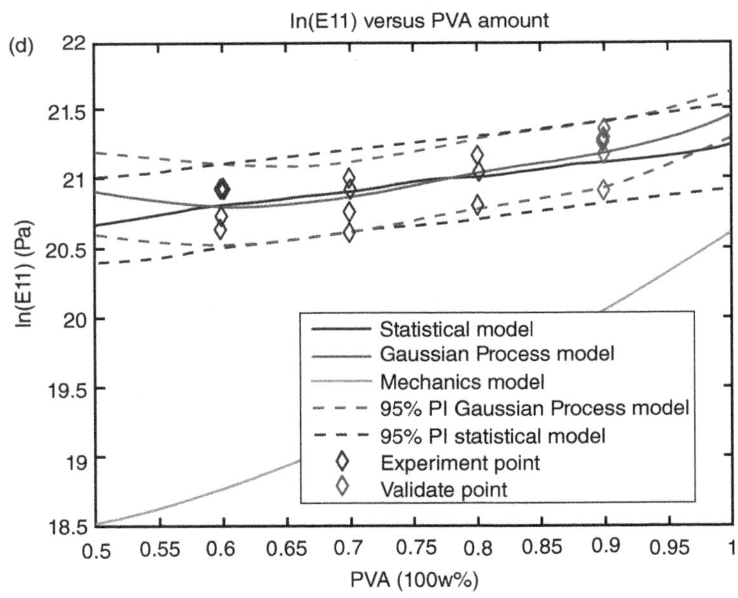

Figure 9.17 *(Continued)*

Table 9.4 Comparison of SRMSE (statistical adjusted physical models versus regression model).

Model	SRMSE			
	PVA level = 0.6	0.7	0.8	0.9
GP model	0.1253	0.1330	0.1285	0.1660
Polynomial model	0.1367	0.1479	0.1301	0.1622
Pure statistical model	0.2041	0.1698	0.1251	0.1850

subject to

$$\sum_{i=1}^{n} \|\alpha_i - g(\text{PVA}_i)\| < M$$

$$g \in G = \{\text{class of monotone decreasing splines}\}$$

where $y_{ij} = P(\text{PVA}_i) + \varepsilon_{ij}$ is the Young modulus for the corresponding PVA for the physical experiment at level $i = 1,2, \dots, n$; the random variable $\varepsilon_{ij} \sim N(0, \sigma^2)$ accounts for the noise associated with the jth replication of the physical experiment ($j = 1,2, \dots, m_i$) at the fixed value PVA_i for each $i = 1,2, \dots, n$.

$S(\text{PVA}_i, \alpha_i)$ is the Young modulus for the corresponding PVA and α for the simulation at the level $i = 1,2, \dots, n$.

B is a constant that reflects the bias that exists between the computer simulations and physical experiments.

M is a constant that can be derived using LOO cross-validation.

The nonlinear problem can be solved by a two-step algorithm:

1) Estimate the functional relationship, g, between PVA and α for the observed physical responses.
2) For any untried value of PVA, denoted PVA*, calculate the associated $\alpha^* = g(\text{PVA}^*)$. Predict the modulus of PVA-treated BP with simulation result $S(\text{PVA}^*, \alpha^*)$.

The procedure is illustrated in Figure 9.18.

In the training stage, for simplicity, assume that only one replication is tested for each $i = 1,2, \dots, n$. The nonlinear problem becomes

$$\min_{\alpha_i, B} \sum_i \left(P\left(\text{PVA}_i\right) - S\left(\text{PVA}_i, \alpha_i\right) - B\right)^2$$

The equation above is solved by dynamic programming (DP)

$$f_i\left(\alpha_i^m\right) = \begin{cases} \min_{\alpha_n \in A_n, \alpha_n < \alpha_n^m, B}\left(P\left(\text{PVA}_n\right) - S\left(\text{PVA}_n, \alpha_n\right) - B\right)^2 \\ \min_{\alpha_i \in A_i, \alpha_i < \alpha_i^m, B}\left(P\left(\text{PVA}_i\right) - S\left(\text{PVA}_i, \alpha_i\right) - B\right)^2 + f_{i+1}\left(\alpha_i\right) \end{cases}$$

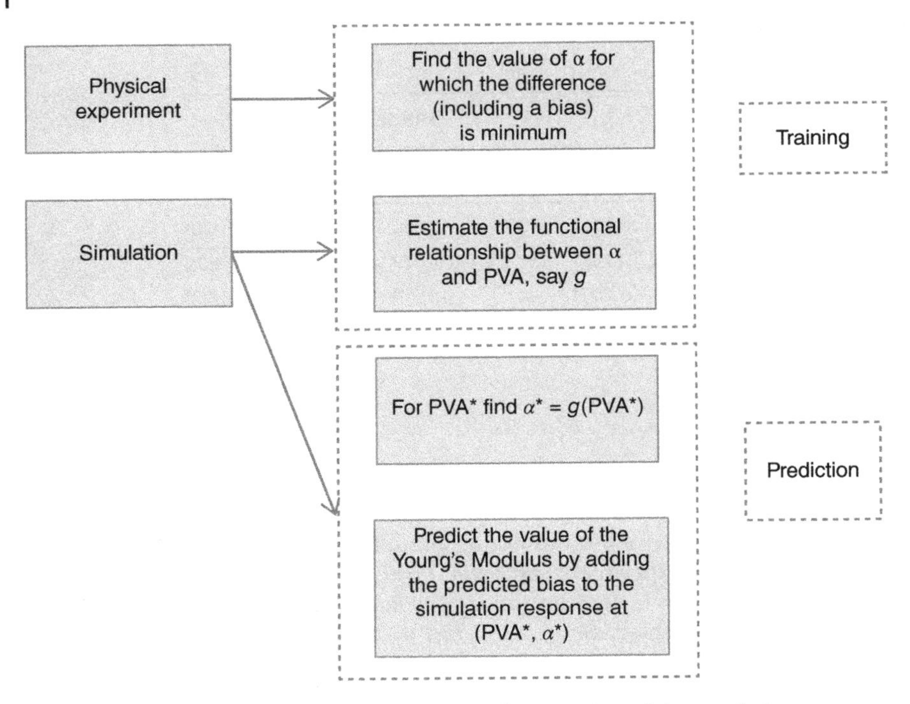

Figure 9.18 The procedure of two-step algorithm for Young's modulus prediction.

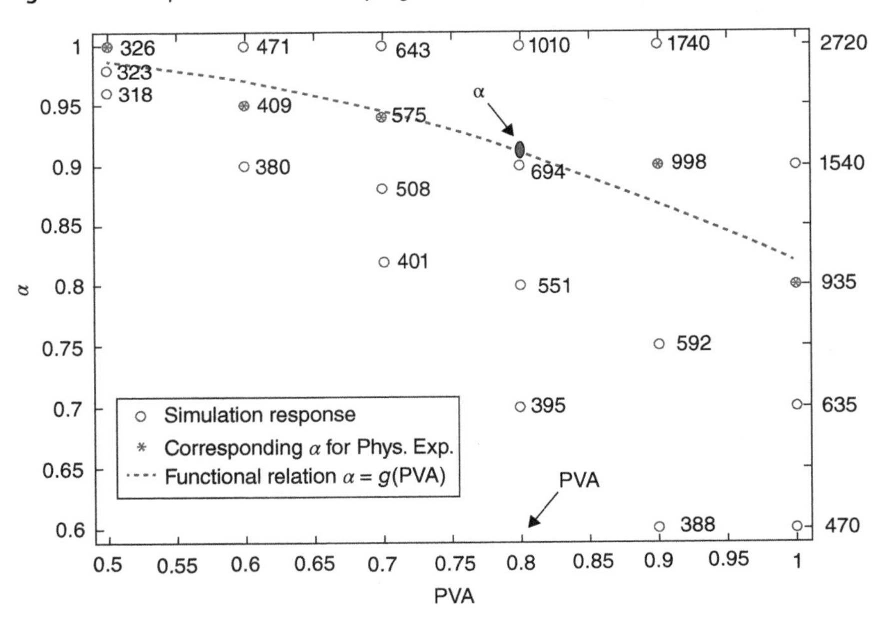

Figure 9.19 The comparison of experimental data and different prediction methods.

Table 9.5 Comparison of SRMSE (latent variable versus other methods).

Method	Proposed method (Interp.; Sim.)*	GP	Weighted multiresolution
SRMSE	0.0188; 0.0190	0.0195	0.1174

*Interp.: interpolated
Sim.: simulated

For PVA = {0.5, 0.6, 0.7, 0.8, 0.9, 1.0}, the solutions of α are {1.00, 0.95, 0.94, 0.80, 0.75, 0.70}.

LOO cross-validation is used to evaluate the performance of the algorithm. An example is given in Figure 9.19.

The calculated SRMSE from LOO cross-validation is compared with other multiresolution methods, and the results are listed in Table 9.5.

9.5 Summary

Buckypaper is a promising platform nanomaterial that exploits the excellent properties of CNTs in bulk scale. To date, the most popular and cost-efficient way to produce BP is through a filtration-based manufacturing process. The obstacles of scaling up the filtration-based manufacturing process of BP include the difficulty of handling BP, the large variation in final products, and the lack of process modeling, monitoring, prediction, and control strategies.

In order to predict the Young modulus of PVA-treated BP, a two-dimensional finite element-based mechanical model was developed. The major difference between this model and other truss/beams models is how it handles the randomness in the input variables. Traditional models usually use the average value of nanostructure characteristics, such as CNT lengths and diameters. The model developed in this procedure can be used to investigate the sources of variations in the manufacturing system. Therefore, in addition to the mean values, the representative distributions of input variables are also considered. By doing this, the effects of variations in the input variable can be evaluated. This is a new capability that no existing models possess.

The accuracy of this physical model is further improved by a few experimental data using either Bayesian-based model calibration and adjustment or latent variable-based multiresolution analysis. Bayesian-based approach is commonly used in other manufacturing processes. However, this particular case requires an input calibration to handle the uncertainty. The uncertainty in the input variables is also a typical issue in nanomanufacturing. An improvement of Bayesian-based model updating method allows the calibration of the input variable and adjustment of the model output to be accomplished at the same time, resulting in a reduction in computational time and resources. The latent variable-based approach was specially designed for this particular case study. The special data structure in this problem makes accurate prediction via existing multiresolution methods difficult. The underlying pattern in the discrepancy between the physical model and experimental data called for introducing a latent variable. This latent variable has a clear physical meaning, the effectiveness of PVA, and is a practical example of handling undetectable variables in the manufacturing system.

In conclusion, this chapter demonstrates an overall methodology, from design and modeling to prediction, for problems in the nanomanufacturing process of a nanomaterial, PVA-treated BP. Using the manufacturing process of PVA-treated BP as a case study, this work embodies the idea of fusing information from both engineering knowledge and the statistical toolset in general industrial engineering problems, especially under the situations that the problem cannot be properly solved by only one of them.

References

1 Iijima, S. (1991) Helical microtubules of graphitic carbon. *Nature*, **354 (6348)**, 56–58.
2 Jones, D.E.H. (1996) Science of fullerenes and carbon nanotubes - Dresselhaus, MS, Dresselhaus, G, Eklund, PC. *Nature*, **381 (6581)**, 384–384.
3 Yu, M.F. *et al.* (2000) Strength and breaking mechanism of multiwalled carbon nanotubes under tensile load. *Science*, **287 (5453)**, 637–640.
4 Demczyk, B.G. *et al.* (2002) Direct mechanical measurement of the tensile strength and elastic modulus of multiwalled carbon nanotubes. *Materials Science and Engineering A: Structural Materials Properties Microstructure and Processing*, **334 (1-2)**, 173–178.
5 Wang, Z. *et al.* (2004) Processing and property investigation of single-walled carbon nanotube (SWNT) buckypaper/epoxy resin matrix nanocomposites. *Composites Part A: Applied Science and Manufacturing*, **35 (10)**, 1225–1232.
6 Ajayan, P.M. and Tour, J.M. (2007) Materials science - nanotube composites. *Nature*, **447 (7148)**, 1066–1068.
7 Sreekumar, T.V. *et al.* (2003) Single-wall carbon nanotube films. *Chemistry of Materials*, **15 (1)**, 175–178.
8 Park, J.G. *et al.* (2008) The high current-carrying capacity of various carbon nanotube-based buckypapers. *Nanotechnology*, **19 (18)**, 185710.
9 Xu, H. *et al.* (2007) Microwave shielding of transparent and conducting single-walled carbon nanotube films. *Applied Physics Letters*, **90 (18)**, 183119.
10 Park, J.G. *et al.* (2009) Electromagnetic interference shielding properties of carbon nanotube buckypaper composites. *Nanotechnology*, **20 (41)**, 415702.
11 Wu, Q. *et al.* (2008) Fire retardancy of a buckypaper membrane. *Carbon*, **46 (8)**, 1164–1165.
12 Kukovecz, A. *et al.* (2007) Controlling the pore diameter distribution of multi-wall carbon nanotube buckypapers. *Carbon*, **45 (8)**, 1696–1698.

13 Endo, M. *et al.* (2005) 'Buckypaper' from coaxial nanotubes. *Nature*, **433** (**7025**), 476–476.
14 Whitby, R.L.D. *et al.* (2008) Geometric control and tuneable pore size distribution of buckypaper and buckydiscs. *Carbon*, **46** (**6**), 949–956.
15 Whitten, P.G., Spinks, G.M., and Wallace, G.G. (2005) Mechanical properties of carbon nanotube paper in ionic liquid and aqueous electrolytes. *Carbon*, **43** (**9**), 1891–1896.
16 Fugetsu, B. *et al.* (2008) Electrical conductivity and electromagnetic interference shielding efficiency of carbon nanotube/cellulose composite paper. *Carbon*, **46** (**9**), 1256–1258.
17 Coleman, J.N. *et al.* (2003) Improving the mechanical properties of single-walled carbon nanotube sheets by intercalation of polymeric adhesives. *Applied Physics Letters*, **82** (**11**), 1682–1684.
18 Wang, S. and Seto, C.T. (2006) Enantioselective addition of vinylzinc reagents to 3,4-dihydroisoquinoline N-oxide. *Organic Letters*, **8** (**18**), 3979–3982.
19 Ashrafi, B. *et al.* (2010) Correlation between Young's modulus and impregnation quality of epoxy-impregnated SWCNT buckypaper. *Composites Part A: Applied Science and Manufacturing*, **41** (**9**), 1184–1191.
20 Liu, J. *et al.* (2011) Nanoparticle dispersion and aggregation in polymer nanocomposites: insights from molecular dynamics simulation. *Langmuir*, **27** (**12**), 7926–7933.
21 Rahmat, M. and Hubert, P. (2011) Carbon nanotube–polymer interactions in nanocomposites: a review. *Composites Science and Technology*, **72**, 72–84.
22 Frankland, S. *et al.* (2002) Molecular simulation of the influence of chemical cross-links on the shear strength of carbon nanotube-polymer interfaces. *The Journal of Physical Chemistry B*, **106** (**12**), 3046–3048.
23 Xie, B. *et al.* (2011) Mechanics of carbon nanotube networks: microstructural evolution and optimal design. *Soft Matter*, **7** (**21**), 10039–10047.
24 Wang, K. (2013) *Statistics-Enhanced Multistage Process Models For Integrated Design & Manufacturing Of Poly (vinyl Alcohol) Treated Buckypaper*. Diss. The Florida State University.
25 Hernandez, A. *et al.* (1997) Surface structure of microporous membranes by computerized SEM image analysis applied to Anopore filters. *Journal of Membrane Science*, **137** (**1**), 89–97.
26 Umbaugh, S.E. (2005) *Computer Imaging: Digital Image Analysis and Processing*, CRC Press.
27 Tsai, C.-H. *et al.* (2012) Predictive model for carbon nanotube–reinforced nanocomposite modulus driven by micromechanical modeling and physical experiments. *IIE Transactions*, **44** (**7**), 590–602.
28 Zaeri, M. *et al.* (2010) Mechanical modelling of carbon nanomaterials from nanotubes to buckypaper. *Carbon*, **48** (**13**), 3916–3930.

10

Fabrication and Fatigue of Fiber-Reinforced Polymer Nanocomposites – A Tool for Quality Control

Daniel C. Davis and Thomas O. Mensah

Georgia Aerospace Systems, Nano technology Division, Georgia Aerospace, Inc., Atlanta, GA, USA

10.1 Introduction

Fiber-reinforced polymer composites are used in many structural applications due to their low density, chemical resistance, and good mechanical properties such as strength, stiffness, and fatigue resistance. For example, new commercial aircraft are designed having over 50% of the primary structural components fabricated with carbon fibers or carbon–glass fiber hybrid reinforced polymer composite materials [1, 2]. Futuristic jet engine fan blades are to be designed using advanced carbon fiber polymer composites. These newer fan blades will be greater in length, thinner in profile, and lighter in weight for achieving greater fuel efficiency, thus reducing operating costs [3–5]. Wind turbine blades are primarily designed using glass fiber-reinforced polymer composites [6, 7]. Lighter weight and high-strength fiber-reinforced polymer composite wind turbine blades are necessary to overcome the gravitational static and cyclic loadings. In particular, for wind turbine rotors diameters greater than 100 m in offshore applications [8]. Aerodynamic structures such as wind turbine blades and aircraft components are subject to high complex mechanical static and cyclic loadings [9, 10]. Fiber-reinforced composite pipelines can provide a specific strength and external surface corrosion resistance not afforded by steel pipes for transporting petrochemical, water, and other fluid products [11]. Glass fiber-reinforced polymer composites are used as overwrap materials for long-term repair of corroded metal transmission pipelines [12, 13]. In this application, the composite materials are subject to elevated temperature creep-fatigue-type loadings [14]. In all of these applications of glass and carbon fiber reinforced polymer composite materials mentioned previously, the expected operational outcomes are lower overall operating costs and greater life endurance due to the benefits of their lower weight, high strength and stiffness, and less detrimental effects due to environment.

Nanotechnology Commercialization: Manufacturing Processes and Products, First Edition.
Edited by Thomas O. Mensah, Ben Wang, Geoffrey Bothun, Jessica Winter, and Virginia Davis.

In composites, the fibers are the high-strength load-carrying components and are achieved without a significant weight penalty. Carbon fibers have a very high strength-to-weight ratio or specific strength in the range 2450 kN m/kg. Glass fibers have good specific strength in the range of approximately 1300 kN m/kg. In comparison, the specific strength of steel and aluminum alloys, 200–300 kN m/kg, is much less. In a composite, the matrix provides bulk and enables transfer of loads between fibers. Fiber-reinforced polymer composites are known to have high in-plane tensile strength and stiffness properties and exceptional resistance to fatigue loadings. An epoxy matrix, in this case, would have good impact resistance, high strength and hardness, and good electrical insulation properties. However, epoxies are known to be subject to brittle fracture that could lead to less than optimal composite properties under mechanical loadings [15]. Epoxies are widely used in the before mentioned structural applications.

Even though fiber-reinforced polymer composites are found to have a high ultimate tensile strength and modulus or stiffness values, and good fatigue resistance, these composites remain susceptible to catastrophic mechanical failures [16, 17]. Continual improvements in the mechanical capabilities of these composite materials are needed to satisfy the ever-increasing requirements of future structural designs and applications in aerospace systems, in particular [18–20]. Now to cite just a few, numerous research studies [21–26] over the past several years demonstrated that nanomaterials integrated into fiber-reinforced polymer composites yield improvements in the mechanical properties of strength, fatigue, and fracture, and also demonstrated that greater life endurances can be achieved for engineering components and systems. This chapter discusses some methods for fabricating nanomaterials in fiber-reinforced composites and their impact on strength, fracture, and fatigue life endurance. The nanomaterials discussed in this article are limited to the widely employed carbon nanotubes (CNTs), carbon nanofibers (CNFs), and nanoclays. The matrix material will be primarily epoxy resins and the fibers will be continuous glass or carbon.

10.2 Materials

The materials considered in this chapter are specifically a glass fiber, a carbon fiber, epoxy resin, CNTs, CNFs, and nanoclays. These high-strength glass or carbon fiber-reinforced polymer composite materials, for example, currently have wide applications in technologically important industries, to include aerospace, automotive, leisure, wind energy, and piping systems. It is the focus in this chapter to discuss the impact nanomaterials can have on the mechanical properties and capabilities of these composites.

10.2.1 Carbon Fabric and Fiber

Carbon fiber is carbon atoms bonded together as crystals and align in the longitudinal length direction. The strong bonding and the carbon crystalline alignment provide the high strength-to-weight ratio of carbon fibers. Thousands of carbon filaments bond together to form a tow, which can be weaved into carbon fiber fabric. A carbon fabric material discussed in this chapter is a four-harness satin weave having identical warp and fill yarns of 6000 filament count [27]. The fiber diameter range is 5–6 μm. This continuous carbon fiber has a reported tensile strength of 5.5–5.6 GPa, an elastic modulus quoted as 276 GPa, elongation at failure as 1.9%, and density of 1.78 g/cm^3 [28]. Carbon fiber surfaces added by the manufacturer typically have a proprietary surface treatment or sizing to improve the interlaminar shear properties for compatibility with the resin and to achieve other material and physical properties. Due to the exceptional high stiffness carbon fiber tends to be brittle and failure shows limited deformation.

10.2.2 Glass Fabric and Fibers

All types of glass fibers are excellent reinforcing materials for polymers and would form a very strong and relatively lightweight fiber-reinforced polymer composite. The most prevalent type of glass fiber in glass fabric is e-glass. Some room temperature physical properties of pristine e-glass fiber are tensile strength = 3.4 GPa, the tensile modulus = 72.3 GPa, strain at failure = 4.8%, and density = 2.58 g/cm^3 [29]. The e-glass fiber has a 10–13 μm diameter. The mechanical properties are less than those of the carbon fibers, especially stiffness. However, glass fiber is much less expensive than carbon and significantly less brittle. The e-glass fiber fabric reinforcement primarily discussed here is of a bidirectional [0°/90°] weave yarn that is approximately 1 mm wide. The glass composition meets the certification for e-glass as defined by ASTM D578-00 Standards Specification for Glass Fiber Strands. According to the supplier, these glass fibers are "a natural, lustrous, white, continuous filament, and are of high stability and durability." With "the exception of surface sizing ingredients, the fibers are inorganic, incombustible and will neither expand nor contract with moisture changes."

10.2.3 Polymer Resin

The polymer resin discussed is a thermoset epoxy resin. Epoxy polymer resins have a wide range of applications in all the aforementioned industry sectors. Epoxies cure by a molecular cross-linking process into a rigid three-dimensional solid polymer structure or gelation [30]. Energy or a heat added to this process can be a catalyst for the molecular chains to more actively link at chemical sites. The resulting thermoset epoxy would have a melting point greater than the ambient temperature. Epoxies cannot be melted, to shape into other forms as thermoplastics, for the decomposition temperature is lower than the melting temperature. Hence, as stated previously, epoxies are subject to brittle fracture that could lead to less than optimal composite strength and fatigue life. Some reported measured properties of commercial epoxies (Simmons Mouldings Ltd, Coventry, UK) are tensile strength = 42.1 ± 1.3 MPa, tensile modulus = 2.4 ± 0.3 GPa, and elongation at break (ductility) = 0.8% [31]. Ultimately, the physical properties of epoxies depend on the cure cycle and hardener. For a low viscosity epoxy resin Araldite LY 5052-1 and curing agent Aradur 5052-1 grade (Huntsman Co., Switzerland), this epoxy/cure system has a tensile strength range 49.0–86.2 MPa; the tensile modulus range 3.0–3.25 GPa; the strain at break ranges from 1.5% to 8.5%; and the glass transition temperature (T_g) ranges from 50 to 133.9 °C [32]. Other physical characteristics include having good chemical and heat resistance and very good electrical insulation properties. Epoxy resins have a low viscosity and can be very favorable for fabricating large complex composite parts that require forming and molding. An aerospace grade epoxy EPON 862 resin [33] is highlighted in this work. Specifically, the commercially available Hexion EPIKOTE™ 862 epoxy, which is cross-linked with Hexion EPIKURE™ curing agent – W. Resin 862, is a difunctional bisphenol F epoxide (diglycidyl ether of bisphenol F). An elevated temperature and pressure cycle was used for EPON 862 curing. Other epoxies can cure at ambient temperature. Some resins other than epoxies used in many of these aforementioned applications are polyester, vinyl ester, and polyurethane.

10.2.4 Carbon Nanotubes

CNTs physically are tubular-like cylinders of carbon atoms having outer diameters ranging from <1 up to 50 nm and lengths up to several microns. Shown in Figure 10.1 [34] are XD-type CNTs, where a single cylinder of CNTs is called single-wall CNTs (SWCNTs); two concentric cylinders of CNTs are called double-wall CNTs (DWCNTs); and CNTs having a greater number of walls of concentric cylinders of carbon atoms are called multiwall CNTs (MWCNTs). CNTs are further distinguished by the chirality or "twist" of the carbon atom chain around the wall along the length of the nanotube. The chirality affects the physical properties, such as the conductance or metal versus semiconductor, density, and other properties [35]. All CNTs exhibit extraordinary mechanical and many functional properties such as electrical, thermal, optical, and chemical resistance properties. SWCNTs have experimentally measured tensile strengths between 13 and 53 GPa and modulus at 1 TPa [36–38]; electrical conductivity between 10^6 and 10^7 S/m [39]; and thermal conductivity greater than 3000 W/(m K) [40]. At these material property values, SWCNTs exhibit 200 times the strength and 5 times the elastic modulus of steel. SWCNTs exceed the electrical conductivities of copper, while the density of CNT forests (0.03–0.11 g/cm^3) [41] is significantly

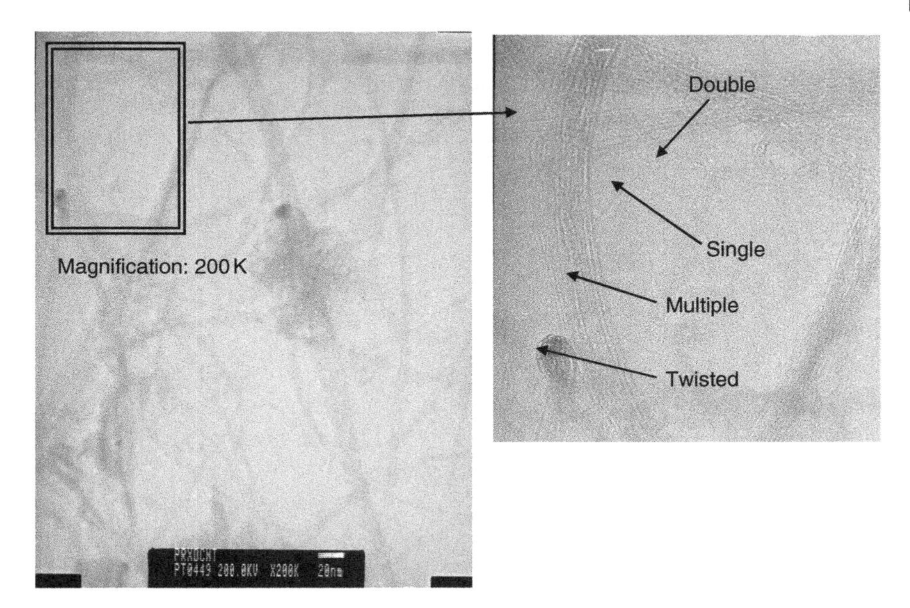

Figure 10.1 XD commercial grade CNTs. Courtesy: Piyush Thrake, Texas A&M University.

less than the density of aluminum (2.70 g/cm^3). As a carbon-based product, CNTs have limited environmental or physical degradation issues such as corrosion, thermal deformation, and sensitivity to light or radiation found in metals. CNTs as produced are in entanglements called ropes because of the strong attractive van der Waals forces between CNTs [42]. With surface functionalization of CNTs, studies demonstrate that CNTs can be untangled and dispersed within glass- or carbon-reinforced polymer composites, resulting in greater performance of the aforementioned system applications [43, 44]. Figure 10.2 shows CNT functionalization routes to oxidation (Ox-CNT), fluorination (F-CNT), and followed-by amination (Am-CNT) [45]. CNTs are expensive. The price of nonpurified, nonsurface-treated CNTs could range as high as US $100–700 per gram. However, efforts to develop new methods for processing CNTs at lower costs are active worldwide [46].

10.2.5 Carbon Nanofibers

Vapor-grown CNFs are cylindrical nanostructures with graphene layers arranged as stacked graphitic cones creating a so-called "herringbone" structure. CNFs diameters reportedly range from 60 to 150 nm and lengths to 100 μm. CNFs also have exceptional mechanical and functional properties [47–51]. CNFs are characterized by a reported tensile strength of 3–4 GPa and Young's modulus varying from 240 to 500 GPa. The density of CNFs (2.0 g/cm^3) is less than that of aluminum. CNFs have good physical properties:

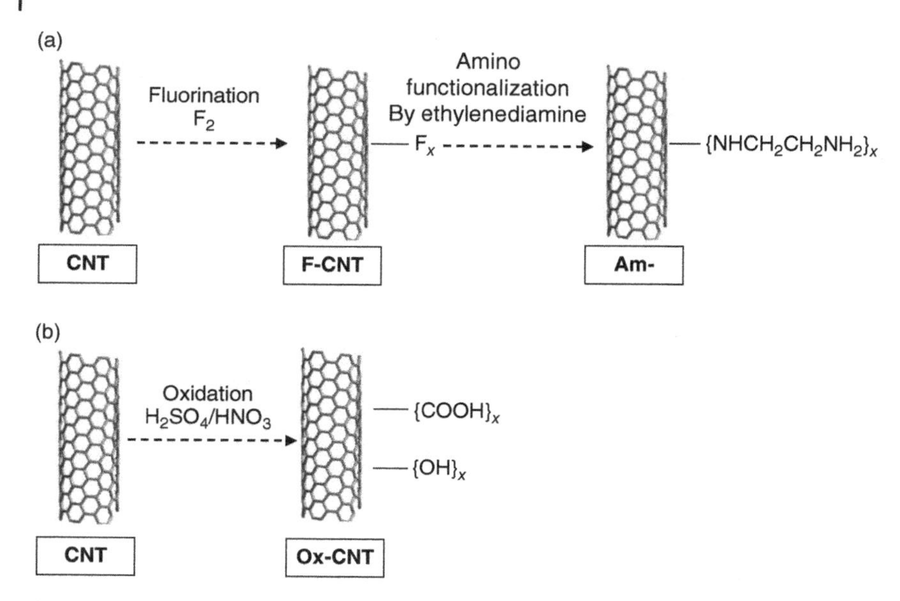

Figure 10.2 CNT functionalization routes. (a) Fluorination – amino route, (b) oxidation route. Courtesy of Dr V.N. Khabashesku, Rice University.

thermal conductivity (20 W/m °K), electrical resistivity (10^{-3} Ω m). It can be observed that the mechanical properties of CNTs are superior to those of CNFs. However, CNFs are an inexpensive nanomaterial that can be purchased in bulk for costs less than US $1 per gram [52], which is significantly less than the costs of CNTs. The stacked cup or herringbone morphology of CNFs results in exposed edge planes on the surfaces of the fiber. These edges along the CNF length provide sites for chemical functionalization of the CNFs surfaces. The functionalization routes as in Figure 10.2 could also apply to CNFs. The functionalized CNFs could then be easily dispersed in the matrix of the composite as CNFs are not greatly hampered by the entanglements such as ropes because the van der Waals forces between CNFs are less than that between CNTs. The surface-functionalized CNFs would allow covalent bonding between the polymer matrix molecular chains and the surfaces of the organically sized glass or carbon fiber reinforcements of the composite. CNFs, as CNTs, are also less affected by environmental conditions as compared to metals. CNFs can have numerous applications in the aerospace, energy, and transportation industrial sectors.

10.2.6 Nanoclays

Nanoclays (NCs) as a raw material is a high aspect ratio layered montmorillonite that can be platelets as thin as 1 nm but with surface dimensions ranging

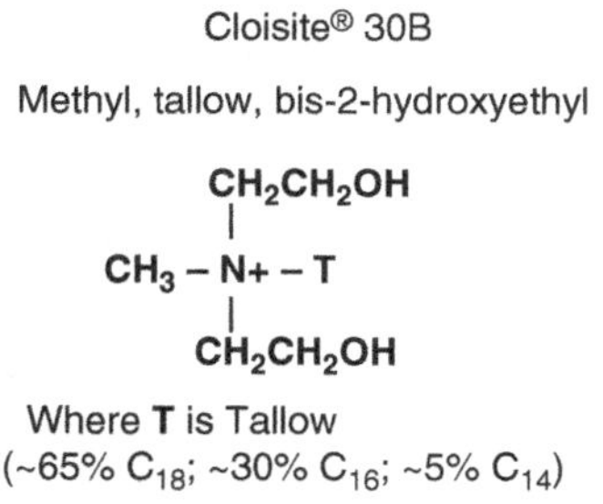

Figure 10.3 Scheme for Cloisite 30B (Southern Clay Products).

Cloisite® 30B

Methyl, tallow, bis-2-hydroxyethyl

$$CH_3-\overset{\displaystyle CH_2CH_2OH \atop |}{\underset{\displaystyle | \atop CH_2CH_2OH}{N^+}}-T$$

Where **T** is Tallow
(~65% C_{18}; ~30% C_{16}; ~5% C_{14})

between 300 and 600 nm [53, 54]. Montmorillonite naturally is hydrophilic and would incur difficulty in attempts to disperse in organophilic polymers. Clays are attractive materials for commercial applications because of their worldwide natural occurrence and low cost [55]. A nanoclay noted in this work is Cloisite 30B, a commercial product that was supplied by Southern Clay Products (Gonzales, TX). According to the supplier, Cloisite 30B has a layered silicate structure built of oval-shaped platelets of 1 nm thickness and 200–300 nm in length. According to the manufacturer, Cloisite 30B is an "organically surface-modified clay by quaternary amines containing a long-chain alkyl tallow and two hydroxyethyl groups" (Figure 10.3). Cloisite 30B is capable of "bonding to epoxy resin compounds either noncovalently through hydrogen bonds or covalently by additional reaction with the resin compound under a moderate heat." In composite applications, Cloisite 30B usually has an intercalated morphology, and with the functional quaternary amines is compatible with polymer matrices.

10.3 Composite Fabrication

This section presents hand layup and resin transfer molding (RTM), the two methods of fiber-reinforced polymer composite fabrication used in laboratory research, testing, and evaluations.

Using either of these two methods, flat composite panels were fabricated for these laboratory studies, but there is mentioning of the application of these fabrication and curing methods for composite component size manufacturing. The test panels are then cured using processes involving combinations of temperature, pressure, and time. From these fabricated composite panels, laboratory tensile-type specimens were water-jet cut to study the influence of nanomaterials on the composite mechanical properties.

10.3.1 Hand Layup

The hand layup is the most basic method for fabricating thermoset composites whether in the laboratory or product manufacturing. This most basic method

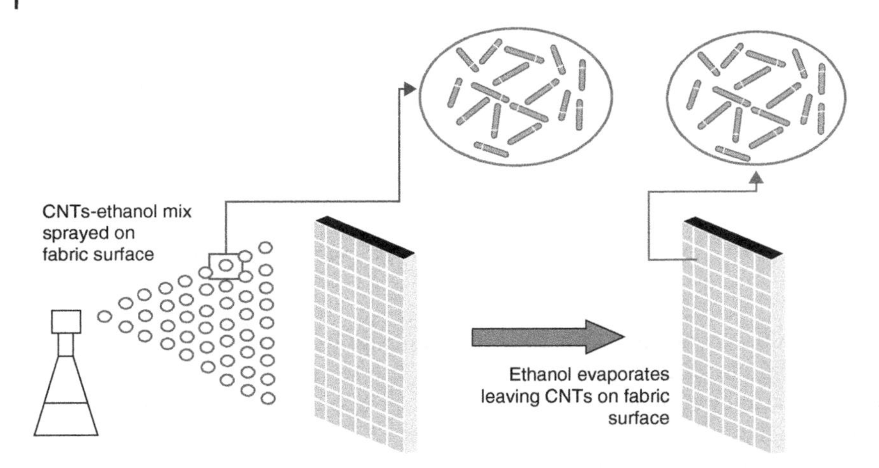

Figure 10.4 Spraying technology schematic. Courtesy of Professor E.V. Barrera, Rice University.

typically consists of stacking fabric plies laboriously by hand onto a tool to form a laminate. The fabric plies may be dry or wet. For dry ply layup, the resin is physically applied by pouring, brushing, or rolling it to "wet-out" the fabric after each ply layup. For nanocomposites, the resin can contain dispersed nanomaterials (nanotubes, nanofibers, nanoclays, or others) in the mixture. For wet-ply layup, the fabric could have been coated with the resin, for example, by dipping the fabric through a resin bath with or without dispersed nanomaterials. A spraying technology [56] has been developed to apply an ethanol–nanomaterial mixture onto the dry fabric; the ethanol evaporates, then leaving the nanomaterial deposits on the fabric. A schematic of the spraying technology approach is shown in Figure 10.4. Images of XD-type CNTs (Figure 10.1) deposited on carbon fabric surfaces from using the spraying technology are shown in Figure 10.5. Afterward, the resin, with dispersed nanomaterials or without, could be applied onto the fabric surfaces as discussed before. When the required numbers of fabric plies are stacked to reach the laminate lateral dimension requirements, typically the laminate is then compressed laterally and cycled through a time, pressure, and temperature cure using, for example, a hydraulic press or vacuum bagging methods [57–59]. This final step in fabrication emulates that obtained through use of the autoclave where the aim is to also mitigate porosity and air pockets in the laminate.

10.3.2 Resin Transfer Molding

In the RTM methods, typically dry fabric (glass, carbon, or others) plies are placed on a heavy thick mold structure and stacked to a required laminate

Figure 10.5 Fluorinated functionalized CNT bundles and ropes deposits on (a) carbon fabric and (b) carbon fiber surfaces.

thickness. RTM is primarily used to mold components with complex shapes and large surface areas, and of high quality [60, 61]. The next step is to attach and torque down the other half of the mold to the bottom half housing the fabric plies. Primarily used to mold components with large surface areas, complex shapes, and smooth finishes, the mold has an inlet opening used for resin injection and an outlet opening. To evacuate or remove air from the fabric stack and porous material in the mold cavity, the inlet opening is closed and the outlet opening is used to vacuum out the air in the mold cavity. The air needs to be removed from the laminate stack plies to allow the resin to fully permeate the laminate and not have micro air pockets to become defects in the composite laminate. In RTM, the resin is pumped through the laminate of stacked fabric plies using an applied pressure at the inlet end to a lower pressured outlet end. Final RTM products will be light in weight and high in strength. However, RTM uses heavy structured tooling to withstand the hydraulic pressure, and hence it has high tooling cost. Finally, RTM produces high-quality and high-strength composite structures with surface quality matching the mold surfaces. However, the heavy-duty mold structure is required to withstand the hydraulic pressure of the resin injection and associated equipment setup is expensive.

Vacuum-assisted resin transfer molding (VARTM) [62] is a variation of RTM where the resin is drawn from a reservoir into and through the stacked fabric laminate by having a vacuum or lower pressure at the outlet end. A schematic in Figure 10.6 [63] shows a special heated-VARTM (H-VARTM) setup used for fabricating laboratory nanocomposite panels. The stack of fabric is covered with a flexible nonpermeable vacuum bag sealed around its edges to the

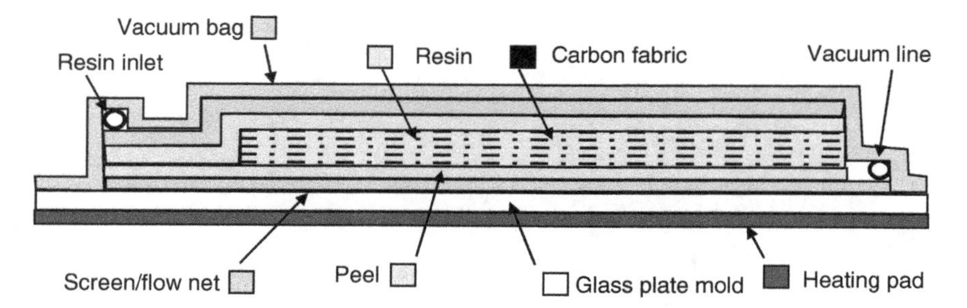

Figure 10.6 Heated vacuum assisted resin transfer molding fabrication (H-VARTM) setup. Courtesy of Drs A. Kelkar and R. Bolick, North Carolina A&T University.

bottom mold. Here, the epoxy resin enters the inlet at temperature and is drawn through the laminate stack by the force of the vacuum at the outlet. Of course, the fabric stack system is vacuumed of air and other impurities before infusion of the resin. In H-VARTM, the resin is housed in a heated sealed reservoir and a heating pad maintains the resin at temperature as it infuses through the fabric stack to the outlet. H-VARTM can be particularly beneficial since a nanomaterial-resin mix is more viscous. The main difference between the VARTM and RTM is that in VARTM resin is drawn into a fabric stack through use of a vacuum only, rather than resin being injected through the stack of fabric layers by an applied pressure pump. VARTM in a general composite fabrication has some advantages over hand layup and RTM. VARTM does not need expensive autoclaves as with hand layup and VARTM does not require heavy-duty equipment to resist the hydraulic pressure differential conditions as with RTM. VARTM can be employed to fabricate flat plat panels for laboratory specimen testing and very large, and as well, complex composite structures such as wind turbine blades, boat hulls, and other shell-like structures of excellent quality that could be too large for the largest autoclaves [64, 65].

10.4 Discussion – Fatigue and Fracture

10.4.1 Fatigue and Durability

The objective of this section is to discuss the fatigue and life endurance of fiber-reinforced polymer composites and the influences nanomaterials can have on improving the mechanical properties. The importance of fatigue of materials, whether metals, ceramics, polymers or their composites, cannot be overstated as most structural component failures are due to fatigue loadings and mechanisms [66]. Here the fatigue loading type being discussed is limited to unidirectional tension–tension. In Figure 10.7a is a schematic of a constant amplitude tension–tension fatigue loading cycle for a fiber-reinforced polymer

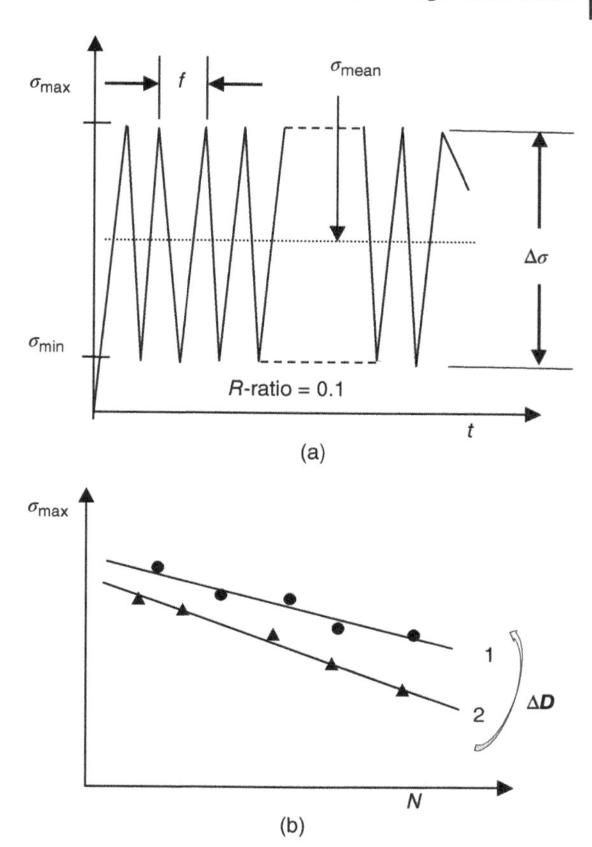

Figure 10.7 (a) Schematic of tension–tension fatigue loading cycle for a fiber reinforced. (b) Schematic plots of the stress (σ) – number of cycles to failure (*N*), or *S–N* plots.

composite. Figure 10.7b shows plots of the stress (σ_{max}) – number of cycles to failure (N), or S–N plots, from laboratory fatigue tests of two assumed different fiber-reinforced polymer composite materials. This tension–tension fatigue testing uses a stress ratio $R = \sigma_{min}/\sigma_{max} = 0.1$, where σ_{max} is the maximum stress level and σ_{min} is the minimum stress level in the cycle. σ_{mean} is the stress level at the midpoint between σ_{max} and σ_{min}, and $\Delta\sigma$ is the stress range of the cycle. From the two plots of data in Figure 10.7b, the material represented by plot 1 would have a greater fatigue strength and life endurance than the material represented by plot 2. The slope of curve-fit line through the data points in plot 1 is less than that through the data points for line 2. Hence, according to a definition given by Dardon *et al.* [67], it could be inferred that the composite represented by line 1 would have a greater durability (D) than the composite material represented by line 2. The difference in durability (ΔD) is measured as the differences in the slopes of the curve-fit lines through the respective data. It is proposed here that greater durability of fiber-reinforced

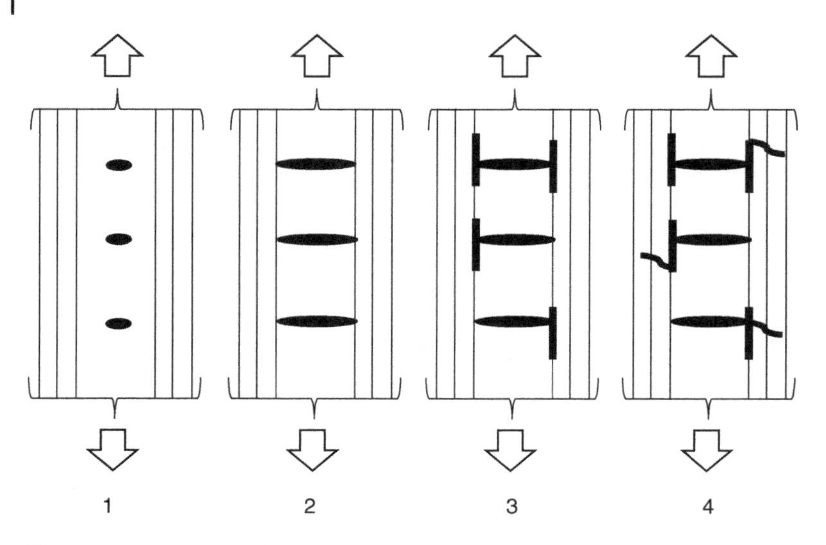

Figure 10.8 Four stages between matrix crack initiation and failure in fiber reinforced polymer composites.

polymer composites results from greater mitigation of crack initiation and propagation due to fatigue loadings [66]. From the illustration of "fatigue damage evolution" in unidirectional fiber-reinforced polymer composites proposed by Talreja [68, 69], a modified version is shown in Figure 10.8 as four stages of fatigue damage and crack propagation in fiber-reinforced polymer composites. At Stage 1, microcracks are assumed to initiate in the matrix in shear or at microscopic defects or discontinuities within the matrix material. In Stage 2, these cracks are proposed to propagate in the matrix normal to the applied cyclic loads, then interact with and damage the fiber–matrix interface. In Stage 3, delamination ensues along the fiber–matrix interface and propagates. In Stage 4 of this scenario, the overloaded fibers rupture, leading to the final failure of the composite cross section [70]. Figure 10.9a micrograph (loading direction in the horizontal) shows Stages 1 and 2 normal matrix cracking and longitudinal fiber/fabric–matrix cracking. Figure 10.9b shows fiber/fabric–matrix delamination and overloaded fabric rupture, and Figure 10.9c shows laminate cross-section failure.

In reference to Figure 10.8, for improved fatigue life of fiber-reinforced composites to be achieved, the matrix must be toughened to mitigate crack initiation and propagation; and as well, the fiber/fabric–matrix interfacial strength must be enhanced to mitigate delamination. It has been implied in the writings of this chapter that chemically functionalized nanomaterials, CNTs, CNFs, and nanoclays, can be dispersed in polymer composite materials for improved fatigue properties and life endurance. The nanomaterial would

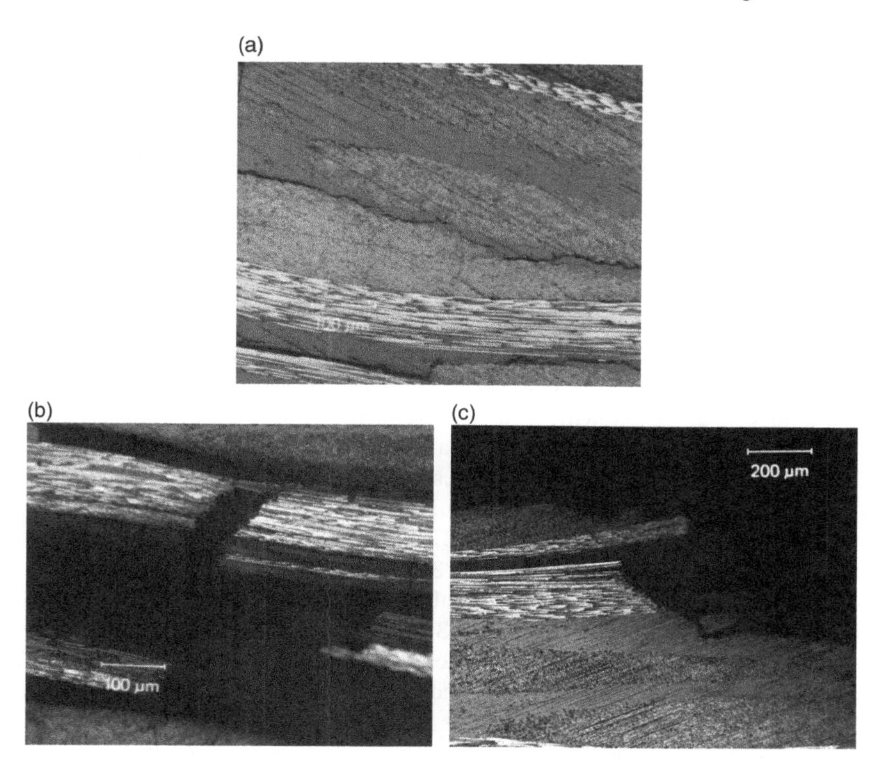

Figure 10.9 (a) Composite normal matrix cracking, fiber–matrix cracking. (b) Fiber/fabric delamination, overloaded fibers and rupture. (c) Composite cross-section failure. (Loading direction – horizontal).

covalently bond with the polymer chains to toughen the matrix. Also, it has been demonstrated that functionalized nanomaterials can be directly deposited on fiber surfaces [71, 72]. The nanomaterials, if covalently bonded to the sizing on fiber surfaces, a stronger fiber–matrix interfacial strength would improve the mechanical properties of the fiber-reinforced composite. In this chapter, the focus will be on the chemical functionalization, as illustrated in Figure 10.2 for CNTs and CNFs, and Figure 10.3 for NCs, as a necessary process for nanomaterial dispersion and nanoreinforcement of composites.

10.4.2 Carbon Nanotube – Polymer Matrix Composites

CNT polymer composites have been the most studied and evaluated nanocomposite due to the extraordinary physical properties of CNTs. However, efficient dispersion of CNTs in a polymer matrix is necessary in order to make use of these properties for composites as CNTs must be untangled from their natural occurrence as ropes due to strong van der Waal forces between nanotubes.

Various types of mechanical methods [73], including "ultrasonication, shear mixing, calendering, ball milling, stirring, and extrusion," are used to achieve good dispersion of nanomaterial in a resin. However, these methods if performed without caution can damage or break the nanomaterials as have been reported happening to CNTs and other high aspect ratio nanomaterials. As mentioned previously, epoxies are brittle, and the expectation is that CNTs, for example, will toughen epoxies for improved mechanical properties. With a combination of acid treatment and subsequent fluorination, Zhu and coworkers [74] achieved good dispersion of chemically functionalized SWCNTs in an epoxy composite after a mechanical mixing method. With 1 wt% SWCNT dispersion in the epoxy matrix, the modulus and tensile strength were enhanced by 30% and 18%, respectively. Guo *et al.* [75, 76] studied the strength, stiffness, and ductility of an epoxy with acid-treated surface-modified MWCNTs. The MWCNTs were dispersed using a sonication method at concentrations 1 wt% through 8 wt% in the composite. Fabricated using a "case molding method," the 8 wt% MWCNT-epoxy nanocomposite showed an 11.7% improvement in tensile strength and significantly a 127.8% improvement in tensile fracture strain at break; however, there was a 30.5% reduction in elastic modulus. Korayem *et al.* [77] studied the effects of CNTs on the ductility of epoxy and found with high-quality dispersion of 3 wt% functionalized CNTs in brittle epoxy matrices, Young's modulus was increased by 12% and tensile strength by 10%. Drescher and coworkers [22] studied the fiber–matrix interfacial strength of a carbon fiber-reinforced polymer composite. In combination, CNTs functionalized with carboxyl groups or amino groups were directly deposited on the carbon fiber surfaces. Functionalized CNTs were also dispersed in the matrix of this composite. Drescher and coworkers concluded that reinforcing the matrix with functionalized CNTs and treating the fibers with a sizing agent containing CNTs would lead to improvements in mechanical strength properties of the fiber-reinforced composite.

Davis and coworkers [78, 79] studied for improvements in quasi-static strength and stiffness, and tension–tension fatigue cycling at stress-ratio (R-ratio) = +0.1 of carbon fiber-reinforced epoxy composite laminates, where functionalized CNTs were incorporated at the fiber/fabric–matrix interfaces over the laminate cross section. This CNT reinforcement strategy was aimed at mitigating the fiber/fabric–matrix damage and delamination during fatigue at Stage 3 of the "Four stages of damage to fatigue failure of fiber-reinforced polymer composites" as shown in Figure 10.8. To fabricate these nanocomposite laminates, first fluorinated or amine functionalized CNTs (Figure 10.2) were dispersed in an organic ethanol solvent using a high shear mix followed by ultrasonication. Second, the CNT-solvent mix is sprayed (Figure 10.4) onto both sides of the fabric layers for the laminate fabrication. Figure 10.5 shows typical images of the CNTs deposited on the fabric and fibers. Third, fiber-epoxy composite flat panels were fabricated using a VARTM process as illustrated in Figure 10.6 and then cured. The composite panels were used in laboratory experiment mechanical testing for the neat case (0.0 wt% CNTs) and the cases of 0.3 and 0.5 wt% CNT reinforcements. Straight bar test specimens were water-jet cut from these fabricated and cured panels. In a comparison between the composite laminate material with and without CNT reinforcements, there are modest improvements in the mechanical properties of strength and stiffness; however, a significant increase is demonstrated in the long-term fatigue life of the functionalized CNT-reinforced composite materials. The improvements in the tensile strength and stiffness, modest as they were, are believed due to toughening of the epoxy matrix by the functionalized CNTs near the fabric/fiber–matrix interface. The improvements in fatigue life is believed due to epoxy matrix toughening plus the CNT-reinforced fiber–matrix interfacial strength improvements, which mitigated the Stage 3 (Figure 10.8) delamination damage and failure. Figure 10.10a is a schematic of a laminate cross section at a fiber/fabric–matrix interface from a VARTM (Figure 10.6) fabricated panel illustrating a network of functionalized CNTs deposited at the interface via the spraying technology (Figure 10.4). The CNTs are shown in Figure 10.10b interacting on the fiber surfaces, within the body of the fabric and within the neighboring polymer chains in the matrix. The Raman microscope (multichannel T64000 Horiba Jobin Yvon), as in Figure 10.10b, was used to visualize and verify the presence of CNTs at the fabric–matrix interface [80, 81]. The Raman image displays the distribution of functionalized CNTs, carbon fibers, and plain epoxy matrix at the interface region obtained by deconvoluting the measured spectra into the Raman spectra of the nanocomposite constituents. As a consequence, the fatigue S–N curves for the neat (0.0 wt% CNTs) material versus the 0.5 wt% amine functionalized SWCNT reinforcement composite material showed a $\Delta D = 55\%$ (per Figure 10.7b), and thus the CNT-reinforced composite would have a significantly greater life endurance. The fatigue testing was conducted consistent with ASTM D3479/D3479M – 12 Standard Test Method for Tension–Tension Fatigue of Polymer Matrix Composite Materials.

10.4.3 Carbon Nanofiber – Polymer Matrix Composites

The objective of this section is to investigate and discuss the effects of well-dispersed vapor-grown CNFs in fiber-reinforced polymer composites on the mechanical properties. As already noted, CNFs have high strength and stiffness mechanical properties and, as a carbon-based material, excellent thermal and electric conductive properties. With the large interfacial surface area, CNFs offer some of the best opportunities to enhance the mechanical and physical properties of these composites through modifying and improving matrix properties [82].

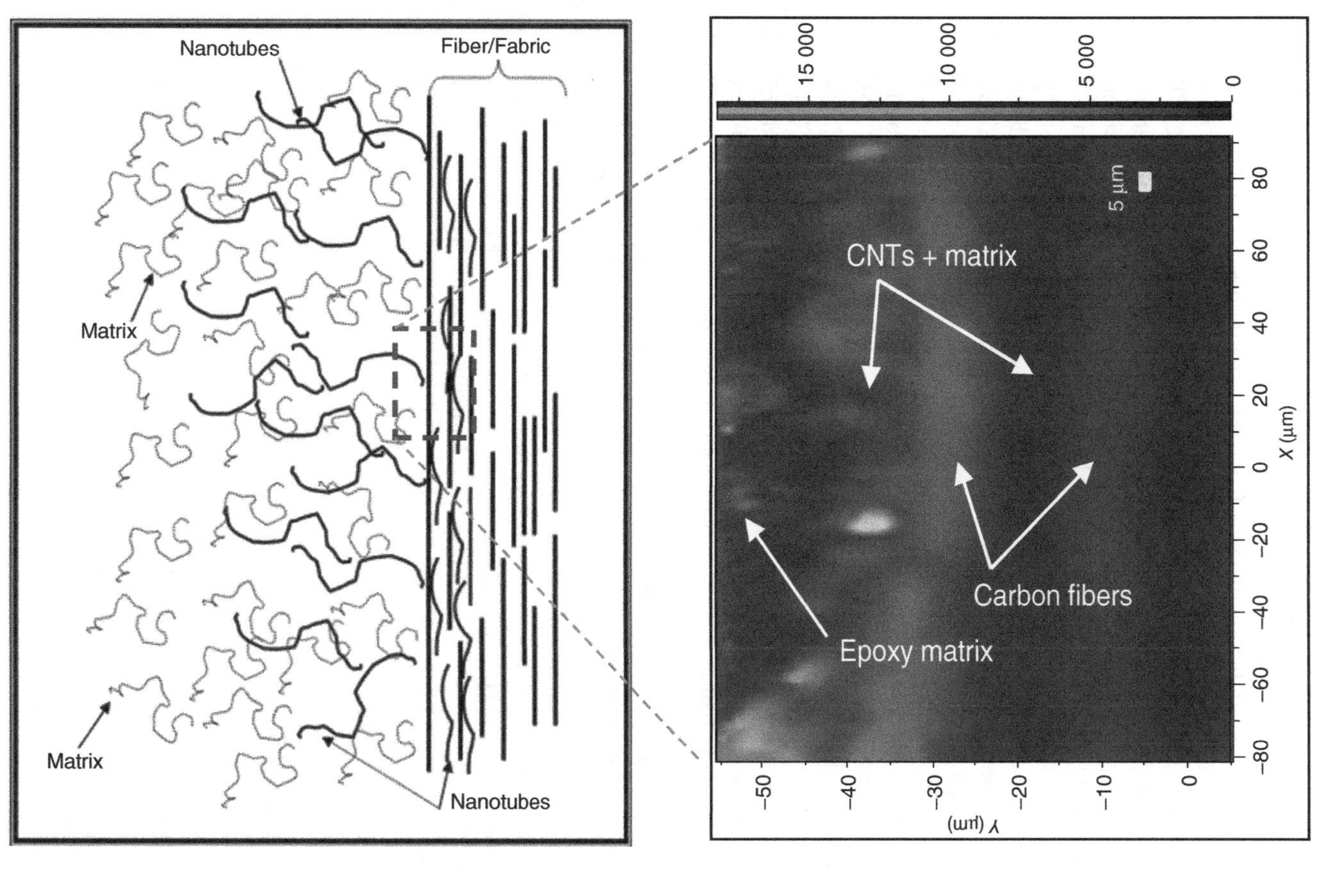

Figure 10.10 (a) Schematic of functionalized CNT reinforced fabric–matrix interface, CNTs strengthen matrix. (b) Raman spectrometry scan of CNTs reinforcing fabric–matrix interface. Courtesy: Dr V. Hadjiev, U. Houston.

Rana *et al.* [83] provided methods for dispersion of CNFs in epoxy-carbon fiber composites. Excellent dispersion is necessary to fully exploit the properties of nanomaterials that would enhance the properties of epoxies or other polymers. Remaining agglomeration of nanoparticles in the composite matrix could generate defects and then deteriorate rather than enhance composite properties. Rana et al. demonstrated several dispersion techniques such as ultrasonication, high-speed mechanical stirring, and also the use of solvents, surfactants, and higher temperatures to aid dispersing CNFs in the resin. This study used an epoxy resin-based "bisphenol A diglycidyl ether (DGEBA) with epoxide equivalent weight of 172–176 supplied by Sigma Aldrich (India). The resin was cured with a triethylene tetramine hardener (HY 951) supplied by Huntsman India Ltd." The "vapor-grown CNFs, Pyrograf III PR24 AGLD grade, were obtained from Applied Sciences, Inc. (USA)." A carbon fiber-epoxy with CNFs dispersed at 0.5 wt% with respect to the epoxy was fabricated and cured for comparison with the neat (or 0.0 wt% CNFs) composite similarly fabricated and cured. The nanocomposite Young's modulus and tensile strength increased by 37% and 18%, respectively. Bortz and coworkers [84] studied an epoxy systems reinforced with a 0.5 and 1.0 wt% commercial grade helical-ribbon CNFs supplied by Grupo Antolín Ingeniería (Burgos, Spain). The CNFs were dispersed in the epoxy resin by hand mixing and then using a high shear laboratory mixer to remove agglomerations. The hardener was subsequently added and further high speed mixed. A neat epoxy sample without CNFs was similarly processed. All epoxy mixes were poured into molds and degassed to remove air bubbles that could result from the mixing. Epoxy-CNF specimens and neat epoxy specimens for the fatigue and the Mode I G_{Ic} fracture toughness testing were cut according to accepted dimensions and prepared. The G_{Ic} improvement over the neat sample at 1.0 wt% CNF was 144%. Uniaxial fatigue tests were conducted. Both the 0.5 wt% CNF epoxy and the 1.0 wt% CNF epoxy had a fatigue life with greater endurance than the neat material as shown in the *S–N* plots in Figure 10.11. From the fatigue *S–N* plots, the % ΔD for the 0.5 wt% CNF epoxy and the 1.0 wt% CNF epoxy composites are less than the neat material plot by 24.4% and 25.9%, respectively. These results would suggest the CNFs in the epoxy matrix acted as inhibitors to the fatigue crack propagation in the matrix. Thus, this toughening mechanism, often characterized as "crack pinning, crack deflection, crack bowing or crack front trapping," are assumed in affect here resulting in the greater fatigue endurance for this epoxy-CNF composite [85, 86]. Zhou *et al.* [87, 88] conducted an experimental study for strength, fracture, and fatigue of a carbon fiber epoxy composite with 2 wt% as-received CNFs. The CNFs were mixed with the epoxy resin by a high-intensity ultrasonic processor. The mixture was then subjected to a high vacuum to remove any air bubble resulting from the mixing and ultrasonication processes. The carbon fiber epoxy flat panel was fabricated using a VARTM setup similar to Figure 10.6

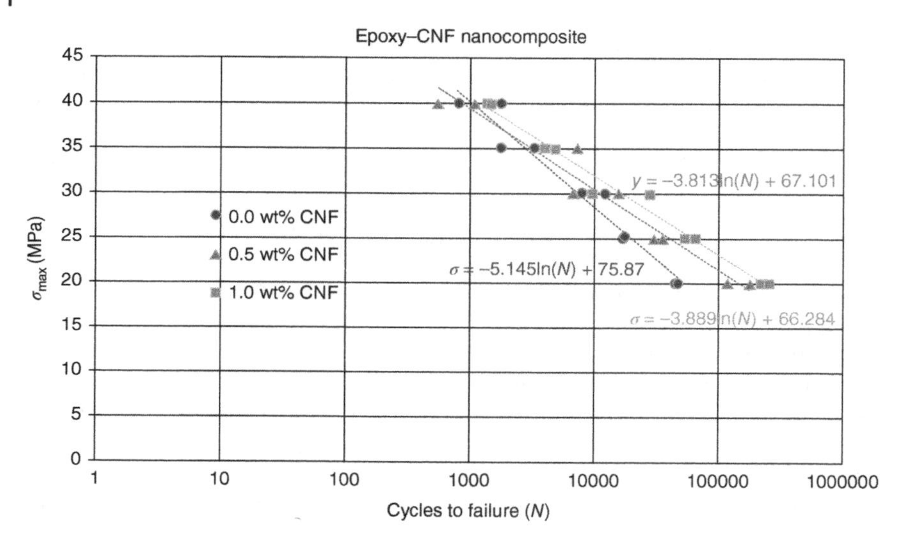

Figure 10.11 Epoxy-CNF, neat epoxy resin fatigue tests.

and then room temperature cured. From tensile testing with the resin modified at 2 wt% CNF, the carbon fabric epoxy composite tensile strength had a 17.4% increase, tensile modulus increase by 19.4%, but there was a 4.2% decrease in tensile strain in comparison to the neat material. The fracture toughness increases by 5.8%. However, the improvements in fatigue were substantial, with a durability change (ΔD) equal to 23.2% determined from the fatigue S–N data. This CNF-reinforced carbon fabric epoxy composite would be projected to have an extended long-term fatigue endurance. "From thermo gravimetric analysis (TGA), the decomposition temperatures for the neat carbon fiber epoxy composite and the CNF-reinforced carbon fiber epoxy composite were the same." "Thus, the CNFs had little effect on the thermal stability of this carbon fiber epoxy composite." The glass transition temperature improved from 111 to 115 °C.

The vast research literature on the subject of chemical functionalization and related dispersion techniques has been in reference to CNTs, as there have been few articles in reference to CNFs on this subject. Varela-Rizo and coworkers [89] present results for polymer composites of a commercial-grade poly(methyl methacrylate) (PMMA) resin and CNFs functionalized with carboxylic and amine groups, and compared with as received CNFs. The helical ribbon CNFs were supplied by Grupo Antolin Ingenieria (Burgos, Spain). The PMMA composites containing 1 wt% pristine CNF, 1 wt% carboxylated CNFs and 1 wt% amino-functionalized CNFs were prepared by procedures as outlined in the reference. The aim of the study was to assess the effects of different functional groups and processing techniques on the properties of

CNF poly(methyl methacrylate) (CNF PMMA) composites. The processing techniques compared are melt-compounding, solvent casting, and *in situ* polymerization in PMMA composites containing pristine, carboxylated, and amino-functionalized CNFs. Varela-Rizo and coworkers use confocal microscopy micrographs of the composites to estimate the degree of dispersion; and at the nanoscale, images from a transmission electron microscope (TEM) were used to estimate the degree of dispersion. A main outcome of this study was the pristine or neat CNF PMMA composite required more aggressive shear mixing for breaking apart and dispersing the agglomerates. The functionalized CNF PMMA composites required less aggressive shear mixing for breaking apart and good dispersion of agglomerates in the matrix. With the less-aggressive shear mixing required, the CNFs were not broken as greatly. Larger aspect ratio (length to diameter) CNFs in the matrix provides for more surface area to bond with the polymer molecular chain.

Withers and coworkers [90] studied and published the research on the mechanical and fatigue properties of a water-activated polyurethane (wPU)-glass fiber (GLF) composite reinforced with 2.0 wt% functionalized CNFs. There have been a number of research studies assessing the properties of polyurethane nanocomposites; however, the nanomaterial has mainly been the CNT. In this case, the CNFs were amine functionalized (Am-CNF) using the route in Figure 10.2, developed by V.N. Khabashesku (University of Houston). The wPU-GLF panel fabrication methods are well described in the Withers and coworkers. Figure 10.12 shows the well-dispersed f-CNFs, measured diameters 70–120 nm, in the wPU matrix on a failure surface. However, there seems to have been a substantial amount of CNF-matrix interface cracking, which could have had an impact on mechanical properties. The aim was to determine improvements in the mechanical properties of the wPU matrix when reinforced with f-CNFs as compared with the neat case without CNFs as reinforcements. All mechanical testing was conducted at 60 °C. As reported in Withers et al., the quasi-static ultimate tensile strength for the f-CNFs reinforced laminate was 8% greater than the neat material. The average % strain-to-failure or ductility was 16% greater for the f-CNF-reinforced composite assumed due to an observed change to an interlaminar shear failure mode. The average stiffness or modulus for the f-CNF-reinforced composite was 11% less than the neat material. The fatigue testing was conducted consistent with ASTM D3479/D3479M – 12 Standard Test Method for Tension–Tension Fatigue of Polymer Matrix Composite Materials. The fatigue cycling of the wPU-GLF composite specimens was at a high mean load level with superimposed smaller amplitude ±vibratory loadings. This cycling was under load control at a frequency of 3 Hz, a stress (or R) – ratio $(\sigma_{min}/\sigma_{max}) = +0.9$, and at a test temperature of 60 °C. It was found that the ΔD between the $S–N$ plots for the reinforced composite (f-CNFs wPU-GLF) and the neat wPU-GLF composite was −7.6%. In this case, the functionalized CNFs dispersed into

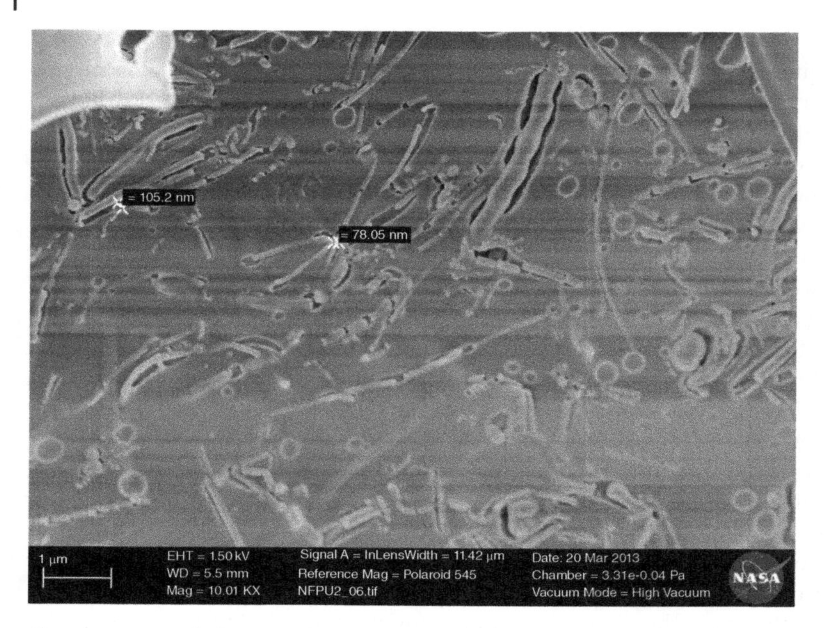

Figure 10.12 Well-dispersed f-CNFs in PU matrix. f-CNF measured diameters, 78.05, 105.2 nm. f-CNF bonding with PU matrix weak. Courtesy of Dr J. Martinez, NASA JSC, Houston, TX.

the wPU-GLF composite did not improve the fatigue durability (*D*). The quasi-static f-CNFs wPU-GLF composite ultimate tensile strength versus the neat material did increase as noted previously, and hence because of the improved tensile strength, at 10^9 cycles the fatigue strength was projected to be 15% greater and the fatigue life was projected to be 800% greater than the neat wPU-GLF composite. In Figure 10.13 are images of the composite failure surfaces, (a) PU-glass fiber composite and (b) CNF-reinforced PU-glass fiber composite. The deformation and fracture are quite similar in both cases and would account for the nearly equal mechanical properties obtained. In this case, it is believed for the composite as fabricated the f-CNFs reinforced the PU matrix mainly and not strategically the fiber/fabric–matrix interface of the composite as the spraying technology was not used to directly deposit the CNFs onto the fabric surfaces as previously demonstrated.

10.4.4 Nanoclay – Polymer Matrix Composites

The nanoclay (NC) being discussed is Cloisite 30B supplied by Southern Clay Products, Gonzales, TX. Cloisite B is an "organically surface modified clay by quaternary amines containing a long-chain alkyl tallow and two hydroxyethyl groups" as described in scheme Figure 10.3. Per the supplier Cloisite 30B is

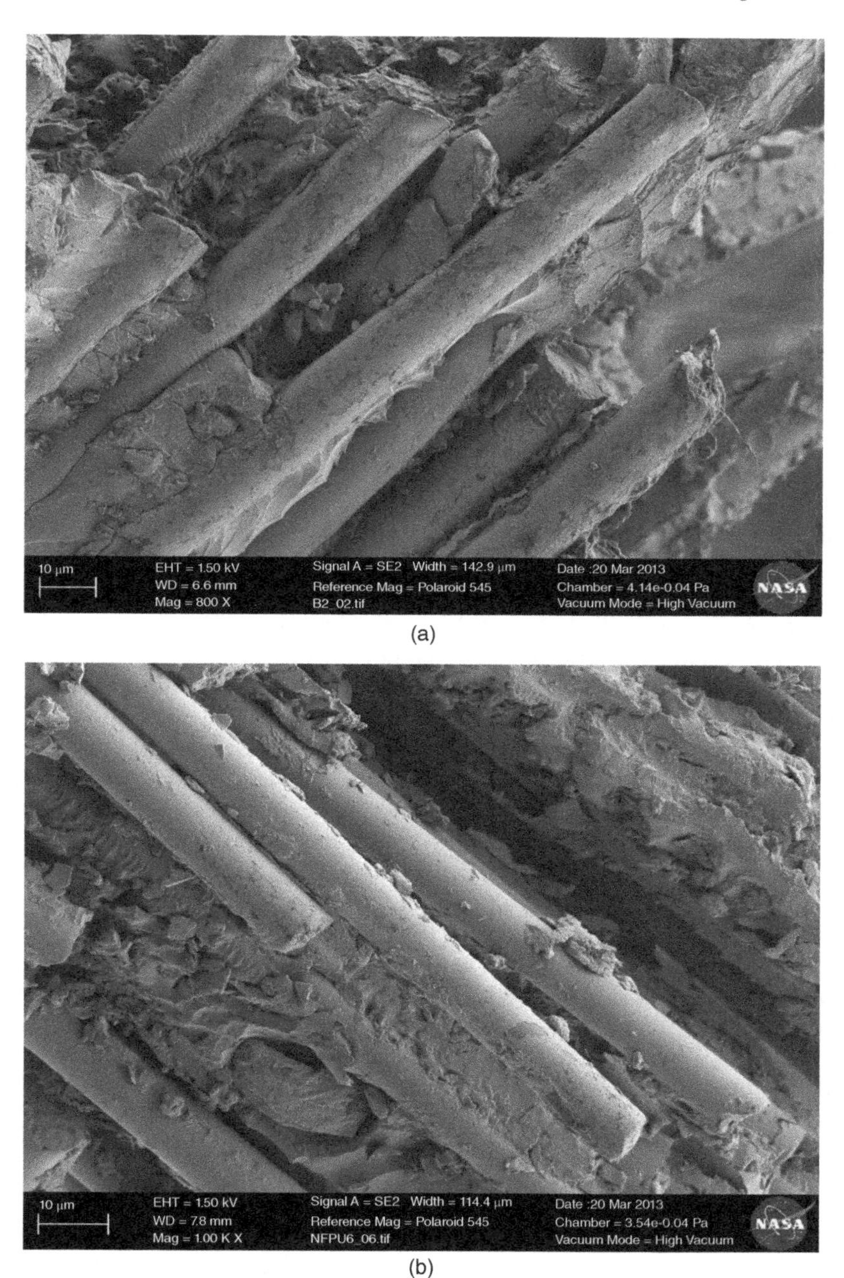

Figure 10.13 Composite failure surfaces. (a) PU-glass fiber composite, (b) CNF reinforced PU-glass fiber composite. Courtesy of Dr J. Martinez, NASA JSC, Houston, TX.

capable of covalently bonding to organic resin compounds and normally has an intercalated morphology.

With a small amount, such as <5 wt%, of nanoclays (NCs) dispersed in an epoxy resin, a substantial improvement in the mechanical properties over the neat material can be achieved [91–95]. These organomodified NCs had an intercalated or the preferred exfoliated morphology where the platelets or layers were well separated with a wide gallery *d*-spacing (>50 Å) and randomly dispersed throughout the polymer matrix. In these cases, the polymer matrix had more surface areas (up to hundreds of m^2/g) to bond with the reinforcing organomodified clay layers, that seemingly act as "nano-level entanglers" with the polymer chains [96]. The degree of an intercalated or exfoliated morphology is dependent on the nanoclay composite curing agent, the surface modification, and the processing method [97]. Using X-ray diffraction (XRD) measurements and TEM observations, the *d*-spacing between organoclay layers can be measured [98].

Only a limited number of studies have been conducted on the effects of adding nanoclays into epoxy-long continuous fiber composite laminate materials. Bozkurt *et al.* [99] evaluated 1 wt% through 10 wt% surface-modified montmorillonite layered aluminoclay-reinforced epoxy-glass fiber composites and a non-surface-modified clay epoxy-glass fiber composite for strength, stiffness, and fracture toughness improvements. Mechanical properties continually increased through 6 wt% surface-modified NC loaded material as XRD and TEM evaluations indicated that clay layer intercalation and dispersion was good. Thereafter, mechanical property values generally declined for both the surface-modified and nonmodified surface NC composites. It is suspected that at high nanoclay loadings, the greater number of agglomerates and air bubble voids in the epoxy matrix between clay layers deteriorated the mechanical properties. On the other hand, in a dynamic mechanical analysis (DMA) of the composite material at 6 wt% NC content, the storage and loss moduli of the laminates increased probably due to the restricted molecular motion. The T_g of the laminates increased by 5% at 10 wt% surface-modified NC loading. It is believed this improvement in T_g over the 1–10 wt% range of nanoclay loadings is due to the restricted mobility of polymer molecules. Khan *et al.* [100] presented a more recent study of tension–tension fatigue of organoclay (octadecylamine-modified montmorillonite)-reinforced epoxy-carbon fiber composites. The organoclay content was at 1, 3, and 5 wt% of the epoxy hardener mixture. The organoclay solution was further high shear mixed, sonicated, and degassed to remove air voids and bubbles. TEM images indicated full intercalation and partial exfoliation of nanoclay dispersion in the mixture. An NC-reinforced epoxy-carbon fiber composite laminate and a similar neat laminate were fabricated each using a hot press for panel dimensions 300 mm square by thickness 2.5 mm. Tensile and fatigue straight-bar specimens were cut from the panels. The ultimate tensile strength and strain-at-failure or ductility of the nanoclay-reinforced composite peaked at 3 wt% NC loading while the tensile stiffness continued to increase through 5 wt% clay loading. The tension–tension fatigue tests were loaded at a stress-ratio (or R-ratio) = $(\sigma_{min}/\sigma_{max}) = +0.1$. The maximum stress ($\sigma_{max}$) levels of the cycle were 90%, 80%, 60%, and 45% of the ultimate tensile strength of the composite, and σ_{min} would be in each case 10% of σ_{max}. The σ_{max} versus N (cycles-to-failure) data showed the NC-reinforced epoxy-carbon fiber composites had a consistently greater fatigue strength and life as compared to the neat composite. SEM images of the nanocomposite fracture surfaces showed features of a more toughened epoxy matrix and greater fiber-epoxy matrix interfacial bonding. Khan *et al.* concluded that these microstructural observations illustrated the primary reasons for the improved tensile strength and fatigue performance in this NC-reinforced composite.

Withers and coworkers [101] studied and published on the mechanical and fatigue properties of the surface organic-modified nanoclay-glass-fiber-epoxy composites being presented as follows. The surface organomodified nanoclays were obtained from Southern Clay Products, Gonzalez, TX. Three formulations of this thermoset glass-fiber-epoxy (GLF-EP) composite were fabricated, a pristine or neat composite with no reinforcing nanoclays (0 wt% nanoclays) and two GLF-EP composites with dispersed surface organomodified Cloisite 30B nanoclays in the matrix at 2.0 and 4.0 wt% levels in respect to the epoxy. There have been many research studies assessing the properties of epoxy nanocomposites; however, the nanomaterial has mainly been the CNT. In this case, the NCs are functionalized as shown in Figure 10.3, per Southern Clay Products. The GLF-EP panel fabrication methods and test specimen configurations are well described in the Withers and coworkers. The aim here is to present improvements in the mechanical properties, primarily tension–tension fatigue life and endurance, of the epoxy matrix when reinforced with Cloisite B as compared with the neat case without NC reinforcements. All mechanical testing was conducted at 60 °C. For the results, the 2 wt% Cloisite B nanoclay-loaded material had the highest ultimate tensile strength, and the 4 wt% Cloisite B NC-loaded specimen showed the ultimate tensile strength to decline. The average σ_{UTS} for the 2 wt% Cloisite B NC-reinforced glass-fiber-epoxy (GLF-EP) composite was 11.7% greater than the GLF-EP neat composite material. The average % strain-to-failure or ductility is 10.5% greater for the GLF-EP NC-reinforced composite versus the GLF-EP neat composite material. The average stiffness or modulus values for the NC-reinforced composite were 10.6% greater than the neat material. The fatigue testing was conducted consistent with ASTM D3479/D3479M – 12 Standard Test Method for Tension–Tension Fatigue of Polymer Matrix Composite Materials: the fatigue cycling was under load control at a frequency of 3 Hz, a stress (or R) – ratio $(\sigma_{min}/\sigma_{max}) = +0.9$, and at a test temperature of 60 °C. It was found that the ΔD between the S–N plots for the Cloisite 30B

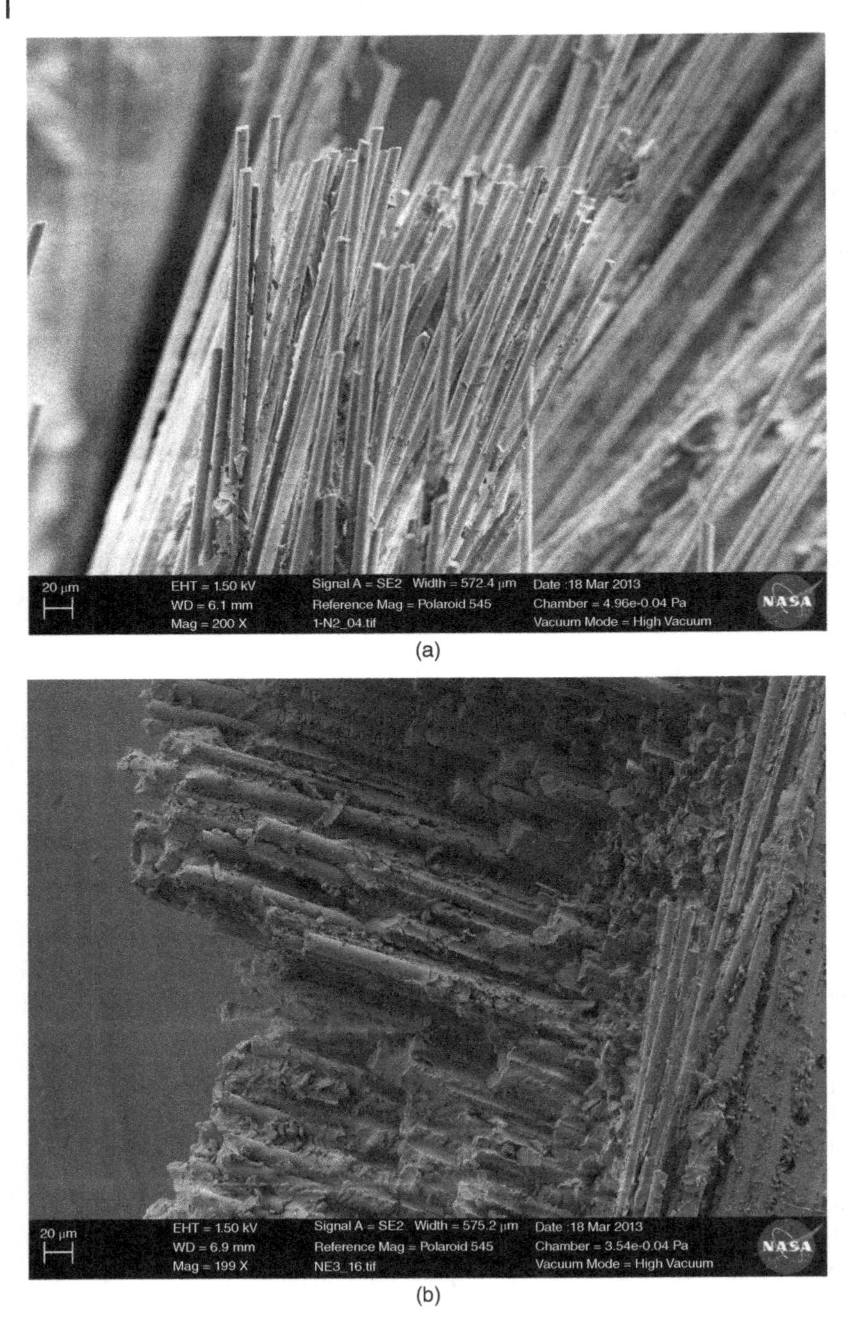

Figure 10.14 Composite failure surfaces. (a) Epoxy-glass fiber composite, (b) nanoclay reinforced epoxy-glass fiber composite. Courtesy of Dr J. Martinez, NASA JSC, Houston, TX.

reinforced composite and the neat EP-GLF composite was 4.9%; hence, the two *S–N* plots were essentially parallel. However, the epoxy matrix was sufficiently toughened such that the GLF-EP NC composite projected at 10^9 cycles a 15% improvement in fatigue strength and a 1000% improvement in fatigue life over the neat GLF-EP composite, as reported in Withers *et al.* A viewing of the composite failure surfaces can possibly provide some explanation for the differences between the NC-reinforced composite and the neat composite. Figure 10.14a shows a brittle-type composite, whereas Figure 10.14b shows the resin to be ductile and appear toughened. The Cloisite B nanoclay-reinforced resin in this case appears to remain more attached to the fibers, thus suggesting a stronger fiber–matrix interfacial strength.

10.5 Summary and Conclusion

Polymer–fiber composites have three primary components, the fibers that generally carry the load imposed on the composite, the matrix that provides bulk to the composite and transfers load between fibers, and the fiber/fabric–matrix interfaces [102] that must have sufficient interfacial strength for loads to transverse the interface without inducing cracking and delamination. To improve the mechanical properties of these composites, the greatest opportunity comes with adjusting the properties of the matrix and the interface between the matrix and fibers. Nanomaterials were initially employed to toughen the matrix by mitigating crack growth. Good dispersion of nanomaterials in the resin, through the use of chemical modifications or functionalization of their surfaces in addition to mechanical dispersion methods, became a necessity in order to toughen the matrix. It was further recognized that the strength of the fiber–matrix interface is most important in hindering the onset of delamination along the interfaces of composites due to fatigue-type cyclic loading.

Figure 10.15a shows a fatigue crack that propagated through a matrix-dominated region of a composite cross section and is seemingly deviated and hindered near the matrix–fabric interface. Figure 10.15b shows that the fatigue crack propagated is apparently deviated and hindered due to the CNTs. The CNT deposition in the matrix and at the fabric–matrix interface was via the spraying technology discussed previously and illustrated in Figures 10.4, 10.5, and 10.10. It is proposed that the spraying technology offers a practical approach to reinforcing fiber-polymer composites with nanomaterials and mitigating the progression of the stages between crack initiation and failure in fiber-reinforced polymer composites. The examples here using the spraying technology have been limited to functionalized CNTs; however, it is suggested that similar studies using other nanomaterials, nanofibers, nanoclays, and graphene platelets, be investigated.

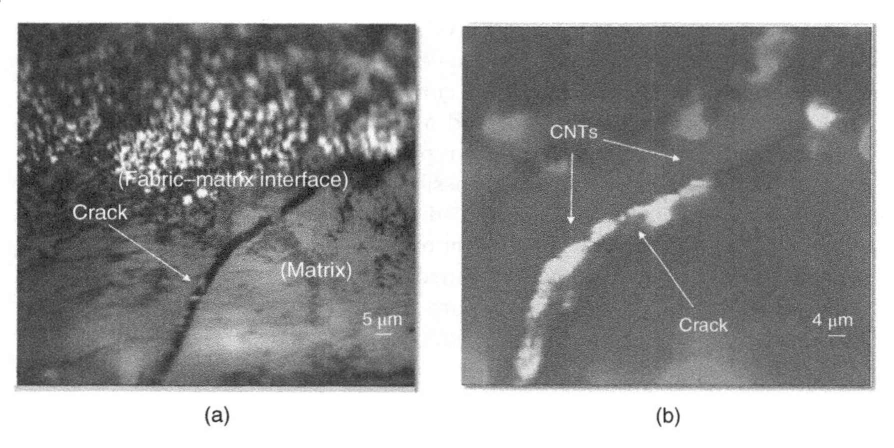

Figure 10.15 Raman images. (a) SEM image of crack propagating in matrix and seemingly deviated and blunted at fabric–matrix interface. (b) Raman spectrometry scan of CNT-reinforced fabric–matrix interface. Crack deviated and blunted due to CNTs. (Courtesy: Dr V. Hadjiev, U. Houston.)

Acknowledgments

The following individuals contributed to the research presented: Dr J.W. Wilkerson (University of Texas, San Antonio, TX), Dr D.C. Lagoudas (Texas A&M University, College Station, TX), Dr J. Zhu (formerly of NanoRidge Materials, Houston, TX), Dr V. Hadjiev (University of Houston, Houston, TX), Dr V.N. Khabashesku (University of Houston and Rice University, Houston, TX), Drs A. Kelkar and R. Bolick (North Carolina A&T University, Greensboro, NC), and Dr J. Martinez (NASA JSC, Houston, TX). The research presented had the sponsorship of the National Aeronautical and Space Administration, Washington, D.C. and the Air Force Research Laboratory (AFRL), Wright Patterson Air Force Base, OH and several awards by the National Science Foundation, Arlington, VA.

References

1 Glover, B.M. (2004) History of development of commercial aircraft and 7E7 dreamliner. *Aviation English*, **592**, 16–21.
2 Pora, J. (2003) Advanced materials and technologies for A380 structure. Flight airworthiness support technology. *Airbus Customer Services*, **32**, 3–8.

3 Red, C. (2015) Composites in commercial aircraft engines, 2014–2023. *Composites World*. http://www.compositesworld.com/articles/composites-in-commercial-aircraft-engines-2014-2023.

4 Brooks, R. (2015) Rolls plans center for carbon-fiber composite fan blades, cases. *American Machinist*. http://americanmachinist.com/shop-operations/rolls-plans-center-carbon-fiber-composite-fan-blades-cases.

5 Anoshkin, A.N., Zuiko, V.Y., Shipunov, G.S., and Tretyakov, A.A. (2014) Technologies and problems of composite materials mechanics for production of outlet guide vane for aircraft jet engine. *PNRPU Mechanics Bulletin*, **2014** (**4**), 5–44.

6 Nijssen, R.P.L. (2013) Chapter 6 – Fatigue as a design driver for composite wind turbine blades. Advances in Wind Turbine Blade Design and Materials, in *Knowledge Centre Wind turbine Materials and Constructions, The Netherlands* (ed. P. Brøndsted), Woodhead Publishing Series in Energy, pp. 175–209.

7 Mohamed, M.H. and Wetzel, K.K. (2006) 3D woven carbon/glass hybrid spar cap for wind turbine rotor blade. *Journal of Solar Energy Engineering*, **128** (**4**), 562–573.

8 Griffith, D.T., Resor, B.R., and Ashwill, T.D. (2012) *Challenges and Opportunities in Large Offshore Rotor Development: Sandia 100-meter Blade Research. AWEA Windpower 2012*, Conference and Exhibition; Scientific Track Paper, Atlanta, GA.

9 Vassilopoulos, A.P. (2013) Chapter 8 – Fatigue life prediction of wind turbine blade composite materials, in *Advances in Wind Turbine Blade Design and Materials*, R. Nijssen, P. Brøndsted, eds., Woodhead Publishing Series in Energy, pp. 251–297.

10 Toft, H.S. and Sørensen, J.D. (2011) Reliability-based design of wind turbine blades. *Structural Safety*, **33**, 333–342.

11 Toutanji, H. and Dempsey, S. (2001) Stress modeling of pipelines strengthened with advanced composites materials. *Thin-Walled Structures*, **39** (**2**), 153–165.

12 Duell, J.M., Wilson, J.M., and Kessler, M.R. (2008) Analysis of a carbon composite overwrap pipeline repair system. *International Journal of Pressure Vessels and Piping*, **85** (**11**), 782–788.

13 Porter, P.C. and Patrick, A.J. (2002) Using composite wrap crack arrestors saves money on pipeline conversion. *Pipeline and Gas Journal*, **229** (**10**), 65–67.

14 da Costa, M.H., Reis, J.M.L., Paim, L.M. *et al.* (2014) Analysis of a glass fibre reinforced polyurethane composite repair system for corroded pipelines at elevated temperatures. *Composite Structures*, **114**, 117–123.

15 Zhang, G., Karger-Kocsis, J., and Zou, J. (2010) Synergetic effect of carbon nanofibers and short carbon fibers on the mechanical and fracture properties of epoxy resin. *Carbon*, **48**, 4289–4300.

16 Rakow, J.F. and Pettinger, A.M. (2006) *Failure Analysis of Composite Structures in Aircraft Accidents, ISASI 2006*, Annual Air Safety Seminar Cancun, Mexico.

17 Wu, H., Xiao, J., Xing, S. *et al.* (2015) Numerical and experimental investigation into failure of T700/bismaleimide composite T-joints under tensile loading. *Composite Structures*, **130**, 63–74.

18 Williams, J.C. and Starke, E.A. Jr. (2003) Progress in structural materials for aerospace systems. *Acta Metallurgica*, **51** (**10**), 5775–5799.

19 Ahmed, K., Noor, A.K., Venneri, S.L. *et al.* (2000) Structures technology for future aerospace systems. *Journal of Computers & Structures*, **74**, 507–519.

20 Mangalgiri PD. Composite materials for aerospace applications (1999). *Bulletin of Materials Science*. **22**(**3**): 657-664, DOI:10.1007/BF02749982.

21 Loos, M.R., Yang, J., Feke, D.L. *et al.* (2013) Enhancement of fatigue life of polyurethane composites containing carbon nanotubes. *Composites Part B: Engineering*, **44** (**1**), 740–744.

22 Dresche, P., Thomas, M., Borris, J. *et al.* (2013) Strengthening fibre/matrix interphase by fibre surface modification and nanoparticle incorporation into the matrix. *Composites Science and Technology*, **74**, 60–66.

23 Böger, L., Sumfleth, J., Hedemann, H., and Schulte, K. (2010) Improvement of fatigue life by incorporation of nanoparticles in glass fibre reinforced epoxy. *Composites Part A: Applied Science and Manufacturing*, **41**, 1419–1424.

24 Isitman, N.A., Aykol, M., and Kaynak, C. (2010) Nanoclay assisted strengthening of the fiber/matrix interface in functionally filled polyamide 6 composites. *Composite Structures*, **92**, 2181–2186.

25 Quaresimin, M., Salviato, M., and Zappalorto, M. (2012) Fracture and interlaminar properties of clay-modified epoxies and their glass reinforced laminates. *Engineering Fracture Mechanics*, **81**, 80–93.

26 Yokozeki, T., Iwahori, Y., Ishibashi, M. *et al.* (2009) Fracture toughness improvement of CFRP laminates by dispersion of cup-stacked carbon nanotubes. *Composites Science and Technology*, **69**, 2268–2273.

27 IM7 Carbon Fiber Data Sheet. http://www.Hexcel.com/products.

28 Hexcel carbon fiber data sheet. http://www.hexcel.com/resources/datasheets/carbon-fiber-data-sheets/im7.pdf

29 Hartman, D., Greenwood, M.E., and Miller, D.M. (1996) *High Strength Glass Fibers*, AGY Technical Paper.

30 Varshney, V., Patnaik, S.S., Roy, A.K., and Farmer, B.L. (2008) A molecular dynamics study of epoxy-based networks: cross-linking procedure and prediction of molecular and material properties. *Macromolecules*, **41** (**18**), 6837–6842.

31 Simmons Mouldings Ltd, Coventry, UK. http://www.epoxyworktops.com/epoxy-resin/mech-properties.html.

32 Huntsman Advanced Materials Araldite® LY 5052/Aradur® 5052. http://www.swiss-composite.ch/pdf/t-Araldite-LY5052-Aradur5052-e.pdf

33 EPIKOTE™ Resin 862 Data Sheet. http://www.Hexion.com/products.

34 Thakre, P.R. (2009) Processing and characterization of carbon nanotubes reinforced epoxy resin based multi-scale multi-functional composites. Ph.D. Dissertation. Texas A&M University, College Station, TX 77873.

35 Wilder, J.W.G., Venema, L.C., Rinzler, A.G., and Smalley, R.E. (1998) Dekker C; Electronic structure of atomically resolved carbon nanotubes. *Nature*, **391** (**6662**), 59–62.

36 Lau, K.-T. and Hui, D. (2002) The revolutionary creation of new advanced materials-–carbon nanotube composites. *Composites Part B*, **33**, 263–277.

37 Demczyk, B.G., Wang, Y.M., Cumings, J. *et al.* (2002) Direct mechanical measurement of the tensile strength and elastic modulus of multiwalled carbon nanotubes. *Materials Science and Engineering*, **A334**, 173–178.

38 Ruoff, R.S., Qian, D., and Liu, W.K. (2003) Mechanical properties of carbon nanotubes: theoretical predictions and experimental measurements. *Comptes Rendus Physique*, **4**, 993–1008.

39 Hong, S. and Myung, S. (2007) Nanotube electronics: a flexible approach to mobility. *Nature Nanotechnology*, **2** (**4**), 207–208.

40 Pop, E., Mann, D., Wang, Q. *et al.* (2005) Thermal conductance of an individual single-wall carbon nanotube above room temperature. *Nano Letters*, **6** (**1**), 96–100.

41 Sakurai, S., Inaguma, M., Futaba, D.N. *et al.* (2013) Diameter and density control of single-walled carbon nanotube forests by modulating Ostwald ripening through decoupling the catalyst formation and growth processes. *Small*, **9** (**21**), 3584–3592.

42 O'Connell, M.J., Bachilo, S.M., Huffman, C.B. *et al.* (2002) Band gap fluorescence from individual single-walled carbon nanotubes. *Science*, 297 (**5581**), 593–596.

43 Grishchuk, S. and Schledjewski, R. (2012) Mechanical dispersion methods for carbon nanotubes in aerospace composite matrix systems, Chapter 4, in *Carbon Nanotube Enhanced Aerospace Composite Materials*, In series: Solid Mechanics and Its Applications, Gladwell, G.M.L., ed., vol. **188**, pp. 99–154.

44 Dimitrios, J., Giliopoulos, D.J., Triantafyllidis, K.S., and Gournis, D. (2012) Chemical functionalization of carbon nanotubes for dispersion in epoxy matrices, Chapter 5, in *Carbon Nanotube Enhanced Aerospace Composite*

Materials, In series: Solid Mechanics and Its Applications, Gladwell, G.M.L., ed., vol. **188**, pp. 155–183.

45 Zhang, J., Loya, P., Peng, C. *et al.* (2012) Quantitative in situ mechanical characterization of the effects of chemical functionalization on individual carbon nanofibers. *Advanced Functional Materials*, **22** (**19**), 4070–7077.

46 Universiti Sains Malaysia (2012) New method for continuous production of carbon nanotubes. ScienceDaily, www.sciencedaily.com/releases/2012/04/120412105109.htm.

47 Tibbetts, G.G., Lake, M.L., Strong, K.L., and Rice, B.P. (2007) A review of the fabrication and properties of vapor-grown carbon nanofiber/polymer composites. *Composites Science and Technology*, **67**, 1709–1718.

48 Huang, X. (2009) Fabrication and properties of carbon fibers. *Materials*, **2**, 2369–2403.

49 Feng, L., Xie, N., and Jing, Z. (2014) Carbon nanofibers and their composites: a review of synthesizing, properties and applications. *Materials*, **7**, 3919–3945.

50 Maruyama, B. and Alam, K. (2002) Carbon nanotubes and nanofibers in composite materials. *SAMPE Journal*, **38** (**3**), 59–70.

51 Bortz, D.R., Merino, C., and Martin-Gullon, I. (2011) Mechanical characterization of hierarchical carbon fiber/nanofiber composite laminates. *Composites: A*, **42** (**11**), 1584–1591.

52 Pyrograf Products, Inc. (PPI) (2013) Study of carbon nanotubes and carbon nanofibers. www.azonano.com/article. aspx?ArticleID=2885 (accessed 3 April 2014).

53 Uddin, F. (2008) Clays, Nanoclays, and montmorillonite minerals. *Metallurgical and Materials Transactions A*, **39A**, 2804–2814.

54 Ploehn, H.J. and Liu, C. (2006) Quantitative analysis of montmorillonite platelet size by atomic force microscopy. *Industrial & Engineering Chemistry Research*, **45** (**21**), 7025–7034.

55 Schut, J.H. (2006) Nanocomposites do more with less. *Plastics Technology*. Article posted: February 2006, http://www.ptonline.com/articles/nanocomposites-do-more-with-less.

56 Zhu, J., Imam, A., Crane, R. *et al.* (2007) Processing a glass fiber reinforced vinyl ester composite with nanotube enhancement of interlaminar shear strength. *Composites Science and Technology*, **67** (**7-8**), 1509–1517.

57 Walczyk, D. and Kuppers, J. (2012) Thermal press curing of advanced thermoset composite laminate parts. *Composites A*, **43**, 635–646.

58 Hou, T.-H., Miller, S.G., Williams, T.S., and Sutter, J.K. (2014) Out-of-autoclave processing and properties of bismaleimide composites. *Journal of Reinforced Plastics and Composites*, **33** (**2**), 137–149.

59 Golzar, M. and Poorzeibolabedin, M. (2010) Prototype fabrication of a composite automobile body based on integrated structure.

International Journal of Advanced Manufacturing Technology, **49** (**9–12**), 1037–1045.

60 Yu, F., Junjiang, X., Chuyang, L., and Xinyao, Y. (2015) Static mechanical properties of hybrid RTM-made composite I- and π-beams under three-point flexure. *Chinese Journal of Aeronautics*, **28** (**3**), 903–913.

61 Roopa, T.S., Murthy, H.N., Sudarshan, K. *et al.* (2015) Mechanical properties of vinylester/glass and polyester/glass composites fabricated by resin transfer molding and hand lay-up. *Journal of Vinyl and Additive Technology*, **21** (**3**), 166–173.

62 Advani, S. and Hsiao, K.-T. (2012) *Manufacturing Techniques for Polymer Matrix Composites (PMCs)*, Elsevier. ISBN: 978-0-85709-067-6.

63 Bolick, R., Kelkar, A.D. (2007). Innovative composite processing by using H-VARTM method. In: SAMPE EUROPE 28th international conference and forums. Paris Porte de Versailles, Paris, France, 2–4 April 2007.

64 Sharma, S. and Wetzel, K.K. (2010) Process development issues of glass-carbon hybrid-reinforced polymer composite wind turbine blades. *Journal of Composite Materials*, **44** (**4**), 437–456.

65 Kong, C., Park, H., and Lee, J. (2014) Study on structural design and analysis of flax natural fiber composite tank manufactured by vacuum assisted resin transfer molding. *Materials Letters*, **130**, 21–25.

66 Dauskardt, R.H., Ritchie, R.O., and Cox, B.N. (1993) Fatigue of advanced materials: Part 1. *Advanced Materials and Processes*, **7**, 26–31.

67 Dardon, H., Fukuda, H., Reifsnider, K.L., and Verchery, G. (2000) *Recent Developments in Durability Analysis of Composite Systems*, CRC Press, p. 500. ISBN 9789058091031 - CAT# RU41095.

68 Talreja, R. (1981) A continuum mechanics characterization of damage in composite materials. *Proceedings of Royal Society of London*, **A378**, 461–475.

69 Talreja, R. (2008) Damage and fatigue in composites – a personal account. *Composites Science and Technology*, **68** (**13**), 2585–2591.

70 Pupurs, A. (September 2012) *Micro-Crack Initiation and Propagation in Fiber Reinforced Composites Doctoral Thesis*, Luleå University of Technology, Luleå, Sweden.

71 Mei, H., Bai, Q., Ji, T. *et al.* (2014) Effect of carbon nanotubes electrophoretically-deposited on reinforcing carbon fibers on the strength and toughness of C/SiC composites. *Composites Science and Technology*, **103**, 94–99.

72 Bekyarova, E., Thostenson, E.T., Yu, A. *et al.* (2007) Multiscale carbon nanotube–carbon fiber reinforcement for advanced epoxy composites. *Langmuir*, **23** (**7**), 3970–3974.

73 Ma, P.-C., Siddiqui, N.A., Marom, G., and Kim, J.-K. (2010) Dispersion and functionalization of carbon nanotubes for polymer-based nanocomposites: a review. *Composites A*, **41**, 1345–1367.

74 Zhu, J., Kim, J.D., Peng, H. *et al.* (2003) Improving the dispersion and integration of single-walled carbon nanotubes in epoxy composites through functionalization. *Nano Letters,* **3** (**8**), 1107–1113.

75 Guo, P., Chen, X., Gao, X. *et al.* (2007) Fabrication and mechanical properties of well-dispersed multiwalled carbon nanotubes/epoxy composites. *Composites Science and Technology,* **67**, 3331–3337.

76 Guo, P., Song, H., and Chen, X. (2009) Interfacial properties and microstructure of multiwalled carbon nanotubes/epoxy composites. *Materials Science and Engineering,* **517**, 17–23.

77 Korayem, A.H., Barati, M.R., Simon, G.P. *et al.* (2014) Reinforcing brittle and ductile epoxy matrices using carbon nanotubes masterbatch. *Composites A,* **61**, 126–133.

78 Davis, D.C., Wilkerson, J.W., Zhu, J., and Ayewah, D.O.O. (2010) Improvements in mechanical properties of a carbon fiber epoxy composite using nanotube science and technology. *Composites Structures,* **92** (**11**), 2653–2662.

79 Davis, D.C., Wilkerson, J.W., Zhu, J., and Hadjiev, V.G. (2011) A strategy for improving mechanical properties of a fiber reinforced epoxy composite using functionalized carbon nanotubes. *Composites Science and Technology,* **71** (**8**), 1023–1182.

80 Hadjiev, V.G., Lagoudas, D.C., Davis, D.C., and Strong, K. (2007) *Spectroscopic Imaging of Polymer Nanocomposites Containing Carbon Fillers,* Proceedings, SAMPE Fall Technical Conference and Exhibition, Cincinnati.

81 Hadjiev, V.G., Waren, G.L., Sun, L.Y. *et al.* (2010) Raman microscopy of residual strains in carbon nanotube/epoxy composites. *Carbon,* **48**, 1750–1756.

82 Fiedler, B., Gojny, F.H., Wichmann, M.H.G. *et al.* (2006) Fundamental aspects of nano-reinforced composites. *Composites Science and Technology,* **66** (**16**), 3115–3125.

83 Rana, S., Alagirusamy, R., and Joshi, M. (2011) Development of carbon nanofibre incorporated three phase carbon/epoxy composites with enhanced mechanical, electrical and thermal properties. *Composites: Part A,* **42**, 439–445.

84 Bortz, D.R., Merino, C., and Martin-Gullon, I. (2011) Carbon nanofibers enhance the fracture toughness and fatigue performance of a structural epoxy system. *Composites Science and Technology,* **71**, 31–38.

85 Kawaguchi, T. and Pearson, R.A. (2003) The effect of particle-matrix adhesion on the mechanical behavior of glass filled epoxies. Part 2. A study on fracture toughness. *Polymer,* **44** (**15**), 4239–4247.

86 Singh, R.P., Zhang, M., and Chan, D. (2002) Toughening of a brittle thermosetting polymer: effects of reinforcement particle size and volume fraction. *Journal of Materials Science,* **37**, 781–788.

87 Zhou, Y., Pervin, F., Jeelani, S., and Mallick, P.K. (2008) Improvement in mechanical properties of carbon fabric–epoxy composite using carbon nanofibers. *Journal of Materials Processing Technology*, **198**, 445–453.

88 Zhou, Y., Jeelani, S., and Lacy, T. (2014) Experimental study on the mechanical behavior of carbon/epoxy composites with a carbon nanofiber-modified matrix. *Journal of Composite Materials*, **48** (**29**), 3659–3672.

89 Varela-Rizo, H., Bittolo-Bon, S., Rodriguez-Pastor, I. *et al.* (2012) Processing and functionalization effect in CNF/PMMA nanocomposites. *Composites: Part A*, **43**, 711–721.

90 Withers, G.J., Souza, J.M., Yu, Y. *et al.* (2016) Improved mechanical properties of a water-activated polyurethane-glass fiber composite reinforced with amino-functionalized carbon nanofibers. *Journal of Composite Materials*. **50** (**6**), 783–793.

91 Ho, M.-W., Lam, C.-K., K-t, L. *et al.* (2006) Mechanical properties of epoxy-based composites using nanoclays. *Composites Structures*, **75**, 415–421.

92 Yasmin, A., Abot, J.L., and Daniel, I.M. (2003) Processing of clay/epoxy nanocomposites by shear mixing. *Scripta Metallurgica et Materialia*, **49**, 81–86.

93 Abot, J.L., Yasmin, A., and Daniel, I.M. (2003) Mechanical and thermoviscoelastic behavior of clay/epoxy nanocomposites. *Materials Research Society Symposium Proceedings*, **740**, © Materials Research Society, **740**, 167–172.

94 Swaminathan, C. and Shivakumar, K. (2011) Thermomechanical and fracture properties of exfoliated nanoclay nanocomposites. *Journal of Reinforced Plastics and Composites*, **30** (**3**), 256–268.

95 Borrego, L.P., Costa, J.D.M., Ferreira, J.A.M., and Silva, H. (2014) Fatigue behaviour of glass fibre reinforced epoxy composites enhanced with nanoparticles. *Composites: Part B*, **62**, 65–72.

96 Yoo, Y. and Paul, D.R. (2008) Effect of organoclay structure on morphology and properties of nanocomposites based on an amorphous polyamide. *Polymer*, **49** (**17**), 3795–3804.

97 Azeez, A.A., Rhee, K.Y., Park, S.J., and Hui, D. (2013) Epoxy clay nanocomposites – processing, properties and applications: A review. *Composites: Part B*, **45**, 308–320.

98 Akbari, B. and Bagheri, R. (2007) Deformation mechanism of epoxy/clay nanocomposite. *European Polymer Journal*, **43**, 782–788.

99 Bozkurt, E., Kaya, E., and Tanoglu, M. (2007) Mechanical and thermal behavior of non-crimp glass fiber reinforced layered clay/epoxy nanocomposites. *Composites Science and Technology*, **67**, 3394–3403.

100 Khan, S.U., Munir, A., Hussain, R., and Kim, J.-K. (2010) Fatigue damage behaviors of carbon fiber-reinforced epoxy composites containing nanoclay. *Composites Science and Technology*, **70**, 2077–2085.

101 Withers, G.J., Yu, Y., Khabashesku, V.N. *et al.* (2015) Improved mechanical properties of an epoxy glass–fiber composite reinforced with surface organomodified nanoclays. *Composites: Part B*, **72**, 175–182.

102 Shahid, N., Villate, R.G., and Baron, A.R. (2005) Chemically functionalized alumina nanoparticle effect on carbon fiber/epoxy composites. *Composites Science and Technology*, **27**, 1123–1131.

11

Nanoclays: A Review of Their Toxicological Profiles and Risk Assessment Implementation Strategies

Alixandra Wagner, Rakesh Gupta, and Cerasela Z. Dinu

Department of Chemical and Biomedical Engineering, West Virginia University, Morgantown, WV, USA

11.1 Introduction

Nanoscale properties and increased ability for processing, along with abundance in soil and low cost, make nanoclays applicable to numerous areas from sorbents in pollution prevention to environmental remediation, paints, and cosmetics. Further, reinforcement of polymer films with nanoclays has led to the implementation of nanoclays in the next generation of nanocomposites with increased mechanical strength, barrier properties, UV dispersions, and fire resistance capabilities to be applied in the food packaging industry.

The first part of this review describes the characteristics of nanoclays and how they can be exploited for synthetic applications, especially food packaging applications, while the second part focuses on the challenges associated with nanoclay integration in consumer products and their potential to induce deleterious effects that could affect humans at the exposure levels of manufacturing, consumption, and disposal. Lastly, the review highlights the potential mechanisms of toxicity resulted upon cellular exposure to nanoclays and proposes that logistical burden associated with risk assessment resulted from such exposures could be circumvented through implementation of tailored strategies to ensure a greener route for nanoclay functionalization and implementation.

11.2 Nanoclay Structure and Resulting Applications

Nanoclays are layered, mineral silicates that originate from the clay fraction of the soil [1, 2]. Carrying platelet thickness of around 1 nm and lengths and

Nanotechnology Commercialization: Manufacturing Processes and Products, First Edition.
Edited by Thomas O. Mensah, Ben Wang, Geoffrey Bothun, Jessica Winter, and Virginia Davis.

widths of up to several microns [3, 4], the smectite group of nanoclays is largely made up of 2:1 phyllosilicates, a silicate–oxygen tetrahedral and an aluminum octahedral sheet [1, 5]. Such features differentiate them from chlorite and kaolinite nanoclays, which consist of two tetrahedral and two octahedral sheets and one tetrahedral and one octahedral sheet, respectively [1, 6]. Montmorillonite, the most common type of phyllosilicate clay [1], has a negative charge due to the substitution of aluminum for silicon in its tetrahedral sheets and magnesium for aluminum in its octahedral sheets [7]. Bentonite, a source of montmorillonite, contains crystalline quartz, cristobalite, and feldspar [8], and due to the presence of inorganic cations in its galleries is hydrophilic in nature [5, 7].

Positively charged ions, such as, sodium, potassium, and calcium, are attracted to the inner galleries of negative nanoclays [7], allowing for cation exchange and organic modification via an ion-exchange reaction [9]. The addition of the organic modifier increases basal spacing to allow for a relatively larger distance between nanoclay platelets, further permitting for the naturally hydrophilic clay to become more hydrophobic [7, 8, 10]. Smectite clays, for instance, have a cation exchange capacity (CEC) of 70–130 meq/100 g [6], where CEC is a measure of the number of positively charged ions that are able to be held by the negatively charged surface on the clay platelets [8].

Due to their nanoscale thickness and longer relative lengths and widths, nanoclays have a high aspect- and surface area-to-volume ratios that lead to an increase of their reactivity [1, 9]. This, along with their abundance in soil [1] and low cost [1], have made nanoclays applicable to numerous areas from sorbents in pollution prevention [7], to environmental remediation [7], wastewater treatment [11, 12], as well as rheological modifiers for oil well drilling fluids [11, 13], paints [11], cosmetics [11], food packaging [14, 15], automotive [16, 17], medical devices [18, 19], and coatings-related industry [20, 21]. A comprehensive list of the numerous organically modified nanoclays currently in use in consumer applications is shown in Table 11.1.

11.3 Nanoclays in Food Packaging Applications

Polymers such as polylactic acid (PLA) [51, 62], polycaprolactone (PCL) [62], methylcellulose, starch, lignin, and poly(vinyl alcohol) have been of interest to replace the petroleum-based, nonbiodegradable packaging materials [23, 51, 62]. However, such polymers do not always have the thermal stability [42], strength [23], or barrier properties [42] of conventional, synthetic polymers, thus making their implementation as effective food packaging materials challenging. Further, gases, such as oxygen, carbon dioxide, water vapor, or ethylene, can penetrate the polymer matrix and diffuse through it [14] in the process, decreasing the matrix quality [14, 30].

Table 11.1 Examples of nanoclays researched along with their organic modifiers, applications, and associated references.

Nanoclay	Organic modifier	Applications	References
Montmorillonite (MMT)	None	Adsorbents in water and wastewater treatment, drilling fluid, paints, cosmetics, coatings, drug delivery	Pluta *et al.* [22], Mondal *et al.* [23], Meera *et al.* [20], Introzzi *et al.* [21], Baek *et al.* [24], Rawat *et al.* [25], Li *et al.* [26], Liu *et al.* [27], Murphy *et al.* [28]
	Octadecyl amine	Food packaging	Barua *et al.* [19], Manikantan *et al.* [29]
	Trimethyl stearyl ammonium	Coatings	Meera *et al.* [20]
Halloysite	None	Coatings, drug delivery, implants, food packaging, composites	Alipoormazandarant *et al.* [30], Sadegh-Hassani and Nafchi [31], Verma *et al.* [32], Vergaro *et al.* [33]
Bentonite	None	Drilling mud, absorbent, groundwater barrier, cosmetics, pharmaceutical	Barua *et al.* [19], Meibian *et al.* [34], Murphy *et al.* [28], Meibian *et al.* [35], Yuwen *et al.* [36]
	3–5 wt% Na_2CO_3		Geh *et al.* [37]
	HCl		Geh *et al.* [37]
	Distearyldimethylammonium chloride		Geh *et al.* [37]
	H_2SO_4		Meibian *et al.* [34], Meibian *et al.* [35]
Quartz	None	Drilling, glass making, foundry sand, electronics, abrasives	Gao *et al.* [38]
Kaolin	None	Coatings, cosmetics, paints, adsorbents in water and wastewater treatment, medical	Gao *et al.* [38]
Amorphous nanosilica particles	None	Cosmetics, food technology, medical	Yoshida *et al.* [39]
	Amine groups		Yoshida *et al.* [39]
	Carboxyl groups		Yoshida *et al.* [39]

(Continued)

Table 11.1 (Continued)

Nanoclay	Organic modifier	Applications	References
Montmorillonite dellite	None	Adsorbents in water and wastewater treatment, drilling fluid, paints, cosmetics, coatings, drug delivery	Janer *et al.* [40]
Cloisite Na^+ (MMT)	None	Adsorbents in water and wastewater treatment, drilling fluid, paints, cosmetics, coatings, drug delivery	Shojaee-Aliabadi *et al.* [41], Rhim *et al.* [42], Houtman *et al.* [43], Maisanaba *et al.* [44], Maisanaba *et al.* [45], Lordan *et al.* [46], Verma *et al.* [32], Maisanaba *et al.* [47], Sharma *et al.* [48]
Cloisite 10A	Dimethyl, benzyl, hydrogenated tallow, quaternary ammonium	Composites, food packaging, paints, coatings	Molinero *et al.* [49]
Cloisite 11B	Benzyl(hydrogenated tallow alkyl)dimethyl	Composites, food packaging, automotive	Dalir *et al.* [17]
Cloisite 15A (MMT)	Dimethyl, dihydrogenated tallow, quaternary ammonium	Composites, drilling fluid, food packaging, medical, automotive	Agarwal *et al.* [13], Krikorian *et al.* [50], Pereira de Abreu *et al.* [14], Plackett *et al.* [51], Dalir *et al.* [17]
Cloisite 20A (MMT)	Dimethyl dihydrogenated tallow quaternary ammonium chloride	Composites, drilling fluid, food packaging, paints, coatings	Agarwal *et al.* [13], Molinero *et al.* [49], Rhim *et al.* [42], Lertwimolnun and Vergnes [52], Lertwimolnun and Vergnes [53], Choi *et al.* [15], Houtman *et al.* [43]
Cloisite 25A (MMT)	Dimethyl, hydrogenated tallow, 2-ethylhexyl quaternary ammonium methyl sulfate	Composites, drilling fluid, medical, food packaging	Agarwal *et al.* [13], Krikorian *et al.* [50], Plackett *et al.* [51]
Cloisite 30B (MMT)	Methyl, tallow, bis-2-hydroxyethyl, quaternary ammonium	Composites, drilling fluid, food packaging, medical, automotive, paints, coatings,	Agarwal *et al.* [13], Krikorian *et al.* [50], Molinero *et al.* [49], Plackett *et al.* [51], Rhim *et al.* [42], Beltrán *et al.* [54], Dalir *et al.* [17], Abreu *et al.* [55], Maisanaba *et al.* [44], Maisanaba *et al.* [45], Sharma *et al.* [48], Sharma *et al.* [56]

Cloisite 93A (MMT)	Methyl dehydrogenated tallow ammonium	Composites, paints, coatings, food packaging	Molinero *et al.* [49], Lordan *et al.* [46]
Montmorillonite dellite (MMTdell 72T, MMTdell 72Ts, MMTdell 67G, MMTdell 67Gs)	Dimethyl dihydrogenated tallow ammonium	Composites, drilling fluids, medical, food packaging	Janer *et al.* [40]
Montmorillonite dellite (MMTdell 43B, MMTdell 43Bs)	Dimethyl benzyl hydrogenated tallow ammonium	Composites, drilling fluids, medical, food packaging	Janer *et al.* [40]
PSAN-MMT	Oligo(styrene-*co*-acrylonitrile)	Drug delivery	Liu *et al.* [27]
Halloysite MP1	None	Coatings, drug delivery, implants, food packaging, composites	Verma *et al.* [32]
Delilite LVF (Bentonite)	None	Composites	Verma *et al.* [32]
Nanomer PGV (Bentonite)	None	Composites, rheological modifier	Verma *et al.* [32]
Clay 1	Quaternary ammonium salt hexadecyltrimethylammonium bromide (HDTA)	Composites, food packaging	Houtman *et al.* [43], Maisanaba *et al.* [57], Jorda-Beneyto *et al.* [58], Maisanaba *et al.* [59], Maisanaba *et al.* [60]
Clay 2	HDTA and acetylcholine chloride (ACO)	Composites, food packaging	Houtman *et al.* [43], Maisanaba *et al.* [57], Jorda-Beneyto *et al.* [58]
Anionic nanoclay	Carbonate	Composites, catalyst, pharmaceuticals, filters	Chung *et al.* [61]
	Chloride	Composites, catalyst, pharmaceuticals, filters	Chung *et al.* [61]
Bentone MA (hectorite clay)	None	Cosmetics, adhesives, paints, cleaners, coatings	Verma *et al.* [32]
ME-100 (Somasif – a synthetic fluoromica clay)	None	Composites	Verma *et al.* [32]

The addition of nanoclays into a polymer matrix at a low silicate content [3, 63] allows for better reinforcement within the polymer plane [50, 64], as well as an increase in its mechanical strength [50, 65], barrier properties [3, 14, 49], UV dispersions [49], and fire resistance [22, 66], and makes it applicable to food packaging industry [14, 15]. As such, nanoclay-enforced polymers were shown to maintain their transparency [14, 30], with further research showing that nanoclay addition into a polymer matrix creates a tortuous path in which the nanoclays are acting as physical barriers to slow down the movement of gases [14] and create a greener route for production and disposal of packaging [14, 30, 42, 67]. For instance, Plackett *et al.* found that nanoclays only caused a slight reduction in light transparency when incorporated into PLA-PCL films [51], with Shojaee-Aliabadi *et al.* showing that the addition of up to 10% weight of nanoclays into polymers can still lead to translucent films when kappa-carrageenan/Cloisite Na^+ (a pristine montmorillonite with Na^+ between the platelets) were used, for instance [41]. However, Rhim *et al.* found that nanoclays significantly reduced the light transmittance when introduced in PLA alone, with a better exfoliation within PLA being observed for Cloisite 20A (a hydrophobic dimethyl dihydrogenated tallow quaternary ammonium chloride nanoclay) [42].

The degree of exfoliation [14] was shown to depend on the organic modifiers being used, as well as the temperature [52], processing time [51], and processing conditions [53]. For example, Pereira de Abreu *et al.* showed that longer processing mixing times generally have a positive influence on the dispersion of nanoclays [14]. Studies also found that between 180 and 200 °C, better exfoliation of Cloisite 20A in polypropylene (PP) was obtained at 180 °C presumably as a result of the polymer stress in this domain temperature [52]. Further, Lertwimolnun and Vergnes showed that exfoliation of Cloisite 20A in PP films increased with decreasing feeding rate from 29 to 4.5 kg/h and increasing the screw speed from 100 to 300 rpm [53].

Other studies showed that the addition of a low percent weight of nanoclays can increase the mechanical properties of polymers [14, 23]. The displayed increase in mechanical strength was shown to help the nanocomposite packaging materials withstand the stresses encountered during handling and transportation of food products [31]. Pereira de Abreu *et al.*, for instance, showed that the addition of Cloisite 15A (nanoclay organically modified with dimethyl, dihydrogenated tallow, quaternary ammonium) into PP increased Young's modulus of the material by 692 MPa relative to the control PP [14] while the addition of halloysite (a 1:1 aluminosilicate) nanoclay to soluble soybean polysaccharide (SSPS) films increased their tensile strength by 4.1 MPa relative to SSPS with no halloysite [30]. Shojaee-Aliabadi *et al.* showed increases in tensile strength of 8.38 MPa when montmorillonite (MMT) was exfoliated in kappa-carrageenan (KC) films relative to KC films without MMT [41]. Sadegh-Hassani and Nafchi showed that tensile strength increased

2.49 MPa for potato starch films containing halloysite nanoclay [31] while Beltrán *et al.* showed that the addition of Cloisite 30B (nanoclay organically modified with methyl, tallow, bis-2-hydroxyethyl, quaternary ammonium) increased elongation at break for PCL [54].

Studies also revealed that addition of nanoclays to polymers such as PLA reduced water vapor permeability (WVP) by 6–33% when compared to control PLA alone [42]. Sadegh-Hassani and Nafchi also showed a decrease in oxygen permeability upon addition of halloysite nanoclay [31], while Alipoormazandarani *et al.* showed that halloysite nanoclay reduced water vapor and oxygen permeability by 56% and 58%, respectively, when incorporated into SSPS [30]. Complementarily, Manikantan *et al.* found that nanoclays added into PP also decreased the WVP with banana chips packaged in 2% nanoclay/PP films showing 22% lower moisture content and banana chips packaged in 4% nanoclay/PP films having 24% lower moisture content, respectively, all relative to the control [29], while Shojaee-Aliabadi *et al.* showed a decrease in WVP by around 78% upon addition of MMT into KC films [41]. Numerous other studies have shown similar results with different nanoclays and polymer matrices [23, 29–31, 41, 54], with further analysis of PLA reinforced with Cloisite 30B also displaying antimicrobial activity against *Listeria monocytogenes* [42] while starch films containing Cloisite 30B decreased microbial growth for *Staphylococcus aureus* and *Escherichia coli* [55].

11.4 Possible Toxicity upon Implementation of Nanoclay in Consumer Applications

The large consumer implementation of nanoclays, especially in food packaging applications as indicated above, has the potential to affect humans at the exposure levels of manufacturing, consumption/usage, and disposal [68, 69]. As such, studies aimed to differentiate toxicity and toxicity-induced mechanisms based on the organic modifier or the size of the nanoclays.

For the organic modifier, for instance, Meibian *et al.* found that activated bentonite particles had a greater cytotoxic response relative to untreated counterparts, indicating that surface characteristics may be playing a large role in mechanisms of toxicity, such as the adsorption capacity, cation exchange, charge interactions, and surface area [34]. Janer *et al.* also observed differences in toxicity based on the organic modifier being used, with the modifier dimethyl benzyl hydrogenated tallow ammonium displaying greater toxic effects relative to the modifier dimethyl dehydrogenated tallow ammonium, for instance [40]. However, Yoshinda *et al.* found that the organically modified silica particles coated with amine or carboxyl groups were less toxic than the unmodified silica particle, as the modified particles reduced the amount of reactive oxygen generated in a human keratinocyte cell line (HaCaT) and a murine hepatocyte cell line (TLR-1), and the amount of DNA damage in HaCaT cells [39].

For the size, Sharma *et al.* showed that samples of Cloisite 30B that had been filtered through a 0.2-μm filter, thus eliminating particles in the micro range, were less cytotoxic than their unfiltered counterparts [48]. However, Janer *et al.* did not observe any differences in cytotoxicity between the small- (100–822 nm) and large- (100–3230 nm) sized pristine MMT particles [40]. Size played a role in the uptake of bentonite particles, with a maximum uptake for particles in the size range 0.4–1.6 μm for the activated bentonite particles and a less selective size range for the unactivated particles [37]. Other *in vitro* studies evaluated the toxicity of nanoclays (both pristine and organically modified nanoclays) and differentiated the observed nanoclay-induced effects based on the exposure levels or the cell type being used, with the majority of studies focusing on understanding toxicity at the consumption level of exposure (i.e., around 48%, 24%, and 28% for consumption exposure, inhalation exposure, and others, respectively) and assuming that nanoclays will eventually migrate out of the polymer matrix into food stocks when used for food packaging applications [43–46, 57, 58]. Detailed below are several investigations aim to identify the deleterious effects that nanoclays can have and possibly propose means to reduce the logistical burden associated with developing meaningful risk assessment strategies for evaluating their potential and feasibility for implementation in food packaging industry.

11.4.1 *In Vitro* Studies Reveal the Potential of Nanoclay to Induce Changes in Cellular Viability

Studies proposed that the small size and platelet morphology of nanoclays have the potential to allow for their inhalation and deposition in the bronchial or alveolar regions of the lung [68, 70–72]. In support of this hypothesis, Verma *et al.* investigated the inhalation toxicity of both platelet- and tubular-shaped nanoclays using *in vitro* models, that is, human alveolar epithelial cells (A549), and a dosage ranging from 1 to 250 μg/ml [32]. Analyses showed that tubular nanoclays did not induce toxicity until doses of 250 μg/ml, which was in contrast with the platelet nanoclays that induced toxicity at only 25 μg/ml and 24 h exposure [32]. Further studies by the same authors showed that the pristine nanoclay, that is, Cloisite Na^+ was internalized by the exposed cells and accumulated in their perinuclear region when compared to control cells (Figure 11.1a). Figure 11.1b, c shows nanoclays localized at the perinuclear region and labeled in light gray using rhodamine dyes and their localization is indicated with white arrowheads [32]. All cells were also counterstained with Alexa Fluor 488 phalloidin (gray) for cellular cytoskeleton and Hoechst (dark gray) for nucleus identification. Analysis showed that the uptake was time dependent with increases in the amount of Cloisite Na^+ occurring

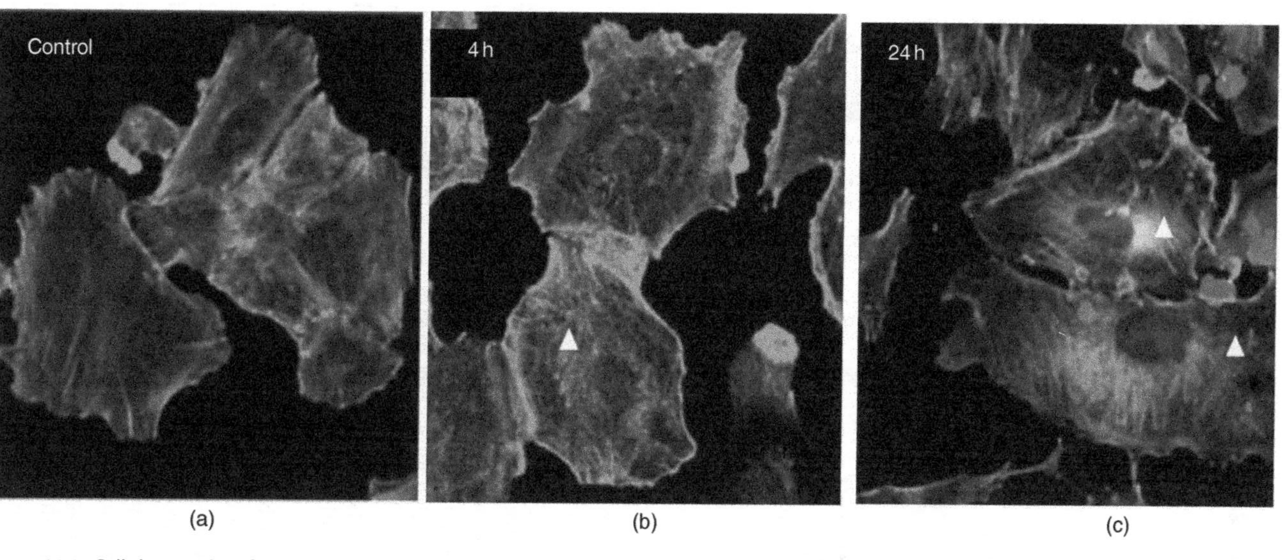

Figure 11.1 Cellular uptake of nanoclays by A549 cells. A549 cells grown onto eight-well chamber slides were (a) fixed (control) or (b) incubated with rhodamine (light gray) labeled Cloisite Na^+ (25 μg/ mL) for 4 h, (c) or 24 h. Cells were also counterstained with Alexa Fluor 488 phalloidin (gray) for their cytoskeletal organization identification and Hoechst (dark gray) for their nuclear localization. Intracellular accumulation of the nanoclays was detected by confocal microscopy; the representative images show the nanoclays localization mostly concurrent with the nuclear regions. Verma *et al*. 2012 [32]. Reproduced with permission of Springer. HepG2 cells uptake (d) 0–62.5 μg/ml Cloisite Na^+ labeled with Neutral red uptake (NR) or (e) 0–500 μg/ml Cloisite 30B labeled with NR. All values are expressed as mean ± SD. *Significantly different from control ($p \leq 0.05$). Maisanaba *et al*. 2013 [44]. Reproduced with permission of Elsevier. Comet assay results of (f) Caco-2 cells after 24 and 48 h of exposure to 8.5, 17, or 34 μg/ml Clay 2; (g) HepG2 cells after 24 and 48 h of exposure to 22, 44, and 88 μg/ml Clay 2. Results from three independent experiments with two replicates/experiment. All values are expressed as mean ± S.D. *Significantly different from control ($p < 0.05$). **Significantly different from control ($p \leq 0.01$). Houtman *et al*. 2014 [43]. Reproduced with permission of Elsevier.

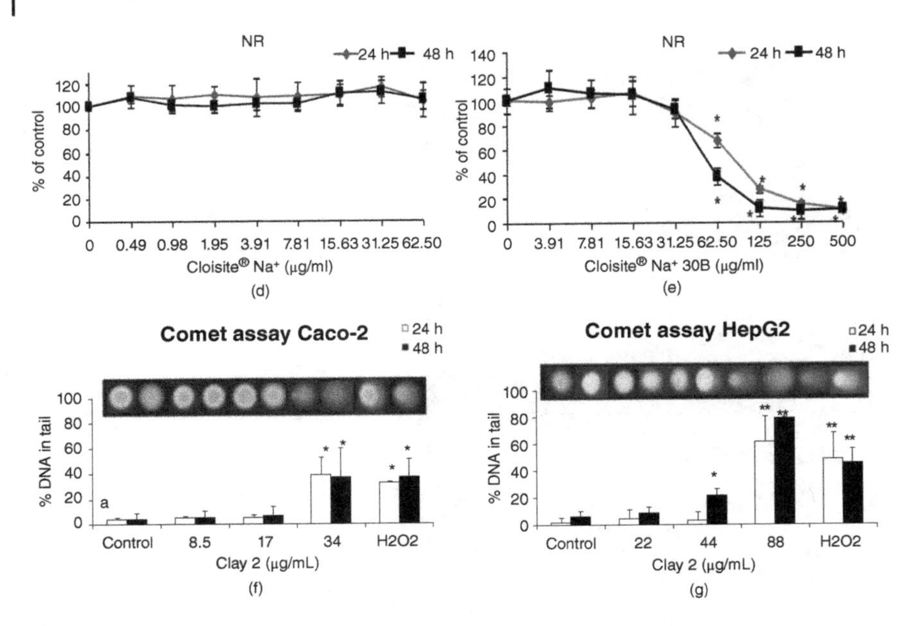

Figure 11.1 (*Continued*)

accumulatively over the 24 h exposure time. Geh *et al.* also observed uptake of bentonite particles into human lung fibroblasts (IMR90 cells) over 24 h exposure at a dose of 10 µg/cm^2, with uptake increasing when the bentonite particles were activated with quartz (5–6%) [37]. Such nanoclays were also more cytotoxic when compared to nonactivated or lower quartz content particles with the observed cytotoxicity being attributed to the lysis of the cell membrane upon translocation of the activated nanoclays [37].

Complementarily, Janer *et al.* found that organically modified nanoclays induced a greater toxicity than the pristine nanoclays [40]. For instance, when A549 cells were treated with pristine clay or nanoclays modified with dimethyl dihydrogenated tallow ammonium or dimethyl benzyl hydrogenated tallow ammonium, respectively (doses <500 µg/ml), a greater loss in cellular viability was experienced even though both the nanoclays were taken up by the cell during the 72 h window of incubation, with the internalization being more prevalent for the pristine clay [40]. Other studies showed that quartz and kaolin dust decreased pulmonary alveolar macrophage viability within 1 day of exposure at doses of 40 and 20 µg/cm^2, respectively [38]; however, quartz and kaolin treated with the surfactant dipalmitoyl phosphatidylcholine (DPPC) did not induce significant decreases in viability even after 3 or 5 days of treatment [38]. Lastly, Chung *et al.* examined both the short- and long-term toxicity of anionic nanoclays (carbonate and chloride forms) on A549 cells and found no toxic effects at high exposure doses of 1000 µg/ml [61]; however, the anionic nanoclay seemed to have inhibited colony formation after 10 days of exposure to doses ranging from 250 to 500 μg/ml [61].

However, due to the nanoclay incorporation into nanocomposites for food packaging applications [14, 15, 73], it is likely that nanoclays will not only come into contact with humans through the route of inhalation but also through the route of ingestion. Studies aimed to examine the toxicity of nanoclays upon exposure to cell lines isolated or belonging to the ingestion track have shown toxic effects induced by Cloisite 30B on human hepatocellular carcinoma epithelial cell line (HepG2), for instance, with significant decreases in both cellular growing rate and viability (Figure 11.1d,e) [44]. In particular, analyses showed that significant decreases in uptake occurred starting at the dose of 62.5 to 500 μg/ml upon both 24 and 48 h cellular exposure to Cloisite 30B (Figure 11.1e). However, analyses on pristine nanoclay Cloisite Na^+ did not reveal any cytotoxic effects on the HepG2 cells up to a dose of 62.5 μg/ml upon the same time of exposure (Figure 11.1d) [44]. Further, Cloisite Na^+ again did not show any cytotoxicity after 24 h of exposure to cells in doses up to 62.5 μg/ml [47]. Similarly, Sharma *et al.* found that Cloisite 30B showed increased toxicity relative to Cloisite Na^+ (40% toxicity in human colon carcinoma cells-Caco-2) after 24 h exposure to a dose of 226 μg/ml [48], with Cloisite 30B inducing greater toxicity than Cloisite Na^+ and also causing both time- and dose-dependent decreases in protein content (PC), 3-(4,5-dimethylthiazol-2-yl)-5-(3-carboxymethoxyphenyl)-2-(4-sulfophenyl)-2*H*-tetrazolium (MTS) reduction [45], all standards for the *in vitro* evaluation of cytotoxicity. A significant toxic effect for Cloisite Na^+ was only obtained using the MTS assay after 48 h of exposure to doses of 31.25 μg/ml, all relative to control cells [45].

Janer *et al.* complemented the above studies showing that the organically modified nanoclays were more toxic than the pristine ones when a colorectal carcinoma cell line (HCT116) and HepG2 cells were assessed, with all organoclays displaying IC_{50}'s at doses below 25 μg/ml, whereas the IC_{50}'s of pristine clays were above 100 μg/ml. Houtman *et al.* treated HepG2 and Caco-2 cells with three other types of clays used in packaging, namely, Cloisite 20A, Clay 1 (modified with quaternary ammonium salt hexadecyltrimethyl-ammonium bromide (HDTA)), and Clay 2 (modified with HDTA and acetylcholine chloride (ACO)), respectively [43]. Analyses showed that Clay 2 caused significant decreases in viability in both HepG2 cells and Caco-2 cells, with more prevalent effects being observed in Caco-2 cells [43]. In another study, Clay 1 was shown to cause significant reductions for HepG2 cells at the same dosage [58]. Both Cloisite Na^+ and Cloisite 93A (modified with methyl dehydrogenated tallow ammonium) caused significant dose-dependent decreases in HepG2 cell viability after 24 h starting from a dose of 1 μg/ml [46]. Similarly, MMT caused significant decreases in the viability of human normal intestinal cells (INT-407) in a dose- and time-dependent manner at doses of 100 μg/ml or above, all over

24–72 h [24]. For instance, Rawat *et al.* found that MMT induced 50% cytotoxicity in the human embryonic kidney cell line at a dose as low as 0.005 μg/ml [25]. When examining the effects on HepG2 or Caco-2 cells after 24 and 48 h of exposure to extracts of PLA-Clay 1 and PLA-Clay 2 nanocomposites, for instance, Maisanaba *et al.* did not observe any cytotoxic effects up to 2.5–100% extracts [57]. Li *et al.* showed that nanosilicate platelets (NSP) originated from MMT produced significant decreases in viability of Chinese Hamster Ovary (CHO) cells after 24 h of treatment with doses from 62.5 to 1000 μg/ml [26], while Meibian *et al.* treated a human B lymphocyte cell line (HMy2.CIR) with active and native bentonite particles and showed that both the active (treated with 10–15% H_2SO_4) and native bentonite particles resulted in dramatic decreases in cellular viability within only 4 h of exposure, with 1000 μg/ml exposure dose causing almost complete loss of cellular viability [34]. Mouse embryonic fibroblast (NIH 3T3) and Human Embryonic Kidney 293 (HEK) cells however did not show a loss in cellular viability when treated with MMT or MMT modified with oligo(styrene-*co*-acrylonitrile) (PSAN-MMT) until exposure doses of 1 g/l, with MMT showing a greater loss in viability relative to PSAN-MMT [27]. Complementarily, human breast cancer cell line (MCF-7) and human cervical cancer cell line (HeLa) displayed decreases in viability after concentrations of 75 μg/ml of the tubular nanoclay, halloysite [33], and halloysite coated with amino-propyl-triethoxysilane (APTES) produced similar trends in toxicity relative to its uncoated counterpart [33].

11.4.2 Proposed Mechanisms of Toxicity for the *In Vitro* Cellular Studies

Due to the observed changes in cellular viability upon exposure to different types of nanoclays, complementary studies aimed to determine the nanoclay-induced mechanisms of toxicity. For this, research has considered the effects of nanoclays exposure to cell morphology, structure, cell signaling, cell–cell and cell–substrate interactions, as well as cell progression through cell cycle and appropriate cellular proliferation since these aspects are known to be influencing cell viability and ultimately determine cellular fate [74, 75]. Further, the evaluations of such changes were meant to provide the means for assessing and differentiating the nanoclay-induced cyto- and/or genotoxic mechanisms in a clay type-dependent manner.

Analyses showed that HepG2 cells had dilated endomembranes after treatment with Cloisite 30B [44], while Caco-2 cells treated with 20 and 40 μg/ml of the same nanoclay displayed changes in cell morphology, intense vacuolization, and euchromatic irregular nuclei [45].

Nanoclays exposure led to membrane damage and changes in cellular structure, likely due to their induced charge interactions with the membrane and resulting membrane lysing ability [28]. For instance, Murphy *et al.* determined that primary murine spinal cord neurons were lysed after 60 min of incubation with 0.1 mg/ml bentonite or MMT, whereas differentiated N1E-115 cells did not appear to be lysed or undergo any morphological damage [28]. Studies by Meibian *et al.* found that both active and native bentonite particles induced significant lactate dehydrogenase (LDH) leakage in human B lymphocyte cells after 4 h of exposure to doses of 60 and 120 μg/ml, respectively [34]. Li *et al.* also observed significant membrane damage in CHO cells after cell treatment with NSP particles in 62.5–1000 μg/ml doses [26]. Both Cloisite Na^+ and Cloisite 93A caused significant increases in LDH release in HepG2 cells, with the organically modified clay inducing the greater response [46]. However, Baek *et al.* only observed significant LDH release in INT-407 cells at the top dose of 1000 μg/ml MMT after 48 and 72 h [24]. The organically modified MMT, that is, PSAN-MMT, showed lower LDH release relative to MMT [27].

Combined with effects on cellular morphology and structure, changes in mitochondrial function were proposed as another viable mean to explain nanoclay-induced toxicity since it is known that the mitochondria regulates redox signaling to cellular cytosol and nucleus [76, 77]. Studies by Maisanaba *et al.* have showed that the mitochondria of cells treated with nanoclays exhibited both matrix and inner membrane degradation [45]. However, when reactive oxygen species (ROS) were assessed in the HepG2 cell line treated with Cloisite Na^+ or Cloisite 30B up to a dose of 88 μg/ml, for instance, no significant increases were recorded [44]. Further both the HepG2 and Caco-2 cells showed no ROS generation when treated with Clay 2 up to a dose of 88 μg/ml for 48 h [43]. However, Lordan *et al.* found that both Cloisite Na^+ and Cloisite 93A produced ROS in HepG2 cells, with Cloisite Na^+ inducing more ROS than the organically modified counterpart and with the significant increases being recorded for smaller doses of 50 μg/ml over 24 h of exposure [46]. This is in contrast with studies performed with both active and native bentonite, which showed significant ROS levels upon only 30 min of exposure to human B lymphocyte cells [34]. ROS was also generated in Caco-2 cells upon treatment with 40 μg/ml of Cloisite 30B [45], while MMT generated ROS in INT-407 cells at concentrations above 50 μg/ml after 48–72 h [24]. One possible explanation for the observed dose and nanoclay-dependent ROS generation was that the toxic effects of nanoclays could be potentially circumvented by changes in the endogenous antioxidant glutathione (GSH), a known regulator of the intracellular redox balance [78]. Maisanaba *et al.* for instance found that Cloisite 30B caused significant decreases in GSH cellular concentration for both HepG2 cells and Caco-2 cells [44, 45], with the GSH concentration being dependent on both the type of clay used (e.g., Clay 2 did not affect the GSH content of Caco-2 cells) and the dose of nanoclay [43].

With changes in cellular structure and energetic activity being known to influence cell cycle and overall fate [75, 79], the role of nanoclays to induce genotoxicity was also investigated. Studies of NIH 3T3 cells exposed to 1 g/l MMT for 24 h identified nuclei fragmentation and condensed chromatin;

however, the changes were minimal for cells exposed to organically modified PSAN-MMT [27]. Maisanaba *et al.* also showed that Cloisite 30B caused significant time-dependent DNA breaks in HepG2 cell [44], while Houtman *et al.* showed that Clay 2 induced DNA changes in both HepG2 and Caco-2 cells, respectively, with Clay 2 having a slightly greater effect on Caco-2 cells (Figure 11.1f,g) [43]. However, exposures to Cloisite 20A and Clay 1 did not induce DNA strand breaks in either Caco-2 or HepG2 cells after 48 h and concentrations up to 62.5 and 8 μg/ml, respectively [43]. In addition, neither of the extracts of the PLA-Clay 1 or PLA-Clay 2 nanocomposites induced any genotoxicity after 24 or 48 h of exposure [57]. Studies by Sharma *et al.* also complemented previous findings and showed DNA damage in a nanoclay dose-dependent manner when Caco-2 cells were treated with Cloisite 30B, with the two highest doses of 113 and 170 μg/ml, respectively, being significantly different from the control (not exposed cells) after 24 h of exposure [48]. However, no change was recorded for the Caco-2 cells treated with Cloisite Na^+ [48] or Cloisite 30B [45]. However, Cloisite Na^+ caused an increase in micronuclei frequency in HepG2 cells after 24 h at exposure levels of 62.5 μg/ml [47]. Similarly, when CHO cells were treated with NSPs from MMT, DNA dose-dependent damage was observed within the exposure dose range 62.5–1000 μg/ml [26]. An increase in DNA damage was observed when human B lymphoblast cells were treated with active or native bentonite particles from doses of 120 μg/ml for 24 h, 60 μg/ml for 48 h, or 30 μg/ml over 72 h, with active bentonites showing a greater effect, and with both particles causing significant increases in micronucleus frequency [35]. Untreated quartz particles induced significant DNA damage in rat pulmonary alveolar macrophages within 1 day of exposure at a dose of 20 $\mu g/cm^2$, whereas quartz/DPPC induced damage after 3 days at double that dose [38]. In addition, untreated kaolin induced DNA damage after 1 day at an exposure dose of 40 $\mu g/cm^2$ and kaolin/DPPC inducing DNA damage after 5 days at a dose of 40 $\mu g/cm^2$ [38]. Lastly, Janer *et al.* observed a slight increase in caspase 3/7 for HepG2 cells exposed to 100 μg/ml nanoclay [40]. Caspases 3 and 7 are cysteine asparate proteases with similar structures [80, 81] that control apoptotic pathways. Briefly, intrinsic apoptosis is mediated by mitochondrial outer membrane permeabilization with cytosol release of proapoptotic proteins such as cytochrome c activating caspase-9 [82, 83]. The resulting activated caspase-9 then cleaves and activates caspase-3 and 7, respectively, to initiate the degradation of cellular structures and cell detachment, eventually leading to cell death [82].

Maisanaba *et al.* proposed to further investigate genotoxicity of nanoclays by evaluating the effect of Cloisite Na^+ on the regulation of genes in HepG2 cells [47]. Cloisite Na^+ was found to deregulate genes associated with cellular metabolism, immediate-early response/signaling, DNA damage response, oxidative stress, and apoptosis/survival [47]. Specifically, four out of the five genes studied for metabolism and two DNA damage responsive genes were found to be upregulated at the tested concentrations of 6.25 or 62.5 μg/ml of Cloisite Na^+ after 24 h of exposure. Catalase, an oxidative stress responsive gene, was downregulated however at both concentrations used. Complementarily, both apoptosis responsive genes and early response/signaling genes were also affected after 24 h, with early response/signaling also having a gene (JUNB) affected at 4 h at the dose of 62.5 μg/ml [47].

11.4.3 *In Vivo* Evaluation of Nanoclay Toxicity

While the mechanism of toxicity is still not completely understood, the deleterious effects observed upon exposure to both pristine and organically modified nanoclays from the *in vitro* studies have prompted increased interest for their *in vivo* evaluations at a consumer level of exposure. Herein the consumer level of exposure is defined as exposures to nanoclays when in use by the public who buys the product containing embedded nanoclay. Analyses by Sharma *et al.* found that Wistar rats orally exposed for two times to Cloisite 30B ranging from 250 to 1000 mg/kg of their body weight, with 24 h apart the exposures, did not induce changes in their organs nor DNA strand breaks in the cells isolated from the colon, liver, or kidney (Figure 11.2a) [56]. Specifically, analyses showed that the Wistar rats treated with Cloisite 30B suspended in water or cell culture media had no significant differences in % tail DNA relative to the controls, even up to a dose of 1000 mg/kg body weight for the liver, kidney, or colon. However, the tracer element aluminum was found in the rat's feces, thus indicating there was no absorption of the clay from the gastrointestinal tract. Complementarily, results by Maisanaba *et al.* administering Clay 1 over 90 days at 40 mg/kg/day showed that rats underwent an adaptive response in result of an increased oxidative stress [59]. Specifically, significant increases in catalase (CAT; responsible in maintaining ROS levels) activity in the kidney (Figure 11.2b) as well as changes in proteins expressions level of CAT in the kidney of the rat were observed, which was in contrary to no significant effects on the antioxidant enzymes, superoxide dismutase (SOD), glutathione peroxidase (GPx), and gluthatione S-transferase (GST) responsible for maintaining ROS levels, respectively (Figure 11.2c) [59].

Maisanaba *et al.* also tested the migrant extract of a PLA-Clay 1 (4%) nanocomposite on the same animal models over an exposure of 90 days and showed no significant effects on their biomarkers, which included the oxidative stress biomarkers of enzymes gluthatione S-transferase (GSH)/glutathione disulfide (GSSG) ratios, lipid peroxidation via thiobarbituric acid (TBA), and antioxidant enzyme activities (CAT, SOD, GPx, and GST) [60]. Further, unlike the previous study dealing with Clay 1, the migrant extract did not cause any changes in CAT activity or in genetic and protein expressions of CAT [59, 60].

There were no major differences observed in the histopathology of the control versus the exposed groups to the migrant extract of PLA-Clay 1

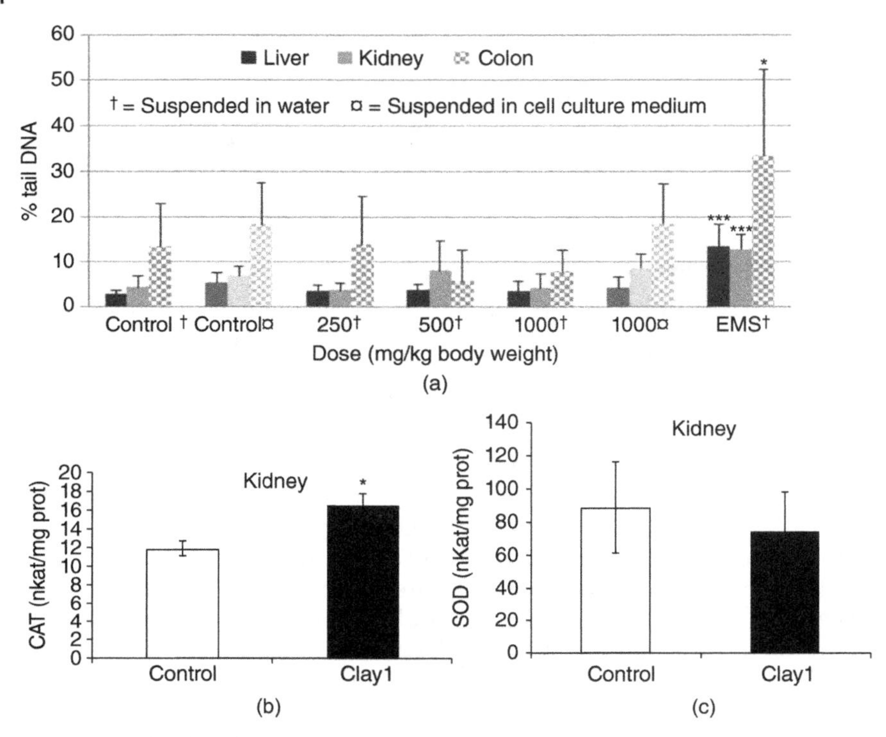

Figure 11.2 (a) Comet assay performed on Wistar rats ($n = 6$) exposed to Cloisite 30B suspended in water (being administered at 250, 500, and 1000 mg/kg body weight of rat) or cell-culture medium (being administered at 1000 mg/kg body weight of rat). Ethylmethane sulfonate (EMS) suspended in water was the positive control. For the experiments, six rats were exposed to 250 mg/kg body weight. Data from liver, kidney, and colon cells of the EMS-exposed group were statistically significantly different ($p < 0.001$, $p < 0.001$: *** and $p < 0.05$: *), respectively, from the values in the corresponding control group (two-tailed, unpaired *t*-test). Internal standards: positive controls (Caco-2 cells exposed to 0.05% ethylmethane sulfonate):16 slides, % tail DNA, mean ± S.D., 21.6 ± 6.6; negative controls (untreated Caco-2 cells): 16 slides, % tail DNA, mean ± S.D., 1.8 ± 0.6. Sharma *et al.* 2014 [56]. Reproduced with permission of Elsevier. (b) Catalase (CAT) and (c) superoxide dismutase (SOD) activities (nKat/mg protein) in kidney of rat exposed to Clay 1. The values are expressed as mean ± SD ($n = 10$). The levels observed are significant at *$p < 0.05$ in comparison to control group values. Maisanaba *et al.* 2014 [59]. Reproduced with permission of Taylor & Francis.

(Figure 11.3). Specifically, Figure 11.3a,b displays the unexposed rat liver and kidney tissues, respectively, stained with hematoxylin and eosin (HE) [60]. Figure 11.3c,d displays the exposed rat liver and kidney tissues, with analyses showing that the liver tissue displayed hepatocytes similar to that of the control and with the kidney tissue displaying a normal parenchyma and normal proximal convoluted tubules (Pct) and distal convoluted tubules (Dct) relative

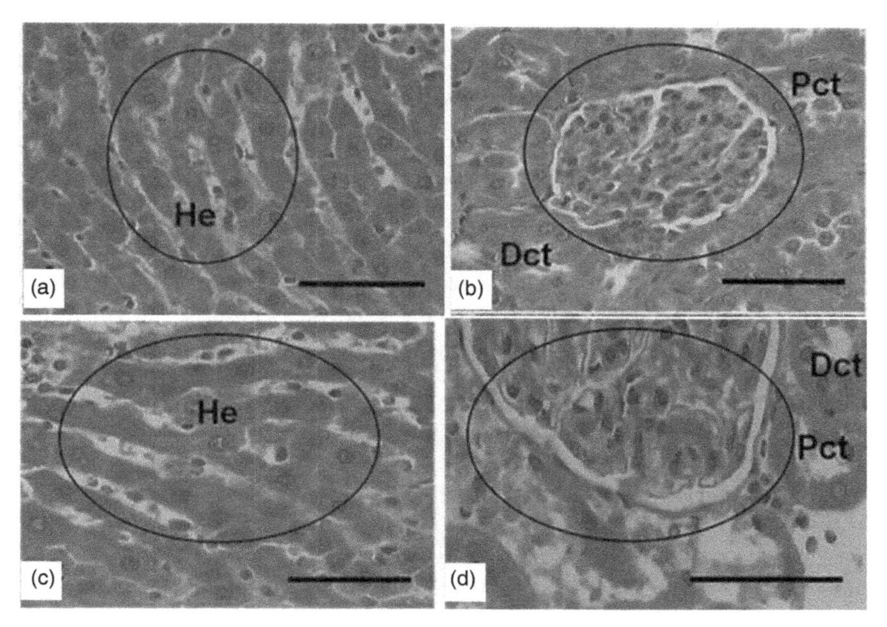

Figure 11.3 Histopathological evaluations of liver of Wistar rats exposed to a PLA-Clay1 extract as beverage for 90 days. (a) He-stained liver section and (b) He-stained kidney sections. Bars, 100 μm. (a,b) Control rats. (a) Liver parenchyma with hepatocytes with normal morphology, central nuclei, and light cytoplasm (He), organized in hepatic cords (circle). (b) Normal structure of kidney parenchyma with glomerulus (circle), proximal convoluted tubules (Pct), and distal convoluted tubules (Dct). (c,d) Exposed rats. (c) Liver parenchyma with hepatocytes with normal morphology, central nuclei, and light cytoplasm (He), organized in hepatic cords (circle). (d) Normal structure of kidney parenchyma with glomerulus (circle), proximal convoluted tubules (Pct), and distal convoluted tubules (Dct). Maisanaba *et al.* 2014 [59]. Reproduced with permission of Taylor & Francis.

to the control group [60]. Li *et al.* also evaluated the LD_{50} of NSP particles being fed to Sprague–Dawley rats at doses of 1500, 3000, and 5700 mg/kg over 14 days and showed no acute oral toxicity of these nanoclays [26], while results by Baek *et al.* obtained in ICR mice exposed to MMT particles orally in single administered doses between 5 and 1000 mg/kg showed that after 14 days there was no significant accumulation of nanoclay in any specific organs, and the LD_{50} was estimated to be over 1000 mg/kg [24].

11.5 Conclusion and Outlook

In the past decades, research aimed to answer critical questions related to the properties and characteristics of nanoclays that could allow for enhancing their consumer-related utility at minimum consumer toxicological risks. However, even though extensive body of evidence from *in vitro* studies supports that nanoclay could be toxic at the exposure level of consumption and inhalation, it is still unclear as to what extent they could potentially affect humans. The isolated study performed by Yuwen *et al.*, for instance, showed that organic bentonite particles could affect workers in two different factories producing such nanoclays. Specifically, group I was exposed to high concentration (around $13\,mg/m^3$), while group II was exposed to moderate concentration (around $8\,mg/m^3$) of bentonites [36]. Preliminary analyses performed in the study identified that group I had a higher frequency of micronuclei, nucleus buds, micronucleated cells, nucleoplasmic bridges, apoptotic cell rate, and necrotic cell rate relative to group II and the control (unexposed individuals) in all the isolated human lymphocytes. Further, group II had higher frequencies of all the above parameters relative to the control group, indicating that genetic damage can occur. Further, both groups showed an increase in lipid peroxidation, responsible for cell damage due to oxidation of lipids in cell membranes, all relative to the control groups, with the age factor further accentuating such correlations.

We propose that circumventing strategies to limit toxicological risks of nanoclays should consider that their toxicity is a combination of both cyto- and genotoxic effects (Scheme 11.1), where effect differentiation is based on the cell and nanoclay type being studied, as well as on the nanoclay physical and chemical characteristics (size, organic modifier, etc.). Such effects could include not only ROS generation or cellular membrane degradation but also possible apoptosis activation through mechanisms involving key proteins responsible for cell cycle or cell proliferation (caspase family), as well as nuclear fragmentation and DNA degradation. For risk prevention, for instance, one could envision controlling the synergistically induced toxicological effects by reducing the nanoclay-induced generation of ROS through direct activation of nanoclay degradation upon their cellular uptake. As such, if the potential of nanoclays implementation in user-directed applications is to be fully reached at the minimum human and environmental logistical burden, both the development of functionalization strategies that allow for such activation strategy to occur, as well as the confirmative body of evidence to demonstrate the feasibility of the activation need to be established. A technical approach in which the addition of user-tailored copolymers to the surface of the nanoclay would facilitate both their cellular uptake as well as their cellular-based autophagy, that is, the lysosome-based degradation could be implemented before consumer integration. With degradative pathways being activated at minimum changes of the lysosome-encapsulated enzymes, the continuous cellular degradation of any uptaken nanoclays will then take place to maintain cellular homeostasis and reduce any cellular toxicity. The underlying functionalization techniques that would ensure such a greener route to nanoclay

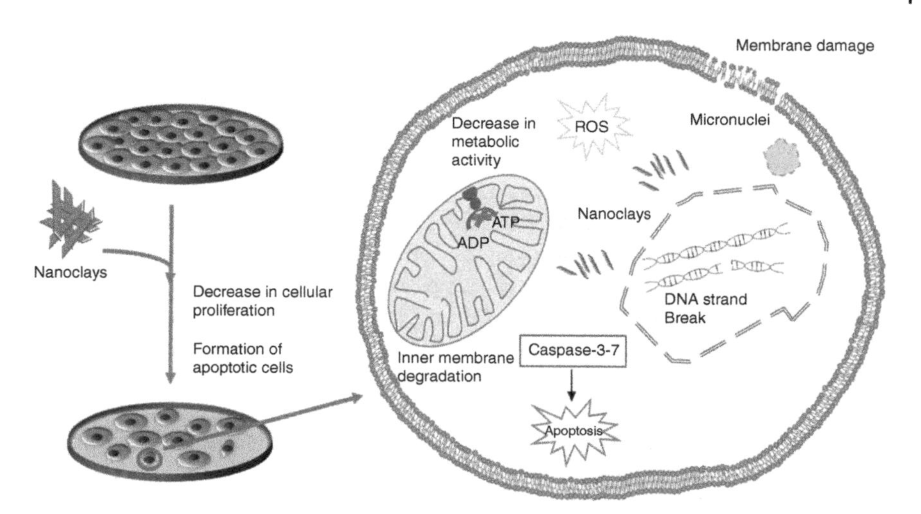

Scheme 11.1 Schematic representation of the proposed mechanisms of toxicity induced by cellular exposure to nanoclays. *In vitro* exposure of cells to nanoclays with different physicochemical properties undergo changes in their structure and functions, with such changes being directly correlated with cellular morphology or proliferation rates controlled at the genetic level.

degradation as well as the lysosome-based degradative pathways will still need to be elucidated. However, until such means are achieved, the "no harm" policy for the worker implementing nanoclays in food packaging or derived products can only account for a proper system of safety measures to be implemented. They could impose limiting the amount of any airborne particles in workplaces or developing personal protective equipment such as respirators capable of removing nanoclays in a reliable and timely fashion, and with a high efficiency. Lastly, high-throughput screening tests to quickly gain an idea of the toxicity of the numerous types of nanoclays in a time efficient manner could be developed and implemented to allow for realistic human and environmentally relevant concentrations rather than excessively high concentrations to be determined and assessed for toxicity of nanoclay risk mitigation.

Acknowledgments

This work was supported by National Science Foundation (NSF) grant EPS-1003907, NSF 1434503, and the National Institute of Health (NIH; R01-ES022968). Alixandra Wagner acknowledges support through the Provost Office.

References

1 Floody, M.C., Theng, B.K.G., Reyes, P., and Mora, M.L. (2009) Natural nanoclays: applications and future trends-a Chilean perspective. *Clay Minerals*, **44**, 161–176.

2 Calabi, M., Jara, A., Bendall, J. *et al.* (2010) *Structural Characterization of Natural Nanomaterials: Potential Use to Increase the Phosphorus Mineralization*, vol. **1–6**, 19th World Congress of Soil Science, Soil Solutions for a Changing World, Brisbane, Australia, pp. 29–32.

3 Patel, H.A., Somani, R.S., Bajaj, H.C., and Jasra, R.V. (2006) Nanoclays for polymer nanocomposites, paints, inks, greases and cosmetics formulations, drug delivery vehicle and waste water treatment. *Bulletin of Material Science*, **29** (**2**), 133–145.

4 Ray, S.S. and Okamoto, M. (2003) Polymer/layered silicate nanocomposites: a review from preparation to processing. *Progress in Polymer Science*, **28**, 1539–1641.

5 Liu, G., Wu, S., van de Ven, M. *et al.* (2010) Characterization of organic surfacant on montmorillonite nanoclay to be used in bitumen. *Journal of Materials in Civil Engineering*, **22**, 794–799.

6 Odom, I.E. (1984) Smectite clay minerals: properties and uses. *Philosophical Transactions of the Royal Society of London. Series A: Mathematical and Physical Sciences*, **311**, 391–409.

7 Xi, Y., Ding, Z., He, H., and Frost, R.L. (2004) Structure of organoclays-an X-ray diffraction and thermogravimetric analysis study. *Journal of Colloid and Interface Science*, **277**, 116–120.

8 Udon, F. (2008) Clays, nanoclays, and montmorillonite minerals. *Metallurgical and Materials Transactions A*, **39A**, 2804–2814.

9 Lewis, D.R. (1955) Ion exchange reactions of clays. *Proceedings of First National Conference on Clays and Clay Technology*, **169**, 54–69.

10 Singla, P., Mehta, R., and Upadhyay, S.N. (2012) Clay modification by the use of organic cations. *Green and Sustainable Chemistry*, **2**, 21–25.

11 Beall, G.W. and Goss, M. (2004) Self-assembly of organic molecules on montmorillonite. *Applied Clay Science*, **27**, 179–186.

12 Beall, G.W. (2003) The use of organo-clays in water treatment. *Applied Clay Science*, **24**, 11–20.

13 Agarwal, S., Tran, P., Soong, Y. *et al.* (2014) Flow behavior of nanoparticles stabilized drilling fluids and effect of high temperature aging. *Journal of Petroleum Science and Engineering*, **117**, 15–27.

14 Pereira de Abreu, D.A., Paseiro Losada, P., Angulo, I., and Cruz, J.M. (2007) Development of new polyolefin films with nanoclays for application in food packaging. *European Polymer Journal*, **43**, 2229–2243.

15 Choi, R., Cheigh, C., Lee, S., and Chung, M. (2011) Preparation and Properties of polypropylene/clay nanocomposites for food packaging. *Journal of Food Science*, **76**, 62–67.
16 Okada, A. and Usuki, A. (1995) The chemistry of polymer-clay hybrids. *Material Science and Engineering: C*, **3**, 109–115.
17 Dalir, H., Farahani, R.D., Nhim, V. *et al.* (2012) Preparation of highly exfoliated polyester-clay nanocomposites: process-property correlations. *Langmuir*, **28**, 791–803.
18 Sahoo, R., Sahoo, S., and Nayak, P.L. (2013) Synthesis and characterization of gelatin-chitosan nanocomposite to explore the possible use as drug delivery vehicle. *European Scientific Journal*, **9**, 135–141.
19 Barua, S., Dutta, N., Karmakar, S. *et al.* (2014) Biocompatible high performance hyperbranched epoxy/clay nanocomposite as an implantable material. *Biomedical Materials*, **9**, 1–14.
20 Meera, K.M.S., Sankar, R.M., Murali, A. *et al.* (2012) Sol–gel network silica/modified montmorillonite clay hybrid nanocomposites for hydrophobic surface coatings. *Colloids and Surfaces B: Biointerfaces*, **90**, 204–210.
21 Introzzi, L., Blomfeldt, T.O.J., Trabattoni, S. *et al.* (2012) Ultrasound-assisted pullulan/montmorillonite bionanocomposite coating with high oxygen barrier properties. *Langmuir*, **28**, 11206–11214.
22 Pluta, M., Galeski, A., Alexandra, M. *et al.* (2002) Polylactide/montmorillonite nanocomposites and microcomposites prepared by melt blending: structure and some physical properties. *Journal of Applied Polymer Science*, **86**, 1497–1506.
23 Mondal, D., Bhowmick, B., Mollick, M.M.R. *et al.* (2013) Effect of clay concentration on morphology and properties of hydroxypropylmethylcellulose films. *Carbohydrate Polymers*, **96**, 57–63.
24 Baek, M., Lee, J., and Choi, S. (2012) Toxicological effects of cationic clay, montmorillonite in vitro and in vivo. *Molecular and Cellular Toxicology*, **8**, 95–101.
25 Rawat, K., Agarwal, S., Tyagi, A. *et al.* (2014) Aspect ratio dependent cytotoxicity and antimcrobial properties of nanoclay. *Applied Biochemistry and Biotechnology*, **174**, 936–944.
26 Li, P., Wei, J., Chiu, Y. *et al.* (2010) Evaluation on cytotoxicity and genotoxicity of the exfoliated silicate nanoclay. *Applied Materials & Interfaces*, **2**, 1608–1613.
27 Liu, Q., Liu, Y., Xiang, S. *et al.* (2011) Apoptosis and cytotoxicity of oligo(styrene-*co*-acrylonitrile)-modified montmorillonite. *Applied Clay Science*, **51**, 214–219.
28 Murphy, E.J., Roberts, E., Anderson, D.K., and Horrocks, L.A. (1993) Cytotoxicity of aluminum silicates in primary neuronal cultures. *Neuroscience*, **57**, 483–490.

29 Manikantan, M.R., Sharma, R., Kasturi, R., and Varadharaju, N. (2014) Storage stability of banana chips in polypropylene based nanocomposite packaging films. *Journal of Food Science and Technology*, **51** (**11**), 2990–3001.

30 Alipoormazandarani, N., Ghazihoseini, S., and Nafchi, A.M. (2015) Preparation and characterization of novel bionanocomposite based on soluble soybean polysaccharide and halloysite nanoclay. *Carbohydrate Polymers*, **134**, 745–751.

31 Sadegh-Hassani, F. and Nafchi, A.M. (2014) Preparation and characterization of bionanocomposite films based on potato starch/halloysite nanoclay. *International Journal of Biological Macromolecules*, **67**, 458–462.

32 Verma, N.K., Moore, E., Blau, W. *et al.* (2012) Cytotoxicity evaluation of nanoclays in human epithelial cell line A549 using high content screening and real-time impedance analysis. *Journal of Nanoparticle Research*, **14**, 1–11.

33 Vergaro, V., Abdullayev, E., Lvov, Y.M. *et al.* (2010) Cytocompatibility and uptake of halloysite clay nanotubes. *Biomacromolecules*, **11**, 820–826.

34 Meibian, Z., Yezhen, L., Xiaoxue, L. *et al.* (2010) Studying the cytotoxicity and oxidative stress induced by two kinds of bentonite particles on human B lymphoblast cells *in vitro*. *Chemico-Biological Interactions*, **183**, 390–396.

35 Meibian, Z., Xiaoxue, L., Yezhen, L. *et al.* (2011) Studying the genotoxic effects induced by two kinds of bentonite particles on human B lymphoblast cells in vitro. *Mutation Research/Genetic Toxicology and Environmental Mutagenesis*, **720**, 62–66.

36 Yuwen, H., Meibian, Z., Hua, Z. *et al.* (2013) Genetic damage and lipid peroxidation in workers occupationally exposed to organic bentonite particles. *Mutation Research/Genetic Toxicology and Environmental Mutagenesis*, **751**, 40–44.

37 Geh, S., Yücel, R., Duffin, R. *et al.* (2006) Cellular uptake and cytotoxic potential of respirable bentonite particles with different quartz contents and chemical modifications in human lung fibroblasts. *Archives of Toxicology*, **80**, 98–106.

38 Gao, N., Keane, M.J., Ong, T., and Wallace, W.E. (2000) Effects of simulated pulmonary surfactant on the cytotoxicity and DNA-damaging activity of respirable quartz and kaolin. *Journal of Toxicology and Environmental Health, Part A: Current Issues*, **60**, 153–167.

39 Yoshida, T., Yoshioka, Y., Matsuyama, K. *et al.* (2012) Surface modification of amorphous nanosilica particles suppresses nanosilica-induced cytotoxicity, ROS generation, and DNA damage in various mammalian cells. *Biochemical and Biophysical Research Communications*, **427**, 748–752.

40 Janer, G., Fernández-Rosas, E., Mas del Molino, E. *et al.* (2014) *In vitro* toxicity of functionalised nanoclays is mainly driven by the presence of organic modifier. *Nanotoxicology*, **8**, 279–294.

41 Shojaee-Aliabadi, S., Mohammadifar, M.A., Hosseini, H. *et al.* (2014) Characterization of nanobiocomposite kappa-carrageenan film with *Zataria multiflora* essential oil and nanoclay. *International Journal of Biological Macromolecules*, **69**, 282–289.
42 Rhim, J., Hong, S., and Ha, C. (2009) Tensile, water vapor barrier and antimicrobial properties of PLA/nanoclay composite films. *LWT-Food Science and Technology*, **42**, 612–617.
43 Houtman, J., Maisanaba, S., Puerto, M. *et al.* (2014) Toxicity assessment or organomodified clays used in food contact materials on human target cell lines. *Applied Clay Science*, **90**, 150–158.
44 Maisanaba, S., Puerto, M., Pichardo, S. *et al.* (2013) *In vitro* toxicological assessment of clays for their use in food packaging application. *Food and Chemical Toxicology*, **57**, 266–275.
45 Maisanaba, S., Gutièrrez-Praena, D., Pichardo, S. *et al.* (2013) Toxic effects of a modified montmorillonite clay on the human intestinal cell line Caco-2. *Journal of Applied Toxicology*, **34**, 714–725.
46 Lordan, S., Kennedy, J.E., and Higginbotham, C.L. (2011) Cytotoxic effects induced by unmodified and organically modified nanoclays in human hepatic HepG2 cell line. *Journal of Applied Toxicology*, **31**, 27–35.
47 Maisanaba, S., Hercog, K., Filipic, M. *et al.* (2016) Genotoxic potential of montmorillonite clay mineral and alteration in the expression of genes involved in toxicity mechanisms in the human hepatoma cell line HepG2. *Journal of Hazardous Materials*, **304**, 425–433.
48 Sharma, A.K., Schmidt, B., Frandsen, H. *et al.* (2010) Genotoxicity of unmodified and organo-modified montmorillonite. *Mutation Research/Genetic Toxicology and Environmental Mutagenesis*, **700**, 18–25.
49 Molinaro, S., Romero, M.C., Boaro, M. *et al.* (2013) Effect of nanoclay-type and PLA optical purity on the characteristics of PLA-based nanocomposite films. *Journal of Food Engineering*, **117**, 113–123.
50 Krikorian, V. and Pochan, D.J. (2003) Poly (L-lactic acid)/layered silicate nanocomposite: fabrication, characterization, and properties. *Chemistry of Materials*, **15**, 4317–4324.
51 Plackett, D.V., Holm, V.K., Johansen, P. *et al.* (2006) Characterization of L-polylactide and L-polyactide-polycaprolactone co-polymer films for use in cheese-packaging applications. *Packaging Technology and Science*, **19**, 1–24.
52 Lertwimolnun, W. and Vergnes, B. (2005) Influence of compatibilizer and processing conditions on the dispersion of nanoclay in a polypropylene matrix. *Polymer*, **46**, 3462–3471.
53 Lertwimolnun, W. and Vergnes, B. (2006) Effect of processing conditions on the formation of polypropylene/organoclay nanocomposites in a twin screw extruder. *Polymer Engineering and Science*, **46**, 314–323.
54 Beltrán, A., Valente, A.J.M., Jiménez, A., and Garrigós, M.C. (2014) Characterization of poly(ε-caprolactone)-based nanocomposites containing

hydroxytyrosol for active food packaging. *Journal of Agricultural and Food Chemistry*, **62**, 2244–2252.

55 Abreu, A.S., Oliveira, M., de Sá, A. *et al.* (2015) Antimicrobial nanostructured starch based films for packaging. *Carbohydrate Polymers*, **129**, 127–134.

56 Sharma, A.K., Mortensen, A., Schmidt, B. *et al.* (2014) In-vivo study of genotoxic and inflammatory effects of the organo-modified Montmorillonite Cloisite 30B. *Mutation Research/Genetic Toxicology and Environmental Mutagenesis*, **770**, 66–71.

57 Maisanaba, S., Pichardo, S., Jordá-Beneyto, M. *et al.* (2014) Cytotoxicity and mutagenicity studies on migration extracts from nanocomposites with potential use in food packaging. *Food and Chemical Toxicology*, **66**, 366–372.

58 Jorda-Beneyto, M., Ortuño, N., Devis, A. *et al.* (2014) Use of nanoclay platelets in food packaging materials: technical and cytotoxicity approach. *Food Additives & Contaminants: Part A*, **31** (**3**), 354–363.

59 Maisanaba, S., Puerto, M., Gutièrrez-Praena, D. *et al.* (2014) In vivo evaluation of activities and expression of antioxidant enzymes in Wistar rats exposed for 90 days to a modified clay. *Journal of Toxicology and Environmental Health, Part A: Current Issues*, **77**, 456–466.

60 Maisanaba, S., Gutiérrez-Praena, D., Puerto, M. *et al.* (2014) In vivo toxicity evaluation of the migration extract of an organomodified clay-poly(lactic) acid nanocomposite. *Journal of Toxicology and Environmental Health, Part A: Current Issues*, **77**, 731–746.

61 Chung, H., Kim, I., Baek, M. *et al.* (2011) Long-term cytotoxicity potential of anionic nanoclays in human cells. *Toxicology and Environmental Health Sciences*, **3**, 129–133.

62 Rhim, J., Park, H., and Ha, C. (2013) Bio-nanocomposites for food packaging and applications. *Progress in Polymer Science*, **38** (**10–11**), 1629–1652.

63 Lebaron, P.C., Wang, Z., and Pinnavaia, T.J. (1999) Polymer-layered silicate nanocomposites: an overview. *Applied Clay Science*, **15**, 11–29.

64 Wang, Z. and Pinnavaia, T.J. (1998a) Nanolayer reinforcement of elastomeric polyurethane. *Chemistry of Materials*, **10**, 3769–3771.

65 Paul, D.R. and Robeson, L.M. (2008) Polymer nanotechnology: Nanocomposites. *Polymer*, **49**, 3187–3204.

66 Zheng, X. and Wilkie, C.A. (2003) Flame retardancy of polystyrene nanocomposites based on an oligomeric organically-modified clay containing phosphate. *Polymer Degradation and Stability*, **81**, 539–550.

67 Majeed, K., Jawaid, M., Hassan, A. *et al.* (2013) Potential materials for food packaging from nanoclay/natural fibres filled hybrid composites. *Material and Design*, **46**, 391–410.

68 Stern, S.T. and McNeil, S.E. (2008) Nanotechnology safety concerns revisited. *Toxicological Sciences*, **101** (**1**), 4–21.

69 Souza, P.M.S., Morales, A.R., Marin-Morales, M.A., and Mei, L.H.I. (2013) PLA and montmorillonite nanocoposites: properties, biodegradation, and potential toxicity. *Journal of Polymers and the Environment*, **21**, 738–759.

70 Hoet, P.H., Brüske-Hohlfeld, I., and Salata, O.V. (2004) Nanoparticles-known and unknown health risks. *Journal of Nanobiotechnology*, **2** (**12**), 1–15.

71 Bakand, S., Hayes, A., and Dechsakulthorn, F. (2012) Nanoparticles: a review of particle toxicology following inhalation exposure. *Inhalation Toxicology*, **24** (**2**), 125–135.

72 Schinwald, A., Murphy, F.A., Jones, A. *et al.* (2012) Graphene-based nanoplatelets: a new risk to the respiratory system as a consequence of their unusual aerodynamic properties. *ACS Nano*, **6**, 736–746.

73 Laufer, G., Kirkland, C., Cain, A.A., and Grunlan, J.C. (2013) Oxygen barrier of multilayer thin films comprised of polysaccharides and clay. *Carbohydrate Polymers*, **95**, 299–302.

74 Frisch, S.M. and Francis, H. (1994) Disruption of Epithelial Cell-Matrix Interactions Induces Apoptosis. *The Journal of Cell Biology*, **124** (**4**), 619–626.

75 Huang, S. and Ingber, D.E. (1999) The structural and mechanical complexity of cell-growth control. *Nature Cell Biology*, **1**, E131–E138.

76 Murphy, M.P. (2009) How mitochondria produce reactive oxygen species. *Biochemical Journal*, **417**, 1–13.

77 Handy, D.E. and Loscalzo, J. (2012) Redox regulation of mitochondrial function. *Antioxidants & Redox Signaling*, **16**, 1323–1367.

78 Trachootham, D., Lu, W., Ogasawara, M.A. *et al.* (2008) Redox regulation of cell survival. *Antioxidants & Redox Signaling*, **10** (**8**), 1343–1374.

79 da Veiga Moreira, J., Peres, S., Steyaert, J. *et al.* (2015) Cell cycle progression is regulated by intertwined redox oscillators. *Theoretical Biology and Medical Modelling*, **12**, 1–14.

80 Lakhani, S.A., Masud, A., Kuida, K. *et al.* (2006) Caspases 3 and 7: key mediators of mitochondrial events of apoptosis. *Science*, **311**, 847–851.

81 Wei, Y., Fox, T., Chambers, S.P. *et al.* (2000) The structures of caspases-1, -3, -7 and -8 reveal the basis for substrate and inhibitor selectivity. *Chemistry & Biology*, 7, 423–432.

82 Brentnall, M., Rodriguez-Menocal, L., Ladron De Guevara, R. *et al.* (2013) Caspase-9, caspase-3 and caspase-7 have distinct roles during intrinsic apoptosis. *BMC Cell Biology*, **13**, 1–9.

83 Mcllwain, D.R., Berger, T., and MAk, T.W. (2013) Caspase functions in cell death and disease. *Cold Spring Harbor Perspectives in Biology*, **3**, 1–28.

12

Nanotechnology EHS: Manufacturing and Colloidal Aspects

Geoffrey D. Bothun[1] *and Vinka Oyanedel-Craver*[2]

[1] *Department of Chemical Engineering, University of Rhode Island, Kingston, RI, USA*
[2] *Department of Civil and Environmental Engineering, University of Rhode Island, Kingston, RI, USA*

12.1 Introduction

Nanotechnology is now ubiquitous in technology development in academic, governmental, and commercial sectors. Advances in analyzing nanoscale structures and phenomena, and the ability to control nanomaterial synthesis, have yielded nano-enabled technologies and products in virtually every industry. We are now on the precipice of a "nanotechnology revolution" akin to the industrial revolution that began in the seventeenth century. This begs the question – can we learn from the past mistakes? The full impact of the industrial revolution was massive and has defined our world today. It also changed the way humans interact with their environment and came at the cost of enormous ecological impacts including the depletion of natural resources and environmental pollution. Certainly, one can argue that, for example, population growth, an increase in life expectancy, and the availability of goods and services outweigh the negative impacts of the industrial revolution. However, many of us would agree that if knew then what we know now about the legacy negative environmental impacts, more environmentally friendly approaches would have been pursued. This concept is the motivation behind the field of nanotechnology environmental health and safety (nano-EHS) that has emerged over the past decade. Potential routes through which engineered nanomaterials (ENMs) enter the environment are shown in Figure 12.1 [1]. The premise behind nano-EHS is simple – can we identify health and safety risks of nanotechnologies, specifically nanomaterials, to enable the rapid pace of nanotechnology at the early stages of development? By doing so, we would

Nanotechnology Commercialization: Manufacturing Processes and Products, First Edition.
Edited by Thomas O. Mensah, Ben Wang, Geoffrey Bothun, Jessica Winter, and Virginia Davis.

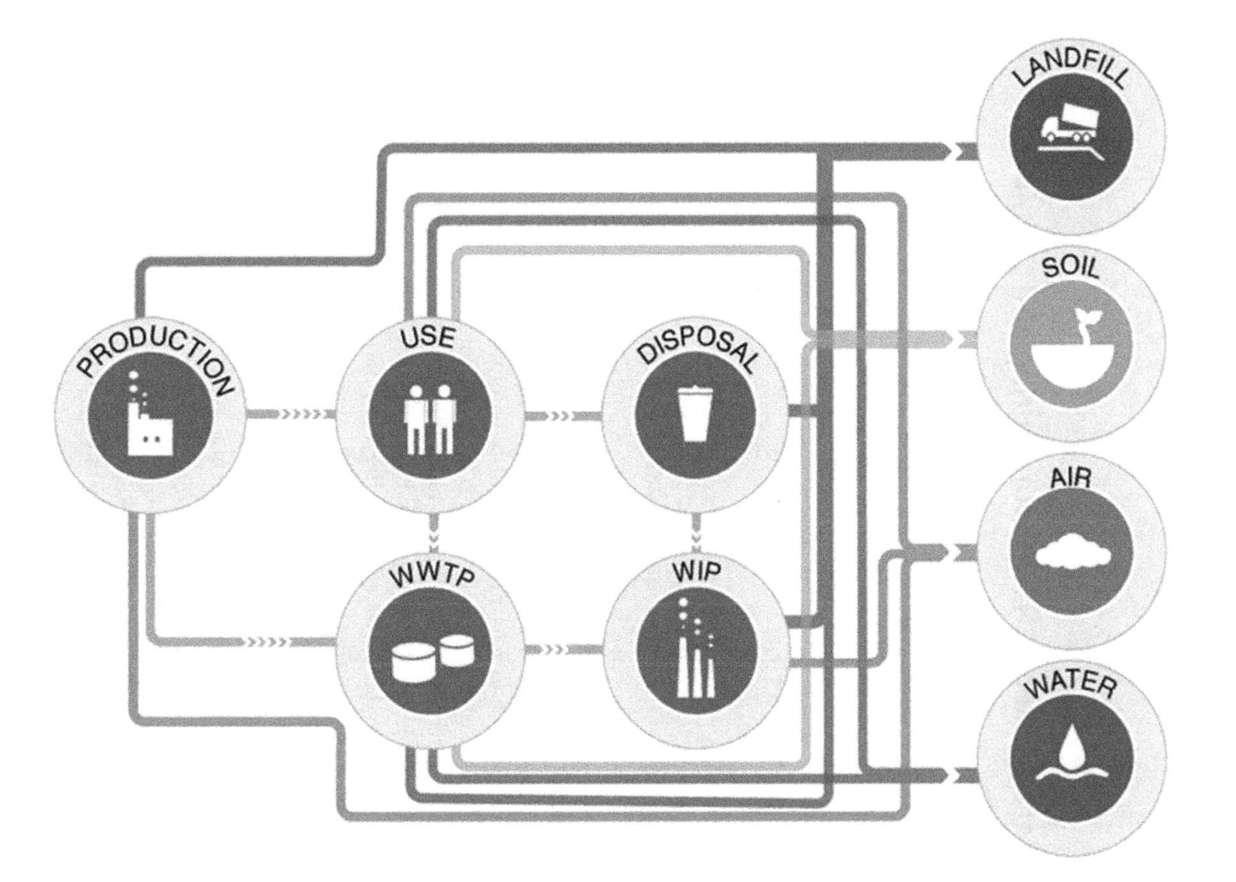

Figure 12.1 Engineered nanomaterials entering the environment. Links are shown between production, consumer use and disposal, wastewater treatment plants (WWTPs), and waste incineration plants (WIPs). Keller and Anastasiya 2014 [1]. Reproduced with permission of American Chemical Society.

be able to identify nanomaterials that pose EHS risks and guide the design of safer technologies. While the premise is simple, achieving this has proved to be as complex and diverse as nanotechnology itself.

12.1.1 Challenges

The *first challenge* to assessing nano-EHS lies in the diverse range of inorganic, organic, and hybrid nanoparticle compositions that are available or being developed for commercial application. In addition to composition, the type of surface coating for a given particle can vary widely – from hydrophilic to hydrophobic, small to large molecule, synthetic to naturally derived, and organic to inorganic – imparting different chemical and physical properties on the surface of the nanoparticle that influence its stability in aqueous or organic media and its interactions with biological and environmental molecules (e.g., proteins or natural organic matter, respectively) and interfaces (e.g., cellular and particulate surfaces, respectively) (Figure 12.2). The type of surface coating also affects the chemical stability and surface reactivity of the core nanoparticle. Changes in the chemical and physical properties of nanoparticles in biological or environmental milieu, which differ based on factors discussed earlier, are referred to as "transformations." A core concept in nano-EHS is that nanoparticles in biological and environmental milieu do not have the same physicochemical properties that were initially engineered with; while the initial materials may still be present, they have been transformed – through dissolution and chemical reactions with dissolved compounds in the media – and can also contain a "corona" composed of molecules and ions that exist in the milieu. These phenomena are particularly impactful to complex *engineered* "nano hybrids" that comprise assemblies of nanoparticles with different compositions, shapes, and sizes; and can further lead to the formation of *incidental* nano hybrids [3]. It should be noted that when we use the term "nanoparticle," we are referring to the core nanoparticle(s) and the surface coating(s), which determine the type of transformation(s) that occurs.

This brings us to the *second challenge* to assessing nano-EHS, which is determining how transformations impact *in vitro* or *in vivo* nanotoxicity studies. *In vitro* studies are conducted with a range of prokaryotic or eukaryotic cells lines in different buffer systems (e.g., pH, ion concentration, and protein concentration), under different culture conditions (e.g., temperature, aerobic versus anaerobic, and two- versus three-dimensional) and under different modes of operation (e.g., batch versus fed-batch versus continuous). It should be noted that the conditions are dependent on whether studies are being conducted using bacteria or mammalian cell lines or plants. *In vivo* studies are conducted using various animal models with different modes of exposure (e.g., intravenous, inhalation, ingestion, and transdermal). Common to both *in vitro* and *in vivo* studies are the range of potential exposure dosing strategies. Early work, and many recent studies, employed high exposure dosage that, while insightful and relevant to workplace safety where nanoparticles are being manufactured, do not reflect a realistic environmental exposure scenario.

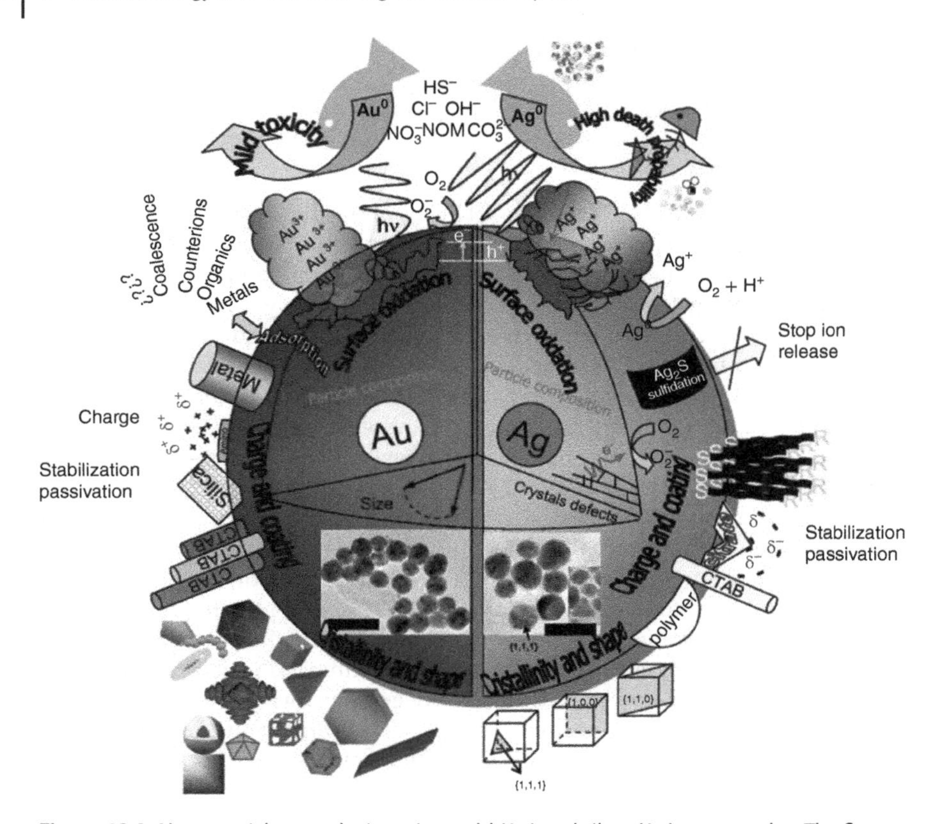

Figure 12.2 Nanoparticle complexity using gold (Au) and silver (Ag) as examples. The figure depicts differences in particle shape, size, crystallinity, and surface coating composition and charge (bottom half). Also shown are different transformations that take place in aqueous media and different toxicity mechanisms. Lapresta-Fernández *et al.* 2012 [2]. Reproduced with permission of Elsevier.

However, high dosages do elicit an inhibitory or toxic response that can be used to better identify toxicity mechanisms. Furthermore, direct comparisons of *in vitro* or *in vivo* toxicity for a given nanoparticle can be made, but the extent and nature of the nanoparticle transformation will be different, which means that the toxicity mechanisms may also be different.

This brings us to the *third challenge* to assessing nano-EHS, which is ability to identify specific toxicity mechanisms, the relative contribution of each mechanism, and the underlying chemical and physical interactions that govern each mechanism. Toxicity mechanisms are generally referred to as "chemical mechanisms" and "physical mechanisms" in the field of nano-EHS. Chemical mechanisms include ion toxicity due to nanoparticle dissolution (e.g., silver and quantum dot nanoparticles) and oxidative stress due to the generation of oxygen radicals by the nanoparticles. Physical mechanisms include nanoparticle binding to biological molecules (e.g., proteins and oligonucleotides) and interfaces (e.g., cellular membranes), which can interfere with the "normal" function of the molecule or interface and/or physically disrupt their structure. Chemical and physical mechanisms depend on the size, specific surface area, and dissolution potential or reactivity of nanoparticles, which in turn vary based on the nanoparticle transformation that has taken place. As a result, nanoparticles that may be considered toxic based on their core composition can experience transformations that restrict ion dissolution and the generation of oxygen radicals, and lead to nanoparticle aggregation. In this case, reduced toxicity may stem from the suppression of chemical mechanisms or from nanoparticle aggregation (and possibly sedimentation) that reduced the bioavailability of nanoparticles and their ability to be internalized by cells.

12.1.2 Recent Initiatives and Reviews

As outlined earlier, a full understanding of nano-EHS concerns is difficult to achieve and results from nano-EHS studies are highly dependent on the materials employed and the experimental conditions. There are efforts underway to develop standard protocols for nano-EHS to address these challenges through initiatives that foster collaboration between agencies, institutions, and disciplines. The National Institute of Environmental Health and Safety (NIEHS) nano-EHS program supports the Centers for Nanotechnology Health Implications Research (NCNHIR) and the Nano Grand Opportunities (Nano GO) program, both of which are focused on assessing the EHS of engineered nanomaterials related to human health [4, 5]. Nanomaterials toxicity is also being investigated through the NIEHS National Toxicology Program (NTP) using standard toxicology protocols. The National Science Foundation is also addressing nano-EHS and is currently supporting the Center for Environmental Implications of Nanotechnology and the Center for Sustainable Nanotechnology.

There have been a number of review articles pertaining to nano-EHS and/or human health effects [6–22], and the transformations and interactions that are involved [23–25] (note, only a subset is referenced here). There are new journals focused on nano-EHS, including *Nanotoxicology* and *Environmental Science: Nano*, and there have been a number of focused issues or thematic collections in journals including *Environmental Science & Technology* (2012) and *Nature Nanotechnology*. In this chapter, we focus on nano-EHS as it relates to environmental systems – how do engineered nanoparticles behave within the environment and what is their environmental fate? This is a precursor to human health effects due to environmental exposure to nanomaterials. The chapter is divided into three parts discussing (i) the colloidal properties and transformations of nanoparticles, (ii) the chemical and physical mechanisms that lead to inhibitory or toxicity effects, and (iii) techniques for assessing environmental impacts.

12.2 Colloidal Properties and Environmental Transformations

Wet chemistry methods are commonly used for the synthesis of organic and metal nanoparticles, with the product of this process being a nanosuspension [26–33]. During synthesis, nanosuspensions are stabilized against particle agglomeration using either steric (polymers or nonionic surfactants) or electrostatic (ionic surfactants) surface functionalization. After synthesis (downstream), nanosuspensions are cleaned and concentrated using methodologies such as centrifugation, filtration (dead-end and cross-flow), and dialysis. During the purification stage, the core-shell nanoparticle complexes can experience agglomeration, increasing the size of the particles and potentially reducing dissolution patterns as well as toxicity effect [34]. Additional stabilization is often sought after synthesis, and this is achieved by providing additional coatings or by exchanging the synthesis coating with an alternative.

In the production of nano-enabled products, nanosuspensions have limited market due to stability issues; therefore, the conversion of a nanosuspension into a dried powder form increases the stability and shelf life of the nanoparticles–stabilizer complex and adds versatility for the incorporation of these materials into polymeric matrices and consumer products [35, 36]. A drying step is needed when possible stability issues, both physical (such as Ostwald ripening and agglomeration) and chemical (such as hydrolysis) can change the properties of the nanosuspension. In some cases, such as the incorporation of nanoparticles in matrices, this step is essential since the solution in which the nanoparticles are suspended may not be compatible with the matrix.

Aqueous nanosuspensions provide the basis for most nano-EHS studies. Furthermore, when nanoparticles are released into the environment, they often eventually become nanosuspensions (stable or unstable). In aqueous environments, the surface charge of a nanoparticle stabilized by electrostatic repulsion varies with pH via protonation or deprotonation, and ion concentration via surface ion binding and the formation of an electrical double layer. Uncharged coatings where steric stabilization is employed may not experience these variations. The stability of nanosuspensions before transformation can be assessed reasonably well using DLVO-based models that can incorporate additional forces, beyond electrostatic and dispersion (van der Waals) interactions, that arise from hydration, hydrophobic, or magnetic interactions. However, it should be noted that the length scale of these interactions are similar to those of the nanoparticles themselves, which means that the interactions are not additive and DLVO-based models are often either qualitative or give erroneous results for small particles (1–20 nm in diameter) [37].

Assessing stability is critical to nano-EHS studies where nanosuspensions experience different pH and ion conditions that can cause aggregation, changing the effective nanoparticle size "experienced" by cells or organisms and

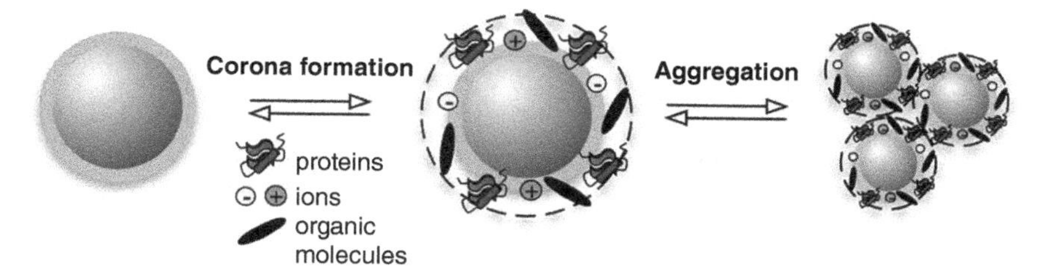

Figure 12.3 Corona formation (biological and/or environmental) around nanoparticles and nanoparticle aggregation. Chen and Bothun 2014 [38]. Reproduced with permission of American Chemical Society.

leading to sedimentation, which can reduce the amount of "available" nanoparticles that remain suspended and to which cells or organisms are exposed. All of these effects are dynamic, particularly when the aqueous environment changes spatially or temporally, which means that the kinetics must also be evaluated. The impact of kinetics is apparent when we consider that nanotoxicology studies occur over a range of timescales and that the lifecycle of a nanomaterial may include many different aqueous environments.

Nanosuspension stability becomes quite complicated when transformations take place. In addition to ion adsorption, biological and natural molecules adsorb onto the surface and form coronas that completely change the effective (hydrodynamic) size of the nanoparticles, the surface topography and compressibility (e.g., "hard" or "soft" coronas), and the surface charge (Figure 12.3). Simplified studies using model biological molecules, such as a single protein, or defined natural organic matter can be used to examine corona formation and the stability can be examined using DLVO-based models. However, corona formation *in vitro* or *in vivo* is dynamic and the composition is complex with multiple species adsorbing. An end-result of transformation due to corona formation is often aggregation. Corona formation can also increase the stability of nanosuspensions, which can have negative consequences when the nanoparticles in question exhibit toxicity and it would be advantageous from an ecological or perspective if they remained aggregated and sedimented. We have only recently begun to develop comprehensive strategies to quantify corona composition and dynamics, and our understanding of how coronas impact nano-EHS is limited.

Several studies have reported the influence of physicochemical properties of nanoparticles such as size, shape, charge, and stabilizers on their toxicological effect on biological systems [39–42]. Most of these studies have focused on evaluating the effect of these parameters at only one stage of the life, usually right after synthesis, of the nanoparticles, while the effects on toxicological properties of different purification and conditioning processes (i.e., sterilization, drying and compression, among others) have not been

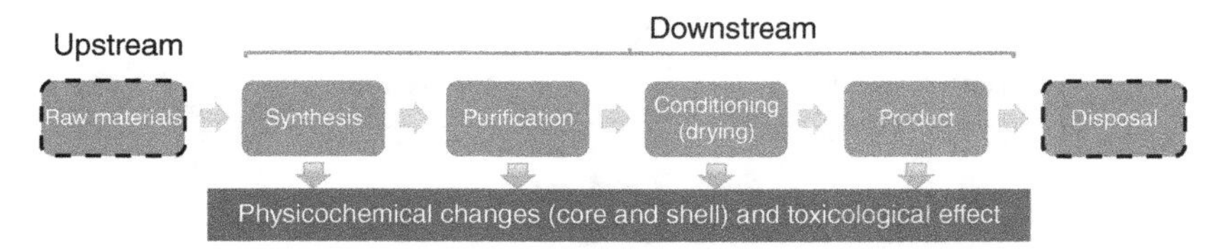

Figure 12.4 Life cycle of nano-enabled products from the raw materials to disposal.

largely considered. A general life cycle model of a nano-enabled product is shown in Figure 12.4. For example, some studies use freshly synthesized and purified nanoparticles at the bench scale; therefore, these nanoparticles have not gone through extensive downstream processes[42–46]. Others used nanosuspensions or nanopowders synthesized and processed by the manufacturers without detailed information of the synthesis and downstream processes applied to the nanoparticles. In addition, studies that have used the same nanoparticle-surface functionalization-water chemistry conditions obtained inconclusive information regarding the cell response to nanoparticle exposure [44, 47, 48]. It is possible that two nanoparticles with the same core-coating composition size and shape could have different toxicological properties due to the different processes applied during their life cycle.

12.3 Assessing Nano EHS

In order to reduce time and cost associated with the screening of specific or a combination of physicochemical properties (Figure 12.5) of nanoparticles using *in vivo* systems, several research are developing *in vitro* methodologies [49]. High-throughput *in vitro* nanotoxicity assays are employed to evaluate several combinations for toxicity end points, cell lines, and composition of nanomaterials [49].

Most of the *in vitro* methods are performed in liquid environments, in which nanoparticles can interact with organic and inorganic compounds in the culture media. Given the difference in culture media composition for studies using cell liners, biofilms or planktonic organism processes such as flocculation, agglomeration, and dissolution of the nanoparticles in these different media will produce different size, charge, and surface properties in each case [50]. These processes can cause discrepancies in terms of the effective dose of nanoparticles in contact with the cell or organism (Figure 12.6). In order to overcome these difficulties, several researches have developed dosimetry protocols to determine the effective dose of nanoparticles in contact to cells [51–53].

In addition to the challenges for assessing ecotoxicological effects of nanoparticles, due to physicochemical changes that nanoparticles can undergo during

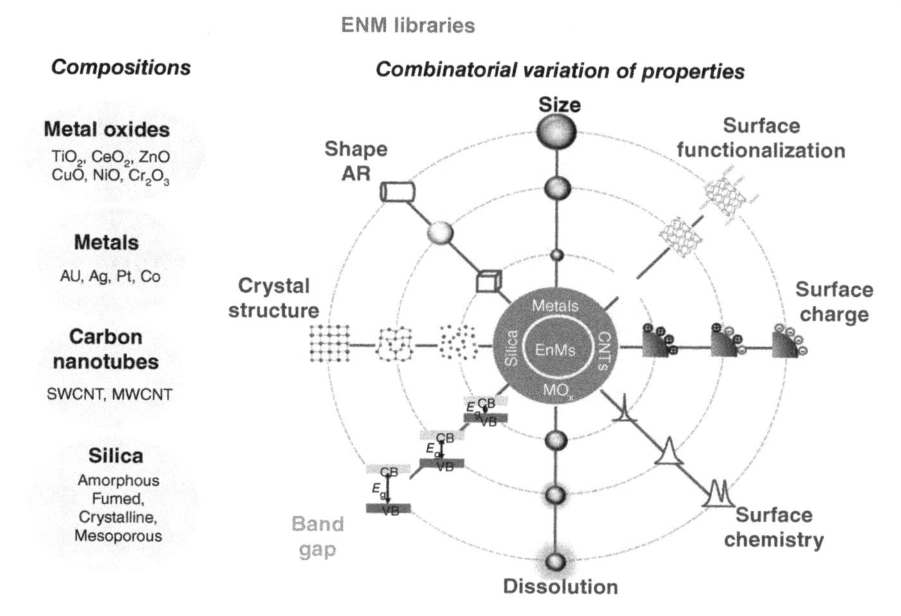

Figure 12.5 Compositional and combinatorial engineered nanomaterials (ENMs) libraries, including metals, metal oxides, carbon nanotubes, and silica-based nanomaterials, to perform mechanism-based EHS and toxicity analyses. Nel *et al.* 2012 [49]. Reproduced with permission of American Chemical Society.

their life cycle, different methodologies to assess these effects provide conflicting results [44]. For example, most ecotoxicological tests are performed in batch reactors of different volumes; in this regime, all parameters such as number of microorganism, substrate concentration, and exopolymeric substances concentration, among others, change with time [54]. Knowing the high reactivity of nanoparticles with dissolved and particulate matter, it could be challenging to obtain reproducible results with batch tests due to the time dependence of the variables. In order to gather reproducible and meaningful information about the effect of the physiological stage of the bacterial population, they must be grown in a defined, ideally constant, and controllable set of physicochemical conditions. Continuous cultures have the advantage that time-independent steady states can be achieved. Another advantage of the continuous growth system is that the specific growth rate can be manipulated by changing the rate of supply of the limiting substrate. These systems have been extensively used to study growth at different environmental conditions, nutrient limitations, and stress response and have a higher reproducibility in interlaboratory testing than batch systems. Finally, continuous tests have been previously used for long-term adaptive experiments in a controlled

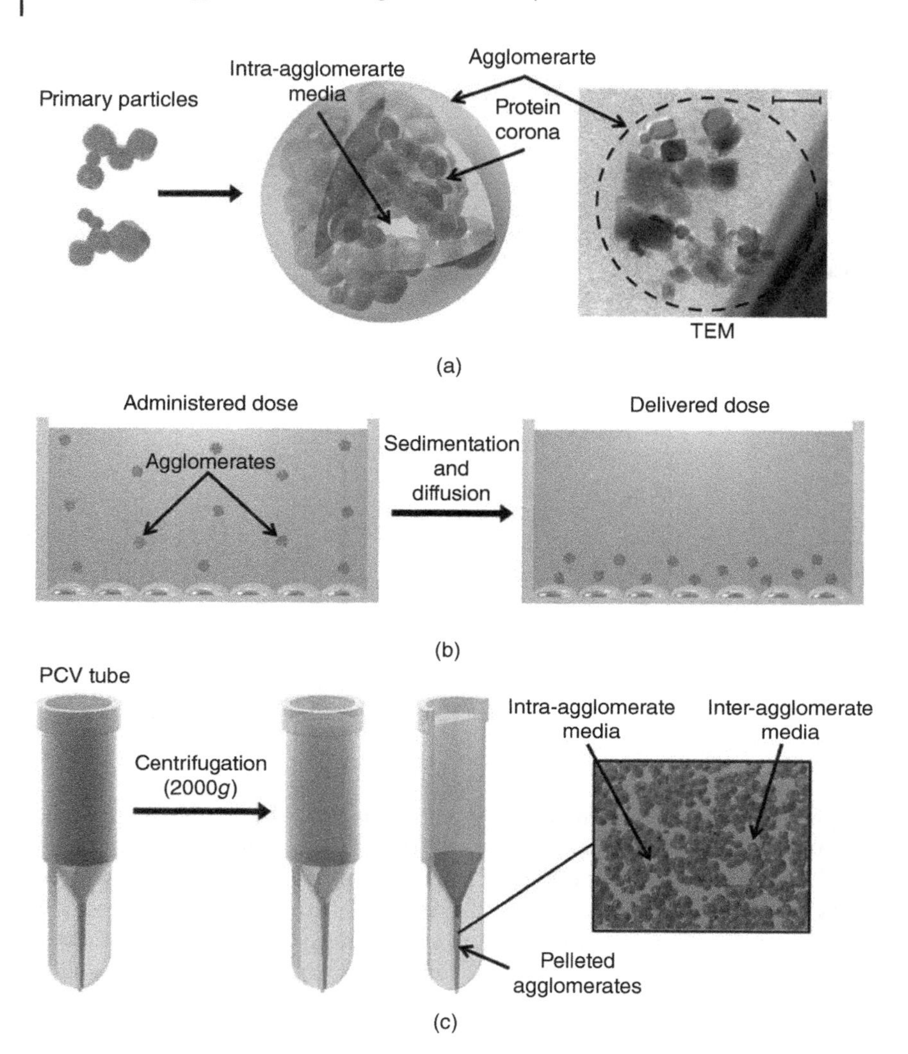

Figure 12.6 Agglomeration and nanoparticle "delivery" to cells *in vitro*. (a) In cell culture media, primary particles form agglomerates with media contained within the agglomerates and protein coronas formed on the particle surfaces. (b) Agglomerates administered to cells settle over time and form a concentrated layer near or deposited onto the cell surface. As a result, the administered dose is less than the delivered dose. (c) The effective density or delivered dose can be estimated by centrifuging the agglomerates and evaluating their packing factor (referred to also as a stacking factor). DeLoid *et al.* 2014 [51]. Reproduced with permission of Nature Publishing Group.

environment due to the capability to maintain the culture at a specific growth rate [55]. Previous studies have reported that organisms growing at slow specific growth rate could adapt easier than those growing at a faster rate [56].

The ability of bacteria to adapt to environmental changes or from exposure to stressors, such as limited nutrient availability or nonlethal concentrations of antibiotics or nanoparticles, is essential for their survival and successful competition with other microorganisms [57–61]. One of the most studied bacterial adaptations to stressors is their resistance to antibiotics. Resistance is the result of the evolutionary process in which bacteria, through either vertical (mutation) or horizontal (e.g., plasmid acquisition) genetic changes, can adapt or counteract the action of a stress agent [62, 63]. Among the mechanisms that bacteria use to respond to metal and antibiotic exposure are membrane permeability [64, 65], metabolic by-pass [66], efflux [67], alteration of cellular targets, and sequestration [68, 69] (Figure 12.7).

In addition, it is well known that coresistance can occur when the genetic element determining the resistance to a particular stressor contains genes related with resistance to another stressor. In this context, several reports have evidenced metal–antibiotic coresistance [68]. A recent publication by Aruguete et al. highlighted a promising application of antimicrobial nanoparticles as alternatives to conventional antibiotics to fight microbial drug resistance; however, it also describes in detail the possible risk that some of the antimicrobial mechanisms of the nanoparticles, such as cell membrane

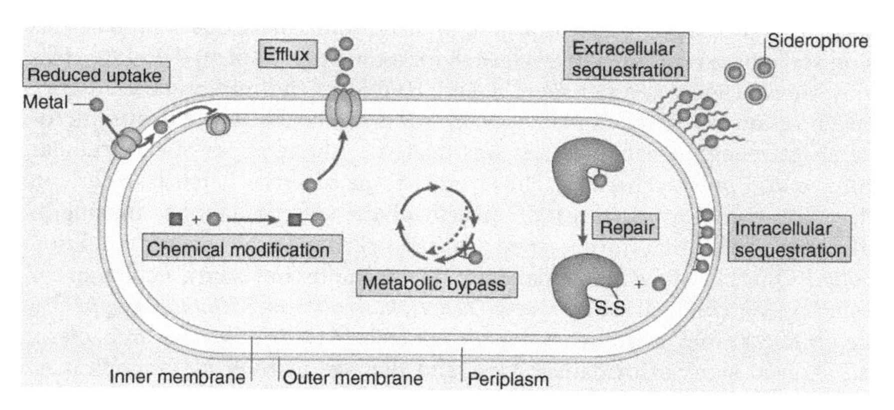

Figure 12.7 Adaptation mechanisms of microorganisms to metals and antibiotics. These can include (i) reduced uptake of metal ions via transporter downregulation; (ii) efflux of metal ions via transporter upregulation; (iii) upregulation of extracellular biomolecular expression, which can in turn influence metal–cell surface binding, uptake, and efflux; (iv) intracellular sequestration through the formation of metal precipitates or metal–protein aggregates; (v) repair in response to ROS; (vi) metabolic bypass via enzymatic disruption; and (vii) chemical modification that reduce metal toxicity (e.g., via precipitation). Lemire *et al.* 2013 [66]. Reproduced with permission of Nature Publishing Group.

damage, generation of reactive oxygen species, and release of toxic ions, could lead to unexpected development of antibiotic resistance [70]. Regarding cell damage, even though most nanoparticles have a nonspecific mechanism to disturb cell membranes and therefore low probability to develop resistance, others, such as silver nanoparticles can selectively interact with specific surface biomolecules (similarly to the action of polymixin) in which case bacteria can inhibit the action of the nanoparticle by changing the composition of the membrane [71–74]. In the case of ROS, bacteria can fight the effect of these agents by increasing extracellular polysaccharides [75, 76], and SoxRS and OxyR regulators can activate the expression of various genes against ROS action [77–79]. However, it should be noted that the rate of mutation increases in bacteria exposed to oxidative stress, possibly due to the action of stress-responsive error-prone DNA polymerase V [63], increasing the possibility to develop resistance to the stressor in the bacterial population. Metal and metal oxide nanoparticles are widely used in several industrial and consumer products. These nanoparticles, when dissolved in aqueous medium (depending on surface functionalization and water chemistry), release metal ions than can impair bacterial functionality. Efflux and metal sequestration are some of the response mechanisms for exposure to metal ions (similar mechanisms have been reported to fight the action of some antibiotics) [67].

Scarce reports about the possible adaptation of bacteria to nanoparticle exposure and links with antibiotic resistance have been published. Dhas *et al.* performed an adaption study using isolates from sewage and soil, which were exposed to increasing concentrations of silver and zinc oxide nanoparticles using serial batch transfers for 24 h each for a total period of 10 days [80]. This study showed increases between 5- and 10-fold of the minimum inhibitory concentration for both nanoparticles. In this case, the resistance to antimicrobial effectiveness to nanoparticles was related to the increase of extracellular proteins and polysaccharides. However, the use of serial batch transfer can select for complex mixtures of growth-phase subpopulations, making it difficult to ensure that the observed adaptation is due to the specific condition applied. Qiu *et al.* showed that nanoalumina promotes horizontal gene transfer occurring via conjugation between *Escherichia coli* and *Salmonella* spp. at a rate up to 200-fold greater than for the controls [81]. Both species of bacteria also showed signs of oxidative stress and damage in their cell membranes. Yang et al. reported that heavy-metal ion efflux and production of extracellular compounds for the precipitation of dissolved metal ions were the main stress-response mechanisms of *Pseudomonas aeruginosa* against quantum dots, using batch reactors [82]. This study also showed the development of simultaneous bacterial resistance to several antibiotics. These results clearly evidence how properties of metal nanoparticles can trigger stress-response mechanisms in bacteria, changing their phenotype and promote unintended negative impacts such as antibiotic resistance.

12.3.1 Example: Silver Nanoparticles (AgNPs)

Due in large part to its antimicrobial properties, silver remains one of the most widely employed nanomaterials in commercial applications. Its toxicological effects stem primarily from silver ion (Ag^+) release, where the ion itself can disrupt the structure and function of biomolecules and cellular membranes. AgNP dissolution and ion release are dependent on the surface area and water chemistry conditions, which ties AgNP colloidal properties (aggregation) to its toxicity. Studies measuring the rate of silver ion release at different dilutions of seawater showed that the salt concentration did not affect the AgNPs' oxidation kinetics; however, high ionic strength increased the size of the particles from approximately 2–200 nm after 24 h [83]. Gao *et al.* also found that fresh water samples with higher ionic strength produced large AgNPs [84]. Jin *et al.* studied the effect of different water matrices on AgNPs size, silver ions release, and antimicrobial activity using a fixed concentration of Ca^{2+} and Mg^{2+} [85]. The study revealed that Ca^{2+} and Mg^{2+} increased AgNPs aggregation in different electrolyte solutions with the same ionic strength in comparison with monovalent ions. The antimicrobial test showed that Gram-negative bacteria *Pseudomonas putida* was more resistant to the AgNPs compared to the Gram-positive bacteria *Bacillus subtilis*. Studies using different natural water sources have shown that the antibacterial performance of AgNPs decreases in the presence of dissolved natural organic matter or divalent ions, such as humic acid and calcium carbonate [86, 87].

12.3.2 Role of Manufacturing

We briefly address the role of manufacturing in nano-EHS. Nanomaterials are manufactured using a variety of bottom-up or top-down processes (Table 12.1). Manufacturing conditions can have an important impact in the ecotoxicological effects of the synthesized nanomaterials because they determine the initial size, shape, aggregation state, and dispersion characteristics (gaseous, liquid, or solid media) of the ENM. For example, carbon nanotubes can be synthesized through arc discharges, laser ablation, chemical vapor deposition, flame synthesis, pyrolysis, and electrolysis, among others [89]. While chemical vapor deposition can produce highly pure carbon nanotubes, those synthesized through flame synthesis and pyrolysis can contain polyaromatic hydrocarbons, which can increase their toxic effects [90]. On the other hand, metal nanoparticles such as silver and gold are mostly synthesized by liquid-phase precipitation in which a variety of well defined reducing or stabilizing agents are present. In this context, the fate of silver or gold nanoparticles manufactured with well-defined reagents can be different in comparison with those produced with phytochemicals from plant extracts as reducing and capping agents.

Table 12.1 Commonly used nanomanufacturing techniques.

Top-down techniques	Bottom-up techniques
Lithography	*Vapor-phase techniques*
Conventional lithography	Deposition techniques
Photolithography	Vapor-phase epitaxy
E-beam lithography (for mask generation only)	Metal organic chemical vapor deposition
Next-generation lithography	Molecular beam epitaxy
Immersion lithography	Plasma-enhanced chemical vapor deposition
Lithography with lower wavelengths than photolithography	Atomic layer deposition
Extreme ultraviolet (soft X-ray) lithography	Pulse laser deposition
X-ray lithography	Sputtering
Lithography with particles	Evaporation
E-beam lithography	*Nanoparticle/nanostructured materials synthesis techniques*
Focused ion-bean lithography	Evaporation
Nanoimprint lithography	Laser ablation
Step-and-flash imprint lithography	Flame synthesis
Soft lithography	Arc discharges
Etching	*Liquid-phase techniques*
Wet etching	Precipitation
Dry etching	Sol-gel
Reactive ion etching	Solvothermal synthesis
Plasma etching	Sonochemical synthesis
Sputtering	Microwave irradiation
	Reverse micelle
Electrospinning	Electrodeposition
Milling	*Self-assembly techniques*
Mechanical Milling	Electrostatic self-assembly
Cryomilling	Self assembled monolayers
Mechanochemical bonding	Langmuir–Blodgett formation

Source: Şengül *et al.* 2008 [88]. Reproduced with permission of John Wiley & Sons.

Summary

As discussed in a recent perspective on nano EHS, it should be pointed out that "no human toxicological disease or major environmental impact has been reported for ENMs" [91]. However, studies over the past 10+ years have clearly shown that ENMs can pose serious EHS risks and have advanced our understanding of toxicity mechanisms, both of which are highly dependent on the chemical and physical properties of the ENMs and the transformations that occur. As such, nano EHS remains a field focused on understanding what risks exist for increasingly complex ENMs and guiding the responsible manufacturing and use of "low-risk" ENMs for consumer products and commercial applications. This can be achieved by addressing the challenges to determining nano EHS, determining molecular mechanisms of toxicity, and creating systematic and standard protocols for nano EHS assessment.

Acknowledgments

The authors acknowledge funding from the National Science Foundations, grant CBET-1056652 (Bothun) and CBET-1350789 (Craver).

References

1 Keller, A.A. and Anastasiya, L. (2014) *Predicted releases of engineered nanomaterials: from global to regional to local. Environmental Science & Technology Letters,* **1** (**1**), 65–70.
2 Lapresta-Fernández, A., Fernández, A., and Blasco, J. (2012) *Nanoecotoxicity effects of engineered silver and gold nanoparticles in aquatic organisms. Trends in Analytical Chemistry,* **32**, 40–59.
3 Saleh, N., Navid, S., Afrooz, A. *et al.* (2014) *Emergent properties and toxicological considerations for nanohybrid materials in aquatic systems. Nanomaterials,* **4** (**2**), 372–407.
4 Carlin, D.J. (2014) *Nanotoxicology and nanotechnology: new findings from the NIEHS and Superfund Research Program scientific community. Reviews on Environmental Health,* **29** (**1–2**), 105–107.
5 Schug, T.T., Nadadur, S.S., and Johnson, A.F. (2013) *Nano GO Consortium—a team science approach to assess engineered nanomaterials: reliable assays and methods. Environmental Health Perspectives,* **121** (**6**), a514–a516.
6 Borak, J. and Jonathan, B. (2009) *Nanotoxicology: characterization, dosing, and health effects. Journal of Occupational and Environmental Medicine,* **51** (**5**), 620–621.

7 Gilbertson, L.M., Wender, B.A., Zimmerman, J.B., and Eckelman, M.J. (2015) *Coordinating modeling and experimental research of engineered nanomaterials to improve life cycle assessment studies. Environmental Science: Nano,* **2** (**6**), 669–682.

8 Grieger, K.D., Linkov, I., Hansen, S.F., and Baun, A. (2012) *Environmental risk analysis for nanomaterials: review and evaluation of frameworks. Nanotoxicology,* **6** (**2**), 196–212.

9 Gunsolus, I.L. and Haynes, C.L. (2016) *Analytical aspects of nanotoxicology. Analytical Chemistry,* **88** (**1**), 451–479.

10 Haynes, C.L. (2010) *The emerging field of nanotoxicology. Analytical and Bioanalytical Chemistry,* **398** (**2**), 587–588.

11 Hussain, S.M., Braydich-Stolle, L.K., Schrand, A.M. *et al.* (2009) *Toxicity evaluation for safe use of nanomaterials: recent achievements and technical challenges. Advanced Materials,* **21** (**16**), 1549–1559.

12 Kahru, A. and Ivask, A. (2013) *Mapping the dawn of nanoecotoxicological research. Accounts of Chemical Research,* **46** (**3**), 823–833.

13 Krug, H.F. and Wick, P. (2011) *Nanotoxicology: an interdisciplinary challenge. Angewandte Chemie International Edition,* **50** (**6**), 1260–1278.

14 Larson, J.K., Carvan, M.J. 3rd, and Hutz, R.J. (2014) *Engineered nanomaterials: an emerging class of novel endocrine disruptors. Biology of Reproduction,* **91** (**1**): 20, 1–8.

15 Mackevica, A. and Hansen, S.F. (2016) *Release of nanomaterials from solid nanocomposites and consumer exposure assessment – a forward-looking review. Nanotoxicology,* **10** (**6**), 641–653.

16 Magdolenova, Z., Collins, A., Kumar, A. *et al.* (2014) *Mechanisms of genotoxicity. A review of in vitro and in vivo studies with engineered nanoparticles. Nanotoxicology,* **8** (**3**), 233–278.

17 Malysheva, A., Lombi, E., and Voelcker, N.H. (2015) *Bridging the divide between human and environmental nanotoxicology. Nature Nanotechnology,* **10** (**10**), 835–844.

18 Maynard, A.D., Warheit, D.B., and Philbert, M.A. (2011) *The new toxicology of sophisticated materials: nanotoxicology and beyond. Toxicological Sciences,* **120** (**Suppl 1**), S109–S129.

19 Maynard, R.L. (2012) *Nano-technology and nano-toxicology. Emerging Health Threats Journal,* **5**: 17508.

20 Peixe, T.S., de Souza Nascimento, E., Schofield, K.L. *et al.* (2015) *Nanotoxicology and exposure in the occupational setting. Occupational Diseases and Environmental Medicine,* **03** (**03**), 35–48.

21 Seabra, A.B., Paula, A.J., de Lima, R. *et al.* (2014) *Nanotoxicity of graphene and graphene oxide. Chemical Research in Toxicology,* **27** (**2**), 159–168.

22 Valsami-Jones, E. and Lynch, I. (2015) *NANOSAFETY. How safe are nanomaterials? Science,* **350** (**6259**), 388–389.

23 Bohnsack, J.P., Assemi, S., Miller, J.D., and Furgeson, D.Y. (2012) *The primacy of physicochemical characterization of nanomaterials for reliable toxicity assessment: a review of the zebrafish nanotoxicology model. Methods in Molecular Biology*, **926**, 261–316.

24 Bruinink, A., Wang, J., and Wick, P. (2015) *Effect of particle agglomeration in nanotoxicology. Archives of Toxicology*, **89** (**5**), 659–675.

25 Louie, S.M., Rui, M., and Lowry, G.V. (2014) *Transformations of nanomaterials in the environment. Frontiers of Nanoscience*, Vol 7: Nanoscience and the Environment, R. Palmer (ed), Elsevier 55–87.

26 Das, S.K. and Marsili, E. (2010) *A green chemical approach for the synthesis of gold nanoparticles: characterization and mechanistic aspect. Reviews in Environmental Science and Bio/Technology*, **9** (**3**), 199–204.

27 García-Barrasa, J., López-de-Luzuriaga, J.M., and Monge, M. (2010) *Silver nanoparticles: synthesis through chemical methods in solution and biomedical applications. Central European Journal of Chemistry*, **9** (**1**), 7–19.

28 Ahamed, M., Khan, M.A.M., Siddiqui, M.K.J. *et al.* (2011) *Green synthesis, characterization and evaluation of biocompatibility of silver nanoparticles. Physica E*, **43** (**6**), 1266–1271.

29 Hatakeyama, W., Sanchez, T.J., Rowe, M.D. *et al.* (2011) *Synthesis of gadolinium nanoscale metal-organic framework with hydrotropes: manipulation of particle size and magnetic resonance imaging capability. ACS Applied Materials & Interfaces*, **3** (**5**), 1502–1510.

30 Valodkar, M., Modi, S., Pal, A., and Thakore, S. (2011) *Synthesis and anti-bacterial activity of Cu, Ag and Cu-Ag alloy nanoparticles: a green approach. Materials Research Bulletin*, **46** (**3**), 384–389.

31 Venkatpurwar, V. and Pokharkar, V. (2011) *Green synthesis of silver nanoparticles using marine polysaccharide: study of in-vitro antibacterial activity. Materials Letters*, **65** (**6**), 999–1002.

32 Poulose, S., Panda, T., Nair, P.P., and Theodore, T. (2014) *Biosynthesis of silver nanoparticles. Journal of Nanoscience and Nanotechnology*, **14** (**2**), 2038–2049.

33 Zhang, T., Song, Y.J., Zhang, X.Y., and Wu, J.Y. (2014) *Synthesis of silver nanostructures by multistep methods. Sensors (Basel)*, **14** (**4**), 5860–5889.

34 Palencia, M., Rivas, B.L., and Valle, H. (2014) *Size separation of silver nanoparticles by dead-end ultrafiltration: description of fouling mechanism by pore blocking model. Journal of Membrane Science*, **455**, 7–14.

35 Chaubal, M.V. and Popescu, C. (2008) *Conversion of nanosuspensions into dry powders by spray drying: a case study. Pharmaceutical Research*, **25** (**10**), 2302–2308.

36 Gamble, J.F., Ferreira, A.P., Tobyn, M. *et al.* (2014) *Application of imaging based tools for the characterisation of hollow spray dried amorphous dispersion particles. International Journal of Pharmaceutics*, **465** (**1-2**), 210–217.

37 Batista, C.A.S., Larson, R.G., and Kotov, N.A. (2015) *Nonadditivity of nanoparticle interactions. Science,* **350** (**6257**), 1242477.

38 Chen, K.L. and Bothun, G.D. (2014) *Nanoparticles meet cell membranes: probing nonspecific interactions using model membranes. Environmental Science & Technology,* **48** (**2**), 873–880.

39 Albanese, A., Tang, P.S., and Chan, W.C. (2012) *The effect of nanoparticle size, shape, and surface chemistry on biological systems. Annual Review of Biomedical Engineering,* **14**, 1–16.

40 Asati, A., Santra, S., Kaittanis, C., and Perez, J.M. (2010) *Surface-charge-dependent cell localization and cytotoxicity of cerium oxide nanoparticles. ACS Nano,* **4** (**9**), 5321–5331.

41 Baek, Y.W. and An, Y.J. (2011) *Microbial toxicity of metal oxide nanoparticles (CuO, NiO, ZnO, and Sb2O3) to Escherichia coli, Bacillus subtilis, and Streptococcus aureus. Science of the Total Environment,* **409** (**8**), 1603–1608.

42 Bondarenko, O., Juganson, K., Ivask, A. *et al.* (2013) *Toxicity of Ag, CuO and ZnO nanoparticles to selected environmentally relevant test organisms and mammalian cells in vitro: a critical review. Archives of Toxicology,* **87** (7), 1181–1200.

43 Hasegawa, G., Shimonaka, M., and Ishihara, Y. (2012) *Differential genotoxicity of chemical properties and particle size of rare metal and metal oxide nanoparticles. Journal of Applied Toxicology,* **32** (**1**), 72–80.

44 Levard, C., Hotze, E.M., Lowry, G.V., and Brown, G.E. Jr. (2012) *Environmental transformations of silver nanoparticles: impact on stability and toxicity. Environmental Science & Technology,* **46** (**13**), 6900–6914.

45 Petrochenko, P.E., Zhang, Q., Wang, H. *et al.* (2012) *In vitro cytotoxicity of rare earth oxide nanoparticles for imaging applications. International Journal of Applied Ceramic Technology,* **9**, 881–892.

46 Tong, T., Binh, C.T., Kelly, J.J. *et al.* (2013) *Cytotoxicity of commercial nano-TiO_2 to Escherichia coli assessed by high-throughput screening: effects of environmental factors. Water Research,* **47** (7), 2352–2362.

47 Kirschling, T.L., Golas, P.L., Unrine, J.M. *et al.* (2011) *Microbial bioavailability of covalently bound polymer coatings on model engineered nanomaterials. Environmental Science & Technology,* **45** (**12**), 5253–5259.

48 Shoults-Wilson, W.A., Reinsch, B.C., Tsyusko, O.V. *et al.* (2011) *Role of particle size and soil type in toxicity of silver nanoparticles to earthworms. Soil Science Society of America Journal,* **75** (**2**), 365–377.

49 Nel, A., Xia, T., Meng, H. *et al.* (2012) *Nanomaterial toxicity testing in the 21st century- use of a predictive toxicological approach and high-throughput screening. Accounts of Chemical Research,* **46** (**3**), 607–621.

50 Martinez, Y.N.I., German, A., Castro, G.R. *et al.* (2014) *Nanostability,* in *Nanotoxicology: Materials, Methodologies, and Assessments* (eds N. Duran, S.S. Guterres, and O.L. Alves), Springer.

51 DeLoid, G., Cohen, J.M., Darrah, T. *et al.* (2014) *Estimating the effective density of engineered nanomaterials for in vitro dosimetry. Nature Communications*, **5**, 3514.
52 Cohen, J., Deloid, G., and Demokritou, P. (2015) *A critical review of in vitro dosimetry for engineered nanomaterials. Nanomedicine*, **10** (**19**), 3015–3032.
53 Cohen, J., Deloid, G., Pyrgiotakis, G., and Demokritou, P. (2013) *Interactions of engineered nanomaterials in physiological media and implications for in vitro dosimetry. Nanotoxicology*, 7 (**4**), 417–431.
54 Moore, T.L., Rodriguez-Lorenzo, L., Hirsch, V. *et al.* (2015) *Nanoparticle colloidal stability in cell culture media and impact on cellular interactions. Chemical Society Reviews*, **44** (**17**), 6287–6305.
55 Miller, A.W., Befort, C., Kerr, E.O., and Dunham, M.J. (2013) *Design and use of multiplexed chemostat arrays. Journal of Visualized Experiments*, **72**, e50262.
56 Hindre, T., Knibbe, C., Beslon, G., and Schneider, D. (2012) *New insights into bacterial adaptation through in vivo and in silico experimental evolution. Nature Reviews Microbiology*, **10** (**5**), 352–365.
57 Kussell, E., Kishony, R., Balaban, N.Q., and Leibler, S. (2005) *Bacterial persistence: a model of survival in changing environments. Genetics*, **169** (**4**), 1807–1814.
58 Fraser, D. and Kaern, M. (2009) *A chance at survival: gene expression noise and phenotypic diversification strategies. Molecular Microbiology*, **71** (**6**), 1333–1340.
59 Arense, P., Bernal, V., Iborra, J.L., and Cánovas, M. (2010) *Metabolic adaptation of Escherichia coli to long-term exposure to salt stress. Process Biochemistry*, **45** (**9**), 1459–1467.
60 Carlson, R.P. and Taffs, R.L. (2010) *Molecular-level tradeoffs and metabolic adaptation to simultaneous stressors. Current Opinion in Biotechnology*, **21** (**5**), 670–676.
61 Zhong, S., Miller, S.P., Dykhuizen, D.E., and Dean, A.M. (2009) *Transcription, translation, and the evolution of specialists and generalists. Molecular Biology and Evolution*, **26** (**12**), 2661–2678.
62 McDonnell, G. and Russell, A.D. (1999) *Antiseptics and disinfectants: activity, action, and resistance. Clinical Microbiology Reviews*, **12** (**1**), 147.
63 Martinez, J.L. and Baquero, F. (2000) *Mutation frequencies and antibiotic resistance. Antimicrobial Agents and Chemotherapy*, **44** (7), 1771–1777.
64 Perez, A., Poza, M., Aranda, J. *et al.* (2012) *Effect of transcriptional activators SoxS, RobA, and RamA on expression of multidrug efflux pump AcrAB-TolC in Enterobacter cloacae. Antimicrobial Agents and Chemotherapy*, **56** (**12**), 6256–6266.
65 Nikaido, H. (2003) *Molecular basis of bacterial outer membrane permeability revisited. Microbiology and Molecular Biology Reviews*, **67** (**4**), 593–656.

66 Lemire, J.A., Harrison, J.J., and Turner, R.J. (2013) *Antimicrobial activity of metals: mechanisms, molecular targets and applications. Nature Reviews Microbiology,* **11** (**6**), 371–384.

67 Nies, D.H. (2003) *Efflux-mediated heavy metal resistance in prokaryotes. FEMS Microbiology Reviews,* **27** (**2-3**), 313–339.

68 Baker-Austin, C., Wright, M.S., Stepanauskas, R., and McArthur, J.V. (2006) *Co-selection of antibiotic and metal resistance. Trends in Microbiology,* **14** (**4**), 176–182.

69 Booth, S.C., Workentine, M.L., Wen, J. *et al.* (2011) *Differences in metabolism between the biofilm and planktonic response to metal stress. Journal of Proteome Research,* **10** (7), 3190–3199.

70 Aruguete, D.M., Kim, B., Hochella, M.F. *et al.* (2013) *Antimicrobial nanotechnology: its potential for the effective management of microbial drug resistance and implications for research needs in microbial nanotoxicology. Environmental Science: Processes & Impacts,* **15** (**1**), 93.

71 Franzel, B., Penkova, M., Frese, C. *et al.* (2012) *Escherichia coli exhibits a membrane-related response to a small arginine- and tryptophan-rich antimicrobial peptide. Proteomics,* **12** (**14**), 2319–2330.

72 Manning, A.J. and Kuehn, M.J. (2011) *Contribution of bacterial outer membrane vesicles to innate bacterial defense. BMC Microbiology,* **11**: 258, 1–14.

73 Naghmouchi, K., Belguesmia, Y., Baah, J. *et al.* (2011) *Antibacterial activity of class I and IIa bacteriocins combined with polymyxin E against resistant variants of Listeria monocytogenes and Escherichia coli. Research in Microbiology,* **162** (**2**), 99–107.

74 Tran, A.X., Lester, M.E., Stead, C.M. *et al.* (2005) *Resistance to the antimicrobial peptide polymyxin requires myristoylation of Escherichia coli and Salmonella typhimurium lipid A. The Journal of Biological Chemistry,* **280** (**31**), 28186–28194.

75 Arciola, C.R., Campoccia, D., Gamberini, S. *et al.* (2005) *Antibiotic resistance in exopolysaccharide-forming Staphylococcus epidermidis clinical isolates from orthopaedic implant infections. Biomaterials,* **26** (**33**), 6530–6535.

76 Khan, W., Bernier, S.P., Kuchma, S.L. *et al.* (2010) *Aminoglycoside resistance of Pseudomonas aeruginosa biofilms modulated by extracellular polysaccharide. International Microbiology,* **13** (**4**), 207–212.

77 Chiang, S.M. and Schellhorn, H.E. (2012) *Regulators of oxidative stress response genes in Escherichia coli and their functional conservation in bacteria. Archives of Biochemistry and Biophysics,* **525** (**2**), 161–169.

78 Lushchak, V.I. (2011) *Adaptive response to oxidative stress: bacteria, fungi, plants and animals. Comparative Biochemistry and Physiology - Part C,* **153** (**2**), 175–190.

79 Wu, Y.X., Vulic, M., Keren, I., and Lewis, K. (2012) *Role of Oxidative Stress in Persister Tolerance. Antimicrobial Agents and Chemotherapy,* **56** (**9**), 4922–4926.

80 Dhas, S.P., Shiny, P.J., Khan, S. *et al.* (2014) *Toxic behavior of silver and zinc oxide nanoparticles on environmental microorganisms. Journal of Basic Microbiology,* **54** (**9**), 916–927.

81 Qiu, Z.G., Yu, Y.M., Chen, Z.L. *et al.* (2012) *Nanoalumina promotes the horizontal transfer of multiresistance genes mediated by plasmids across genera. Proceedings of the National Academy of Sciences,* **109** (**13**), 4944–4949.

82 Yang, Y., Mathieu, J.M., Chattopadhyay, S. *et al.* (2012) *Defense mechanisms of Pseudomonas aeruginosa PAO1 against quantum dots and their released heavy metals. ACS Nano,* **6** (7), 6091–6098.

83 Liu, J.Y. and Hurt, R.H. (2010) *Ion release kinetics and particle persistence in aqueous nano-silver colloids. Environmental Science & Technology,* **44** (**6**), 2169–2175.

84 Gao, J., Youn, S., Hovsepyan, A. *et al.* (2009) *Dispersion and toxicity of selected manufactured nanomaterials in natural river water samples: effects of water chemical composition. Environmental Science & Technology,* **43**, 3322–3328.

85 Jin, X., Li, M., Wang, J. *et al.* (2010) *High-through screening of silver nanoparticle stability and bacterial inactivation in aquatic media: Influence of specific ions. Environmental Science & Technology,* **44**, 7321–7328.

86 Zhang, H.Y., Smith, J.A., and Oyanedel-Craver, V. (2012) *The effect of natural water conditions on the anti-bacterial performance and stability of silver nanoparticles capped with different polymers. Water Research,* **46** (**3**), 691–699.

87 Zhang, H.Y. and Oyanedel-Craver, V. (2012) *Evaluation of the disinfectant performance of silver nanoparticles in different water chemistry conditions. Journal of Environmental Engineering,* **138** (**1**), 58–66.

88 Şengül, H., Theis, T.L., and Ghosh, S. (2008) *Toward sustainable nanoproducts. Journal of Industrial Ecology,* **12** (**3**), 329–359.

89 Yan, Y., Miao, J., Yang, Z. *et al.* (2015) *Carbon nanotube catalysts: recent advances in synthesis, characterization and applications. Chemical Society Reviews,* **44** (**10**), 3295–3346.

90 Plata, D.L., Gschwend, P.M., and Reddy, C.M. (2008) *Industrially synthesized single-walled carbon nanotubes: compositional data for users, environmental risk assessments, and source apportionment. Nanotechnology,* **19** (**18**), 185706.

91 Nel, A.E., Brinker, C.J., Parak, W.J. *et al.* (2015) *Where are we heading in nanotechnology environmental health and safety and materials characterization? ACS Nano,* **9** (**6**), 5627–5630.

Index

Nanotechnology Commercialization: Manufacturing Processes and Products, First Edition.
Edited by Thomas O. Mensah, Ben Wang, Geoffrey Bothun, Jessica Winter, and Virginia Davis.

e

f

O